chapter 6

chapter 7

chapter 8

chapter 9

chapter 10

chapter 11

INTRODUCTION TO statistical reasoning

INTRODUCTION TO

statistical reasoning

Gary **Smith**
Pomona College

Boston, Massachusetts Burr Ridge, Illinois Dubuque, Iowa
Madison, Wisconsin New York, New York San Francisco, California St. Louis, Missouri

For my students.

WCB/McGraw-Hill

A Division of The **McGraw-Hill** *Companies*

Introduction to Statistical Reasoning

This book is printed on acid-free paper.

1 2 3 4 5 6 7 8 9 0 DOC DOC DOC 9 0 0 9 8 7

ISBN 0-07-059276-4

Publisher: Thomas L. Casson
Sponsoring editor: Maggie Rogers
Marketing manager: Michelle Sala
Project manager: Terri Wicks
Production supervisor: Michelle Lyon
Designer: Deborah Chusid
Cover designer: Deborah Chusid
Compositor: GTS Graphics, Inc.
Typeface: Times Roman
Printer: R. R. Donnelley & Sons Company

Grateful acknowledgment is made for use of the following:
Realia and Cartoons *Page 8* Reprinted with permission of Carole Cable; *16* Illustrator, by Rob Cross. Reprinted with permission from the *Ottawa Citizen.; 17* Adapted from *Mathematics: A Human Endeavor,* second edition, by Harold Jacobs © 1982 by W.H. Freeman and Company. Used by permission.; *31,* From *The Visual Display of Quantitative Information* by Edward R. Tufte (Cheshire, Conn.: Graphics Press, 1983); *32* Reprinted with permission of *American Political Science Review; 41 The Visual Display of Quantitative Information; 48* Drawing by Mankoff; © 1984 The New Yorker Magazine, Inc.; *50 The Visual Display of Quantitative Information; 55* By permission of Johnny Hart and Creators Syndicate, Inc.; *62 The Visual Display of Quantitative Information; 128* By permission of Johnny Hart and Creators Syndicate, Inc.; *231* THE BORN LOSER reprinted by permission of Newspaper Enterprises Association, Inc.; *365* Cartoon by Rob Pudin used by permission of *The Skeptical Inquirer; 458* From *Statistics: A Guide to the Unknown,* ed. Judith M. Tanur (Oakland, Calif.: Holden Day, 1972); *535* Reprinted by permission of the *Detroit Free Press; 536* Copyright © 1994 New York Times Company.

Library of Congress Cataloging-in-Publication Data

Smith, Gary, 1945-
Introduction to statistical reasoning / Gary Smith.
p. cm.
Includes bibliographical references and index.
ISBN 0-07-059276-4
1. Mathematical statistics. I. Title.
QA276.12.S577 1998 97-1143
519.5—dc21 CIP

http://www.mhhe.com

brief contents

PART 1 Statistical Data 1

PART 2 Statistical Inference 199

contents

*Optional

*Optional

*Optional

*Optional

chapter 11 simple regression 520

*Optional

preface

This textbook is intended for an introductory statistics course that will help students to develop their statistical reasoning. Too many students mistakenly believe that statistics courses are too abstract, mathematical, and tedious to be useful or interesting. To demonstrate the power, elegance, and even beauty of statistical reasoning, this book provides hundreds of interesting and relevant examples, and discusses not only the uses but also the abuses of statistics. These examples show how statistical reasoning can be used correctly to answer important questions and it also exposes the errors—accidental or intentional—that people often make with statistics. As indicated by the selected list inside the front cover, examples are drawn from many areas to show that statistical reasoning is not an irrelevant abstraction, but an important part of everyday life.

I have tried to write a book that students will find clear, interesting, and engaging—not dense, dull, and intimidating. Toward that end, I have tried to avoid the obscure and frivolous and focus instead on topics that students might remember after the final exam. For example, Chapter 2 (Displaying Data) spends more time on the use and misuse of graphs than do most textbooks, but I think that this attention is needed today for visual literacy. Similarly, this book is full of realistic examples and exercises from a wide variety of disciplines that are intended to persuade students that they are learning a powerful set of tools that can be applied in almost any career.

Learning by Doing

In place of rote memorization, I have tried to emphasize critical thinking and problem-solving skills. This is not a plug-and-chug book. It emphasizes statistical reasoning and deemphasizes formulas for doing calculations, as calculations will generally be completed by computer, not hand. Because I believe that memorizing formulas is less important than recognizing and applying the appropriate statistical procedures the book is also relatively nonmathematical. Some instructors may want to give students a page of formulas to use during examinations, or even use open-book tests.

Every chapter has exercises of varying type and difficulty. Some are simple computations to reinforce the chapter material. Many are real-world examples that offer students opportunities to apply important statistical principles. Every chapter also has a set of projects that can be used to involve the students in hands-on data collection and nontrivial writing projects. I use these in my classes as group projects in teams of 3 to 5 persons (which encourages teamwork and reduces the grading burden), but they can also be assigned to individuals. In small classes, students can also make brief oral presentations of their results. I've found that students learn best by doing—by collecting, analyzing, writing, and speaking.

Teaching by Example

The traditional way of teaching statistics is to begin with a definition, theorem, or formula and then use a series of brief (usually hypothetical) examples to illustrate the point. I have turned this approach around because students are more motivated to learn the definition, theorem, or formula if we begin with a real-world example of an interesting question we would like to answer. After posing this question, we look at the statistical tools that will help us. The tools are then applied and the answers discussed. Going from example to tools really persuades students that the tools are useful. After answering the initial question, we look at other examples and see how the tools can be used in a wide variety of contexts—not just to answer a single, narrowly defined question but to answer all sorts of real-world questions.

It is important that the motivating examples be real. Students are more easily convinced of the power of statistical reasoning if they see it applied to questions that are important, interesting, and real. Tools that are used to answer artificial questions will seem artificial too. In addition, students will remember the real-world question and how we answered it much more easily than they will remember some made-up example.

Thus each chapter has a theme that continues through the chapter. The opening page, "Have you ever wondered?," introduces a topic worthy of exploration. Several questions are raised, which the chapter then answers using a variety of statistical techniques. The techniques developed in the chapter are of course applied to a great many other examples too, but the sustained exploration of the theme topic provides continuity to the discussion. In addition, it helps students realize that questions can be addressed in a variety of ways and also gives them what I hope are satisfying answers to some of the important issues raised by the theme topic. Students not only learn more about statistical techniques, but are also convinced that statistical reasoning can be used to answer interesting questions.

special features

Have You Ever Wondered? Each chapter (except for Chapter 1) begins with a discussion of an important topic and related questions that can be explored using statistical techniques. Each topic is explored in detail in the chapter and students are provided with appropriate tools that can be used to help answer the questions raised.

How To Do It Boxes. Although mathematical computations are kept to a minimum in this book, some instructors may want to teach their students how to perform important statistical calculations by hand. To address this need, I have included steps for most statistical calculations used in the book in separate

How To Do It boxes throughout each chapter. These boxes are set apart from the main body of the text and can be omitted without loss of continuity or topic coverage.

Student Projects. At the end of each chapter, I have included projects for the students to complete either individually, or working in groups. These projects are excellent opportunities for students to learn about data collection, data analysis, and report-writing.

Real Data. The majority of the data used in the examples and the exercises is real, not fictional. I have included references to the original data source, where appropriate.

Technology. Introduction to Statistical Reasoning can be used with any computer package or spreadsheet program. Sample output from Minitab and SAS is included in the text. Some instructors use commercial software in their course because of this software's power or because they believe that learning to use such software is an important goal of the course. For instructors who wish to minimize expenses and/or the time students must spend learning to use software, I have written a very user-friendly statistic program that students can use immediately, with no instructions. The statistical programs on this disk are intended to avoid the drudgery of hand calculations and the complexity of large computer packages, so that students can concentrate on understanding and applying statistical reasoning. This software is free to adopters.

supplements

Instructor's Manual. This manual contains lecture suggestions, additional examples and exercises, and detailed solutions to all of the textbook exercises.

Student Study Guide. This manual provides detailed step-by-step calculations for the odd-numbered exercises in the textbook and additional exercises for homework assignment or student practice. Also included are a brief chapter review and worked out examples that illustrate the chapter's main statistical techniques.

Testbank. A print and computerized testbank is available to adopters containing 1000 test questions written by the author.

Software. As described above, an easy-to-use computational software program is available free to adopters.

Minitab Student Version. The Student Version of Minitab is available for purchase when packaged with this textbook.

Please contact your WCB/McGraw-Hill sales representative for more information about any of the supplements described above.

acknowledgments

Many have contributed to this book. I sincerely appreciate your advice, assistance, and encouragement.

Neil Alper, *Northeastern University*
Martin Appel, *University of Iowa*
Sergio Antiochia, *Eastern Michigan University*
Dan Brick, *College of St. Thomas*
George Briden, *University of Rhode Island*
William Burdon, *Claremont Graduate School*
Deborah Burke, *Pomona College*
Ann Cannon, *Cornell College*
Philip Carlson, *Bethel College*
Peyton Cook, *University of Tulsa*
Allin Cottrell, *Elon College*
Michael H. Criqui, *University of California, San Diego*
Joyce Curry Daly, *Cuesta College*
Anirban DasGupta, *Purdue University*
Stephen DeCanio, *University of California, Santa Barbara*
Paul Dussere, *SUNY Oswego*
Eugene A. Enneking, *Portland State University*
Roy V. Erickson, *Michigan State University*
Chris Franklin, *University of Georgia*
H. E. Frech III, *University of California, Santa Barbara*
Joseph Glaz, *University of Connecticut*
Myron F. Goodman, *University of Southern California*
Marvin Jay Greenberg, *University of California, Santa Cruz*
David Harrington, *Kenyon College*
Iftekhar Hasan, *University of Wisconsin*
Catherine Hayes, *University of Mobile*
John Hoffman, *Humboldt State University*
Burt S. Holland, *Temple University*
Linda Host, *University of Wisconsin, LaCrosse*
Rick Hutchinson, *Yellowstone National Park*
Bernard Isselhardt, *Rochester Institute of Technology*
Raj Jagannathan, *The University of Iowa*
Robert E. Johnson, *Virginia Commonwealth University*
Shaun Johnson, *Australian Bureau of Meteorology*
Mario L. Juncosa, *University of California, Los Angeles*
Kevin Keay, *University of Melbourne*
Manfred Keil, *Claremont McKenna College*
Hanhan Kim, *South Dakota State University*
Ronald S. Koot, *The Pennsylvania State University*
Michael Kuehlwein, *Pomona College*
Bret R. Larget, *Duquesne University*
Edward Leamer, *University of California, Los Angeles*
Belinda Lees, *Wynn Institute for Metabolic Research*
Dennis Lendrum, *Nottingham University*

Jerry Lenz, *St. John's University*
Lonnie Magee, *McMaster University*
Linda Malone, *University of Central Florida*
Stephen Marks, *Pomona College*
Robert L. Mason, *Southwest Research Institute*
Fred S. McChesney, *Cornell University*
Lauren McIntyre, *North Carolina State University*
Don Miller, *Virginia Commonwealth University*
Tom Moore, *Pomona College*
Michael Murray, *Bates College*
David Noone, *University of Melbourne*
Harold Petersen, *Boston College*
John R. Pickett, *Georgia Southern College*
Frank W. Puffer, *Clark University*
Svetlozar T. Rachev, *University of California, Santa Barbara*
Larry J. Ringer, *Texas A&M University*
Ronald E. Shiffler, *University of Louisville*
Ian Simmonds, *University of Melbourne*
Bob Smidt, *Cal Poly San Luis Obispo*
James Stapleton, *Michigan State University*
Paul A. Thompson, *Ohio State University*
Suzanne Thompson, *Pomona College*
James Vanderhoff, *Rutgers University*
Ivan Valiela, *Boston University*
Mary Beth Walker, *Emory University*
Steven S. Wasserman, *University of Maryland*
Stefan N. Willich, *Free University of Berlin*
Thomas O. Wisley, *Western Kentucky University*
Barbara Yanoska, *Humboldt State University*
Mary Sue Younger, *University of Tennessee at Knoxville*
Andrew Zimbalist, *Smith College*

I am grateful to the Literary Executor of the late Sir Ronald A. Fisher, F.R.S., to Dr. Frank Yates, F.R.S., and to Longman Group Ltd, London for permission to reprint Table III from their book *Statistical Tables for Biological, Agricultural, and Medical Research* (6th Edition, 1974). I also thank Maggie Lanzillo Rogers, Terri Wicks, and the many others at McGraw-Hill who helped so much with this project.

Most of all, I am grateful to all of the students that I have had during the 25 years that I have been teaching statistics—for their endless goodwill and contagious enthusiasm and, especially, for teaching me how to be a better teacher.

Gary Smith

about the author

Gary Smith is Fletcher Jones Professor of Economics at Pomona College in Claremont, California. Gary received a B.S. in mathematics from Harvey Mudd College and M.A., M/Phil, and Ph.D. degrees in economics from Yale University. He has taught statistics courses as an assistant Professor at Yale University, an associate professor at the University of Houston and Rice University, and a professor at Pomona College.

Gary is the author of six college textbooks and dozens of scholarly articles. His paper on ridge regression was selected as the invited theory and methods paper for 1978 JASA meetings, and he has received Pomona College's Wig Award for teaching excellence. He is particularly interested in encouraging active learning throughout the use of computer simulations, writing exercises, and speaking assignments.

PART 1

Statistical Data

chapter 1

introduction

It is remarkable that a science which began with the consideration of games of chance should become the most important object of human knowledge. . . . The most important questions of life are, for the most part, really only problems of probability.
—Pierre Simon, Marquis de Laplace

H. G. WELLS ONCE WROTE that "Statistical thinking will one day be as necessary for efficient citizenship as the ability to read and write." And today it is. We live in a complex society in which informed decisions require an intelligent interpretation of statistical data. Statistics can help us decide which foods are most healthful, which careers are likely to be lucrative, which cars are unsafe, and which investments are risky. Businesses use statistics to gauge manufacturing precision, monitor production, schedule activities, estimate demand for their products, and design marketing strategies. Government statisticians measure corn production, air pollution, unemployment, inflation, and much more. Each year, the tip of the government's statistical iceberg is published in the *Statistical Abstract,* a surprisingly interesting—some say fascinating—source of information about the United States. Some government data, such as weather patterns and crop production, help citizens and businesses make intelligent plans. Other data, such as unemployment rates and the results of tests of new medications, are used by government policy makers.

Statistical data also provide ammunition for advertising and election campaigns, and citizens must decide whether businesses and politicians are using or abusing statistics. In 1984, Ronald Reagan was so upset by data presented by his political opponent that he used the immortal retort of the 19th-century British politician Benjamin Disraeli: "There are three kinds of lies—lies, damn lies, and statistics." And then President Reagan proceeded to cite some statistics of his own!

Used correctly, statistical reasoning can help us distinguish between informative statistics and damnable lies—to recognize valid uses and identify invalid abuses—so that we can judge the competing claims of advertisers and politicians. Because so many real-world situations require statistical analysis, this book relies heavily on examples to teach its lessons. The list inside the front cover is just a sampling of the hundreds of examples that are used to show the breadth of statistical reasoning, illustrate how these tools are used, and even teach something about the examples themselves.

Every chapter has dozens of exercises of varying type and difficulty. Most are real-world examples that offer opportunities to reinforce the chapter material and to apply important statistical principles. The exercises are an absolutely essential part of the book. We cannot become good racquetball players simply by watching, and we cannot learn statistical reasoning just by reading. We have to hit a few balls and do a few exercises. Everyone, whether playing racquetball or applying statistical reasoning, will miss a few at first, but soon will acquire the necessary skills and experience. Practice is the surest way to strengthen our physical and mental abilities, and it is the surest way to develop the statistical reasoning that will be needed after the final examination, when statistics are encountered almost daily.

To provide an overview of this book's organization, we briefly consider the nature of statistical data and statistical inference.

1.1 Statistical Data

Statistics involves the collection, display, analysis, and interpretation of numerical information. Most statistical data comprise a *sample* from a much larger *population.* To test a product's reliability, quality control people inspect a small sample of the product. To gauge the potential benefits and side effects of a medication, it is administered to a sample of volunteers. To predict the outcome of a presidential election, pollsters interview a few thousand voters, who represent a sample from a population of tens of millions of voters.

Data collection involves the selection of a sample and the design of the experiment or observational study—the product tests that will be conducted, the medical dosages that will be administered, the survey questions that will be asked. Once we have our data, we might use a descriptive number, table, or graph to summarize the most salient features and tell the audience—in clear and unbiased terms—what the data say. In this book, you will learn how to select samples, how to display your data fairly, and how to recognize samples, statistics, and graphs that are misleading.

1.2 Statistical Inference

Probabilities can be used to help us analyze uncertain situations. If only death and taxes are certain, there are plenty of potential targets for probability analysis. Gamblers use probabilities when they say that the odds are 100 to 1 that the Chicago Cubs will be in baseball's World Series. The Educational Testing Service uses probabilities when it calculates standardized Scholastic Aptitude Test (SAT) scores. Colleges use probabilities when they decide how many students to admit, how many professors to employ, and how many classrooms to build. Probabilities are involved when an army decides to attack or retreat, when a business decides to expand or contract, and when you decide whether to carry an umbrella. Uncertainties are all around us, and so are probabilities.

Historically, the first rigorous application of probability theory involved games of chance, and games are still a fertile field for probability analysis. Gambling casinos use probabilities when they set the payoffs for roulette, craps, and slot machines. Governments use probabilities when they set the payoffs for state lotteries. Mathematicians have used probability calculations to devise optimal strategies for blackjack, backgammon, Monopoly, poker, and many other games of chance.[1] Devoted players can also learn these probabilities firsthand from long and sometimes expensive experience.

Another early use of probabilities was in setting insurance rates. This is why life insurance premiums depend on whether a person is aged 18 or 98, in good health or using an artificial heart, a college professor or a soldier of fortune. Probabilities are also used to price disability insurance, medical insurance, car insurance, boat insurance, home insurance, business insurance, and even insurance against a baseball strike or a singer getting laryngitis.

Probabilities are used to describe the anticipated events; empirical data are the observed outcomes. Statistical inference involves the use of probabilities and data together to draw conclusions about the underlying population. For instance, insurance companies use mortality data and probability theory to estimate life expectancies. Political pollsters use survey results and probability calculations to predict election outcomes and to assess how much confidence we can have in these predictions.

The analysis of sample data involves probabilities because there is chance involved in the selection of a sample from the population. Suppose that 50 percent of the population prefer John Bigsmile. In a random sample of 1000 voters, it may turn out, by luck of the draw, that exactly 500 people (50 percent of those sampled) prefer Bigsmile; or it may turn out that a somewhat larger or smaller number in the sample prefer Bigsmile. Probability calculations can tell us how likely it is that the sample proportion will be close to the population proportion. For instance, if 1000 people are randomly selected from a large population in which exactly one-half of the people prefer Bigsmile, then Table 1.1 shows the probability that the number surveyed who prefer Bigsmile will be greater than or equal to various specified values. (You will soon learn how to make these simple calculations.)

Table 1.1

Probability that x or more people prefer a candidate in a random sample of 1000 people from a large population in which 50 percent prefer this candidate.

X	PROBABILITY
500	.5126
510	.2740
520	.1087
530	.0310
540	.0062
550	.0009

Probabilities help us predict the results of sampling, given assumptions about the population that yielded the sample. Statistical inference uses the results—the observed data—to draw conclusions about the underlying population. For example, if it turns out that 380 of the 1000 people sampled prefer Bigsmile, then we might estimate that 38 percent of the entire population prefer Bigsmile. Because the luck of the draw is involved in the selection of the 1000 people who are surveyed, we can gauge the confidence that we have in our prediction by reporting our estimate as "38 percent, plus or minus 3 percent." In later chapters, you will learn where this plus-or-minus figure comes from and how to interpret it.

Statistical inference can also be used to confirm or discredit theories about the underlying population. For example, the theory that John Bigsmile is going to receive more than 50 percent of the votes would be discredited by a random sample of 1000 voters in which only 380 preferred Bigsmile. Similarly, 22,000 doctors participated in an experiment in the 1980s testing the effects of aspirin on the incidence of heart attacks. One-half of these doctors took a single aspirin tablet every other day; the other half did not. After 5 years, 171 of the doctors who had not been taking aspirin had suffered heart attacks, 18 of them fatal, while only 99 of the doctors taking aspirin had heart attacks, of which 5 were fatal. The hypothesis that aspirin has no effect on the incidence of heart attacks was refuted statistically by these results. In fact, the results were so convincing that the study was stopped after 5 years of what was intended to be a 7-year test. In later chapters, you will learn exactly how to show that these data refuted the hypothesis that aspirin has no effect on the incidence of heart attacks. You will also see how to use data to test such theories as smoking is dangerous to your health, vitamin C fights colds, Volvos last longer than Fords, and the unemployment rate influences the outcomes of presidential elections. Statistical inference is used in election polls and unemployment surveys, in marketing tests and product safety tests, in constructing forecasting models and interpreting psychological test scores, and in testing and refining theories about the human body and the nature of the universe.

We begin in Part 1 of this book by seeing how data are displayed, summarized, and produced. Part 2 explains how probabilities can be interpreted and determined, and shows how data can be used to make statistical inferences.

1.3 Uses and Misuses of Statistics

Over the years, the uses of statistics have multiplied in surprising ways as people have recognized the prevalence of uncertainty and the power of informed analysis. For example, the American Society of Composers, Authors, and Publishers (ASCAP) now allocates hundreds of millions of dollars in performance-rights royalties based on a statistical analysis of radio, television, jukebox, and even background music.

Statistical reasoning has long been used to decipher secret codes and is now used in computer programs that compress text for storage, scan written documents, and translate speech to writing. These speech recognition programs use test periods to learn a user's pronunciation. Then, when the computer hears a sound, it selects a set of similar-sounding candidate words and, based on the context, computes probabilities and chooses the most likely word intended by this sound. For instance, if the user says, "Two computers are better than none," the machine considers *to, too,* and *two* for the first word and *none* and *nun* for the last word. Using grammatical patterns and word frequencies, the computer settles on *two* and *none* by deciding that, statistically, the sentence is most likely to make sense this way.

"What we had in mind was more along the lines of a 'A Year in Provence' approach to statistical analysis."

Statistical reasoning now abounds in sports. Basketball plays are designed to give players opportunities to take "high-percentage shots," from a distance and an angle where a particular player has a high probability of making a basket—probabilities that have been estimated from detailed statistical records. Football teams use statistical models to evaluate prospective players, based on not only measurements of speed and strength, but also answers to psychological test questions. Football coaches use statistical reasoning when they must choose between punting, trying for a field goal, and trying for a first down. Football coaches ranging from Plano High School to the Dallas Cowboys estimate their opponents' "tendencies," for instance, the plays a particular team most often uses when it has the ball in the first quarter inside its own 20-yard line on third down with 3 yards to go for a first down. Football teams also estimate their own tendencies, to avoid becoming too predictable.

Baseball is probably the most statistical sport, starting with lifetime batting averages and running through a certain pitcher's record against a specific batter with runners in scoring position in the late innings of a close game played at night in August in a certain ballpark. Good baseball managers are continually "playing the percentages." They bring in a left-handed pitcher to face a left-handed batter. They instruct a .245 hitter to make a sacrifice bunt to advance a runner to second base. They put their good hitters at the start of the lineup so that, on average, over the course of a season they will bat more often than the weak hitters at the end of the lineup. In the late innings, they take out a good-hitting, weak-fielding player who has just batted and bring in a weak-hitting, good-fielding replacement. In all these cases and many more, the manager is trying to use statistics to make victory more probable.

Fallible humans also misuse statistical reasoning. We see skill where there is luck. We see psychic phenomena where there is coincidence. We confirm flimsy theories with anecdotes. We distort and misreport data to support our prejudices. We take foolish chances, counting on an erroneous "law of averages" to protect us from our own folly. We waste money on worthless gambling systems. We play inferior backgammon, bridge, and poker. We buy too much insurance and either too few or too many stocks. We neglect many important risks and exaggerate others.

I wrote this textbook to introduce you to statistical reasoning. It is not encyclopedic, and it is not a collection of arcane formulas and crunched numbers. I emphasize some important principles that you can use all your life. You do not need to memorize 200 formulas or add 2,000,000 six-digit numbers. You can always look up the formulas and have computers do your arithmetic. What you really need is to understand what you are doing.

You will see how data and statistical inferences can be used to help you make informed decisions in the face of otherwise daunting uncertainty. You will also see how data can be used and misused and how to recognize the everyday errors in the assertions of others. You will learn why the unemployment rate is plus or minus .2 percent and why an election prediction is plus or minus 3 percent. You will understand what it means to say that the evidence supports someone's theory. You will learn how to produce and display data and how to test theories yourself. You will find out what is meant by the commonplace report that researchers have found a "statistically significant relationship" between smoking and lung cancer. You will see why expert studies sometimes disagree and how statistical tests can be abused. By learning these things and more, you will make this one of the most interesting and useful courses that you will ever take.

HAVE YOU EVER WONDERED?

Mark Twain once joked, "October. This is one of the peculiarly dangerous months to speculate in stocks. The others are July, January, September, April, November, May, March, June, December, August, and February." Another old joke asks, "How do you make a small fortune in the stock market?" The answer is, "Start with a large fortune."

Stock market zigs and zags are frequently front-page news—especially when the news is bad. The worst day ever for the U.S. stock market was October 19, 1987, when a staggering 98 percent of all stock prices fell, with the market overall down 23 percent and investors $500 billion poorer at the end of the day than they were at the beginning. Investors who lived through this stockmarket meltdown will never forget it. Many decided to follow the old advice that the safest way to double your money is to fold it over once and put it in your pocket. They sold their stocks and vowed never to return to the market.

However, the stock market is not always bad news. The market was up for 1987 as a whole and for 8 of the next 9 years. In 1995 the average return in the stock market was a stunning 37.4 percent, which is a whole lot better than money in a pocket or money in a bank paying 1 or 2 percent interest.

How well has the stock market done historically? In how many years has the market gone up, and how often has it gone down? Which was the most profitable year in recent memory and which was the worst, and how much did investors make or lose in these best and worst years? Are there predictable patterns in stock prices? For example, if the market goes up one year, is it more likely to go up the next year, too, or less likely? In this chapter, we will use some statistical tools to answer these questions.

chapter 2

displaying data

When you can measure what you are speaking about and express it in numbers, you know something about it; but when you cannot measure it, when you cannot express it in numbers, your knowledge is of a meager and unsatisfactory kind.

—Lord Kelvin

IT IS VERY difficult to absorb and analyze substantial amounts of data merely by looking at a seemingly endless list of numbers. What sense could you make of pages and pages of a computer printout showing the gender and height of every student at your school? Instead, we need to organize and present data in ways that will let us readily grasp what the numbers can tell us, if only we let them speak clearly. In the next chapter, we will see how a few summary statistics can help us make sense of large amounts of data. In this chapter, we will focus on visual displays—graphs and charts—that can help us to understand and interpret data and to see patterns and relationships. A picture can be worth not only a thousand words, but a thousand numbers.

One of the most helpful ways to make sense of a large body of data is to divide our observations into meaningful categories or intervals. How many students are

female, and how many are male? How many heights are between 60 and 65 inches? Between 65 and 70 inches? Some data, such as whether a person is male or female, do not have natural numerical values—a person cannot be 3.5 male—and are said to be *qualitative* or *categorical* data, since they can be grouped into nonnumerical categories, such as male or female. Heights, weights, and other data that have natural numerical values—a person can be 63.5 inches tall—are called *quantitative* data and can be separated into numerical intervals, such as from 60 to 65 inches.

Once the data have been sorted into categories or intervals, we can then calculate the **relative frequencies**—the fraction (or percentage) of the observations that fall into each category or interval. By looking at how the data are distributed across a small number of meaningful categories or intervals, we can learn a great deal about the data that would not be apparent from staring at hundreds or thousands of unorganized numbers. What percentage of the students is female? What fraction is between 65 and 70 inches tall? Relative frequencies can be listed in a table, or they can be displayed visually by using some of the techniques described in this chapter. We will also look at graphical techniques that can be used to show how data vary over time or to reveal how two variables may be related.

Because visual displays are intended to communicate information, it is not surprising that they, like other forms of communication, can also be used to distort and mislead. Regardless of whether words or graphs are used, the uninformed can make mistakes and the unscrupulous can lie. In this chapter you will see how to use visual displays to convey information accurately and fairly and how to recognize distortions—innocent and otherwise.

To identify practices to be avoided, I will show you several poorly done graphs. You should not conclude that bad graphs are as prevalent in academic journals and the popular press as they are in this chapter. Most graphs do not have serious problems. By studying some that are flawed, we can recognize these pitfalls and learn how to draw effective, accurate graphs.

2.1 Bar Charts and Pie Charts

One of the simplest ways to display qualitative data is with a **bar chart,** in which the heights of bars are used to show either the number or the relative frequency of observations in each category. For instance, Fig. 2.1 shows a bar chart of these 1994 data on the purchases of four kinds of athletic footwear in the United States:[1]

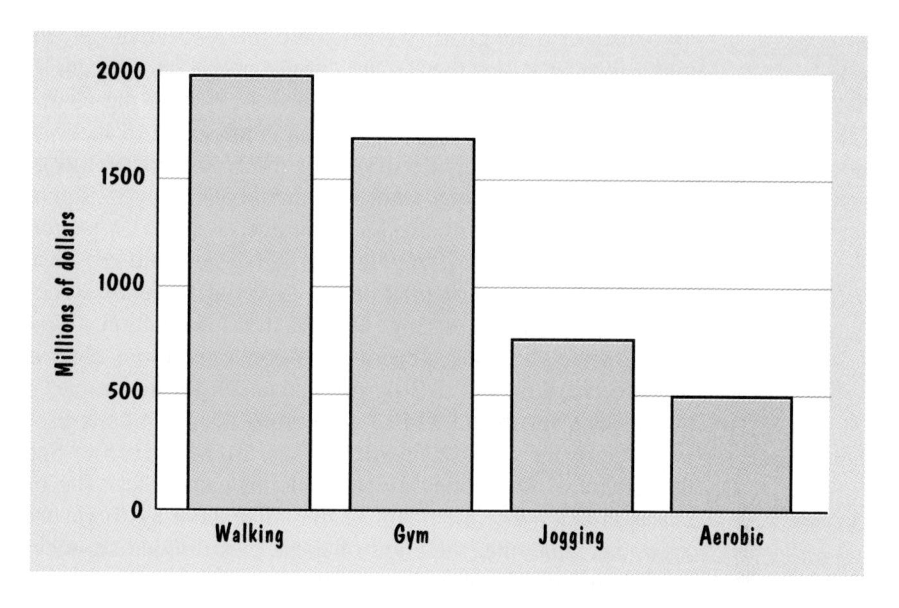

Figure 2.1 A bar chart of purchases of athletic footwear in the United States, 1994.

	MILLIONS OF DOLLARS
WALKING SHOES	1850
GYM SHOES AND SNEAKERS	1457
JOGGING AND RUNNING SHOES	702
AEROBIC SHOES	226

Because the human eye tends to gauge size by area rather than by height, each bar should be the same width; otherwise the wider bars will look disproportionately large. The height of each bar shows the number of dollars spent on that type of footwear. Alternatively, we could let the heights be either the fraction or the percentage of total dollars spent on each type of shoe. Such a graph would look exactly like Fig. 2.1, but with different values on the vertical axis.

The primary advantage of a bar chart over a table is that some people find the visual impression conveyed by the bars to be more accessible and memorable than numbers in a table. The arrangement of the categories from largest to smallest may also help the reader absorb the information in the graph.

Another way to depict categorical data is with a **pie chart,** which resembles a pie that has been cut into slices. For example, here are the contributions (in millions of dollars) by political action committees (PACs) to incumbents and challengers during the 2-year cycle preceding the 1994 congressional elections:[2]

Senate incumbents	26.3	House incumbents	101.4
Senate challengers	5.7	House challengers	12.7
	32.0		114.1

To construct a pie chart, we calculate the respective percentages and slice the pie accordingly. For the Senate campaigns, a fraction 26.3/32.0 = .822 of the PAC money went to incumbents; thus the Senate pie chart in Fig. 2.2 is sliced so that 82.2 percent of the total area is allocated to incumbents and the remaining 17.8 percent to challengers. For the House races, a fraction 101.4/114.1 = .889 of the PAC money went to incumbents; thus 88.9 percent of the House pie chart in Fig. 2.2 is allocated to incumbents and 11.1 percent to challengers.

Pie charts are not particularly enlightening, but are nonetheless often displayed during oral presentations to give the audience something to look at while the speaker is talking. Frankly, the information in a simple pie chart with two or three slices can generally be conveyed more accurately in words. The two pie charts in Fig. 2.2 do not add much to the simple, unadorned statement that PACs contributed $146.1 million to the 1994 congressional elections, with incumbents receiving 82 percent of the money given to Senate candidates and 89 percent of the money given to House candidates. The only advantage of pie charts is that some members of the audience may forget the specific percentages, but remember the sight of these pies divided so unevenly.

When a pie chart is sliced into more than a few pieces, it is usually difficult to keep the various slices in perspective, especially if they have been shaded with the distracting patterns provided by many computer software packages. Figure 2.3 shows a press release by the College Board, explaining the changes in the composition of the verbal reasoning component of the Scholastic Aptitude Test (SAT). The antonyms slice of the current SAT has been separated from the rest of the pie to dramatize the fact that these questions are being

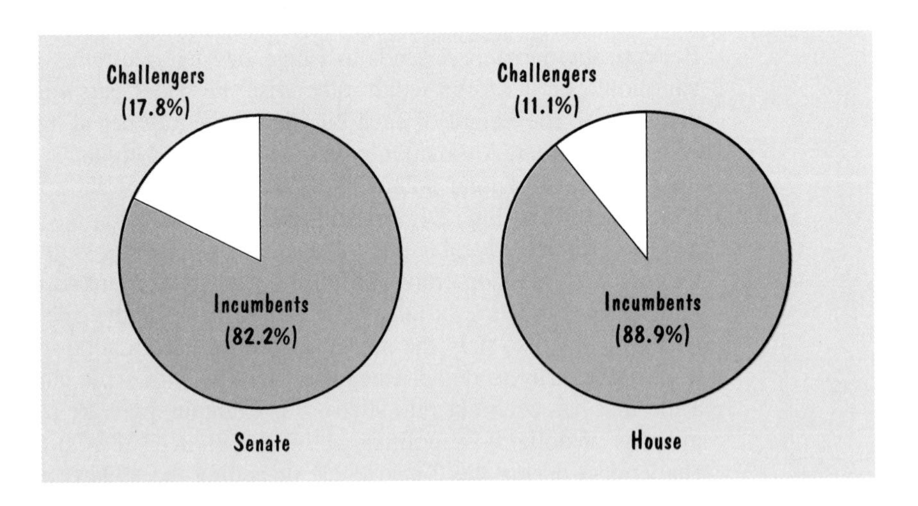

Figure 2.2 Pie charts of PAC contributions to 1994 U.S. congressional election campaigns.

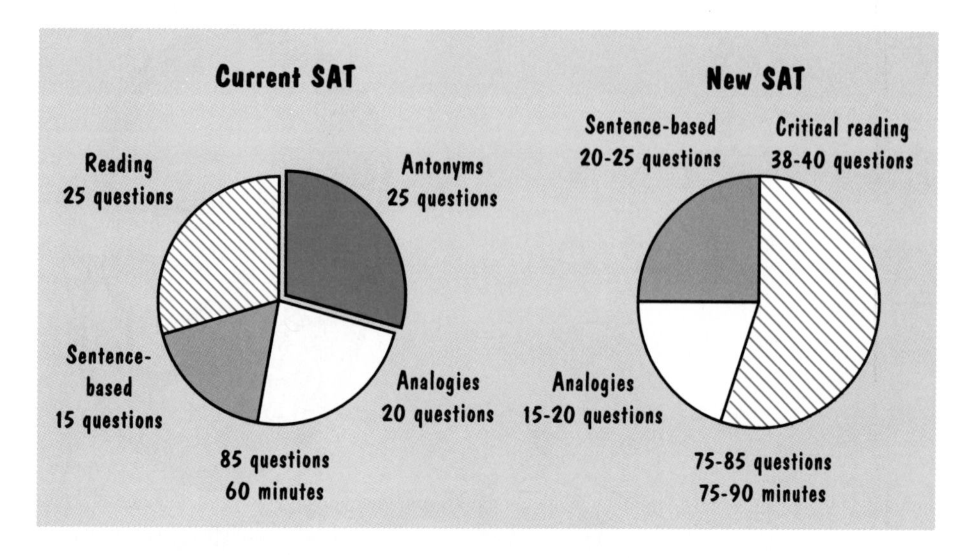

Figure 2.3 SAT I: Reasoning.

removed. The same technique can be used to emphasize the addition of a new category. These pie charts are not much more complicated than the PAC-money charts in Fig. 2.2. Yet it is hard to believe that the audience will take home any visual memories other than the removal of the antonym questions. Without looking, what do you remember about these pie charts? Pie charts that are sliced more finely than Fig. 2.3 are even more difficult to interpret and remember and should usually be avoided. Figure 2.4 shows a grotesque example from a Canadian newspaper. Not only are there far too many slices to digest, but there is no relationship between the size of the slices and the numbers they are meant to represent.

Pie charts are generally not very useful for analyzing data, in that the charts do not reveal relationships or patterns that are not self-evident simply from looking at a table. They may even be confusing to the extent that we have trouble gauging the relative sizes of the slices. Pie charts are used primarily to convey a visual image to audience members who have trouble interpreting numbers.

Two-Dimensional Graphs of One-Dimensional Data

To enliven the bar charts printed in newspapers and popular magazines, graphic artists often use pictures instead of plain bars. For example, the figure below was part of a newspaper advertisement boasting that "*The Times* has 2,244,500

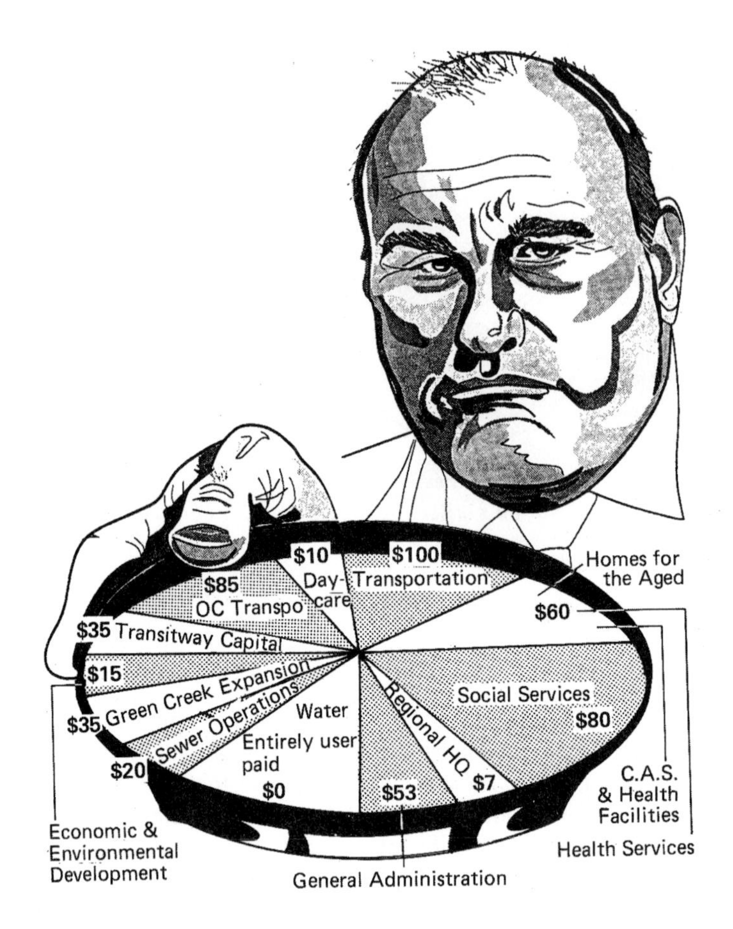

Figure 2.4 An ineffective pie chart.

readers every weekday—more than the next four area newspapers combined."[3] This circulation advantage is made even more impressive by the use of two-dimensional trucks to construct the bar chart.

When the scale of a truck or other graphic is increased, it is made both taller and wider. A figure that is twice as high as another is also twice as wide and therefore covers 4 times as much area on the page. While the numbers are 2 to 1, the picture appears to be 4 to 1. This is why artists who want to exaggerate differences use two-dimensional representations of one-dimensional data. An even greater exaggeration can be created with three-dimensional graphs, since a figure that is twice as high, wide, and deep as another has 8 times the volume.

In the figure below, our attention is drawn to the area occupied by the trucks, not their heights. The truck of *The Times* is about 3.6 times taller than that of the nearest competitor, reflecting the fact that the circulation of *The Times* is 3.6 times that of *The Examiner.* Because it is also 3.6 times wider, the truck of *The Times* fills an area 3.6(3.6) = 13.0 times that of *The Examiner.* The circulation numbers say that *The Times* has 27 percent more readers than the other four newspapers combined. The picture says that *The Times* dwarfs the competition. This visual image is very effective, but misleading.

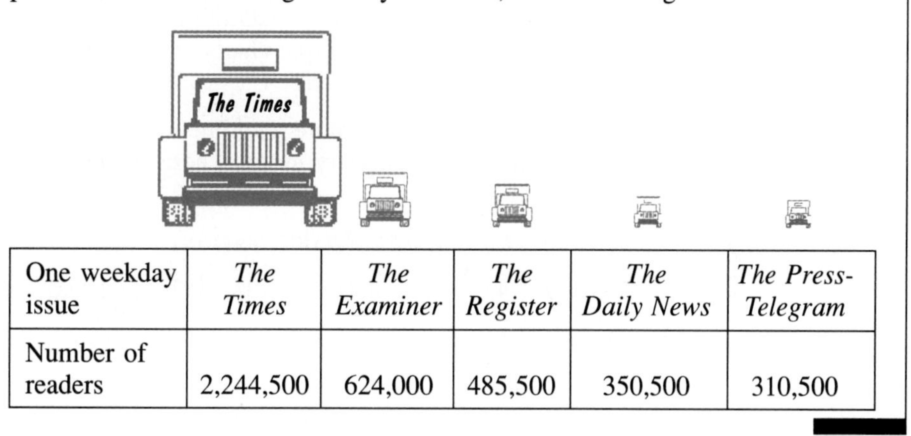

One weekday issue	*The Times*	*The Examiner*	*The Register*	*The Daily News*	*The Press-Telegram*
Number of readers	2,244,500	624,000	485,500	350,500	310,500

2.1 Figure 2.1 shows a bar chart of purchases (in millions of dollars) of four kinds of athletic footwear. Use the data given in the text to compute the percentage of total athletic footwear purchases spent on each type of shoe. Now show these percentages in a bar chart. In what ways does your graph look different from the bar chart in Fig. 2.1?

2.2 Use a bar chart to display these data from a 1996 survey of 64 college students (roughly half female and half male) on the number of biological children they expect to have during their lifetimes.[4]

NUMBER OF CHILDREN	0	1	2	3	4	5	6
NUMBER OF RESPONSES	7	5	31	17	2	1	1

Looking at your graph, write a one-sentence summary of your data.

2.3 Here are 1995 data on the U.S. population and mobile home sales, grouped by geographic region:[5]

	POPULATION (MILLIONS)	MOBILE HOME SALES (THOUSANDS)
NORTHEAST	51.5	14.4
MIDWEST	61.8	55.5
SOUTH	91.9	195.2
WEST	57.6	41.0

a. Use a pie chart to show mobile home sales by region.
b. Use a pie chart to show the population by region.
c. Solely on the basis of these pie charts, do each region's sales seem roughly proportional to its population?

2.4 The 1980 census found that more than one-half of the people in Massachusetts identified with a single ancestry group. These single-ancestry people were divided as follows:

IRISH	ENGLISH	ITALIAN	FRENCH	PORTUGUESE	POLISH	OTHER
665,119	459,249	430,744	313,557	190,034	161,259	947,003

Use a pie chart to display these data and then explain why a pie chart may not be effective.

2.5 What is misleading about the following graphic comparing the 1980 populations of three large eastern cities?

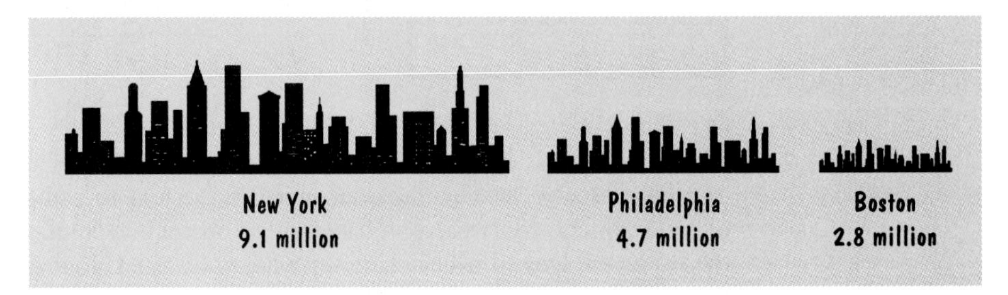

2.2 Histograms

Table 2.1 shows the annual returns from four different types of investments. Treasury bills and Treasury bonds are securities issued by the U.S. Treasury to finance the federal deficit. Treasury bills mature within 1 year of issuance; Treasury bonds are for a longer term. Corporate bonds are long-term bonds issued by U.S. corporations. Corporate stock shareholders are the legal owners

Table 2.1

Annual percentage returns from four investments, 1971–1995

	TREASURY BILLS	TREASURY BONDS	CORPORATE BONDS	CORPORATE STOCK
1971	4.4	13.2	11.0	14.3
1972	3.8	5.7	7.3	19.0
1973	6.9	−1.1	1.1	−14.7
1974	8.0	4.4	−3.1	−26.5
1975	5.8	9.2	14.6	37.2
1976	5.1	16.8	18.7	23.8
1977	5.1	−.7	1.7	−7.2
1978	7.2	−1.2	−.1	6.6
1979	10.4	−1.2	−4.2	18.4
1980	11.2	−4.0	−2.6	32.4
1981	14.7	1.9	−1.0	−4.9
1982	10.5	40.4	43.8	21.4
1983	8.8	.7	4.7	22.5
1984	9.9	15.5	16.4	6.3
1985	7.7	31.0	30.9	32.2
1986	6.2	24.5	19.9	18.5
1987	5.5	−2.7	−.3	5.2
1988	6.4	9.7	10.7	16.8
1989	8.4	18.1	16.2	31.5
1990	7.8	6.2	6.8	−3.2
1991	5.6	19.3	20.0	30.6
1992	3.5	8.1	9.4	7.7
1993	2.9	18.2	13.2	10.0
1994	3.9	−7.8	−5.8	1.3
1995	5.6	31.7	27.2	37.4

Source: Roger G. Ibbotson and Rex A. Sinquefield, *Stocks, Bonds, Bills and Inflation: 1996 Yearbook,* Ibbotson Associates, Inc., Chicago, 1996.

of a company and receive dividends paid by the company out of its profits. The corporate stock data in Table 2.1 refer to the average return on the 500 prominent companies monitored by the Standard & Poor's 500-stock index. The returns in Table 2.1 include both income (bond coupons or stock dividends) and capital gains or losses caused by changes in the market prices of these securities.

For now, we focus on the annual stock returns. One way to summarize these 25 observations is to separate the data into intervals. For instance, we might divide the data into years when investors made money and years when they lost money. A simple count of the data in Table 2.1 shows that stock returns were positive in 20 years and negative in 5 years. Thus, one useful summary of these stock return data is that investors had positive returns in 20 of 25 years, with the best year a 37.4 percent profit in 1995 and the worst year a 26.5 percent loss in 1974.

For more detailed information, we can separate the data into more intervals. In how many years were the annual returns between 0 and 10 percent? Between 10 and 20 percent? Table 2.2 shows the results of grouping the returns into seven intervals and calculating the relative frequency with which observations fall into each interval. For example, in .04 (or 4 percent) of the years returns were between −30 and −20.1 percent. The intervals have been written as 10.0 to 19.9 and 20.0 to 29.9, rather than 10 to 20 and 20 to 30, to make it clear that a borderline value, like 20.0, is placed in the upper interval.

Drawing Histograms

Figure 2.5 shows a bar chart for the stock return data in Table 2.2, with the height of each bar equal to the relative frequency. If, as here, the intervals are equally wide, a bar chart depicts the data accurately. If, however, the intervals are not equally wide, a bar chart is misleading. Suppose, for example, that we combine the intervals from 20.0 to 29.9 and from 30.0 to 39.9, getting the six intervals shown in Table 2.3. If we graph these relative frequencies, as in Fig. 2.6, we get a misleading picture of the data. The wider interval, from 20.0 to 39.9, contains 36 percent of the data but occupies 53 percent of the area covered by these bars. Because we tend to gauge size by area rather than by height, the bar spanning the wider interval gives the misleading impression that 53 percent of the years had returns of 20 percent or more, when in fact only 36 percent did.

To correct this optical illusion, we can use a **histogram,** in which the relative frequency of the observations in each interval is shown by the *area* of a

Table 2.2

Annual stock returns, 1971–1995

PERCENTAGE RETURN	NUMBER OF YEARS	RELATIVE FREQUENCY
−30.0 to −20.1	1	.04
−20.0 to −10.1	1	.04
−10.0 to −.1	3	.12
.0 to 9.9	5	.20
10.0 to 19.9	6	.24
20.0 to 29.9	3	.12
30.0 to 39.9	6	.24
	25	1.00

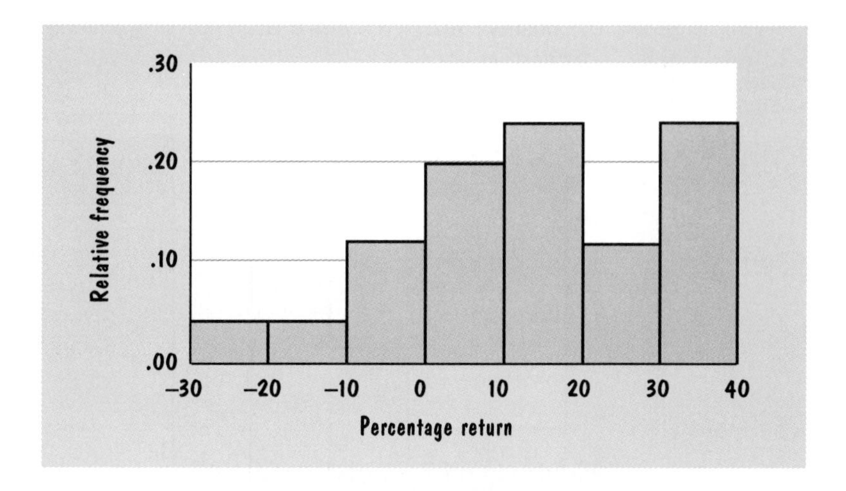

Figure 2.5 A bar chart of annual stock returns, 1971–1995.

bar spanning the interval. Remembering that the area of a bar is equal to its width times its height, we want

$$\text{Relative frequency} = (\text{bar width})(\text{bar height})$$

Solving for the bar's height gives

$$\text{Bar height} = \frac{\text{relative frequency}}{\text{bar width}}$$

These requisite heights are shown in Table 2.3. The last column in this table

Table 2.3

Annual stock returns, in intervals of unequal widths

PERCENTAGE RETURN	NUMBER OF YEARS	RELATIVE FREQUENCY	HISTOGRAM BAR HEIGHT (FREQUENCY/WIDTH)	BAR AREA (WIDTH × HEIGHT)
−30.0 to −20.1	1	.04	.04/10 = .004	10(.004) = .04
−20.0 to −10.1	1	.04	.04/10 = .004	10(.004) = .04
−10.0 to −.1	3	.12	.12/10 = .012	10(.012) = .12
.0 to 9.9	5	.20	.20/10 = .020	10(.020) = .20
10.0 to 19.9	6	.24	.24/10 = .024	10(.024) = .24
20.0 to 39.9	9	.36	.36/20 = .018	20(.018) = .36
	25	1.00		1.00

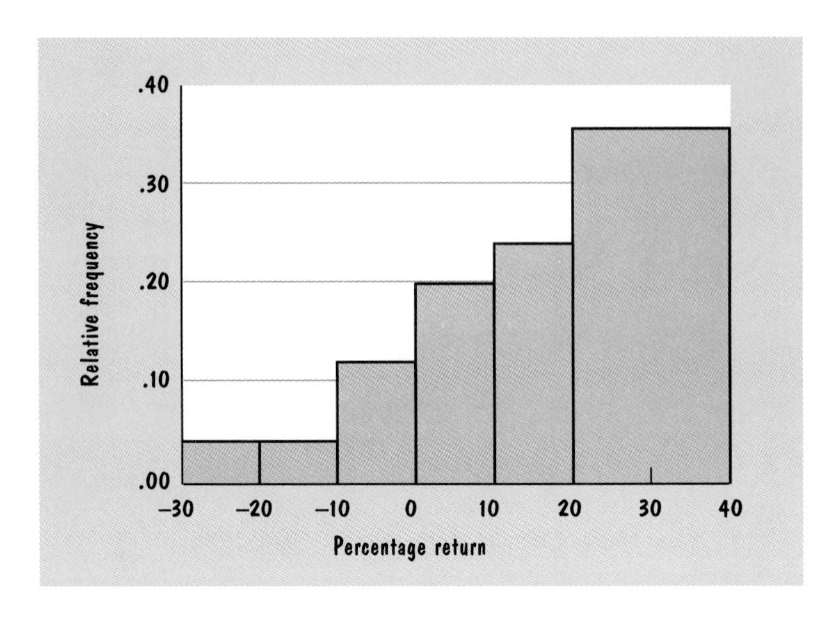

Figure 2.6 A misleading bar chart.

shows that, by using these bar heights, the area of each bar is equal to that interval's relative frequency and that the total area covered by the histogram is equal to 1.

A histogram's bar height is a "density" in that it measures the relative frequency per unit of the variable we are analyzing. The first interval in Table 2.3, from −30.0 to −20.1, spans 10 percentage points. A fraction .04 of the data is in this interval, an average of .04/10 = .004 per percentage point. The last interval, from 20.0 to 39.9, spans 20 percentage points. A fraction .36 of the data is in this interval, an average of .36/20 = .018 per percentage point. If the data were measured in years, the histogram bar heights would be frequencies per year. If the data were measured in pounds, the bar heights would be frequencies per pound. The general term *density* covers all these cases. Because the term is admittedly difficult to grasp and because histograms are standardized to have a total area equal to 1, histograms are often drawn with no label on the vertical axis or no vertical axis at all!

Figure 2.7 shows a correctly drawn histogram, with the bar heights adjusted for the interval widths. If we compare the two bars in Fig. 2.5 for the intervals from 20.0 to 29.9 and from 30.0 to 39.9 with the single bar in Fig. 2.7 for the combined interval from 20.0 to 39.9, we see that the single bar is just an average of the two separate bars. The single bar loses some of the detail provided by two separate bars, but does not misrepresent how often stock returns were 20 percent or higher.

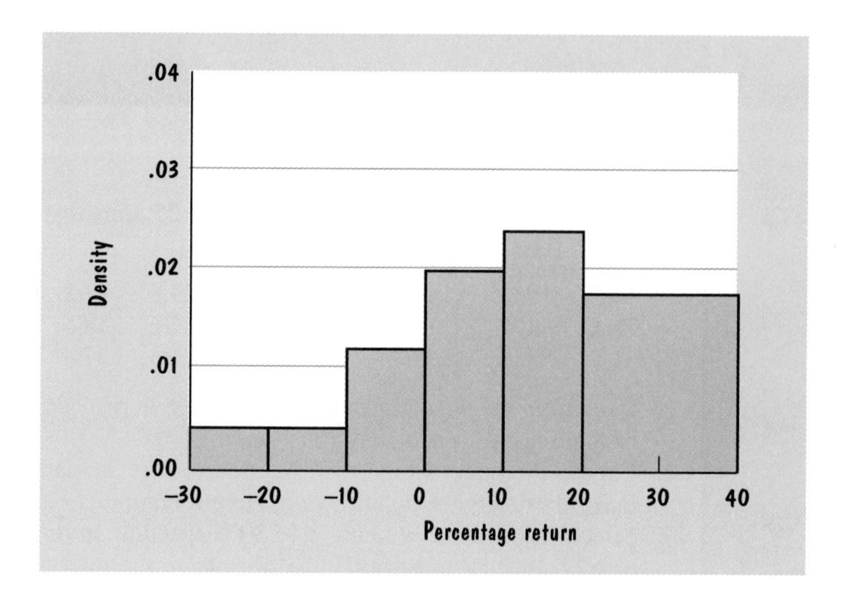

Figure 2.7 A histogram of annual stock returns, 1971–1995.

Within reasonable bounds, the number of intervals used in a histogram is a matter of taste. However, a few guidelines can be given. Histograms with fewer than four intervals usually do not convey enough information to warrant drawing a figure at all. If all we want to say is that 20 of 25 annual stock returns were positive and 5 were negative, we can express this in words more easily than in graphs. At the other extreme, if the number of intervals is too large relative to the number of observations, a histogram can turn into a jumble of narrow spikes, with each representing only a few observations. Usually data are divided into 5 to 15 intervals, and we use more intervals when we have more data.

In choosing the width and location of the intervals, the simplest approach is equal-width intervals that begin with round numbers, such as 10 and 20, and cover all the data. Some intervals may be intrinsically interesting; for example, 0 is often an informative place to end one interval and begin another. With household income data, some of the intervals might coincide with tax brackets. With data on years of schooling, there are natural dividing lines between primary education, high school, college, and graduate school.

If our data have an open-ended interval, such as "more than $100,000 income," it is best to omit this interval entirely. Alternatively, we can close the interval with an assumed value and hope that it represents the data fairly. For instance, if we are working with education data that have an interval "more than 16 years," we might make this interval from 16 to 20 years and assume that this encompasses almost all these data.

HOW TO DO IT

Drawing a Histogram

We want to use a histogram to describe the 25 annual stock returns in Table 2.1:

14.3	19.0	−14.7	−26.5	37.2	23.8	−7.2	6.6	18.4
32.4	−4.9	21.4	22.5	6.3	32.2	18.5	5.2	16.8
31.5	−3.2	30.6	7.7	10.0	1.3	37.4		

1. Determine the smallest and largest values; here, −26.5 and 37.4. The histogram must encompass this range.
2. Choose the intervals that will be used to sort the data; if possible, make the intervals equally wide. Here a natural choice is seven 10 percent intervals, beginning at −30 and ending at 40. To illustrate the case of unequal interval widths, we merge the intervals from 20.0 to 29.9 and from 30.0 to 39.9.
3. Determine how many observations are in each interval, and then calculate the relative frequencies. The histogram bar heights are equal to the interval frequencies divided by the interval widths:

INTERVAL	NUMBER	FREQUENCY	BAR HEIGHT
−30.0 to −20.1	1	1/25 = .04	.04/10 = .004
20.0 to −10.1	1	1/25 = .04	.04/10 = .004
10.0 to −.1	3	3/25 = .12	.12/10 = .012
.0 to 9.9	5	5/25 = .20	.20/10 = .020
10.0 to 19.9	6	6/25 = .24	.24/10 = .024
20.0 to 39.9	9	9/25 = .36	.36/20 = .018
	25	1.00	

Conveying Information

A histogram is intended to help us understand the data. One of the first things we notice in a histogram is the center of the data. Figure 2.8 shows how a histogram for stock returns might be centered at 0 to 20 percent, or at 30 to 50 percent, or might have multiple peaks. A second consideration, as in Fig. 2.9, is whether the data are bunched closely around the center of the data or are widely dispersed.

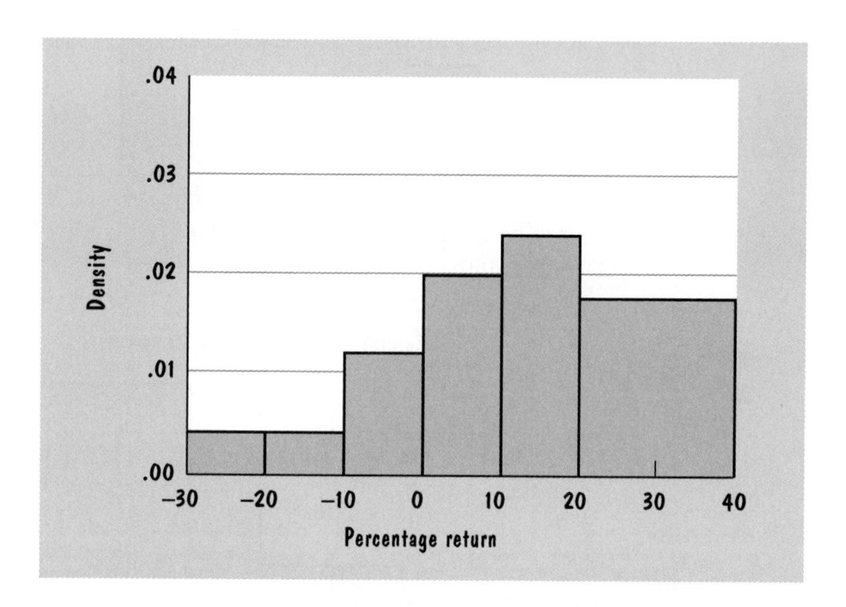

4. If the intervals are equally wide, we can draw a simple bar chart with the bar heights showing the relative frequency of data in each interval. If the interval widths are unequal, we need to draw a histogram with the height of each bar equal to the relative frequency divided by the width of the interval, so that the relative frequencies are given by the areas of the bars.

We also notice if the histogram is roughly symmetric about the center, as in Fig. 2.10*a*. This histogram is perfectly symmetric in that the bars to the right of 10 percent are an exact mirror image of the bars to the left of 10 percent. Data are seldom perfectly symmetric, but are often roughly so. However, histograms sometimes have an asymmetric shape that resembles a playground slide: steep on the side with the ladder and then declining more gradually on the side with the slide. The asymmetric data in Fig. 2.10*b* are said to be *positively skewed,* or skewed to the right, in that the upper half of the data runs off to the right, while the bottom half of the data is bunched together. Data that are negatively

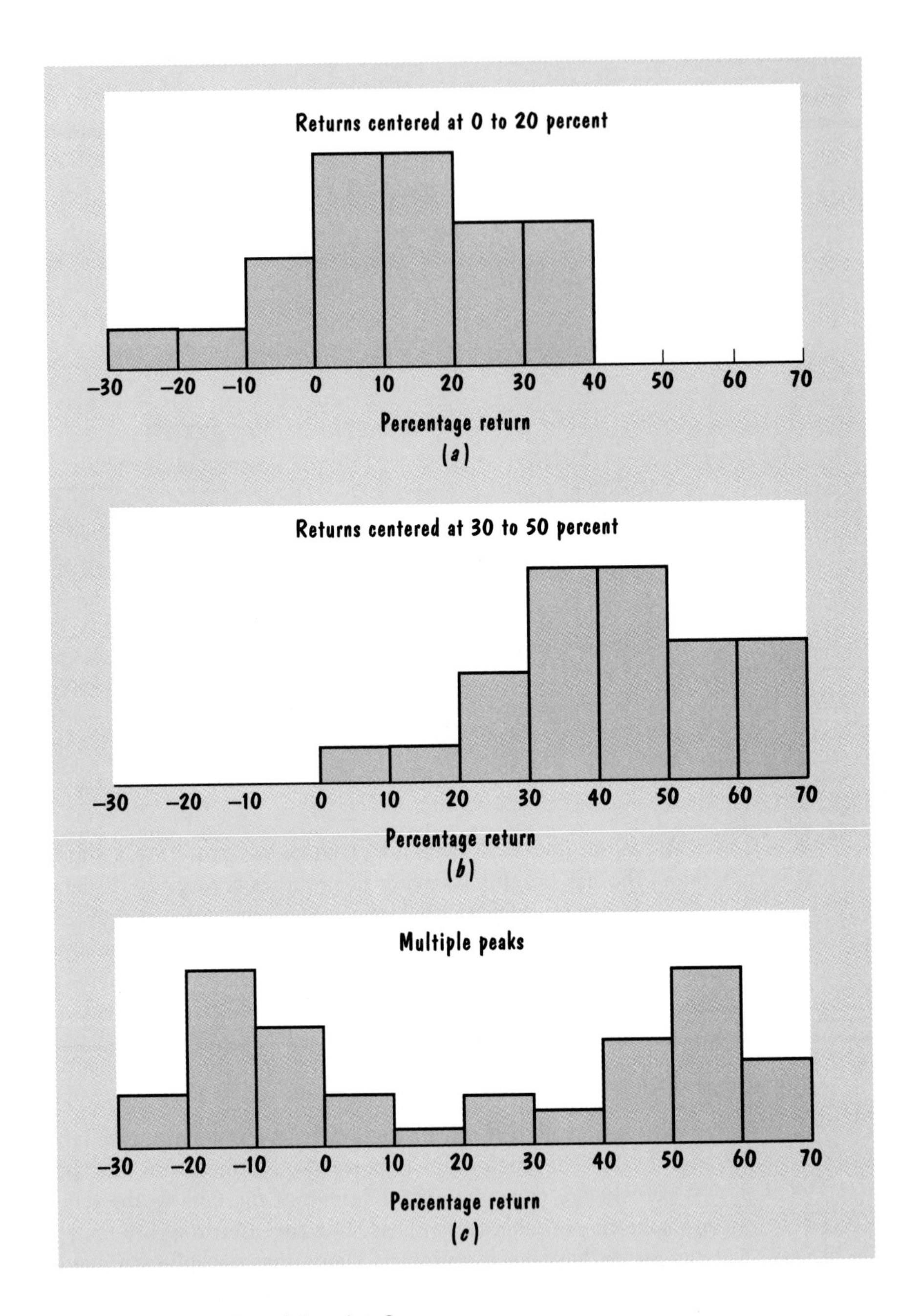

Figure 2.8 Where is the center of the data?

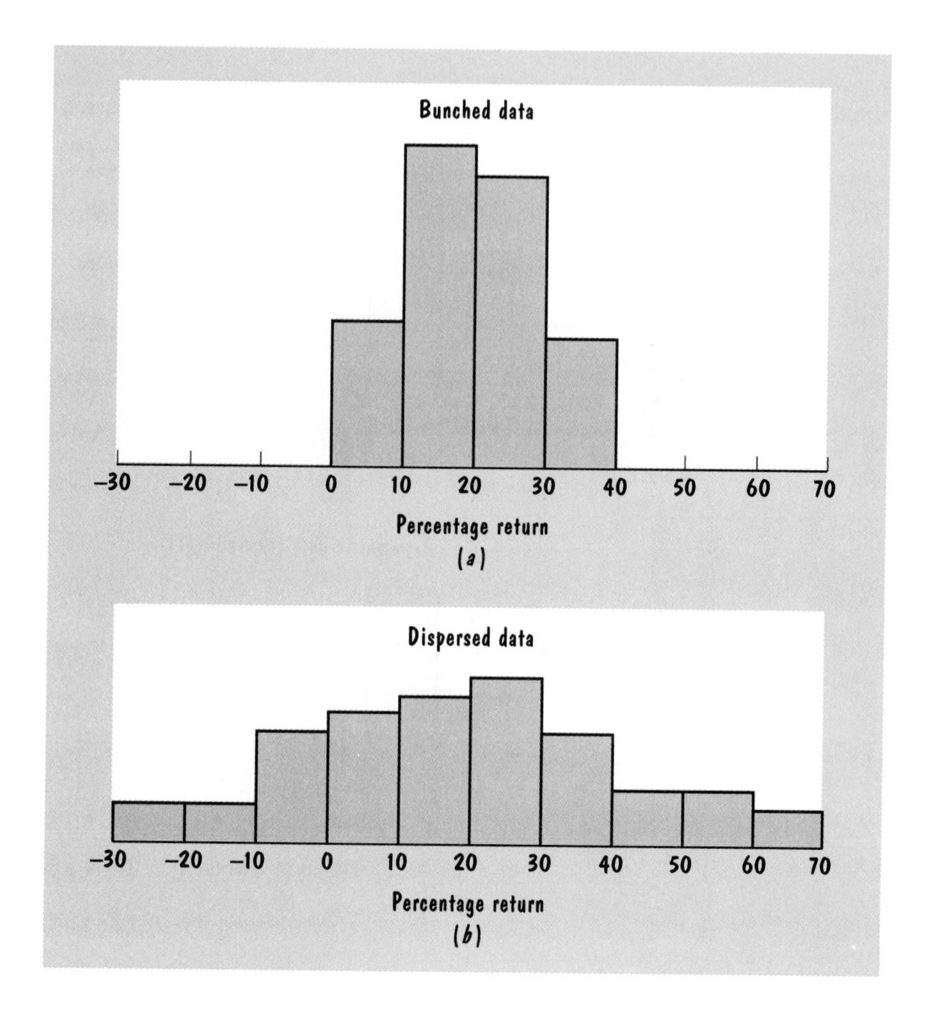

Figure 2.9 Are the data bunched or dispersed?

skewed run off to the left, for example, the stock returns in the how-to-do-it box.

We also notice any *outliers*—values that are very different from the other observations. If one of the values is very different from the rest, as in Fig. 2.11, we should check the data to see which value is so unusual and why. Sometimes an outlier is just a clerical error—the misplacement of a decimal point. Sometimes, it reflects a unique situation—perhaps the year that an immense natural disaster occurred—and it can be discarded if we are not interested in such unusual events. In other cases, the outlier may be a very interesting observation

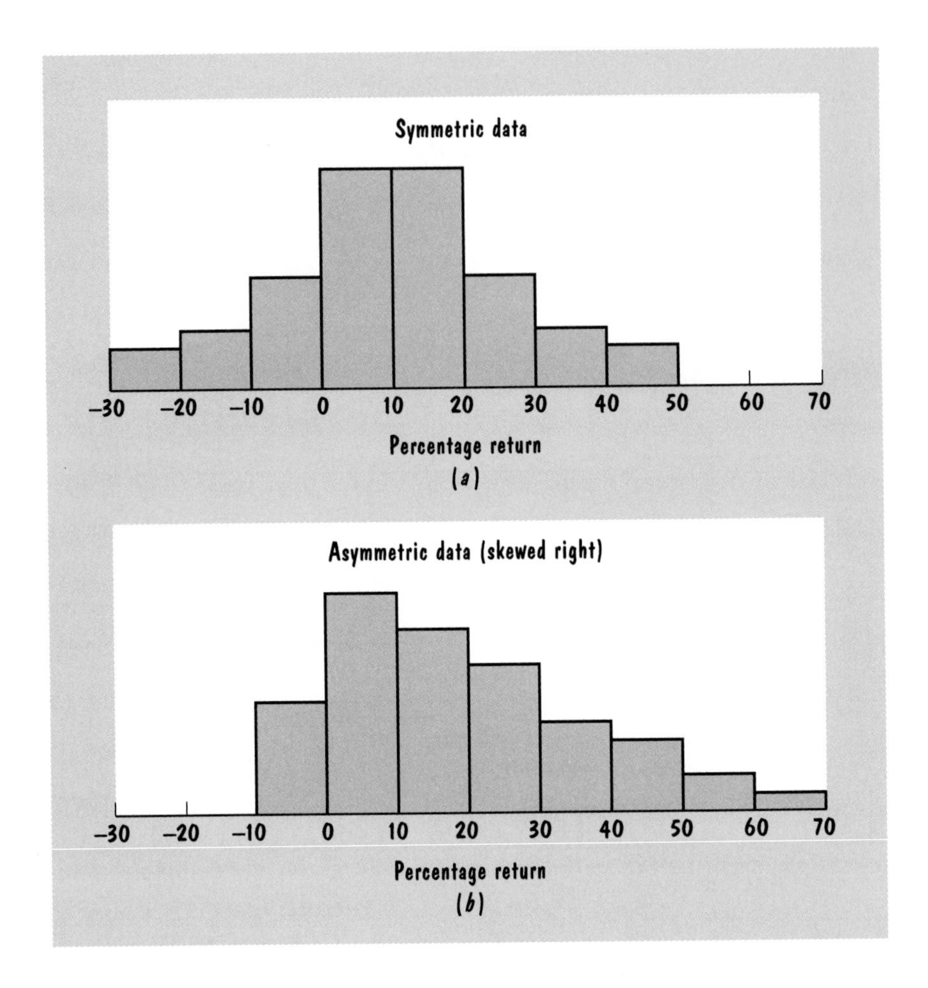

Figure 2.10 Are the data symmetric?

that should be investigated further. It might be very interesting to see how the stock market reacts to a natural disaster, a Presidential assassination, or the end of a war.

A dramatic example of outliers occurred in the 1980s when scientists discovered that the computer software analyzing satellite ozone readings over the South Pole had been automatically omitting a substantial number of very low readings because these were outliers in comparison to readings made in the 1970s. The computer program assumed that readings this far from what had been normal in the 1970s must be mistakes. When scientists reanalyzed the ozone data, including the previously ignored outliers, they found that an omi-

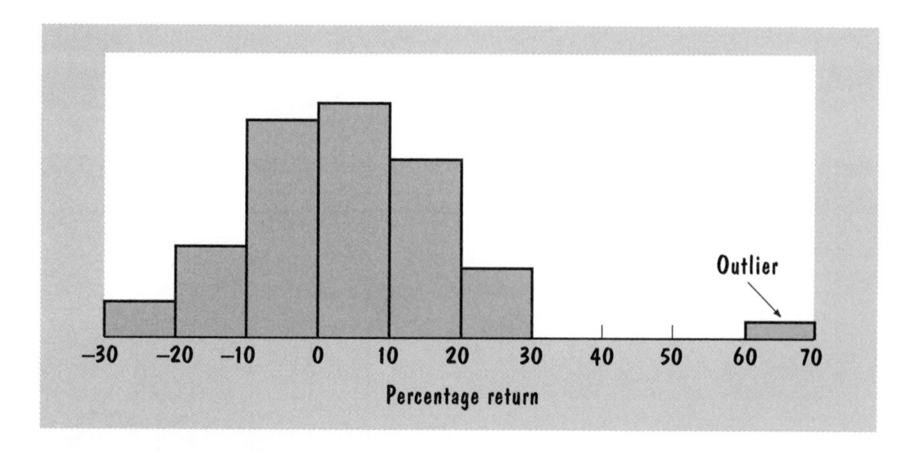

Figure 2.11 Look for outliers.

nous hole in the ozone layer had been developing for several years, with Antarctic ozone declining by 40 percent between 1979 and 1985. Susan Solomon, of the National Oceanic and Atmospheric Administration's Aeronomy Laboratory, said, "This is a change in the ozone that's of absolutely unprecedented proportions. We've just never seen anything like what we're experiencing in the Antarctic."[6] Outliers may be clerical errors, measurement errors, or flukes that, if not corrected or omitted, will distort the analysis of the data; sometimes, they are the most interesting observations.

Often histograms turn out to be roughly bell-shaped, with a single peak and the bars declining symmetrically on either side of the peak—first gradually, then more quickly, and then gradually again. The histogram in Fig. 2.12 of annual stock returns going back to 1926 (the earliest year that data are available) has a very rough bell shape. An even more pronounced bell shape is shown in Fig. 2.13, which is a histogram of the chest measurements (in inches) of 5738 Scottish militiamen in the early 19th century.[7] This bell-shaped curve is known as the *normal distribution* and is encountered frequently in later chapters.

Chart Junk

The bars that are used in bar charts and histograms do nothing more than show relative frequencies. Some well-meaning people add lines, blots, and splotches, apparently intending to enliven the graphs, but too often creating what Edward Tufte calls "chartoons" or "chartjunk," as in Fig. 2.14—unattractive graphics that distract the reader and strain the eyes.[8] Another way to create chart junk is to use multiple colors, preferably bright and clashing. Computers also make

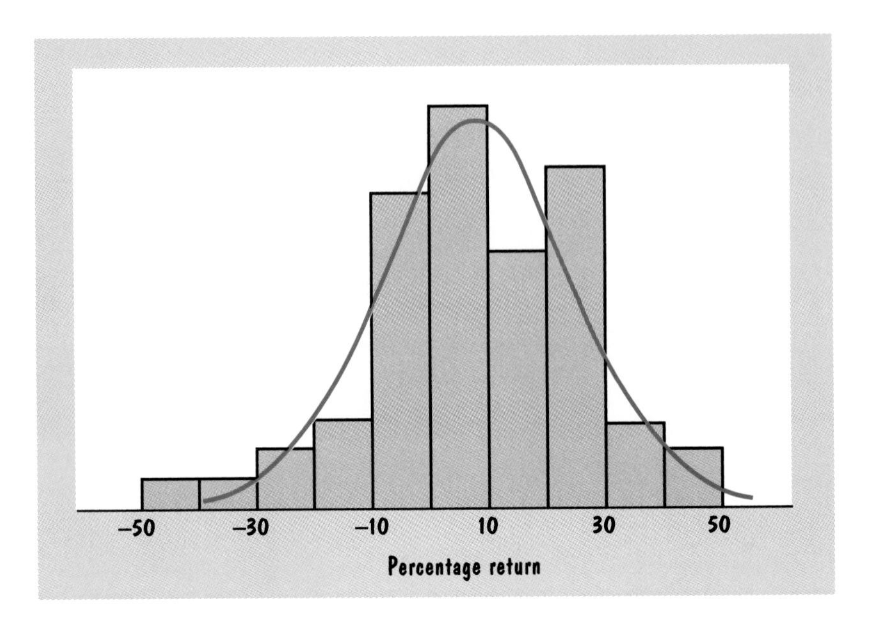

Figure 2.12 A histogram of annual stock returns, 1926–1995.

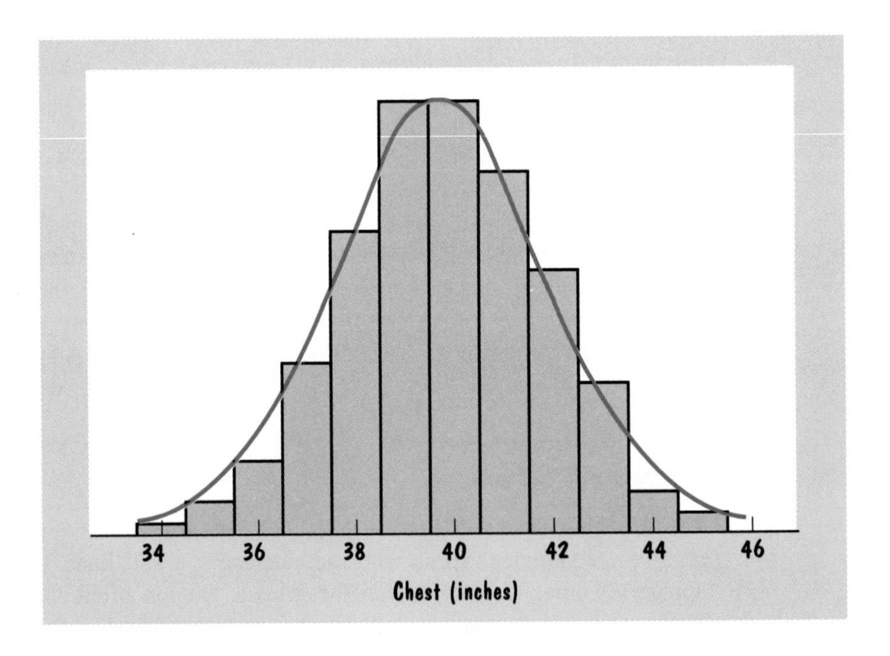

Figure 2.13 Chest measurements of Scottish militiamen, early 19th century.

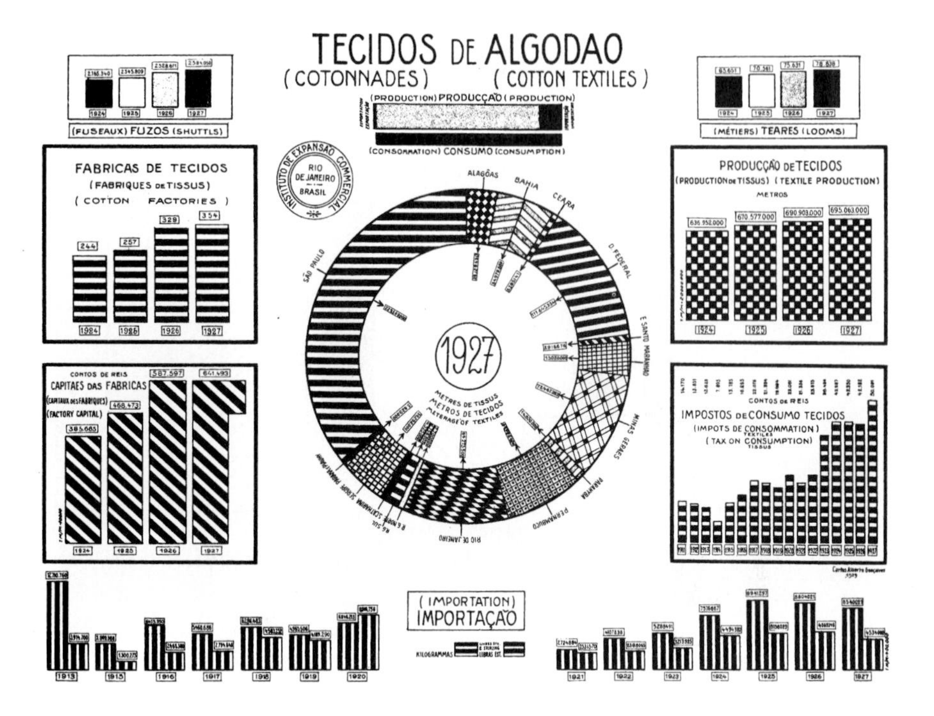

Figure 2.14 Chart junk.

possible text junk—printed documents that look like ransom notes that are made by pasting together characters in mismatched sizes, styles, and fonts. I recently received a two-page newsletter that used 32 different fonts, not counting variation caused by characters that were bold, italic, or in different sizes. It was painful to read.

The purpose of a bar chart or histogram is to reveal information that would not be as apparent or as easily remembered if it were presented in a table. Figure 2.15 is an example from a political science journal that does neither.[9] The nine bars refer to nine types of articles from the front pages of U.S. newspapers, with the heights of the bars showing the percentage of the articles in each category that were critical of a specific person or policy. The order in which the categories are presented is arbitrary; there is no reason, for example, why the inflation bar should precede rather than follow the unemployment bar. The pattern in the bars—dipping, rising, than dipping again—reveals no useful information. It would be more sensible to arrange the bars from tallest to smallest.

The labels for the bars along the horizontal axis are abbreviated and jumbled together, making it difficult to tell which label goes with which bar and what some of the labels mean. Longer versions of the labels appear in a legend to the right of the bar chart, but the similarity in the bar patterns makes

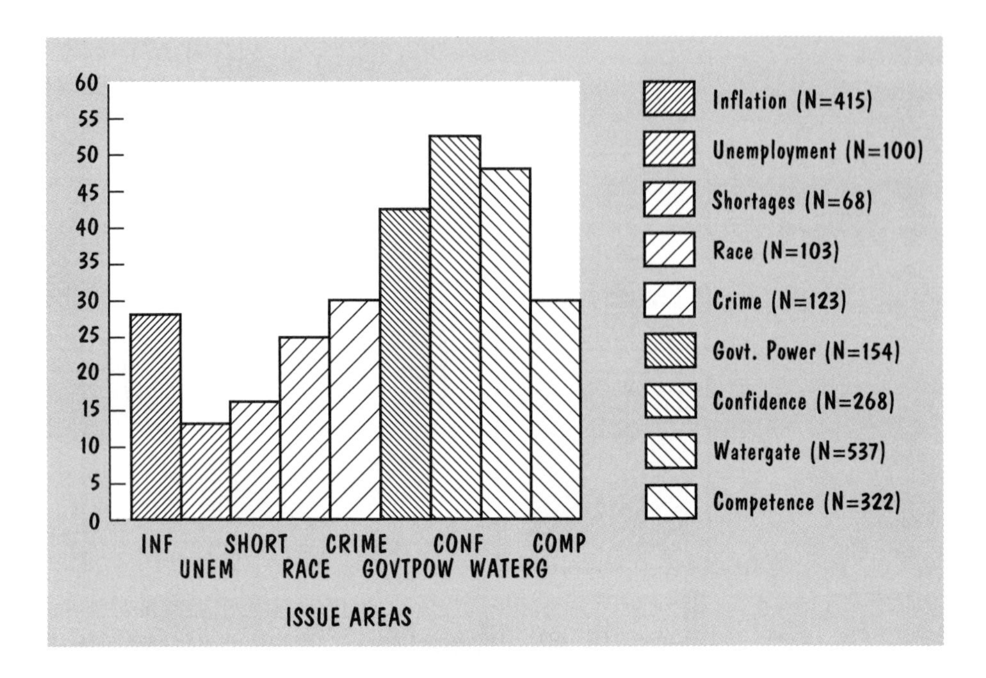

Figure 2.15 An unhelpful bar chart.

it tedious to look back and forth from the bars to the legend and difficult to remember which label goes with which pattern. The bar patterns themselves are distracting and make it hard to focus on the heights of the bars. Finally, if the reader does manage to concentrate on the heights of the bars, it is not easy to line up these heights with the numerical values on the vertical axis. This bar chart is not an improvement over a simple table, such as Table 2.4, showing the percentage of the articles in each category that were critical of a specific person or policy.

Omitting Origins to Stretch the Truth

Sometimes zero is omitted from the vertical axis of a bar chart or histogram to emphasize variations in the bar heights. For example, the figure on the left below appeared in an advertisement that claimed to tell "the simple truth" about the frequency with which insurance companies paid off claims.[10] The vertical axis runs only from 89 to 100, rather than from 0 to 100, to emphasize the fact that this company's 97.5 percent payoff is higher than that of nine

Table 2.4

A table showing the information in Fig. 2.15

TOPIC OF FRONT-PAGE ARTICLES	NUMBER OF ARTICLES	PERCENTAGE CRITICAL
Government honesty	266	52
Watergate scandal	537	49
Government power	154	42
Government competence	322	30
Crime	123	30
Inflation	415	28
Race	103	25
Energy and food shortages	68	16
Unemployment	100	13

competitors. With this interrupted axis, 97.5 percent seems twice the size of 94 percent.

The figure on the right is what the bar chart would look like if zero were not omitted. The payoff rates appear to be essentially the same for all 10 companies, which is not at all the message that the advertising agency intended to convey.

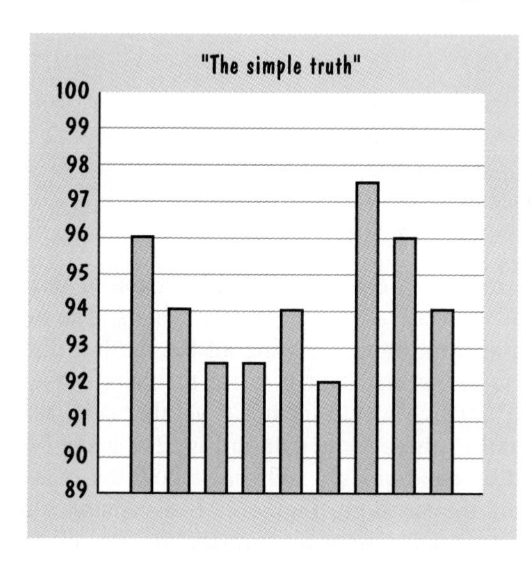

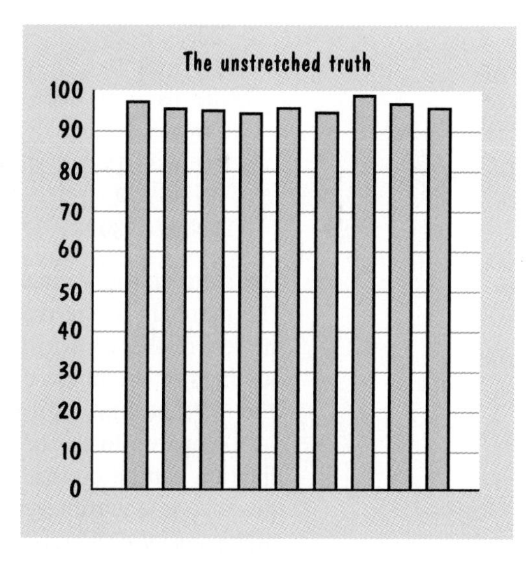

A legitimate reason for omitting zero is to help the reader detect differences that might otherwise be ambiguous. We have to look carefully at the "unstretched truth" figure to determine which bar is the tallest and which is the shortest. By omitting zero, the "simple truth" magnifies the top part of the graph and makes it much easier to identify the tallest and shortest bars. However, once zero has been omitted, we can no longer gauge the relative magnitudes (how much taller and shorter) by comparing the heights of the bars.

Stem-and-Leaf Diagrams*

Instead of cluttering up histogram bars with confusing geometric patterns, John Tukey, a distinguished statistician, invented the **stem-and-leaf diagram,** also called a **stem plot,** in which the bars are replaced by numerical digits, so that they convey more, not less, information about the underlying data. In Tukey's words, "If we are going to make a mark, it may as well be a meaningful one. The simplest—and most useful—meaningful mark is a digit."[11]

Earlier in this chapter, Fig. 2.5 showed a bar chart for the 25 annual stock returns in Table 2.1 grouped into the seven intervals shown in Table 2.2. Here are these 25 returns, rounded to the nearest percent:

14	19	−15	−26	37	24	−7	7	18	32	−5	21	23
6	32	19	5	17	32	−3	31	8	10	1	37	

Now, we put the data in order, from smallest to largest, collecting the returns in seven intervals:

−20 to −29:	−26					
−10 to −19:	−15					
−0 to −9:	−3	−5	−7			
0 to 9:	1	5	6	7	8	
10 to 19:	10	14	17	18	19	19
20 to 29:	21	23	24			
30 to 39:	31	32	32	32	37	37

The stem-and-leaf diagram in Fig. 2.16 is a condensed version of this display. The numbers to the right of the vertical line are the last digits. The numbers to the left of the vertical line label the stem, showing all but the last digit. (For numbers in the range of 0 to 9, the first digit is 0; for numbers in the range of 0 to −9, the first digit is −0, which would include a number such as −3.)

Compared to the bar chart of these same data in Fig. 2.5, a stem-and-leaf diagram can be constructed by hand quickly and shows both the relative frequencies in each interval and the last digit. Figure 2.16 also shows a *back-to-back stem plot,* which uses a common stem to compare two sets of data. In this example, we can see that the annual stock returns have generally been higher but more dispersed than the returns on corporate bonds, although there was a 44 percent bond return in one year.

*Sections marked with an asterisk are optional.

Corporate stock	
−2	6
−1	5
−0	3 5 7
0	1 5 6 7 8
1	0 4 7 8 9 9
2	1 3 4
3	1 2 2 2 7 7

Stem-and-leaf diagram
(*a*)

Corporate bonds		Corporate stock
	−2	6
	−1	5
6 4 3 3 1 0 0	−0	3 5 7
9 7 7 5 2 1	0	1 5 6 7 8
9 6 6 5 3 1 1	1	0 4 7 8 9 9
7 0 0	2	1 3 4
1	3	1 2 2 2 7 7
4	4	

Back-to-back stem plot
(*b*)

Figure 2.16 Stem-and-leaf diagrams for annual returns, 1971–1995.

Stem-and-leaf diagrams can also show the last two digits of each number and can be modified in other ways to handle various kinds of data. Tukey's intention is to facilitate what he calls *exploratory data analysis*—"looking at data to see what it seems to say."[12] For small data sets, stem-and-leaf diagrams are an easy way of sorting the data. As with a histogram, among the things statisticians look for when examining a stem-and-leaf diagram is the center of the data, the number of peaks, whether the data are compact or dispersed, whether the figure is symmetric, and whether there are outliers.

Identifying Authors

Authorship is sometimes in doubt because of a pseudonym or inaccurate claim. Did Conan Doyle really write the Sherlock Holmes novels? Some suspect Dr. Watson. Were the plays attributed to Shakespeare written by Christopher Marlowe? Did Cervantes write novels about Don Quixote, or did he take credit for the work of his servant, Sancho Panza? Did Mark Twain write letters signed *Quintus Curtius Snodgrass,* chronicling a southern soldier's adventures during the Civil War?

Two statisticians used the Federalist papers to show how such claims might be evaluated.[13] In 1778 and 1788, a series of 77 essays urging New Yorkers to ratify the U.S. Constitution, signed with the pseudonym *Publius,* appeared in newspapers and then were published as a book, entitled *The Federalist.* Five of these essays were written by John Jay; the rest were written by Alexander Hamilton and James Madison, although in 12 cases Hamilton and Madison (and later historians) disagreed as to which of these two was the author.

The statisticians compared the frequency with which certain words were used in papers known to have been written by Hamilton and Madison with the frequency in the disputed Federalist papers. Hamilton almost always used *while* instead of *whilst,* and Madison did the opposite; *while* appears in none of the 12 disputed papers, and *whilst* appears in 5 of them, suggesting that these 5 papers were written by Madison. Similarly, *upon* was used often by Hamilton and seldom by Madison and appears in only one of the disputed papers. A less extreme case concerns *by,* which was used by both, but more often by Madison, as shown in the bar charts below. After analyzing similar data for 30 key words, the statisticians concluded that the evidence strongly points to Madison as the author of all 12 papers.

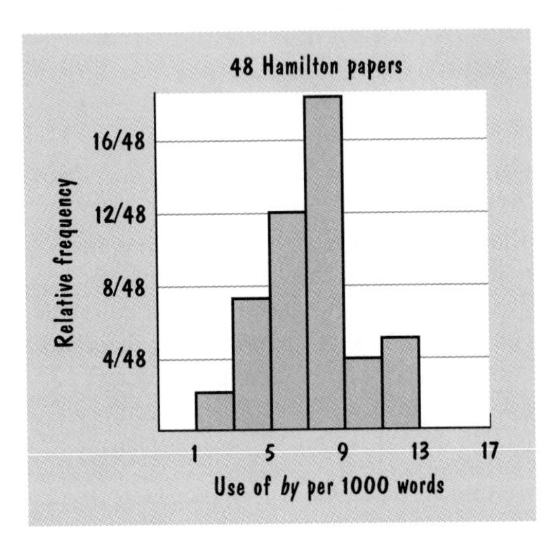

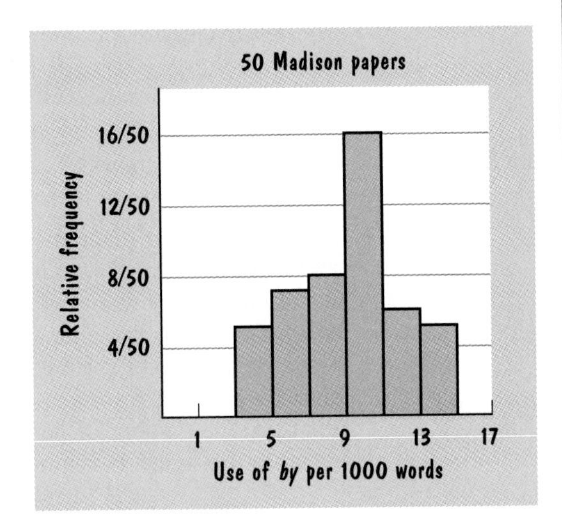

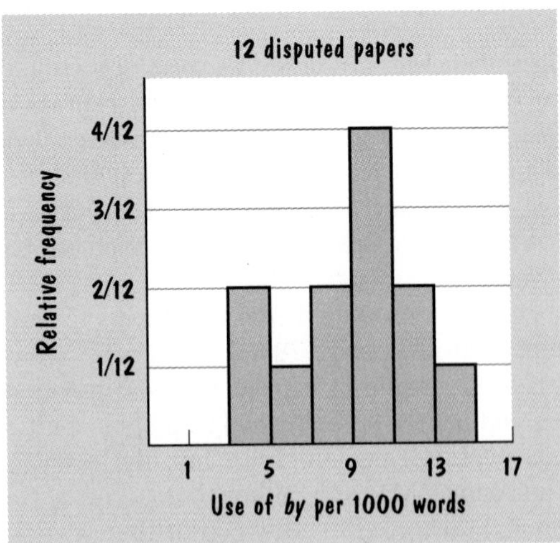

Detecting Cheating

In 1984 a student (whom we will call C, for cheater) was accused of cheating on an examination after the proctor reported that this student seemed to be copying the answers of an adjacent student (I, for innocent).[14] There were 40 multiple-choice questions on this test, and at the student's judiciary hearing, the prosecution showed that of the 16 questions missed by both C and I, the same wrong answer was given in 13 cases. A probability calculation showed that there was very little chance that guessing would result in so many matches, and student C was found guilty of academic dishonesty.

The student appealed, arguing (among other things) that the probability calculation used at the trial incorrectly assumes that each wrong answer to a question is equally likely to be chosen. In fact, some wrong answers are more appealing than others, and a student who makes a mistake on one question because of a misunderstanding of the material may make a similar mistake on other questions.

At a second trial, the prosecution tried an ingenious alternative method for demonstrating that, in comparison with the rest of the class, there was an unusually close correspondence between the answers chosen by C and I. Using I's answers as an answer key, all student tests were regraded. The number of "right" answers in this regrading gives the number of matches between each student's test and I's test. The bar chart below shows the results. Student C gave the same answers as I for 32 questions. Of the 86 other students in this class, the next closest match is 27 questions and most students had fewer than 21 matches. The bar chart reveals student C to be an outlier with an unusually (and suspiciously) close correspondence with I's answers. Nonetheless, C was acquitted after the defense pointed out that this close statistical match could have been

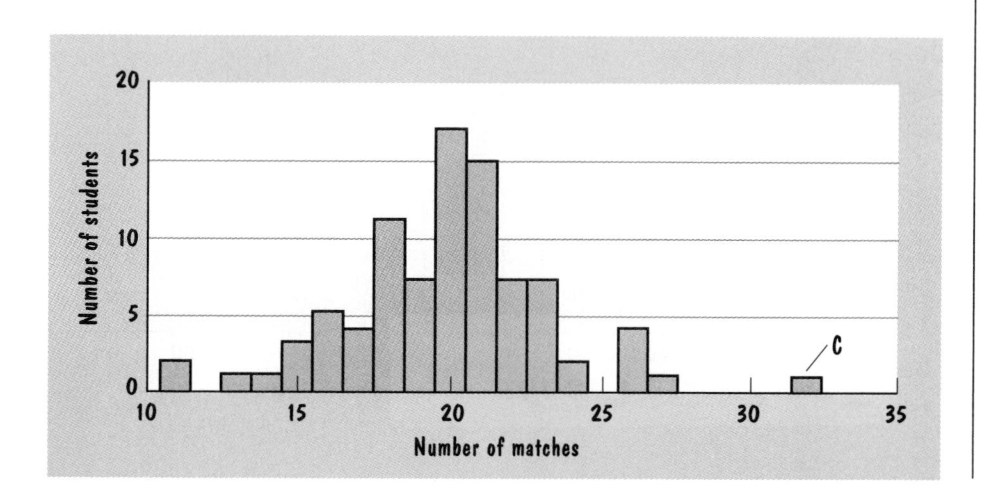

due to I's copying from C, rather than the other way round. The jury evidently forgot or was not impressed by the fact that the initial accusation was based on the proctor's report that C seemed to be copying from I.

Old Semifaithful

Yellowstone National Park has more than 10,000 natural geysers and hot springs. The most famous geyser is Old Faithful, which several times each day has a stunning eruption that lasts from 1 to 5 minutes and sends hot water and steam more than 100 feet in the air. Local tourism promoters used to claim that Old Faithful erupted like clockwork, every hour on the hour. The eruptions have slowed down over the years, but the time between eruptions has never been constant.

The National Park Service rangers in Yellowstone use an infrared eye to record the time between eruptions. Shown below is a histogram summarizing 112 observations made during the daylight hours for 2 weeks in September 1995.[15] This histogram shows considerable variation, with some eruptions less than 50 minutes apart and others nearly 2 hours apart. The histogram also reveals two peaks, one around 60 minutes and one around 90 minutes. Clearly, the time between eruptions is not constant.

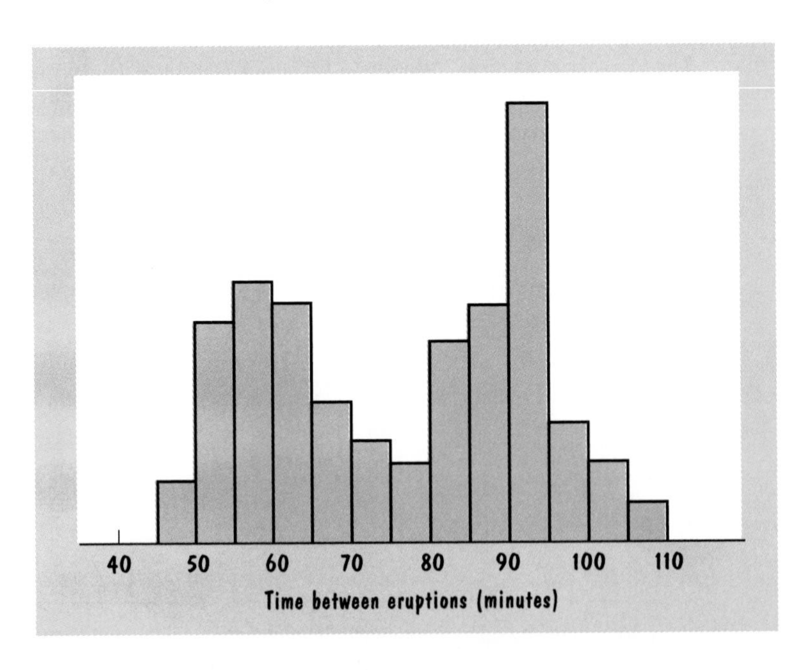

2.6 Use a bar chart with relative frequencies on the vertical axis to display these data on the age of U.S. women who gave birth in 1993:

AGE OF WOMEN	15–19	20–24	25–29	30–34	35–39	40–44
NUMBER (THOUSANDS)	501	1038	1129	901	357	61

Now write a sentence or two summarizing the most important information conveyed by this bar chart.

2.7 Use a histogram with intervals of 0 to 9.99, 10 to 19.99, 20 to 29.99, and 30 to 39.99 to summarize these data on annual inches of precipitation from 1961 to 1990 at the Los Angeles Civic Center. Show your work and write a brief summary of what your histogram tells you about these data.

5.83	15.37	12.31	7.98	26.81	12.91	23.66	7.58	26.32	16.54
9.26	6.54	17.45	16.69	10.70	11.01	14.97	30.57	17.00	26.33
10.92	14.41	34.04	8.90	8.92	18.00	9.11	9.98	4.56	6.49

2.8 Based solely on your own personal opinion, draw a rough sketch of a histogram of the ages of the faculty at your college, using the intervals 20-40, 40-50, 50-60, and 60-80 years. Be sure to explain your reasoning and label the vertical axis clearly.

2.9 What is misleading about this 1953 newspaper advertisement showing the increase in the average weekly sales of *Collier's* magazine?[16]

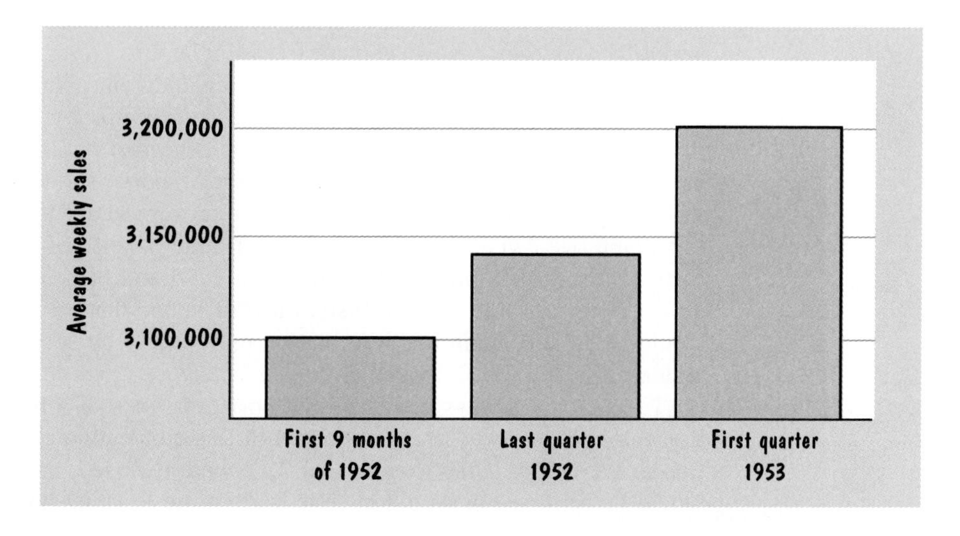

2.10 Draw a back-to-back stem plot, using the annual returns in Table 2.1 for Treasury bills and Treasury bonds. What important differences does this diagram reveal between the annual returns on Treasury bills and those on Treasury bonds?

2.3 Time-Series Graphs

Bar charts, pie charts, and histograms are graphical techniques for showing the frequencies with which data have certain values or fall into specified intervals. *Time series data* are collected at different times, and our concern is often with how these data vary over time, for example, whether there are trends or seasonal patterns. In a **time-series graph,** the variation in data over time is shown by a plot with the data on the vertical axis and time on the horizontal axis.

William Playfair, an English political economist, was one of the first to construct time-series graphs using economic data. In a book of these graphs published in 1786, he argued that someone

> who has carefully investigated a printed table finds, when done, that he has only a very faint and partial idea of what he has read; and that like a figure imprinted on sand, is soon totally erased and defaced. . . . [However] on inspecting any of these Charts attentively, a sufficiently distinct impression will be made to remain unimpaired for a considerable time, and the idea which does remain will be simple and complete.[17]

Figure 2.17 shows one of the graphs in his remarkable book. The line that is generally uppermost shows exports from England to North America from 1770 to 1782; the other line shows imports to England from North America over this same time period. The shaded area labeled *ballance in favour of England* identifies those periods when England's exports exceeded its imports. The small area in the middle, labeled *ballance against England,* shows a brief period when England's imports from North America were larger than its exports. From this graph, we can see that England had a large balance-of-trade surplus in the early 1770s, which turned into a deficit in 1775 and then returned to a surplus, although not as large as previously. Playfair hoped that an attentive reader would grasp and retain this information more easily by seeing a graph than by reading a table.

Time should always be put on the horizontal axis of a time-series graph, as this helps readers who are accustomed to this convention absorb the information in the graph. If the axes in Fig. 2.17 were reversed, the audience either would be confused or would have to waste time learning how to read a graph

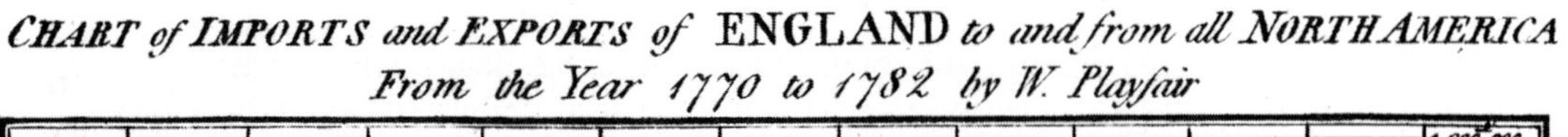

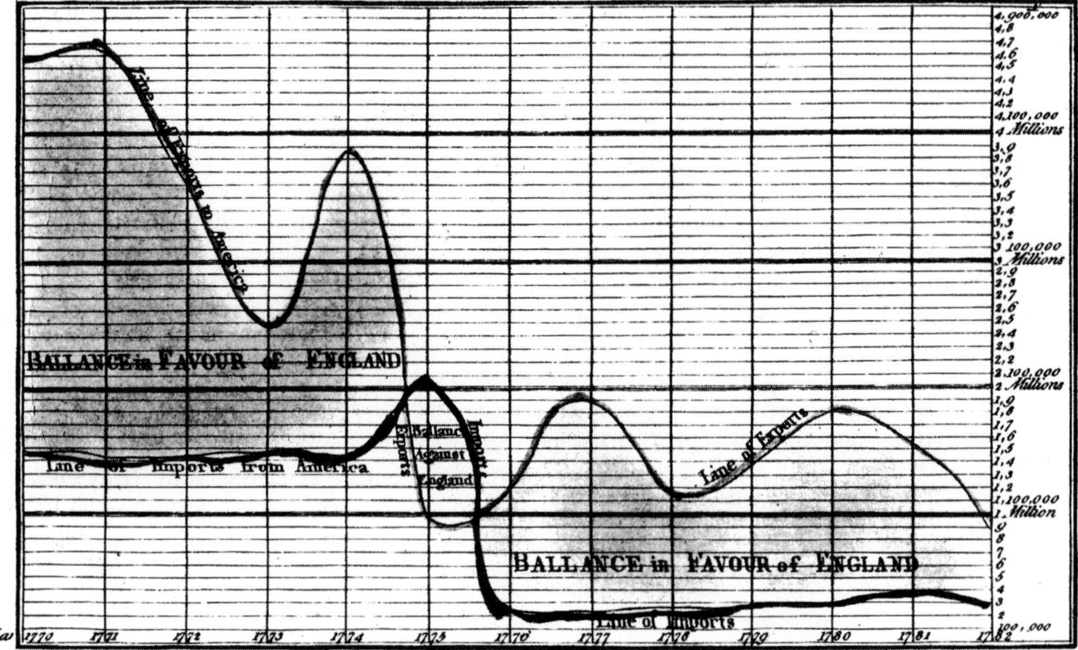

Figure 2.17 A Playfair balance-of-trade graph.

with time on the vertical axis. No doubt, the first reaction of audience members would be to turn the graph on its side, so that time was at the bottom of the page, where they are accustomed to seeing it.

Years ago, time-series graphs had to be drawn by hand, using special graph paper that was covered with vertical and horizontal grid lines to help the artist plot each point in approximately the right place. After each observation was plotted, a ruler was used to draw a series of straight lines connecting these points; for this reason, time-series graphs are sometimes called *line graphs.* Playfair connected his annual data points with curved lines, apparently intending to give a more aesthetically pleasing appearance, but incorrectly suggesting that he had information about the bumps and wiggles in between the data points. Straight lines represent a simple interpolation that has not been embellished with speculative twists and turns.

Today, we usually construct a graph by entering the data into a statistical software package that plots and connects the data points for us. For example, Fig. 2.18 is a software-generated time-series graph of quarterly values of the Dow Jones Industrial Average of stock prices from 1920 to 1940.[18] The Dow is based on the average stock price of 30 prominent corporations, including Coca Cola, Exxon, and General Motors, and is a widely followed barometer of movement in stock prices. This graph shows that the Dow nearly quadrupled between 1925 and late 1929, but then crashed, losing nearly one-third of its value in a few months, made a brief but inadequate recovery, and then slid straight downhill, touching bottom in 1932, down nearly 90 percent from its 1929 peak. This figure gives a vivid, memorable picture of the stock market boom in the late 1920s and the subsequent collapse.

Many software packages allow the user to put grid lines in a graph and to identify each data point with a circle, square, or other design. Grid lines sometimes help the reader identify the axis values that correspond to certain interesting points on the figure, for example, the levels of the Dow Jones average at its 1929 peak and its 1932 bottom. Alternatively, the values of particularly interesting data points can be shown with simple labels, as in Fig. 2.18. Grid lines can

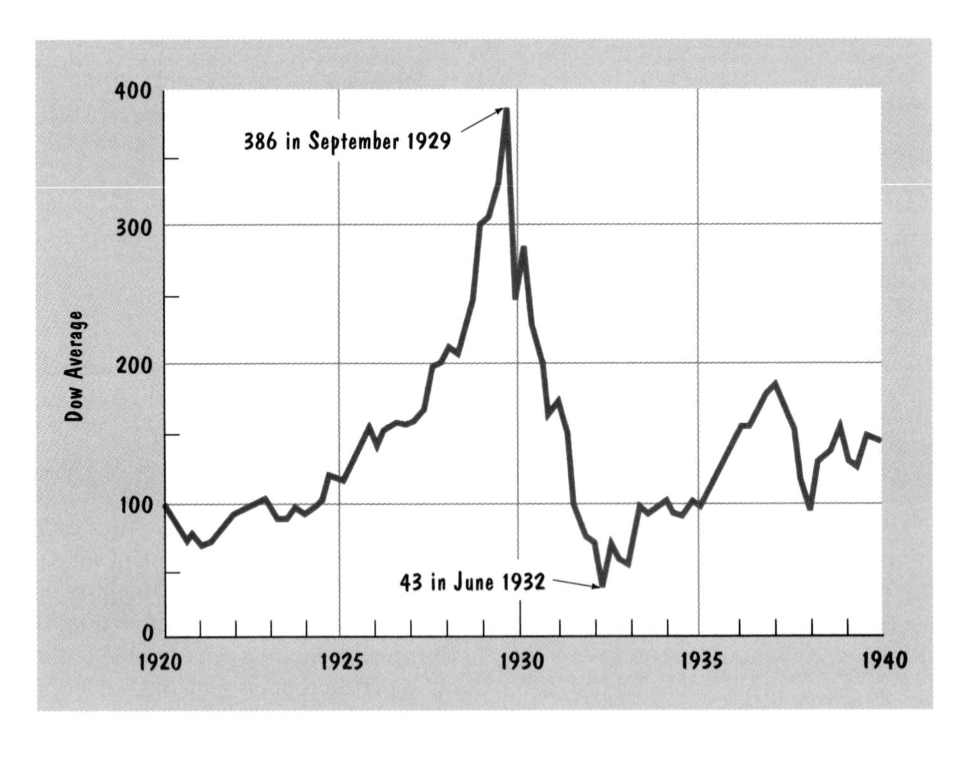

Figure 2.18 The Dow Jones Industrial Average, 1920–1940.

be distracting, as in the Playfair graph in Fig. 2.17, and should generally be either drawn lightly or avoided entirely.

The identification of each data point with a circle or other design is occasionally useful and is often done when two separate time series are on the same graph. However, designs can distract the reader's attention from the essential information conveyed by the data. Figure 2.19 shows how the placement of circles

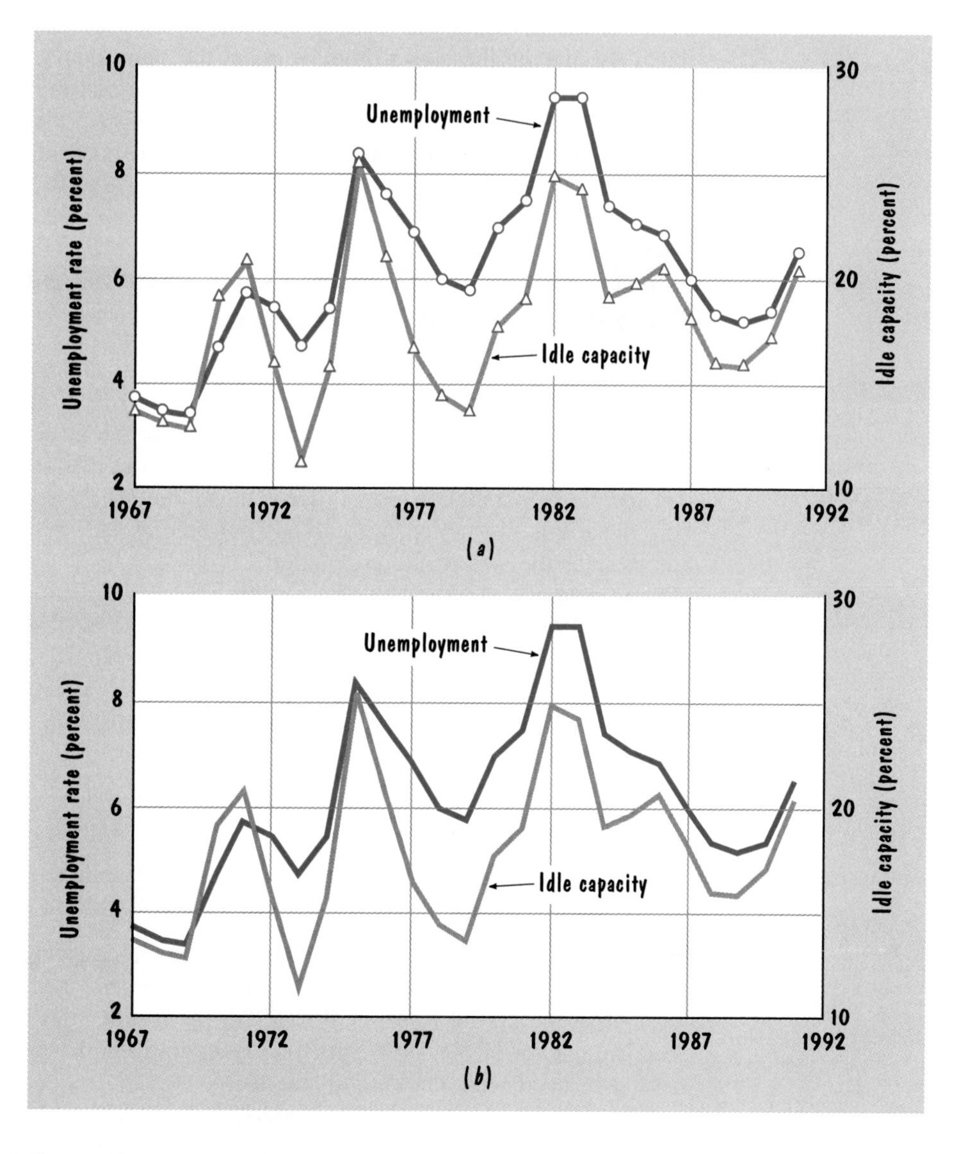

Figure 2.19 Cluttered and uncluttered figures.

and triangles on two overlapping time series can clutter a graph. It is much better to make one of the lines thicker than the other (or to use different colors).

Figure 2.19 also shows how data with different values can be placed on the same graph by using two vertical axes, one for each variable. Here, the left axis is for the unemployment rate, an estimate of the percentage of the nation's labor force that does not have, but is actively seeking, paid employment. The right axis is for the idle capacity rate, an estimate by the Federal Reserve Board of the percentage by which the nation's industrial factories are underutilized. By placing these two time series on the same graph, we can see the close relationship between the underutilization of people and machines. During economic booms, the unemployment rate declines and factories operate at close to capacity; during recessions, workers and machines both sit idle.

Adjusting Data for Population Growth and Inflation

A nation's population generally increases over time, and as a consequence, so do many of the things that people do—marry, work, eat, and play. If we look at time-series data on such human activities without taking into account changes in the size of the population, we will not be able to distinguish changes that are due merely to population growth from those that reflect people's behavior. To help make this distinction, we can use *per capita* data, which have been adjusted for the size of the population.

For example, the number of cigarettes sold in the United States totaled 506 billion in 1960 and 710 billion in 1990, an increase of 40 percent. To put these numbers into perspective, we need to take into account the fact that the population more than doubled during this period, from 120 million to 250 million people. We can do so by dividing each year's sales by the population, to obtain per capita data:

$$\text{1960 per capita cigarette sales: } \frac{506 \text{ billion}}{120 \text{ million}} = 4200$$

$$\text{1990 per capita cigarette sales: } \frac{710 \text{ billion}}{250 \text{ million}} = 2800$$

Total sales increased by 40 percent, but per capita consumption fell by 33 percent. Figure 2.20 shows that while total cigarette sales generally increased throughout the period from 1950 to 1970, per capita sales declined steadily after the release of the 1964 Surgeon General's Report warning of the health risks associated with smoking cigarettes.

Data that are denominated in dollars, such as wages and profits, need to be adjusted for changes in the price level so that we can measure changes in

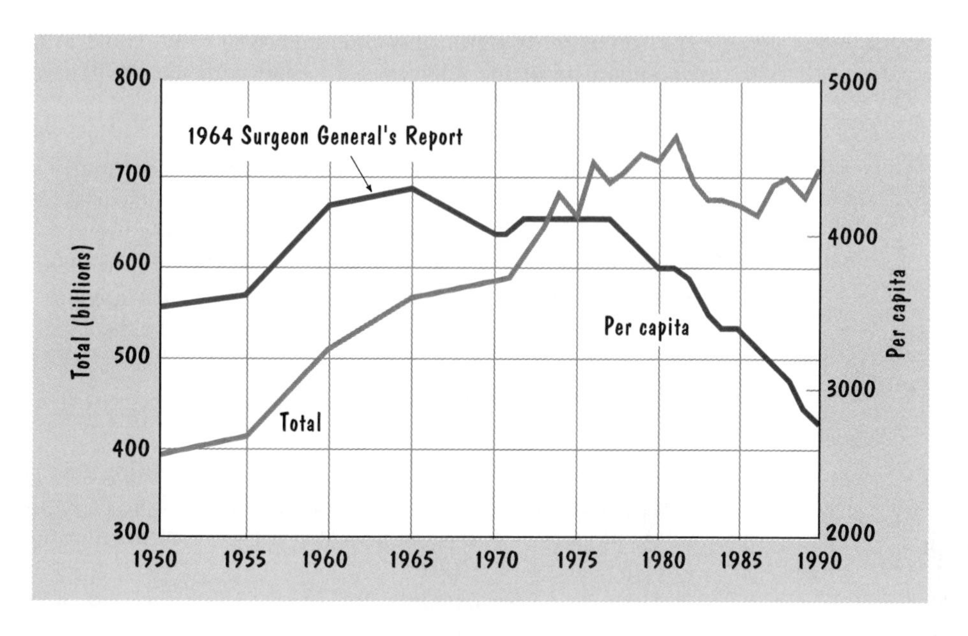

Figure 2.20 Total and per capita U.S. cigarette consumption.

purchasing power. Data that have been adjusted for changes in the price level are known as *real data*. The real values of household wages, income, and wealth are usually calculated by using the *consumer price index* (CPI), which is based on the cost of a representative market basket of the goods and services purchased by households. Every 10 years or so, the Department of Labor's Bureau of Labor Statistics (BLS) surveys thousands of households to learn the details of their buying habits. Based on this survey, the BLS constructs a market basket of approximately 80,000 goods and services, and each month 400 agents call or visit stores in more than 50 cities to collect current price data on these items. These prices are used to calculate the consumer price index, which measures the current cost of the market basket relative to the cost in a base period in which the CPI has been scaled to equal 100. Thus the CPI is a measure of current prices relative to average prices during the base period. Here, for example, are the CPI figures for 1970, 1980, and 1990 using a base of 100 from 1982 to 1984:

YEAR	CPI (100 IN 1982–1984)
1970	38.8
1980	82.4
1990	130.7

The CPI is meaningful only in comparison to its value in another period. Thus a comparison of the 82.4 value in 1980 with the 38.8 value in 1970 shows that prices more than doubled during this 10-year period.

To illustrate how price indexes can be used to adjust data measured in dollars for inflation, consider the fact that the average hourly earnings of nonsupervisory workers in U.S. manufacturing industries were \$7.27 in 1980 and \$10.83 in 1990. These workers earned an additional \$3.56 for each hour of work, but were they able to buy more with \$10.83 in 1990 than they could buy with \$7.27 in 1980? We can convert the \$7.27 earnings in 1980 to 1990 dollars by multiplying by the 1990 price level relative to the 1980 price level:

$$\text{1980 wage in 1990 dollars: } \$7.27\left(\frac{130.7}{82.4}\right) = \$11.53$$

Because prices increased by nearly 60 percent during this period, it would have taken \$11.53 in 1990 to buy as much as \$7.27 bought in 1980. Thus, the actual 1990 earnings of \$10.83 represented a 6 percent decline in real earnings: $(10.83 - 11.53)/11.53 = -.061$.

Here is another example. In January 1995, the cost of mailing a first-class letter in the United States was increased to 32 cents. Fifty years earlier, in 1945, the cost was 3 cents. Was there an increase in the real cost of mailing a first-class letter, that is, an increase relative to the prices of other goods and services? The value of the consumer price index was 17.3 in 1945 and 152.3 in 1995. Thus 3 cents in 1945 bought as much as 26.4 cents in 1995:

$$\text{1945 cost in 1995 dollars: 3 cents}\left(\frac{152.3}{17.3}\right) = 26.4 \text{ cents}$$

Over this 50-year period, the real cost of mailing a first-class letter increased by 21 percent: $(32 - 26.4)/26.4 = .21$.

Some data should be adjusted for both population growth and inflation, thereby giving *real per capita* data. The dollar value of a nation's aggregate income, for example, is affected by the population and the price level, and unless we adjust for both, we will not be able to tell whether there has been a change in living standards or just changes in the population and price level.

Other kinds of data may require other adjustments. An analysis of the success of domestic automakers against foreign competitors will not learn much from a time-series graph of the sales of domestically produced automobiles, even if these data are adjusted for population growth and inflation. A more telling graph might show the ratio of domestic sales to total automobile sales.

A study of motor vehicle accidents over time needs to take into account changes in the number of people traveling in motor vehicles and how far they travel. Thus, studies of motor vehicle deaths generally divide the number of deaths each year by an estimate of the total number of miles driven that year.

Because these adjusted data are quite small, they can be multiplied by 100 million to give the number of motor vehicle deaths per 100 million miles driven. In 1990, for example, there were 44,500 motor vehicle deaths in the United States, and it was estimated that 2147 billion miles were driven that year. The number of deaths per mile driven was 44,500/2,147,000,000,000 = .000000021. Instead of working with so many zeros (which most people have trouble understanding), we can multiply this number by 100 million to obtain a simpler (and more intelligible) figure of 2.1 deaths per 100 million miles driven.

A time-series graph showing the number of votes cast for Democratic party candidates for president since 1900 would not be very informative, since there have been changes in the population and in the fraction of the population that votes in presidential elections. A more informative graph might show the votes for Democratic candidates as a percentage of the total number of votes cast for president (or of the votes cast for the two major-party candidates). This graph would show how the Democrat party's share of the votes for president changed over time.

For one final example, a spokesperson for the New Jersey Department of Environmental Protection boasted, "It speaks well of our enforcement," when the federal government reported that a total of 38.6 million pounds of toxic chemicals had been released into the air above New Jersey in 1982, less than had been released in 21 other states.[19] However, this number does not reflect the fact that New Jersey is a very small state, with relatively little air in which to release chemicals. When the amount of toxic chemicals is divided by New Jersey's area (in square miles), giving the pounds released per square mile, there were only three states that polluted more than New Jersey.

Seasonal Patterns

Sometimes a time-series graph helps us identify regular seasonal patterns in the data. Ice cream sales tend to go up in the summer and down in the winter. Turkey sales increase each Thanksgiving. Sales of toys, books, and jewelry rise in December and fall in January. Bank deposits increase at the end of each week and month, when paychecks are received. Movie attendance goes up on Friday; job attendance goes down on Monday.

Many data are *seasonally adjusted* to remove seasonal patterns. The actual calculations are quite complex and are best left to computer software packages and professional statisticians. The general objective is to take into account the regular patterns in the data so that we can determine whether a particular fluctuation is noteworthy or is just something to be expected for that time of year. For example, because of the holiday season and winter weather, production in the United States tends to increase substantially in the fourth quarter of the calendar year and fall sharply in the first quarter. If this regular pattern were not taken into account, government policy makers might mistakenly believe that a drop in production in the first quarter was the beginning of an economic recession, when it might be just a seasonal blip that would soon be corrected by the arrival of spring. To prevent this error, government statisticians publish seasonally adjusted data on the *gross domestic product* (GDP), the most widely used measure of the nation's production. In the first quarter of each year, the seasonally adjusted GDP is higher than the unadjusted data.

In the first quarter of 1981, the unadjusted GDP fell by $15 billions, but the seasonally adjusted GDP rose by $20 billion. Because of the usual seasonal pattern, government statisticians expected the unadjusted GDP to fall by $35 billion in the first quarter of 1981; when it fell by only $15 billion, this was evidence of a robust economy. In the first quarter of 1982, in contrast, unadjusted GDP fell by $45 billion and seasonally adjusted GDP fell by $10 billion. Government policy makers knew that, even taking into account the usual first-quarter sag, the economy was deteriorating.

It has been said that "With seasonally adjusted temperatures, you could eliminate winter in Canada."[20] That's funny, but (unfortunately) not true. A seasonal adjustment does not eliminate seasonal patterns; it merely takes the usual patterns into account so that we can see if anything unusual is happening. Seasonally adjusted temperatures would not eliminate winter, but they would tell us whether a particular winter was unusually cold.

"Yes, I'm somewhat depressed,
but seasonally adjusted I'm probably happy enough."

Some regular patterns are visible in daily, weekly, and monthly data. For others, you need a longer horizon. For example, there seems to be a presidential political business cycle in the United States, in that the unemployment rate has increased in only two presidential election years since the Great Depression in the 1930s. This is no doubt due to the efforts of incumbent presidents to manipulate the economy so as to avoid the wrath of voters suffering through an economic recession. The exceptions were the reelection bids of Jimmy Carter in 1980 (the unemployment rate went up 1.3 percentage points) and George Bush in 1992 (the unemployment rate rose .7 percentage point). In each case, the incumbent was soon unemployed, too.

Evidence of another political cycle was provided by a 1973 federal lawsuit that challenged the way members of Congress use their franking privilege—the right to send mail for official government business at the taxpayers' expense. This lawsuit argued that too often these free mailings were thinly disguised political advertisements, giving incumbents an advantage over challengers who had to pay for their own campaign mailings. Senate Republicans had created government jobs for two persons to advise them on how best to use the franking privilege to win votes. Senate Democrats had prepared a reelection manual advising members that a model campaign should take well-timed advantage of the franking privilege. The most striking evidence is the data shown in Fig. 2.21, documenting that the volume of congressional mail rises in election years, peaking just before the election and then declining sharply after the election.

Omitted Origins

Sometimes, it is helpful to omit zero from the vertical axis of a time-series graph so that we can see the final detail. For example, Fig. 2.22 shows two graphs of weekly data on the assets of money market mutual funds in early 1990. In Fig. 2.22*a,* which includes zero, the values (ranging from $373 billion to $392 billion) are so far from zero that any patterns in the data have been ironed out of the graph and cannot be detected. The graph in Fig. 2.22*b,* which omits zero, enables us to see that money market assets peaked on March 25 and then declined. We must realize, however, that the omission of zero magnifies the wiggles in the data to help us see them more clearly. Because the data take up a larger fraction of the graph, the fluctuations not only are easier to see, but also appear bigger than they really are. A graph similar to Fig. 2.22*b* (with zero missing) was printed in the *Los Angeles Times* and labeled "After peaking at $392 billion in March, money market fund assets have fallen sharply."[21] With zero omitted, the height of the line drops 30 percent, which does suggest that fund assets fell sharply; however, the actual numbers show that there was only a 2 percent decline, from $392 billion to $384 billion.

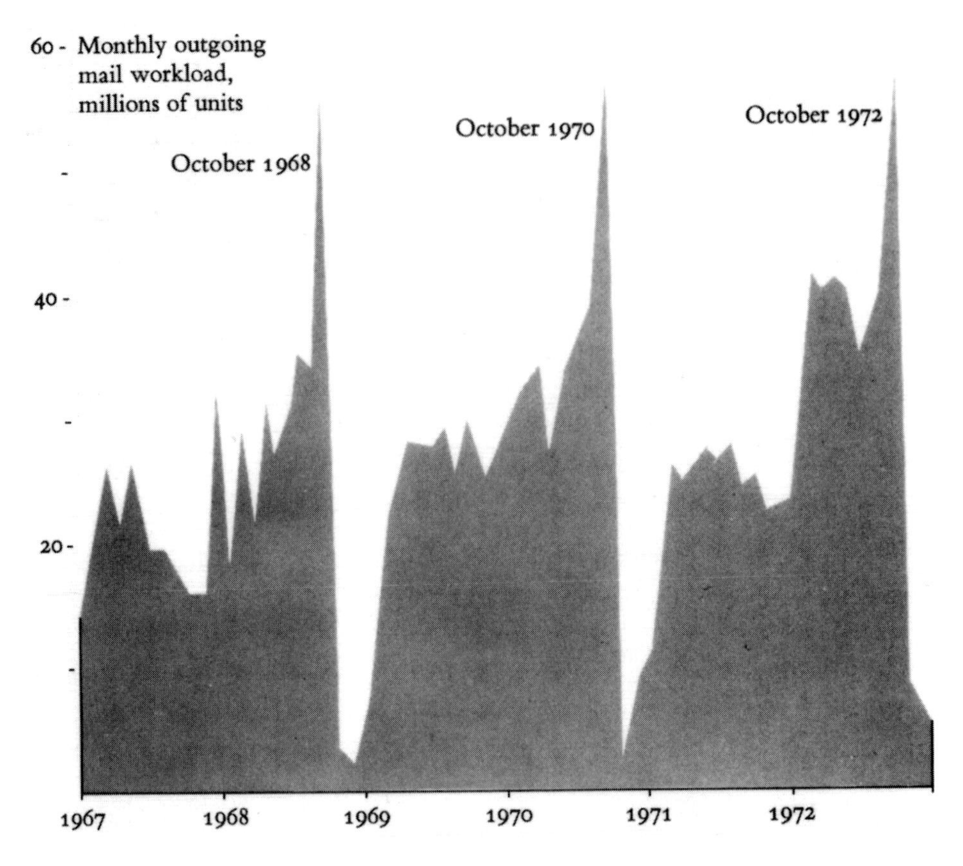

Figure 2.21 Congressional mail.

Omitting zero magnifies the graph so that we can get a closeup look at the data. However, we can no longer judge relative magnitudes by visually inspecting the heights of lines. Before concluding that the changes are large or small, we need to look at the actual numbers. Figure 2.23 gives a famous example in which President Ronald Reagan showed how much a family with a $20,000 income would pay in taxes under the administration's tax proposal, in comparison to a plan drafted by the House Ways and Means Committee. Because the vertical axis does not show any numbers (just a very large dollar sign), we have no basis for gauging the magnitude of the gap between the lines. The 1986 gap in fact represented about a 9 percent tax reduction, from $2385 to $2168. The omission of zero magnifies this 9 percent difference to 90 percent of the height of the line labeled *Their bill;* the omission of numbers from the

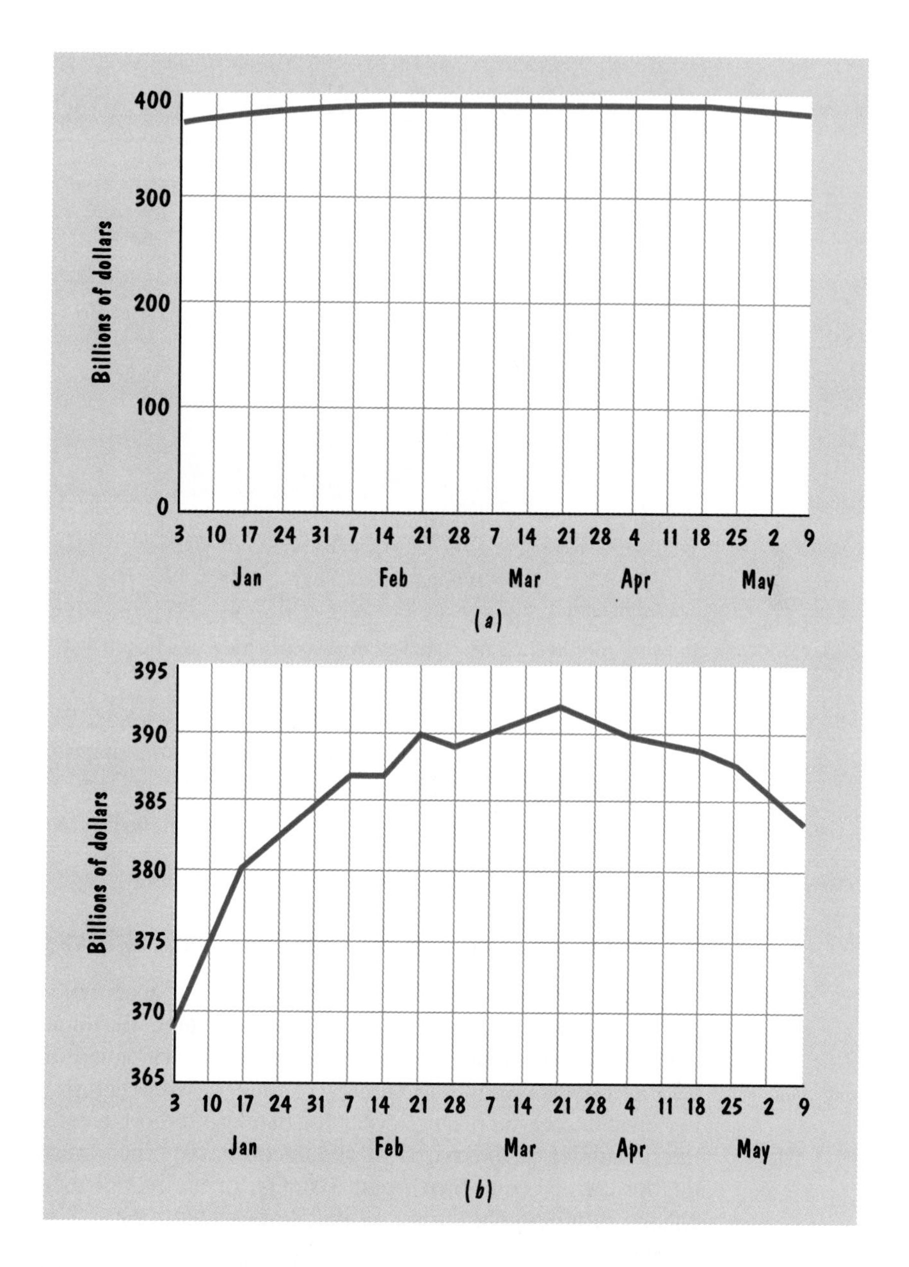

Figure 2.22 Money market mutual fund assets, with and without zero.

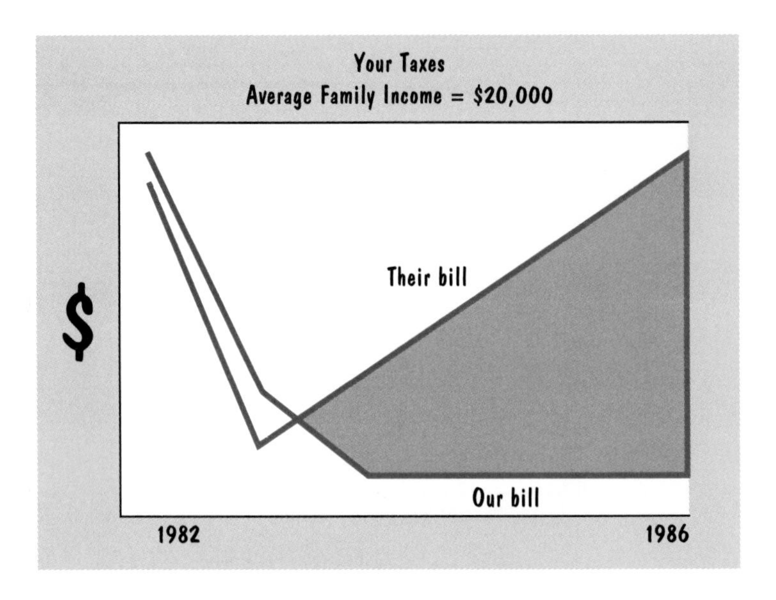

Figure 2.23 Showing middle-income families how much they will be saving in taxes.

vertical axis prevents the reader from detecting this magnification. Afterward, David Gergen, a White House spokesman, told reporters, "We tried it with numbers and found that it was very hard to read on television, so we took them off. We were trying to get a point across."[22]

Changing the Units in Midgraph

For a graph to represent the data accurately, the observations must be comparable and the units on each axis must be consistent. We should not have one observation be the number of times something happened in a year, while another observation is the number of times it happened in a week. We should not have 1 inch represent 1 year for part of the horizontal axis and 10 years for the remainder of the axis. Nor should we have 1 inch mean $1000 for part of the vertical axis and then mean $10,000 for the rest of the axis. If the units change in midgraph, the figure will surely distort the information contained in the data.

In 1976 the National Science Foundation (NSF) published the graph in Fig. 2.24 showing the number of science Nobel Prizes that had been awarded to U.S. citizens.[23] (The graph also included data for four other countries, which were removed to unclutter the figure.) Can you find the trickery in this figure? Look at the time axis. Each of the first seven data points gives the number of

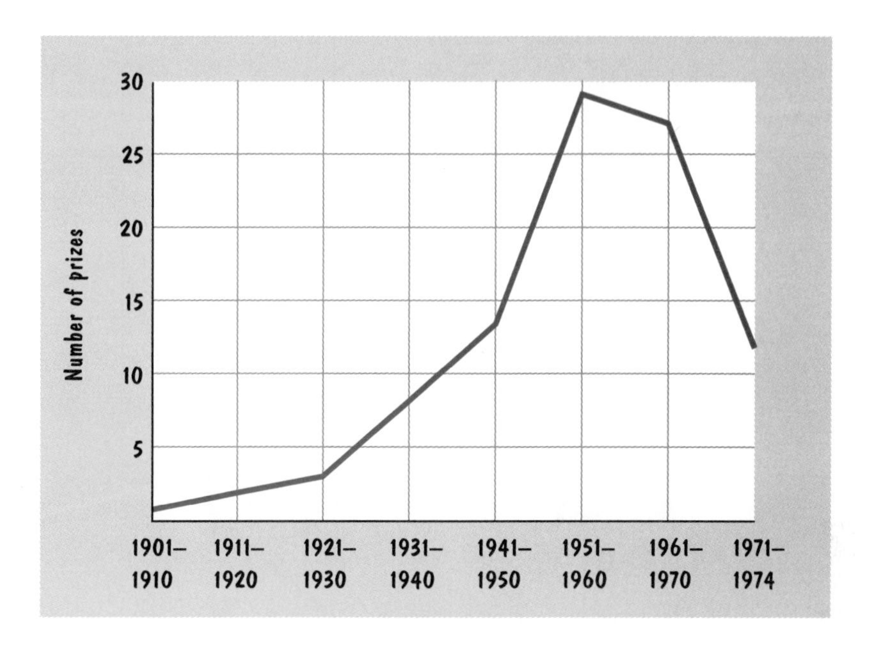

Figure 2.24 A misleading picture of the Nobel Prizes in science awarded to U.S. citizens.

Nobel Prizes during a 10-year-period; but the eighth data point is for a 4-year period, from 1971 to 1974. The NSF, which certainly knows better, altered the time axis in midgraph! The 4-year period from 1971 to 1974 should not have the same spacing on the horizontal axis as do the 10-year periods. Even more seriously, the first seven observations are prizes per decade; the eighth observation is prizes per 4 years. For this eighth observation to have the same units as the other data, it should be multiplied by $10/4 = 2.5$. Alternatively, we could graph annual data.

Because the number of prizes awarded during a 4-year period is less than that during a 10-year period, the NSF was able to create the illusion of an alarming drop in the number of science prizes awarded to U.S. citizens. Taking into account the prizes that were awarded during the subsequent 6 years, Fig. 2.25 shows that the number of science Nobel Prizes given to U.S. citizens continued to increase. (Of course, this trend could not continue forever, unless the Nobel Committee increased the total number of prizes; in the 1970s, the United States won more than one-half of the science Nobel Prizes.)

An even more bizarre variation in the time axis is shown in Fig. 2.26, a version of which (embellished with pictures of doctors and other chart junk) was published in the *Washington Post*.[24] The first unit on the time axis is 8 years, the next two units are 4 years, then another 8 years, then 2 years, and so

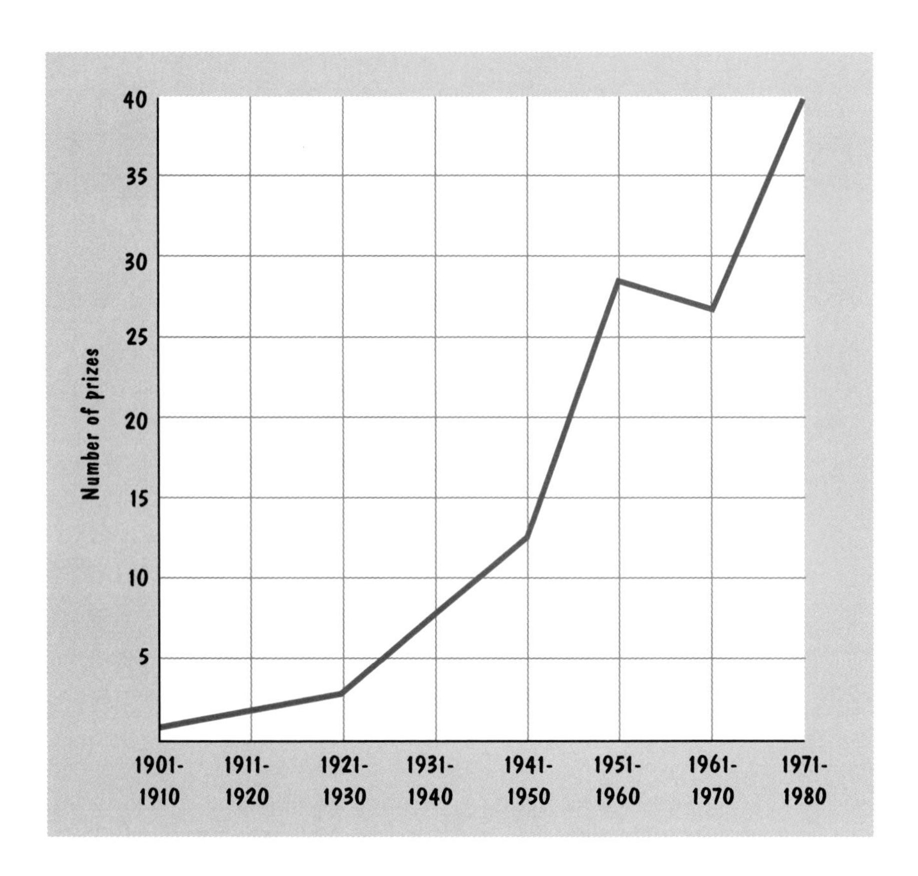

Figure 2.25 A better picture of the Nobel Prizes in science awarded to U.S. citizens.

on. The artist who drew this graph evidently chose the time units so that the graph would turn out to be an approximately straight line, probably reasoning that a straight line looks good. As a consequence, the figure does not tell us anything at all about the data! Figure 2.27 graphs these data with a consistent time axis. Now we can see that doctor incomes did not move upward in a straight line and that there may have been a significant change in doctor salaries after 1964, which is when Medicare began. Besides using a consistent time axis, it would be helpful to adjust the data for changes in the CPI. The purpose of a time-series graph is not simply to draw something that looks good, but to show us what the data say.

Figure 2.28, prepared by the Associated Press, combines several poor graphical practices.[25] First, note that the axes are reversed. Readers who do not look

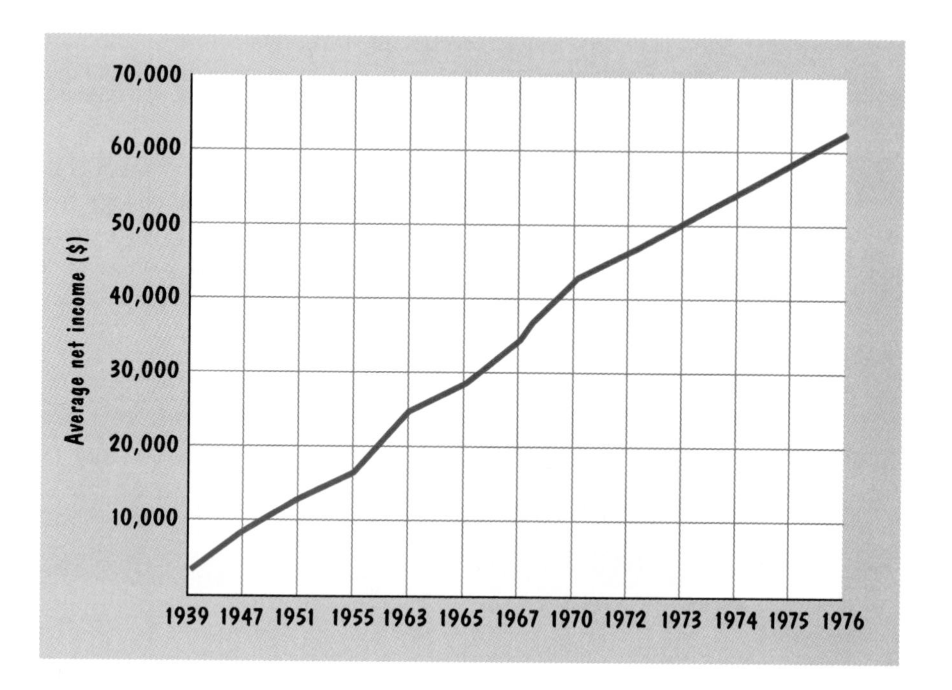

Figure 2.26 Doctors' income with an inconsistent time axis.

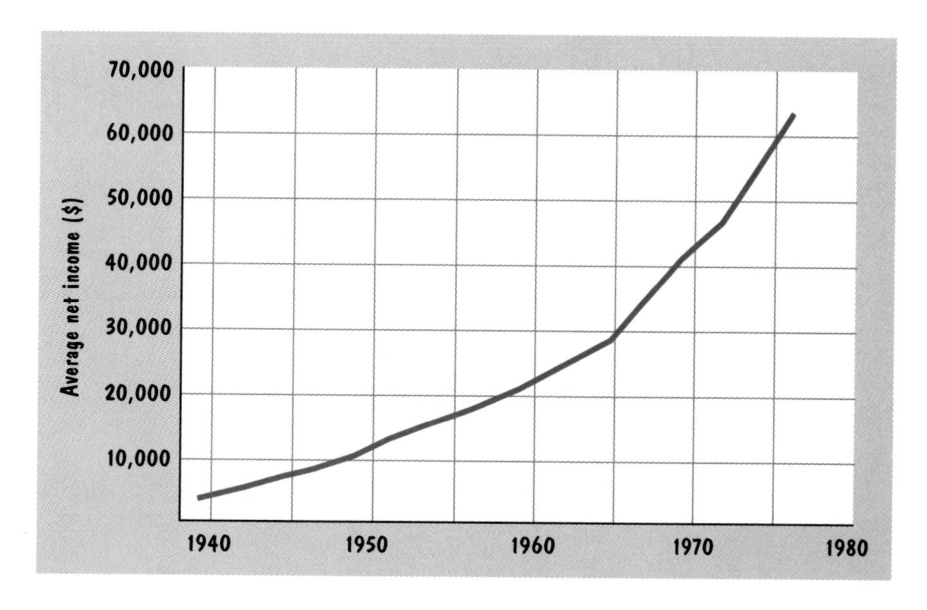

Figure 2.27 Doctors' income with a consistent time axis.

too carefully and are accustomed to seeing time on the horizontal axis may mistakenly conclude that state and local property taxes have been rising by a constant amount each year, or perhaps even leveling off. Second, the time units change several times, representing anywhere from 2 to 6 years. Figure 2.29 shows that using the usual convention for the axes and consistent time units completely changes the graph's appearance. The change from single years (such as 1962) to double years (such as 1964–1965) is also confusing; perhaps it represent a change from calendar-year to fiscal-year accounting. Finally, the data should be adjusted for the growth of the population and the increase in the general price level.

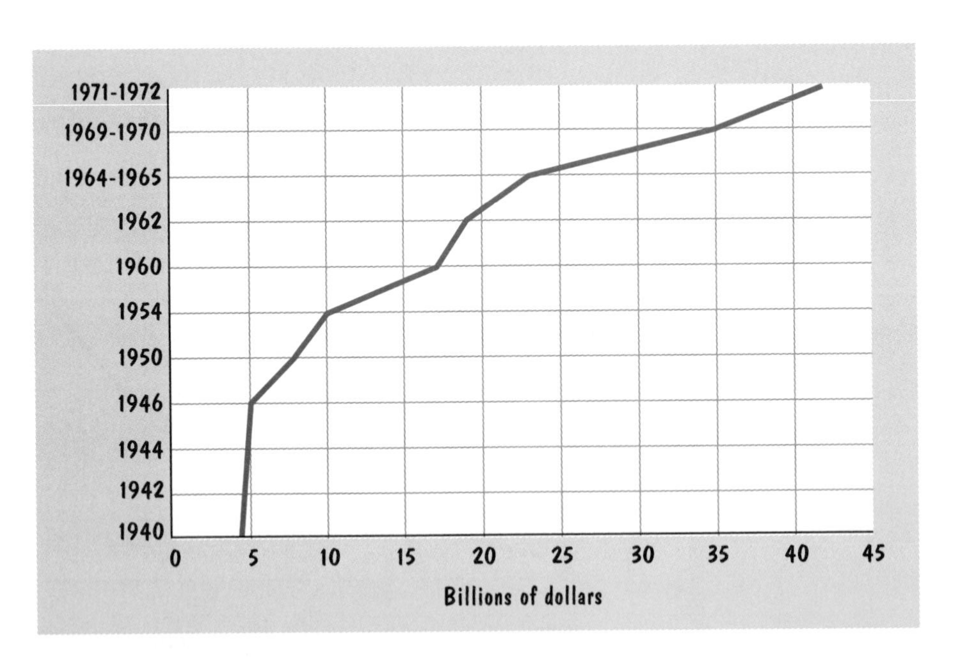

Figure 2.28 State and local property taxes level off.

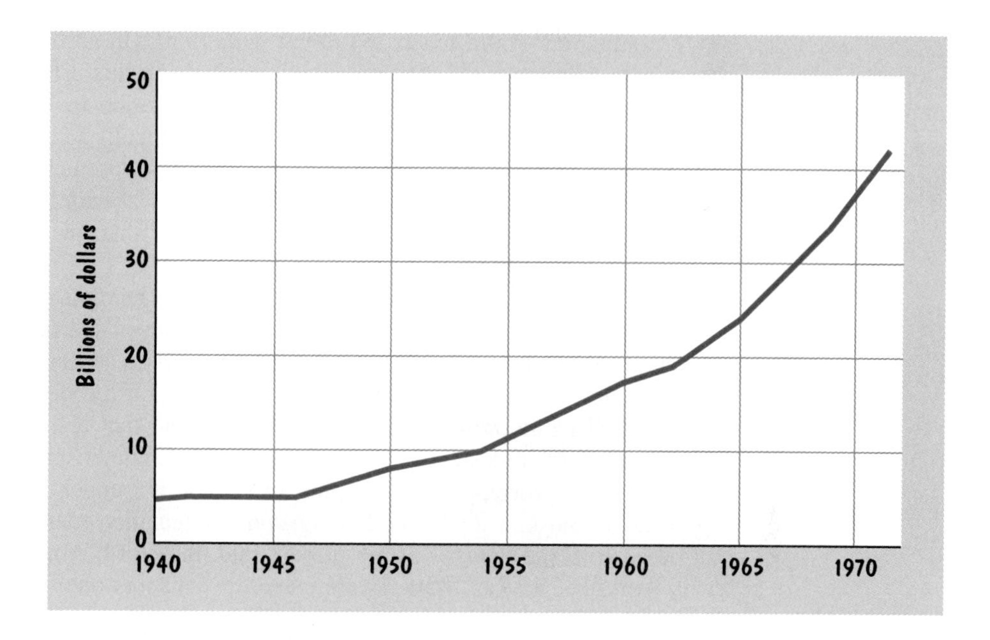

Figure 2.29 State and local property taxes accelerate.

Incautious Extrapolation

While time-series graphs provide a simple and useful way of describing the past, we must not forget that they are merely descriptive. Graphs may reveal patterns in the data, but they do not explain why these patterns are occurring. We, the users, must provide these logical explanations. Sometimes, explanations are easily found and provide good reasons for predicting that patterns in past data will continue. Because of the Thanksgiving holiday, the consumption of turkey has shown a bulge in November for decades and will probably continue to do so for years to come. Because of the cold weather, construction activity declines in winter in the northern part of the United States and will most likely continue to do so.

If we have no logical explanation for the patterns in the data and nonetheless assume they will continue, then we have made an *incautious extrapolation* that may well turn out to be embarrassingly incorrect. The farther into the future we look, the riskier the extrapolations because there is then more time for the underlying causes to weaken or evaporate. We can be more certain of turkey sales next November than in the year 5000.

In his second State of the Union Address, Abraham Lincoln predicted, based on an extrapolation of data for the years from 1790 to 1860, that the

U.S. population would reach 251,689,914 in 1930. (In addition to the incautious extrapolation, note the unjustified precision in this estimate of the population 70 years in the future.) The actual U.S. population in 1930 turned out to be 123 million, less than one-half the size that Lincoln had predicted. A 1938 presidential commission erred in the other direction, predicting that the U.S. population would never exceed 140 million. Just 12 years later, in 1950, the U.S. population was 152 million. Forty years after that, in 1990, the U.S. population was 249 million.

Other examples are so ludicrous that everyone recognizes the error in incautious extrapolation. In 1940, there were an average of 3.2 people in each car on the highway in the United States. By 1960 this average had dropped to 1.4. At this rate, by the year 2000, three out of four cars on the highway will be empty! Other humorous extrapolations have found that if the number of microscope specimen slides acquired by a certain St. Louis hospital continues to increase at the current rate, St. Louis will be buried under 3 feet of glass by the year 2224, and that if *National Geographic* magazines continue to accumulate in basements and garages at the present rate, the North American continent will sink beneath the seas. Mark Twain came up with this one:

> In the space of one hundred and seventy-six years the Lower Mississippi has shortened itself two hundred and forty-two miles. This is an average of a trifle over one mile and a third per year. Therefore, any calm person, who is not blind or idiotic, can see that in the Old Oolitic Silurian Period, just a million years ago next November, the Lower Mississippi River was upward of one million three hundred thousand miles long, and stuck out over the Gulf of Mexico like a fishing rod. And by the same token, any person can see that seven hundred and forty-two years from now the Lower Mississippi will be only a mile and three-quarters long. . . . There is something fascinating about science. One gets such wholesale returns out of a trifling investment of fact.[26]

As a general rule, we should be extremely wary of extrapolations into the distant future and cautious even about near-term predictions, unless we have a reasonable explanation for the patterns that we see in a time-series graph.

Choosing the Time Period

One of the most bitter criticisms of statisticians is "Figures don't lie, but liars figure." The complaint is that an unscrupulous statistician can prove anything by carefully choosing favorable data and ignoring conflicting evidence. Time-series graphs can be a vehicle for such chicanery. By a clever selection of when to begin and end a graph, a dishonest person may show a trend that will be absent in a more complete graph. This is particularly true with a brief series of volatile data. Figure 2.30 illustrates this point with data on the S&P 500

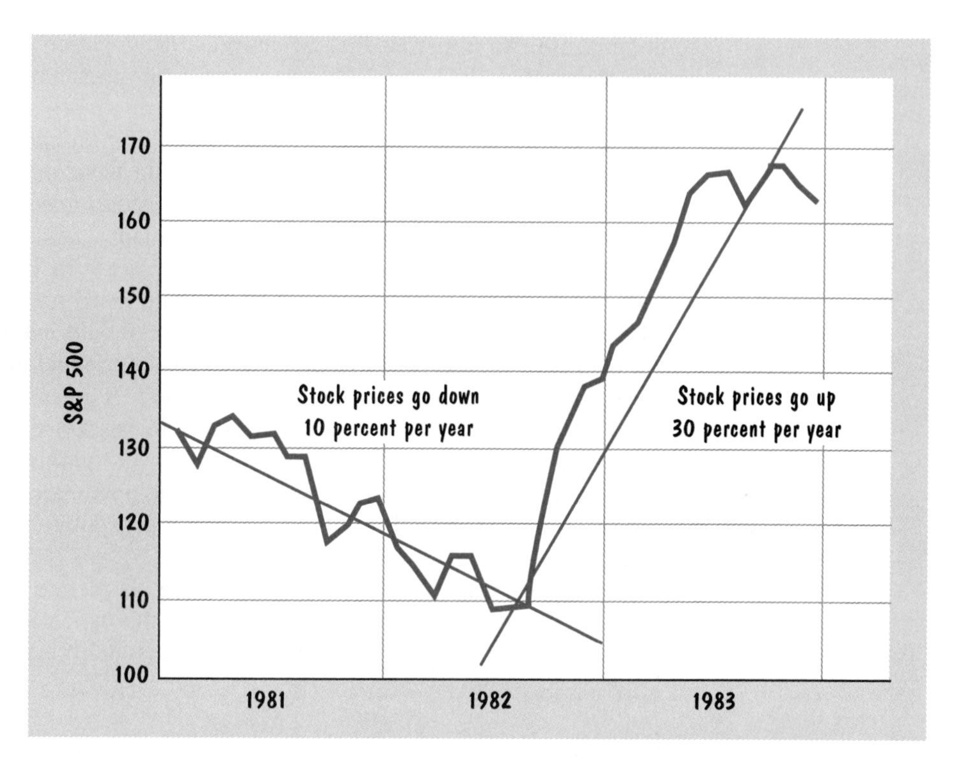

Figure 2.30 What is the trend in stock prices?

index of stock prices. If we show the data only for 1981 through 1982, it appears that stock prices fall by about 10 percent per year; if we only graph the data for mid-1982 through 1983, it appears that stock prices increase by 30 percent per year. Which is correct? Both time periods are too brief to describe meaningful trends in something so volatile. A much longer perspective is needed for a more balanced view. Over the past 100 years, stock prices have increased, on average, by about 4 percent per year.

Would anyone be so naive as to extrapolate stock prices on the basis of only 1 or 2 years of data? Investors do it all the time, over even shorter periods, hoping that the latest zig is the prelude to many happy profits or fearing that a recent zag is the beginning of a crash. The deceitful choose their data carefully to give the trend they want. In December 1985, an investment advisory service touting gold as an inflation-proof investment fit a line to two points—an 11 percent inflation in 1974 and a 13.5 percent inflation in 1979—and concluded that we will have at least 20 percent inflation by 1995! The best antidotes for misleading trends are a visual inspection of the data and some common sense. An impartial look at the pre-1985 inflation data in Fig. 2.31 clearly shows that 1974 and 1979 were chosen not to summarize the data fairly, but to create a misleading trend.

We should be suspicious whenever the data begin or end at peculiar times. If someone uses data for 1853 to 1894, we should wonder why they did not use data for 1850 to 1900, a more natural choice for someone who is not massaging the data. If the data start or stop at important dates (for example, financial data that begin or end with the stock market crash in 1929), we should consider the possibility that a trend is being built based on outliers. If someone in 1995 presents data for 1980 through 1989, we should inquire about more recent data. If someone refers to the last seven presidential elections, we should ask why the eighth, ninth, and tenth were omitted. A general rule is that if the beginning and ending points for the data seem to be choices that one would make only after scrutinizing the data, then they were probably chosen to distort the historical record. There may be a perfectly logical explanation, but we should insist on hearing this explanation before accepting conclusions based on data from a suspicious time period.

A time-series graph is essentially descriptive—a picture meant to tell a story. As with any story, bumblers may mangle the punch line, and dishonest people may tell lies. Those who are conscientious and honest can use

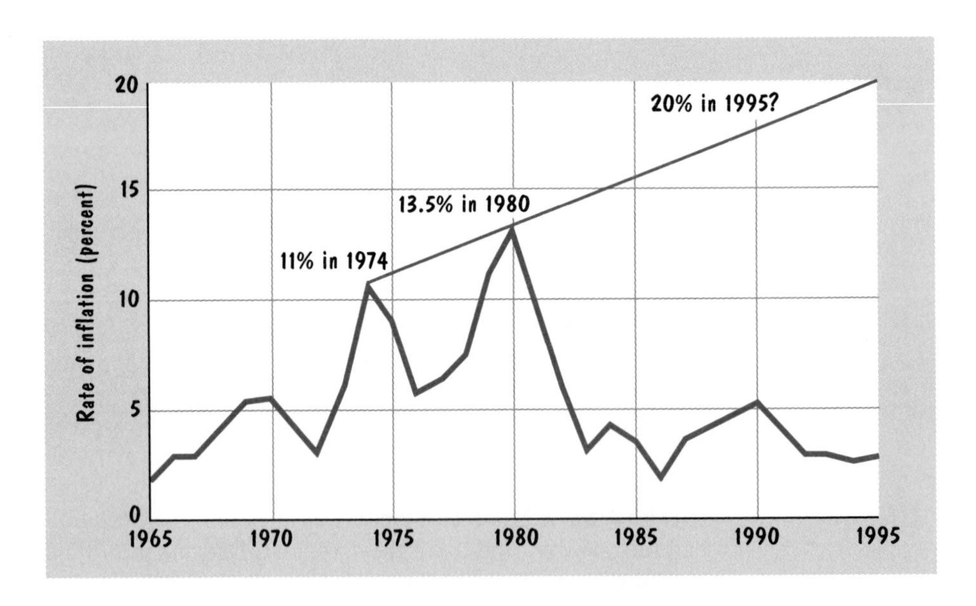

Figure 2.31 A selective use of data predicts 20 percent inflation in 1995.

time-series graphs to describe in a simple and revealing way the behavior of a variable as time passes. There is room for mistakes and chicanery, but there is also room for enlightenment.

Unnecessary Decoration

The first full-time graph specialist at *Time* magazine was an art school graduate who asserted that "the challenge is to present statistics as a visual idea rather than a tedious parade of numbers."[27] Too often graphs are drawn to be artistic, rather that informative. Graphs are not mere decoration, to enliven numbers for the easily bored. They are intended to summarize and display complex data in a way that can facilitate analysis and interpretation. Unnecessary ornamentation that distracts the reader should be avoided. A useful graph displays data accurately and coherently, does not distort the data's implications, and encourages the reader to consider the data's substance rather than to admire the graphical style.

The axes should show enough labels that an interested reader can make rough estimates of the numerical values of points on the graph, but should not be crowded with numbers that are distracting and difficult to read. The labels on the vertical axis of Playfair's balance-of-trade graph (Fig. 2.17) are unnecessarily cluttered. An alternative is to use large hashmarks to identify well-spaced numerical values and to put smaller, unlabeled hashmarks in between these labeled values. Playfair could have eliminated all the labels on the lighter grid lines in between the millions values and let the reader infer that these refer to hundreds of thousands. Figure 2.18 illustrates how small hashmarks can also be used to identify time values in between larger, labeled hashmarks.

Artists sometimes use three-dimensional graphs that are superficially interesting but, on closer inspection, misleading. Figure 2.32 is a graph of the age structure of college enrollment that was prepared by the federal government's National Center for Education Statistics.[28] The top half of the graph is redundant, since the under-25 percentages are just 100 percent minus the 25-and-over percentages. There are really only five data points in this graph; the curvature between these five annual percentages is ornamentation added by an overly imaginative artist that erroneously suggests daily bumps and wiggles in college enrollment. The three-dimensional representation of these two-dimensional data adds to the confusion. Even more garishly, this graph was printed in five colors by a person concerned more with decoration than with information. Edward Tufte includes this graph in one of his wonderful books on displaying data, and he speculates, "This may well be the worst graphic ever to find its way into print."[29] The unembellished graph of the five data points in Fig. 2.33 says about all that these data have to say.

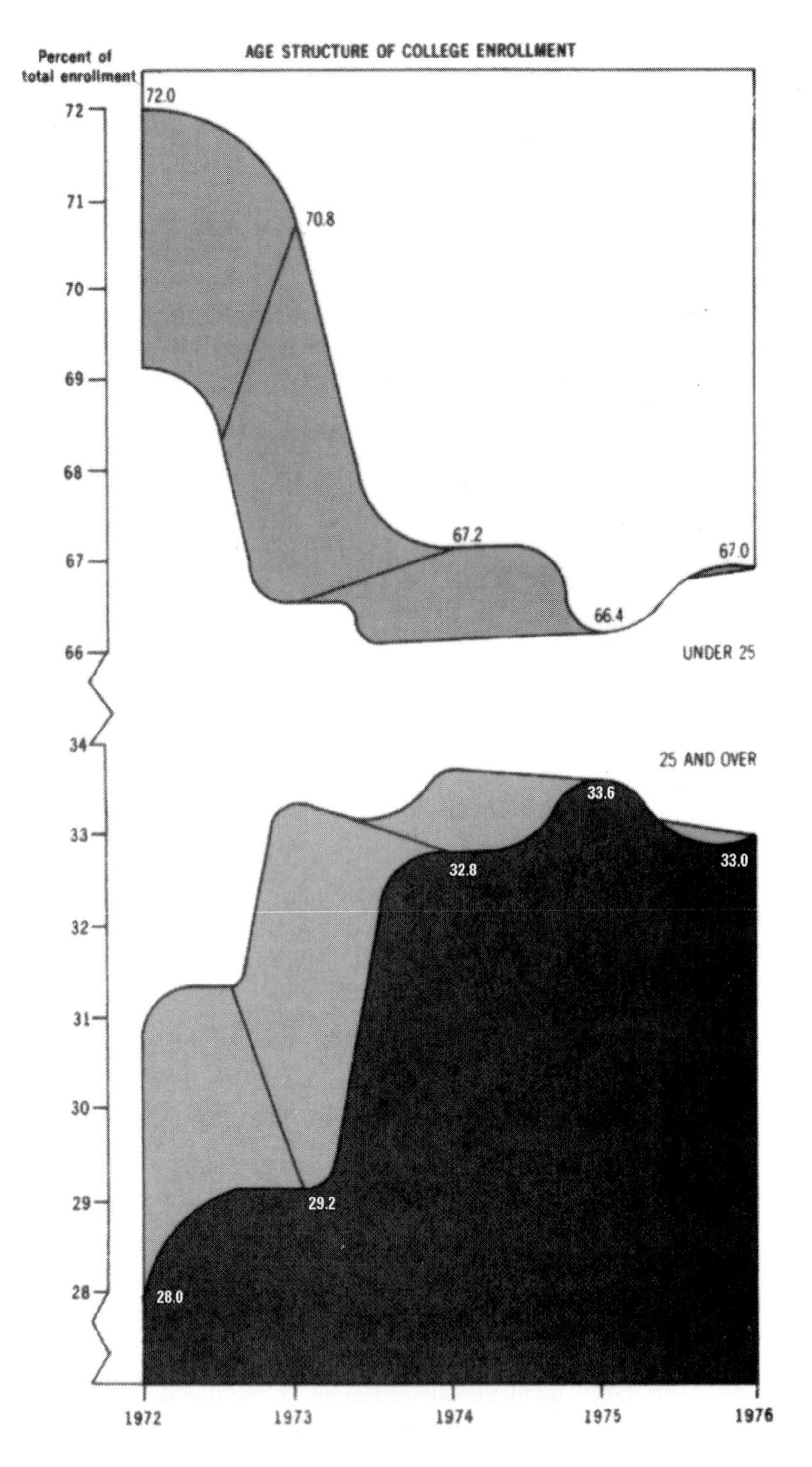

Figure 2.32 The worst graph ever?

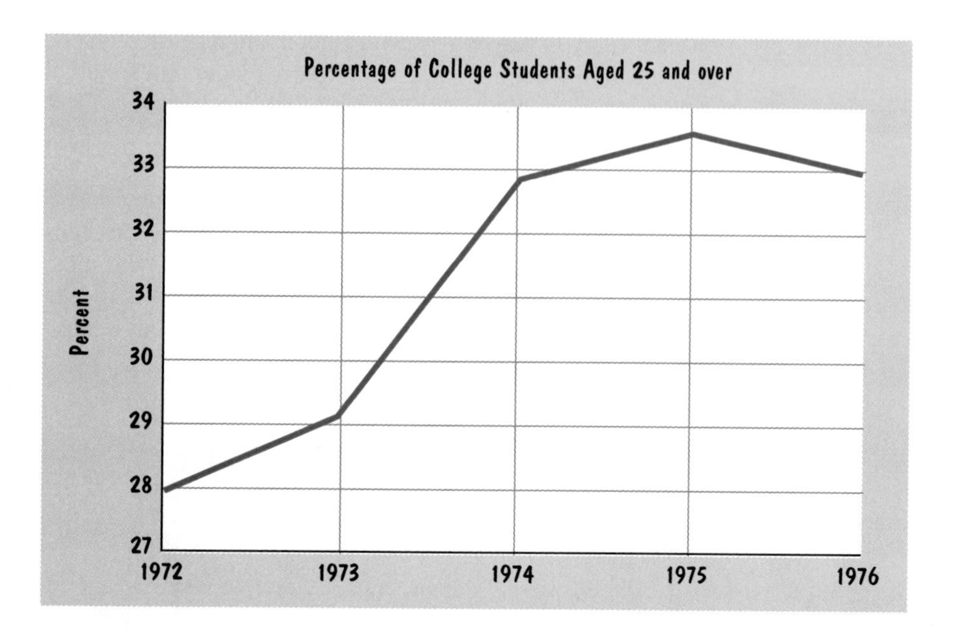

Figure 2.33 A plain-vanilla graph.

Hurricane Bob

Hurricane Bob hit Cape Cod on August 19, 1991, with winds gusting to 100 miles per hour.[30] The casual observer could easily see the terrible damage to buildings, boats, beaches, trees, telephone poles, and power lines. However, relatively little is known about the less obvious effects of hurricanes on land and on water. For Hurricane Bob, it happened that a team of researchers had been monitoring many aspects of the Cape Cod environment for years and were able to assess the effects of Bob by comparing measurements made after the storm with data gathered beforehand.

For instance, among the populations disturbed by this hurricane were hornets whose tree nests were damaged and who became disoriented by the smell of sap from tree limbs that were broken or defoliated. The first figure that follows shows a time-series graph of the number of persons treated for hornet stings at the emergency room of a Cape Cod hospital. This graph shows quite clearly that the number of patients increased by a factor of 5 or 6 in the aftermath of Hurricane Bob and then returned to normal over the next 4 weeks. The second figure shows a quite different consequence of Bob. Wind-borne salt spray temporarily, but only temporarily, increased the amount of sodium in the soil, particularly at sites closer to the shoreline. These researchers also documented how the salt spray damaged some vegetation and disturbed others, for example, causing the fall blooming of cherry trees, lilacs, and other plants that normally would not have bloomed until the following spring.

These researchers also gathered a wealth of data on the release of ammonium from the soil as well as the temperature, surface and bottom salinity, amount of dissolved oxygen, and production of phytoplankton in ponds. One of their most interesting findings was that while the ecosystem experienced many dramatic shocks, it recovered very quickly with most readings returning to pre-Bob levels after only a few days or weeks. In each case, time-series graphs were used to show vividly the temporary, but intense, disturbance to the ecosystem, followed by a quick recovery.

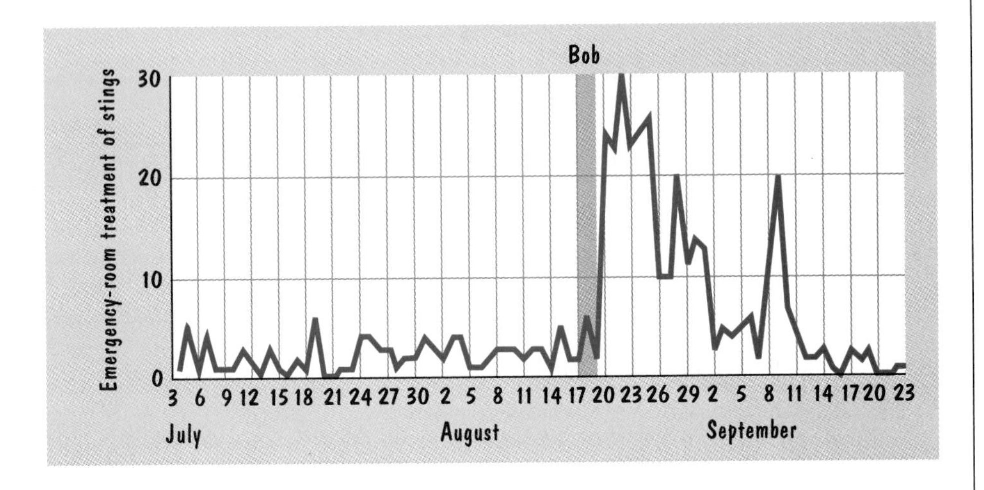

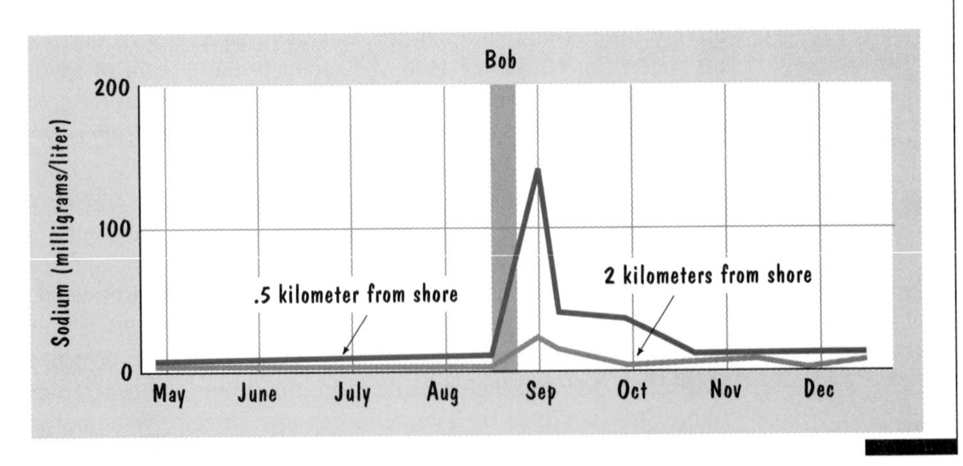

A Projection Gone Awry

In 1924, Computing-Tabulating-Recording Company changed its awkward name to something more ambitious—International Business Machines (IBM)—and proceeded to become the premier growth stock of all time. By 1978, its revenues had been growing (in real terms, adjusted for inflation) by about 16 percent per year for more than 50 years. Some financial analysts looked backward and decided that you could never go wrong buying IBM stock. Others looked forward and cautioned that this phenomenal growth could not continue forever.

The figure below shows how a security analyst in 1978 might have projected IBM's earnings per share over the next 10 years, using data for the current year and the preceding 10 years (1968 to 1978). A smooth curved line fits these data almost perfectly and suggests that IBM's earnings per share will rise to $20 in 1988.

This figure, like all time-series graphs, is merely descriptive and, by itself, provides no underlying rhyme or reason why things should continue as they have. To extrapolate a past trend into a confident prediction of the future, we need to look behind the numbers to understand what has been causing this trend and then think seriously about whether these forces will continue to operate. Sometimes they will. But at other times there are good reasons for anticipating changes in the underlying factors that cause the patterns we observe in time-series graphs.

From its small beginnings, IBM could grow rapidly as the use of computers spread throughout the economy. But by 1978, IBM was one of the largest companies in the United States, with limited room for continued rapid growth.

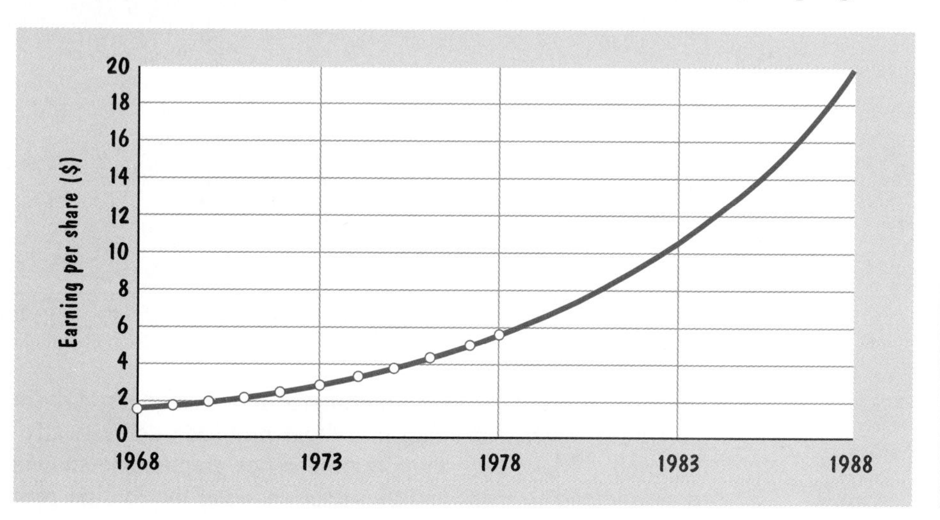

If IBM's 16 percent real growth rate had continued and the overall U.S. economy had continued to grow at its long-run 3 percent rate, then by 2003 one-half of U.S. gross domestic product would be IBM products and in 2008 everything would be made by IBM! At this stage in this fanciful exercise, something would have to give—either IBM's growth rate would have to drop to 3 percent, or the economy's growth rate would have to rise to 16 percent. The latter scenario of sustained 16 percent growth for the entire economy is plainly implausible, since aggregate production is constrained by the rates of growth of the labor force and productivity.

In IBM's case, its 16 percent growth rate did not persist, and a simple extrapolation of its pre-1978 trend in earnings per share proved to be an incautious extrapolation that was recklessly optimistic. Instead of earning $20 per share in 1988, IBM earned less than half that: $9.27. Subsequent years turned out to be even more disappointing. Investors who bought IBM stock in the 1970s because they were confident that IBM's remarkable growth rate would continue forever were disappointed to learn that you cannot always see the future by looking in a rear-view mirror.

exercises

2.11 Summarize the information contained in this graph of the U.S. birthrate, the number of live births per 1000 women 15 to 44 years old.[31]

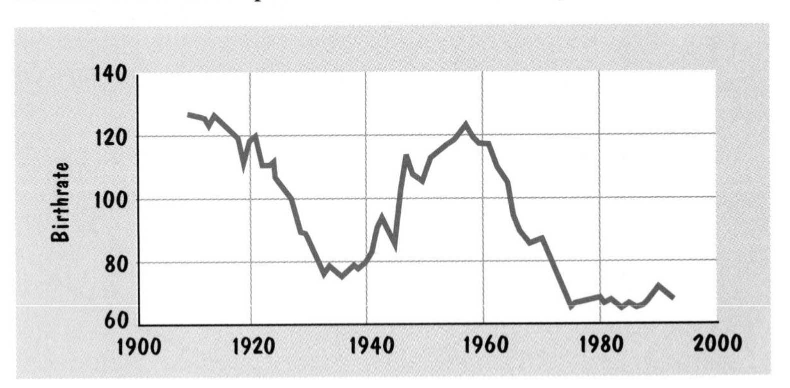

2.12 Shown next are data on the number of new passenger cars (in millions) sold in the United States, separated into those produced domestically and those imported.[32] Make two separate time-series graphs, one showing total car sales (domestic and foreign) and the other showing the foreign percentage of total car sales. Show your data. Do you see any patterns in either of these graphs?

	DOMESTIC	FOREIGN		DOMESTIC	FOREIGN
1980	6.581	2.398	1988	7.526	3.004
1981	6.209	2.327	1989	7.073	2.699
1982	5.759	2.224	1990	6.897	2.404
1983	6.795	2.387	1991	6.137	2.038
1984	7.952	2.439	1992	6.277	1.937
1985	8.205	2.838	1993	6.742	1.776
1986	8.215	3.245	1994	7.255	1.735
1987	7.081	3.196	1995	7.129	1.506

2.13 The Soviet Antarctica Expeditions at Vostok analyzed an ice core that was 2083 meters long in order to estimate the atmospheric concentrations of carbon dioxide during the past 160,000 years. Shown below are some of their data, where D is the depth of the ice core, in meters; A is the age of the trapped air, in years before present; and C is the carbon dioxide concentration, in parts per million by volume.[33] Draw a time-series plot of atmospheric carbon dioxide concentration over the past 160,000 years. (Show the most recent period on the right-hand side of the horizontal axis.) What patterns do you see?

DEPTH D	AGE A	CARBON C	DEPTH D	AGE A	CARBON C
126.4	1,700	274.5	1,225.7	80,900	222.5
302.6	9,140	259.0	1,349.0	90,630	226.0
375.6	12,930	245.0	1,451.5	98,950	225.0
474.2	20,090	194.5	1,575.2	110,510	233.5
602.3	30,910	223.0	1,676.4	119,500	280.0
748.3	42,310	178.5	1,825.7	130,460	275.0
852.5	50,150	201.0	1,948.7	140,430	231.0
975.7	59,770	201.0	2,025.7	150,700	200.5
1,101.4	70,770	243.0	2,077.5	159,690	195.5

2.14 The height of Lake Michigan varies from month to month and from year to year. Here are data over a 100-year period on the highest monthly mean levels (in feet) every 10 years. Graph these data twice, first letting the vertical axis go from 0 to 600 and then letting the vertical axis go from 579 to 584. What advantages and disadvantages do you see to omitting zero from the vertical axis of this graph?

1860	1870	1880	1890	1900	1910	1920	1930	1940	1950
583.3	582.7	582.1	581.6	580.7	580.5	581.0	581.2	579.3	580.0

2.15 In 1992, two professors at UCLA's Physiology Laboratory published a paper comparing the trends over time in male and female world records in several different running events.[34] Here are their data (in minutes) for the marathon, a race covering 42,195 meters:

	1910s	1920s	1930s	1940s	1950s	1960s	1970s	1980s
MALES	156.12	149.03	146.70	145.65	135.28	128.57	128.57	126.83
FEMALES					217.12	187.43	147.55	141.10

Convert each of these times to velocity, in meters per minute, by dividing each time into 42,195 meters. Then plot these velocity data on a single time-series graph, with the velocity axis running from 0 to 400 and the time axis running from 1910 to 2010. (Plot each point at the midpoint of the decade; for example, plot the data for the 1910s at 1915.) The professors say that "Despite the potential pitfalls, we could not resist extrapolating these record progressions into the future." Draw one freehand straight line that fits the female data and another that fits the male data. Do these lines cross before 2010? By extrapolating the female line backward, what was the implied female world record in 1900?

2.16 The following figure was published in *Pravda* in 1982. What is misleading about this graph? How would you graph these data?

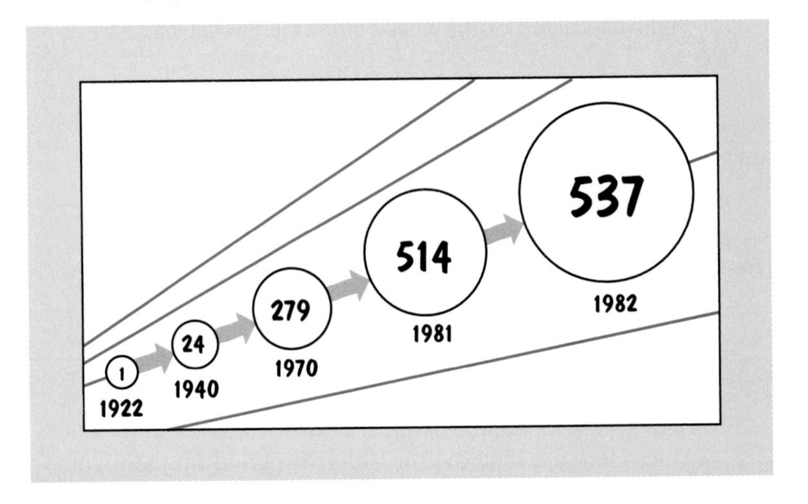

2.4 Relational Graphs

So far we have looked at graphical displays of a single variable, either to see how the data fall into categories or intervals or to examine patterns over time. Graphs can also be used to see if there is a relationship between two different variables: nitrogen and crop yields, cigarette smoking and lung cancer, the unemployment rate during the year preceding a presidential election and the number of votes received by the incumbent political party. In each of these three examples, the relationship is meant to be causal, not coincidental. One of the variables—the *explanatory variable*—influences the value of the other variable—the *dependent variable.* Nitrogen is the explanatory variable, and crop yield is the dependent variable. Cigarette smoking is the explanatory variable, and lung cancer is the dependent variable. The unemployment rate is the explanatory variable, and the number of votes the incumbent receives is the dependent variable.

There is a positive relationship between two variables if when the value of the explanatory variable increases, the value of the dependent variable tends to increase, too. There is a negative (or inverse) relationship between two variables if when the explanatory variable increases, the dependent variable tends to decrease. We might anticipate that there is a positive relationship between cigarette smoking and lung cancer and a negative relationship between the unemployment rate and the votes received by the incumbent party. The relationship between nitrogen and crop yields might be positive for modest levels of nitrogen and then turn negative when nitrogen reaches excessive levels.

Scatter Diagrams

One way to explore whether the data show the presence of a positive relationship, a negative relationship, or no relationship at all is with a **scatter diagram,** in which the explanatory variable is plotted on the horizontal axis and the dependent variable is shown on the vertical axis. Figure 2.34 shows a scatter diagram for per capita cigarette consumption in each of 11 countries in 1930 and the male death rates from lung cancer in these same countries 20 years later, in 1950.[35] The U.S. observation, for example, shows that the per capita cigarette consumption was slightly more than 1250 and that there were almost 200 deaths per million.

This scatter diagram shows a rough positive relationship in that countries with relatively high cigarette consumption also tend to have relatively high death rates from lung cancer. A similar graph was part of the evidence presented in the landmark 1964 U.S. Surgeon General's report that warned of the health risks associated with cigarette smoking. The relationship is imperfect because there are other factors, in addition to cigarette consumption, that can cause lung cancer.

The data in Fig. 2.34 are *cross-sectional* data showing observations (in this case, from different countries) at comparable times: all the cigarette consumption data are from 1930, and all the lung cancer data are from 1950. Scatter diagrams can also be used with *time-series* data. Figure 2.35 is a scatter diagram of the duration of an eruption of Old Faithful and the time until the next

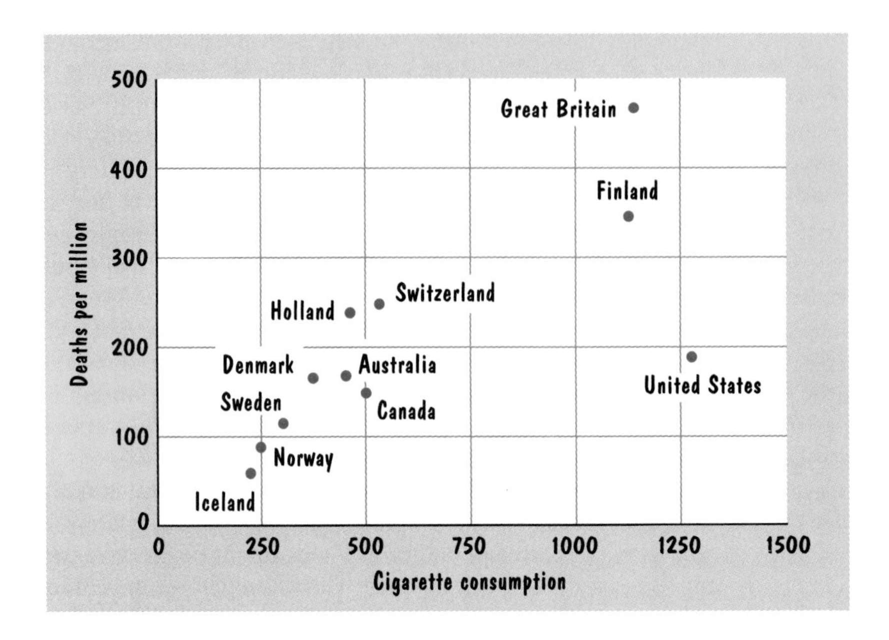

Figure 2.34 Per capita cigarette consumption in 1930 and male death rate from lung cancer in 1950.

eruption. Each of the 112 points in this graph is a different observation, taken at a different time, for example, at 8:55 a.m. on September 10, 1995, and at 7:04 p.m., on September 23, 1995.

Eruptions occur when some of the water in the geyser reaches the boiling point. If the eruption is brief, then more heated water is left in the geyser hole and it takes less time to reach the boiling point for the next eruption. If it is a long eruption, then more hot water is spewed from the geyser and it takes longer for the remaining water to be heated to the boiling point. The scatter diagram in Fig. 2.35 confirms that the longer the eruption lasts, the longer the wait until the next eruption. In a later chapter, you will see how to quantify this relationship to predict when Old Faithful will erupt.

Scatter diagrams can also be used to describe the statistical relationship between two variables that may not be causally related. For instance, in Fig. 2.36 the dependent variable is the annual rate of return from stocks (using the data in Table 2.1), and the explanatory variable is the annual return from stocks in the preceding year. We do not believe that annual stock returns are *caused* by the returns the preceding year, but we may want to investigate whether stock returns can be predicted from the return the preceding year. A positive relationship would indicate that when the stock market does well one

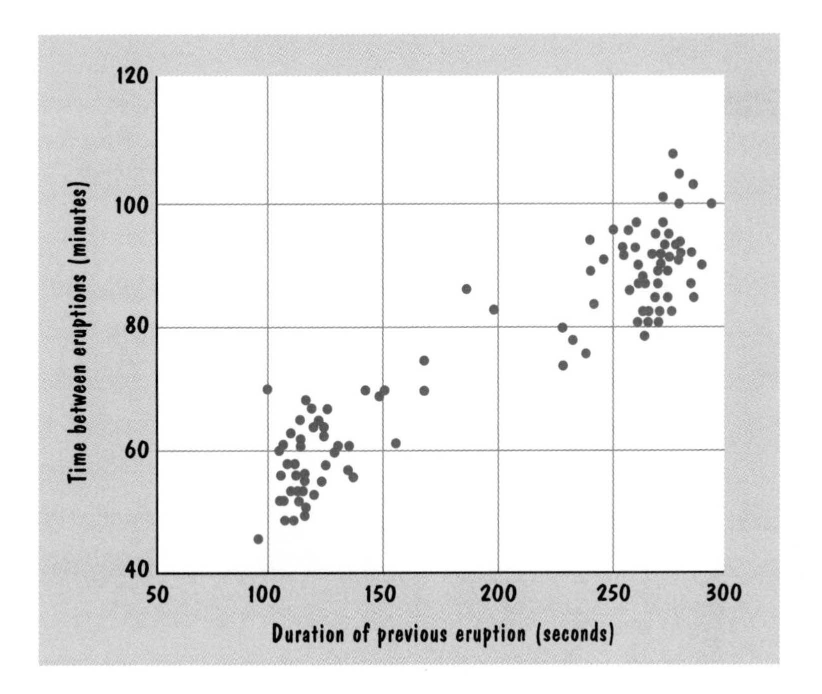

Figure 2.35 Old Faithful's time between eruptions and duration of previous eruptions.

year, it is likely to do well the next year, too. A negative relationship would mean that good years in the stock market tend to be followed by bad years. The haphazard scatter of points in Fig. 2.36 suggests that stock market performance in any given year is unrelated to performance the preceding year, just as the outcome of a coin flip is not related to the outcomes of earlier flips. In later chapters we will see how to quantify the notion of a positive or negative relationship and how to gauge whether a relationship is strong or weak.

Using Categorical Data in Scatter Diagrams

Sometimes, we want to separate the data in a scatter diagram into two or more categories. We can do this by using different symbols or colors to identify the data in the different categories. The two examples in Fig. 2.37 are data on the percentage of students who use computers in school, grouped by household income.[36] Figure 2.37*a* shows data for both 1984 and 1993, separated into these two categories by different symbols. By showing these categorical data in the same scatter diagram, we can see that the use of computers at school

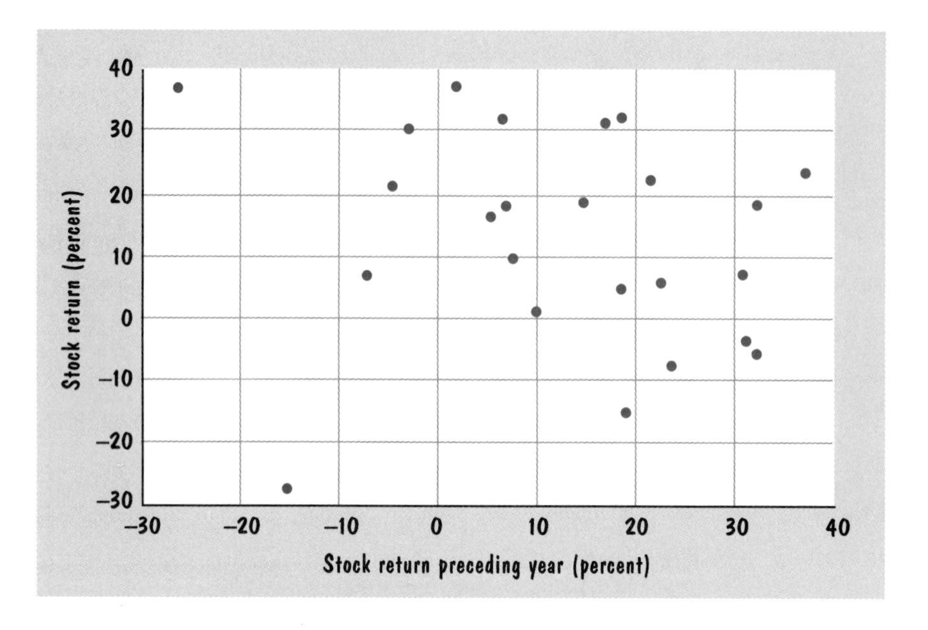

Figure 2.36 Can stock returns be predicted from returns the preceding year?

increased substantially between 1984 and 1993, and that computer usage is positively related to household income in both years—though not as strongly in 1993. Figure 2.37*b* shows 1993 data for two different types of schools: grades 1–8 and college. Here, the use of categorical data in a scatter diagram shows that the percentage of students using computers at school was generally larger for grades 1–8 than for college students and that there was a positive relationship between household income and computer usage at school for students in grades 1–8, but not for college students.

Are Mutual Funds Consistent?

Mutual funds buy stocks and other securities on behalf of their shareholders and pass along the investment profits, less management fees and other expenses. Many investors pour money into recently successful mutual funds, paying substantial fees and other expenses, in the belief that a fund's recent performance ensures future success. Unfortunately, the data show that there is little or no consistency to mutual fund performance. Funds with above-average records are no more likely to do well in the future than funds that have had disappointing records.

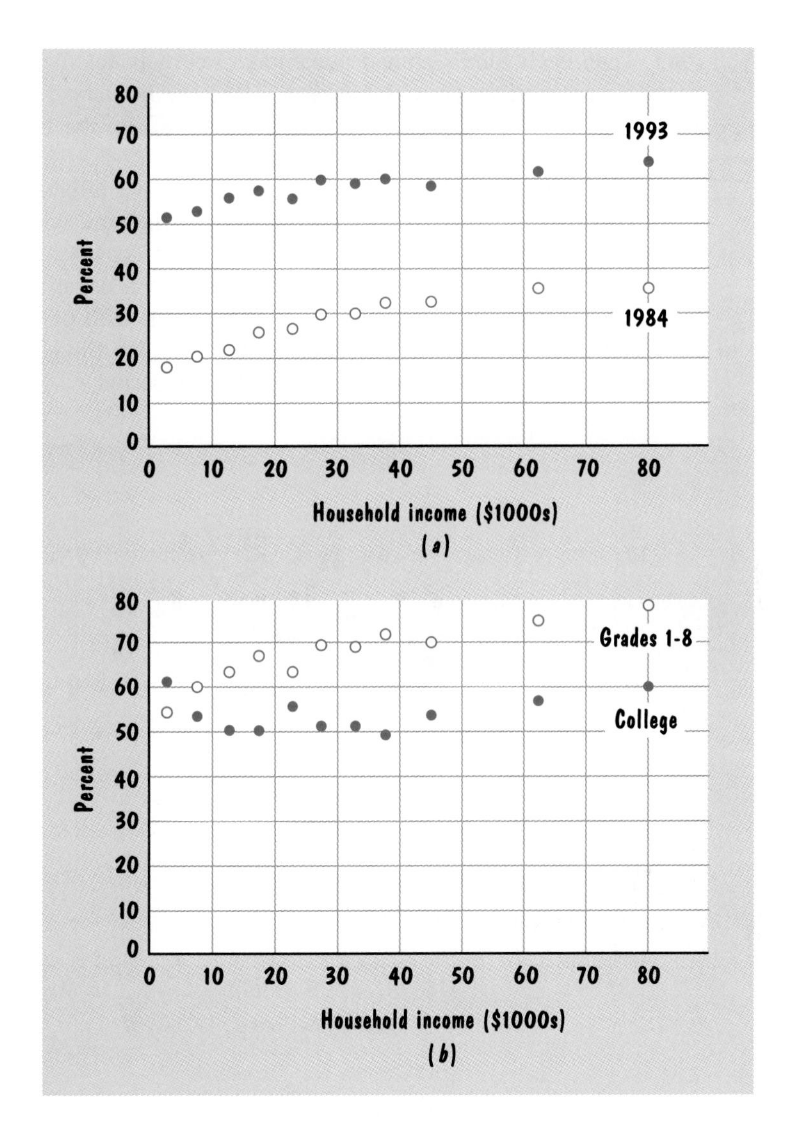

Figure 2.37 The percentage of students who use computers in school.

For example, I looked at the records of the 64 mutual funds that, according to the Wiesenberger Service, seek to maximize capital gains.[37] Each fund's annual percentage return, dividends plus capital gains, is calculated over the 5-year period from January 1, 1977, to January 1, 1982. These returns are the performance record that investors might have used to choose a mutual fund in

1982. Then each fund's annual percentage return is calculated over the subsequent 5-year period, from January 1, 1982, to January 1, 1987. This is the 5-year return for an investor who bought in 1982 on the basis of each fund's performance during the preceding 5 years.

If relatively good performance in the past were a reliable predictor of relatively good performance in the future, a scatter diagram would show a positive relationship. If funds that did relatively well in the past tended to do relatively poorly in the future, a scatter diagram would show a negative relationship. The actual scatter diagram has one or two outliers but, with or without them, indicates that there is no systematic relationship between fund performance in one 5-year period and that in the subsequent 5-year period.

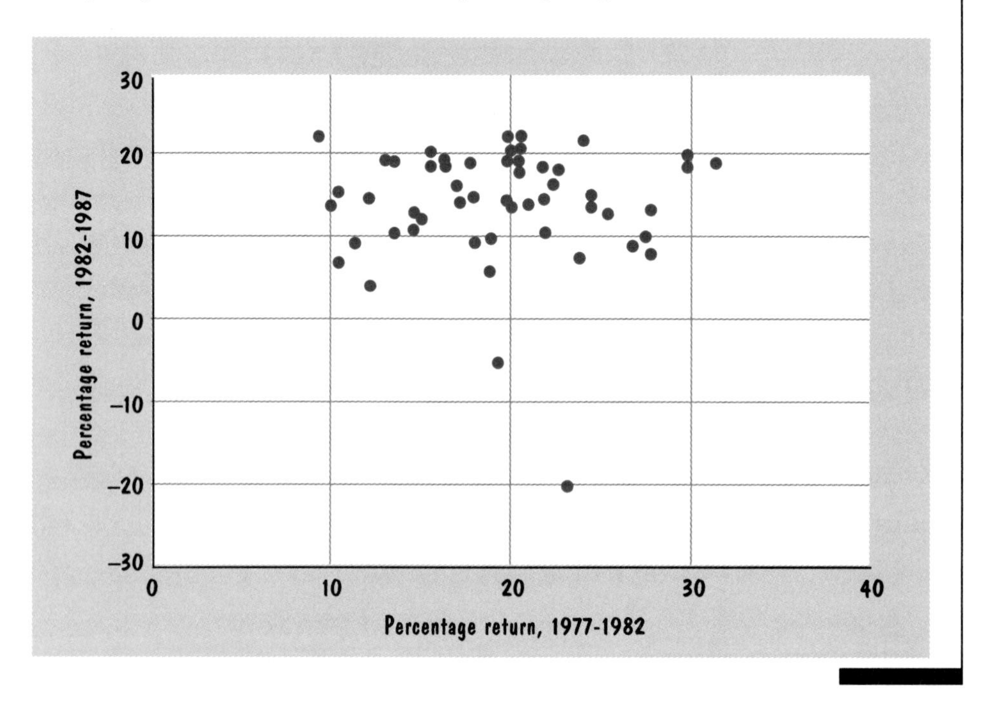

2.17 Explain why each of the following pairs of data is positively related, negatively related, or essentially unrelated. If you think there is a causal relationship, which is the explanatory variable?

a. The height of a father and the height of his oldest son

b. The age of a mother and the age of her oldest child

c. The number of scalp hairs and the age of a man between 30 and 40 years
d. The height and age of a woman between the ages of 30 and 40
e. A woman's age and her cost for a 1-year life insurance policy

2.18 Plutonium has been produced in Hanford, Washington, since the 1940s, and some radioactive waste has leaked into the Columbia River. A 1965 study of cancer incidence in nearby communities compared an exposure index and the cancer mortality rate per 100,000 residents for nine Oregon counties:

EXPOSURE INDEX	8.34	6.41	3.41	3.83	2.57	11.64	1.25	2.49	1.62
CANCER MORTALITY	210.3	177.9	129.9	162.3	130.1	207.5	113.5	147.1	137.5

Which variable is the explanatory variable, and which is the dependent variable? Make a scatter diagram and describe the relationship as positive, negative, or nonexistent.

2.19 After its staff ran more than 5000 miles in 30 different running shoes, *Consumer Reports* rated these shoes for overall performance on a scale of 0 to 100, with 100 the highest possible rating.[38] Below is a scatter diagram of each shoe's rating and its price (in dollars). Does there seem to be a strong positive relationship, strong negative relationship, or essentially no relationship?

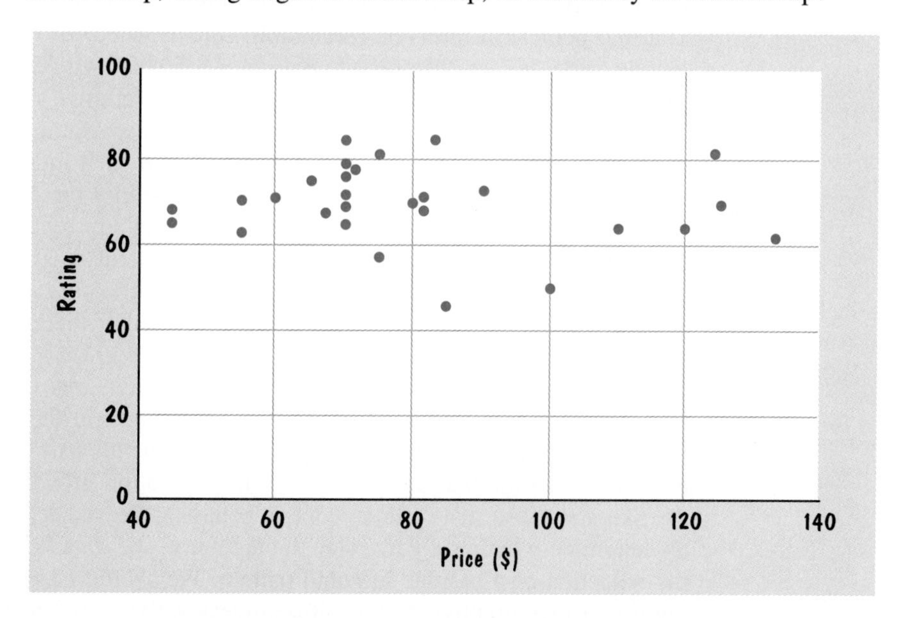

2.20 Use the annual data in Table 2.1 to draw a scatter diagram with the return on Treasury bills as the explanatory variable and the return on Treasury bills the following year as the dependent variable. (You will have 24 observations, since

Table 2.1 does not show the return for the year following 1995.) Does your scatter diagram suggest a positive relationship, a negative relationship, or no relationship?

summary

To analyze categorical data, such as male and female, that do not have natural numerical values, we can sort the data into a small number of meaningful categories and calculate either the number of observations in each category or the relative frequencies (the fraction of the observations that fall into each category). These results can be displayed in a bar chart (using the heights of bars to show either the number or the relative frequency of observations in each category) or in a pie chart (using slices of a pie to show relative frequencies). Bar charts are sometimes distorted by the replacement of simple bars of equal width with two- or three-dimensional pictures. Pie charts are sometimes confusing because it is hard to gauge the relative sizes of the slices.

To analyze quantitative data such as height and weight, we can separate the data into numerical intervals (preferably equally wide). If the intervals are equally wide, we can display the data in a bar chart with the height of each bar equal to either the number of observations or the relative frequency. If the interval widths are not equal, we need to use a histogram, with the bar areas showing the relative frequency of observations in each interval; the height of each bar is equal to the relative frequency divided by the width of the interval.

In a stem-and-leaf diagram (or stem plot), bars are replaced by numerical digits. When examining a histogram or stem-and-leaf diagram, we look for the center of the data, the number of peaks, whether the data are compact or dispersed, whether the figure is symmetric, and whether there are outliers.

In a time-series graph (sometimes called a *line graph*), the variation in data over time is shown with the data on the vertical axis and time on the horizontal axis. Time-series data are often adjusted for changes in the population, price level, or other appropriate variables to help us distinguish between changes in the data over time that are due merely to population growth, price changes, and the like and those that reflect people's behavior. Seasonally adjusted data help us determine whether a particular fluctuation in the data is noteworthy or just the reflection of a regular seasonal pattern. We should be cautious in extrapolating time-series graphs into the future unless we have thought carefully about the underlying causes of these observed patterns and whether the patterns are likely to continue. To detect purposely deceptive time-series graphs, we should ask ourselves whether the beginning and ending points for the data appear to be peculiar choices that one would make only after scrutinizing the data.

The omission of zero magnifies variations in the heights of bars in bar charts and histograms and in the heights of lines in a time-series graph, allowing us to detect differences that might otherwise be ambiguous. However, once zero has been omitted, we can no longer gauge relative magnitudes by comparing the heights of bars or lines. Instead, we need to look at the actual numbers.

A scatter diagram can be used to investigate whether there is a positive or negative relationship (or no relationship at all) between an explanatory variable (on the horizontal axis) and a dependent variable (on the vertical axis). Often, these relationships are thought to be causal—hence the terms *explanatory* variable and *dependent* variable. However, sometimes we just want to investigate the statistical relationship between two variables that may not be causally related. Although we do not believe that annual stock returns are *caused* by the returns of the preceding year, we may want to investigate whether stock returns can be predicted from returns of the preceding year.

Visual displays of data are intended to facilitate analysis and interpretation. We should avoid chart junk—unnecessary ornamentation that distracts the reader from the information conveyed by the data. We should display data accurately and coherently, not distort the data's implications, and encourage the reader to think about the data rather than to admire our artwork.

review exercises

2.21 Do not collect any data, but use your general knowledge to draw a rough sketch of a histogram of the first-year salaries of last year's graduates from your school. Be sure to label the horizontal and vertical axes.

2.22 Shown below are the estimated 1994 U.S. expenditures on four kinds of athletic footwear, separated into female and male shoes. Construct two pie charts, one for females and one for males. Based solely on these pie charts, what relationship does there seem to be between gender and the kinds of shoes purchased? If you did not construct pie charts, what arithmetic calculations could you use to show that shoe purchases vary by gender?

	FEMALES	MALES
WALKING SHOES	1143	707
GYM SHOES AND SNEAKERS	678	779
JOGGING AND RUNNING SHOES	295	407
AEROBIC SHOES	203	23

2.23 Languages differ in the number of syllables used in words. Use the relative frequencies below to sketch bar charts for each of these four languages.[39] (Omit the more-than-4 category.) Summarize the major differences in these charts.

SYLLABLES	ARABIC	ENGLISH	GERMAN	JAPANESE
1	.23	.71	.56	.36
2	.50	.19	.31	.34
3	.22	.07	.09	.18
4	.05	.02	.03	.09
More than 4	.00	.01	.01	.03

2.24 Use two histograms, one for 1980 and one for 1994, to display these data on the number of passenger cars (in millions) of various ages in use in the United States. (Assume the oldest interval is 12 to 17 years, and note that two of the intervals are 3 years wide and two are 6 years wide.) Summarize these data in words.

AGE (YEARS)	0–5	6–8	9–11	12 and up
NUMBER IN 1980	52.3	25.2	14.6	12.5
NUMBER IN 1994	45.4	27.7	25.1	31.4

2.25 Shown below is a bar chart for data on the difference between the sophomore and rookie batting averages for 39 nonpitchers who were selected as major league baseball rookies of the year between 1967 and 1992.[40] Based on this bar chart, summarize these data.

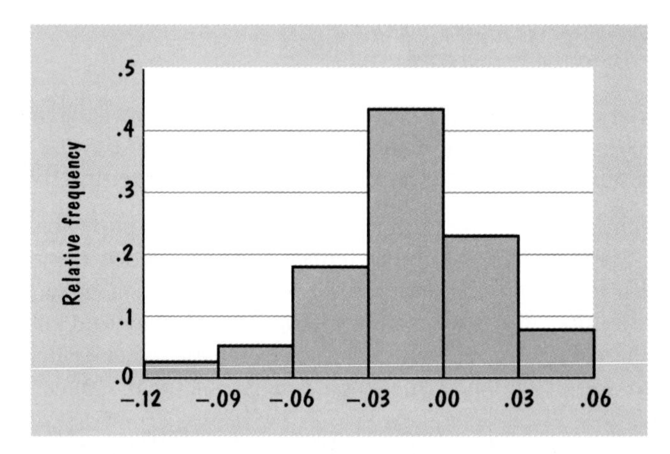

2.26 A study of the average sentence length of British public speakers used the data that follows.[41] Plot these data, using a horizontal axis from 1500 to 2200 and a vertical axis from 0 to 80. Draw a freehand line that fits these data, and then extend this line to obtain a rough prediction of the average number of words per sentence in the year 2200.

SPEAKER	YEAR	WORDS PER SENTENCE
Francis Bacon	1598	72.2
Oliver Cromwell	1654	48.6
John Tillotson	1694	57.2
William Pitt	1777	30.0
Benjamin Disraeli	1846	42.8
David Lloyd George	1909	22.6
Winston Churchill	1940	24.2

2.27 The U.S. Center for Health Statistics estimates the following relative frequencies of death by age for persons born in 1975:

AGE AT DEATH	WHITE FEMALES	WHITE MALES	BLACK FEMALES	BLACK MALES
0–19	.020	.030	.032	.044
20–39	.016	.036	.035	.086
40–59	.025	.044	.050	.083
50–64	.107	.196	.168	.244
Over 64	.832	.694	.715	.543

Draw two histograms, comparing black males and black females. (Assume that the over-64 category is 65 to 89, and note that the interval widths are 20, 20, 20, 15, and 25 years.) Based on your histograms, briefly compare these two data sets.

2.28 Identify what is misleading about this graphic that accompanies a story in *The New York Times* about how discount fares were reducing travel agent commissions:[42]

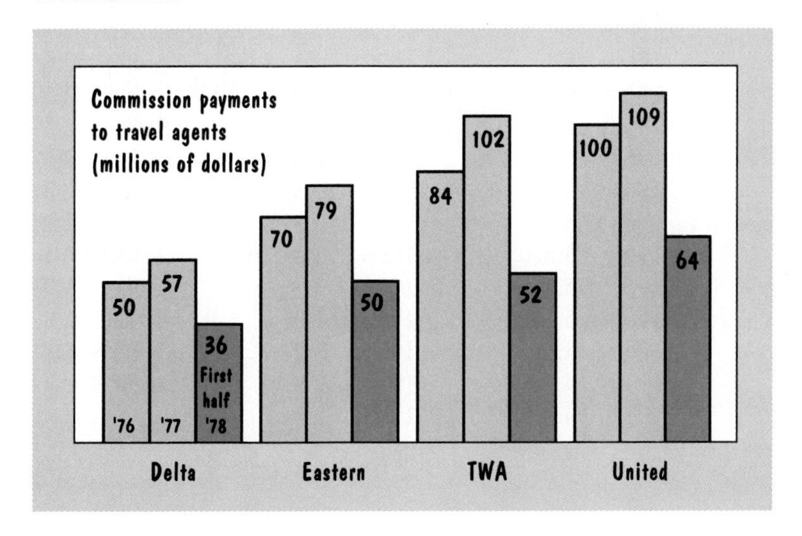

2.29 Below is a properly done graph of data from 21 countries on annual per capita cigarette consumption and deaths from coronary heart disease per 100,000 persons aged 35 to 64. Which variable is on the vertical axis? Explain your reasoning. Does there seem to be a positive relationship, a negative relationship, or no relationship at all?

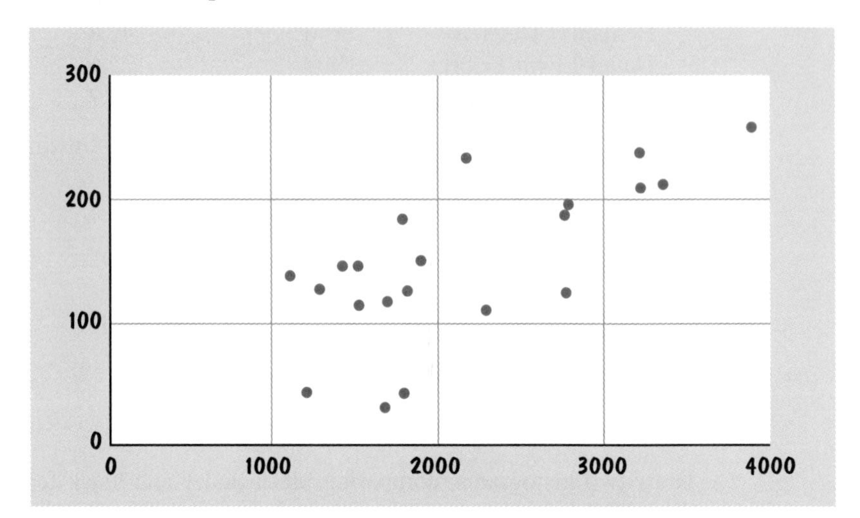

2.30 In 1974 Congress tried to reduce fuel use by imposing a nationwide speed limit of 55 miles per hour. As time passed, motorists increasingly disregarded this speed limit, and in 1987 the speed limit on rural interstate highways was increased to 65 miles per hour. Plot these data on U.S. motor vehicle deaths per 100 million miles driven, and see if anything unusual happened in 1974 and during the next several years.

1960	5.1	1972	4.3	1984	2.6
1961	5.1	1973	4.1	1985	2.5
1962	5.1	1974	3.5	1986	2.5
1963	5.3	1975	3.4	1987	2.4
1964	5.4	1976	3.3	1988	2.3
1965	5.3	1977	3.3	1989	2.2
1966	5.5	1978	3.3	1990	2.1
1967	5.4	1979	3.3	1991	1.9
1968	5.2	1980	3.3	1992	1.8
1969	4.9	1981	3.2	1993	1.8
1970	4.7	1982	2.8	1994	1.7
1971	4.5	1983	2.6		

2.31 A 1993 national survey on drug abuse estimated the number of persons, by age, who had smoked at least one cigarette during the preceding month.[43] The data below are shown in two ways: the number of persons in each age group who smoked a cigarette and the percentage of persons in each age group who had smoked. Draw the two figures that you would use if you were a government employee asked to display these two sets of data.

AGE	NUMBER	RATE
12–17	2,104,000	9.6
18–25	8,415,000	29.0
26–34	11,317,000	30.1
35–64	22,026,000	23.8

2.32 Below is a bar chart of the estimated 1996 market value of the 30 teams in the National Football League.[44]

a. Approximately what percentage of these teams is valued at $160 million to $200 million?

b. In what range is the lowest-value team?

c. Write a one-sentence summary of the information provided by this bar chart.

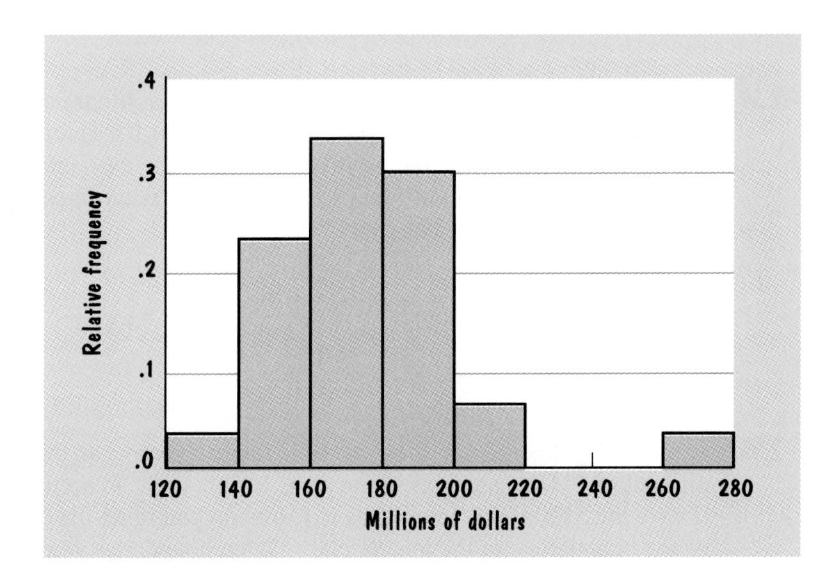

2.33 Explain what is misleading about the following graph.[45]

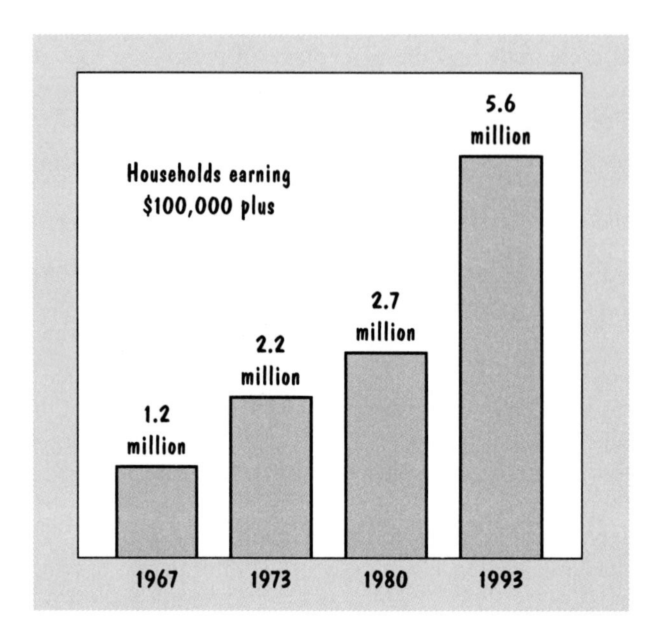

2.34 Use these data on the number of households (in millions) of different sizes to calculate the relative frequencies and to draw two bar charts, one for 1890 and one for 1990. (Calculate the relative frequency for the 7-or-more category, but do not include this category in your bar charts.) What important differences do you detect in these two bar charts?

SIZE	1	2	3	4	5	6	7 OR MORE
1890	.457	1.675	2.119	2.132	1.916	1.472	2.919
1990	23.0	30.2	16.1	14.6	6.2	2.2	1.5

2.35 Explain why this graphic showing 1993 family incomes in the United States is misleading.[46] Using an axis of the same length, draw an accurate representation. (Leave the $100,000+ interval as is.) How do you think the original artist decided on the boundaries for the middle class? What boundaries would you use?

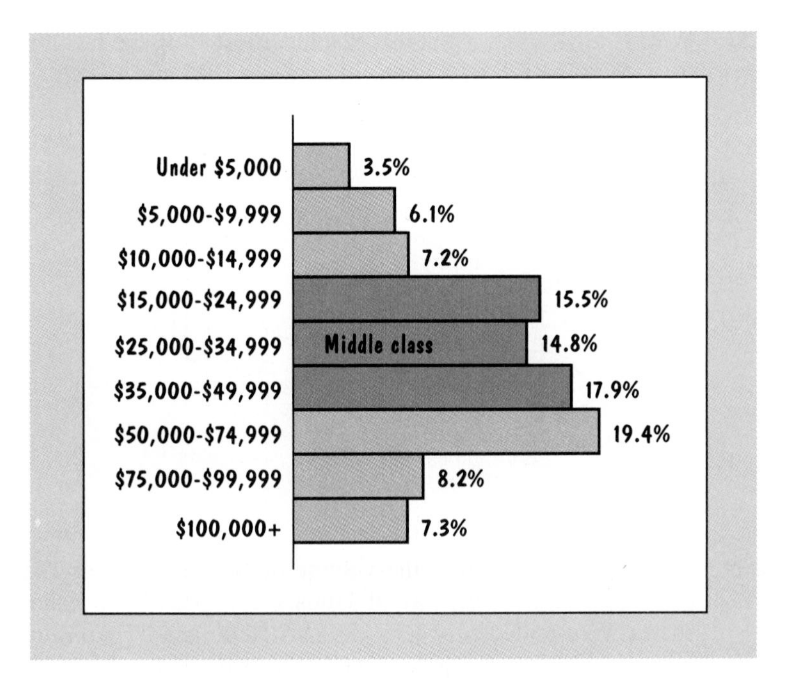

2.36 Here is a graph of the calcium and fat in 1-cup servings of nine foods.[47]

a. Which of these foods has the most fat?

b. Which has the least calcium?

c. Which combines the lowest fat with the most calcium?

d. Approximately how much fat and how much calcium are in the product selected for part c?

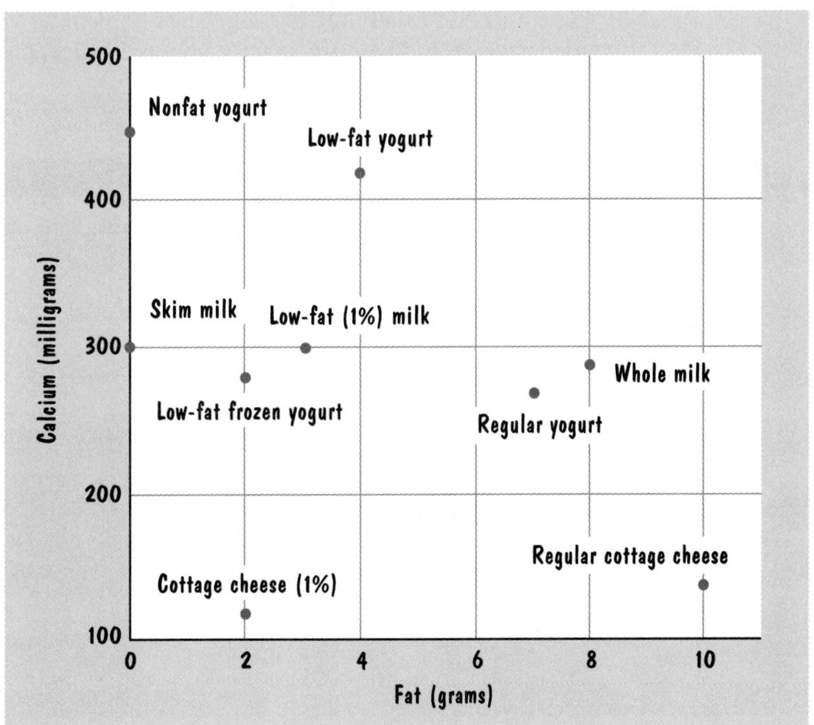

2.37 What is misleading about this graphic, illustrating the fact the purchasing power of the U.S. dollar fell by 50 percent between 1968 and 1978?

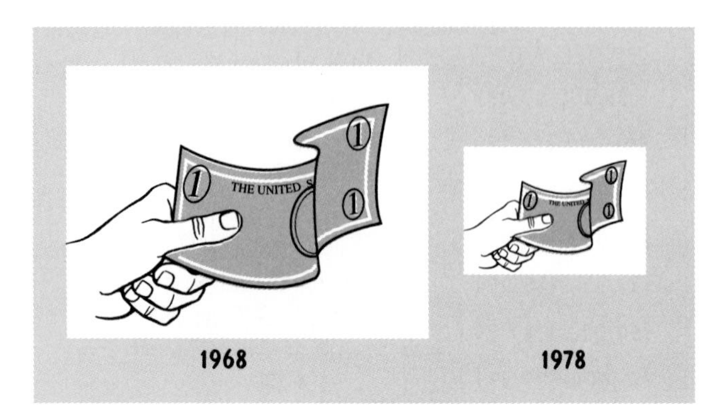

2.38 The graph below shows the volume of New York Stock Exchange (NYSE) trading throughout the day of January 27, 1997.[48] For example, 54.4 million shares were traded between 9:30 and 10:00 and 77.1 million shares were traded between 10:00 and 11:00. Identify the two serious deficiencies that cause this graph to be visually misleading. Then explain how you would correct these deficiencies and identify exactly how the appearance of the graph would be altered dramatically.

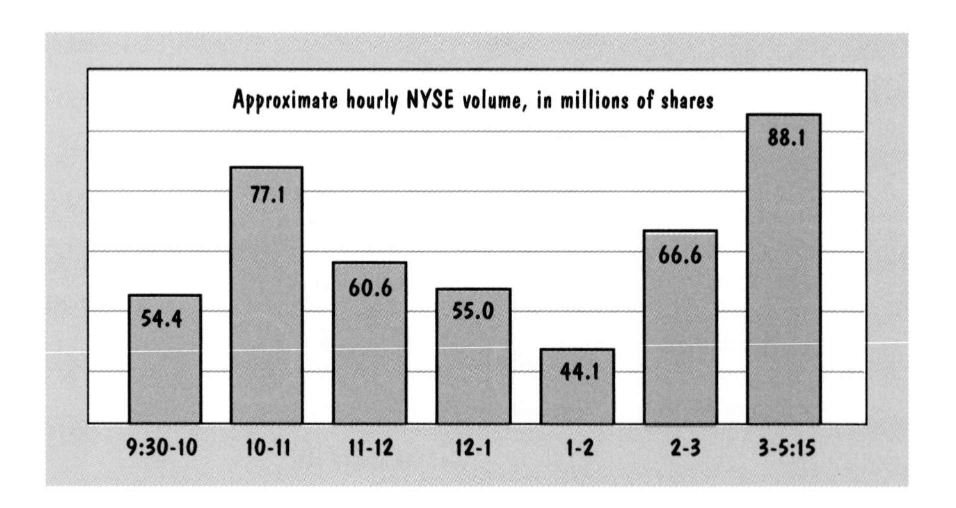

2.39 Babe Ruth was paid $80,000 in 1931. (When asked about the fact that he was being paid more than Herbert Hoover, the President of the United States, Ruth replied, "I had a better year than the president.") Darryl Strawberry was paid $4 million in 1991. The consumer price index was 15.2 in 1931 and 136.2 in 1991. Which of these two baseball players was paid more in real terms? What would Darryl Strawberry's 1991 salary have been if he had been paid an amount with the same purchasing power as Ruth's 1931 salary?

2.40 When a child loses a baby tooth, a U.S. tradition is for the tooth to be put under the child's pillow, so that the tooth fairy will leave money for it. A survey by a Northwestern University professor found that the tooth fairy paid an average of 12 cents for a tooth in 1900 and $1 for a tooth in 1987.[49] The consumer price index was 25 in 1900 and 340 in 1987. Did the real value of tooth fairy payments rise or fall over this period? If tooth fairy payments had kept up with inflation, how large would the 1987 payment have been?

2.41 Explain why this graphic does a poor job of showing that "Clothing accounts for the greatest portion of merchandise purchased through catalogs."[50]

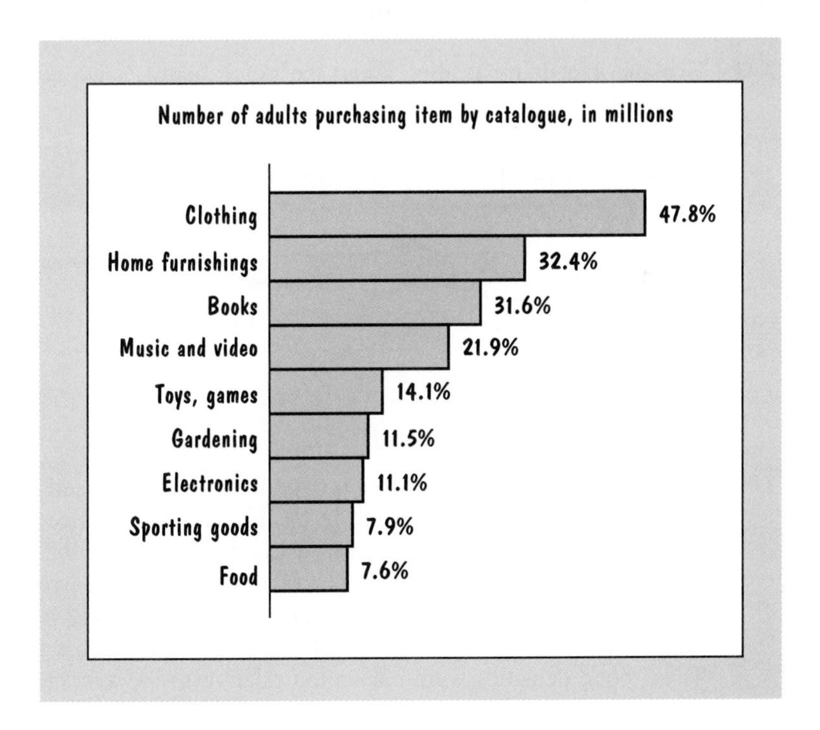

2.42 One hundred and thirty college students were asked if they had had a serious romantic relationship in the past 2 years and, if so, to identify the month in which the most recent relationship began.[51] Write a paragraph interpreting these results.

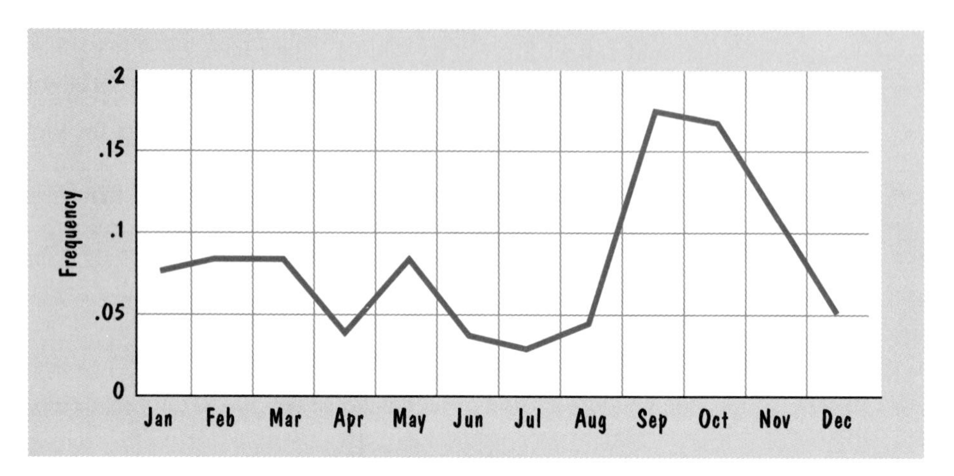

2.43 Explain what is misleading about the graph below.[52] Do not redraw it, but do explain how you would have drawn this graph.

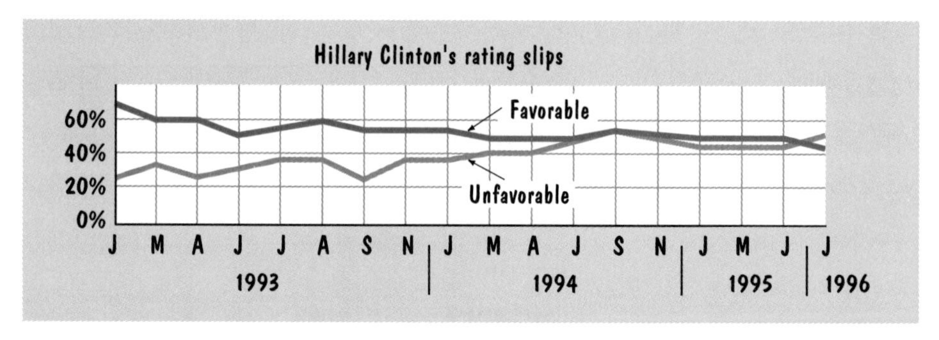

2.44 Modern-day adult British women experience a pronounced loss of bone density as they grow older, contributing to an increasing incidence of osteoporotic hip fractures. During the restoration of a London church in the 1990s, a crypt was opened that contained the skeletons of more than 100 persons buried between 1729 and 1852.[53] Shown next is a graph of bone density in the femoral neck region of the skeletons of 25 females who died between the ages of 15 and 45. These bone densities were calculated relative to the average bone density of women in the sample aged 18 to 30 years. Thus a figure of 120 indicates that this person's bone density was 20 percent larger than that of the average 18- to

30-year-old woman in the sample. Does there seem to be a clear positive relationship, clear negative relation, or essentially no relationship? What is the implication?

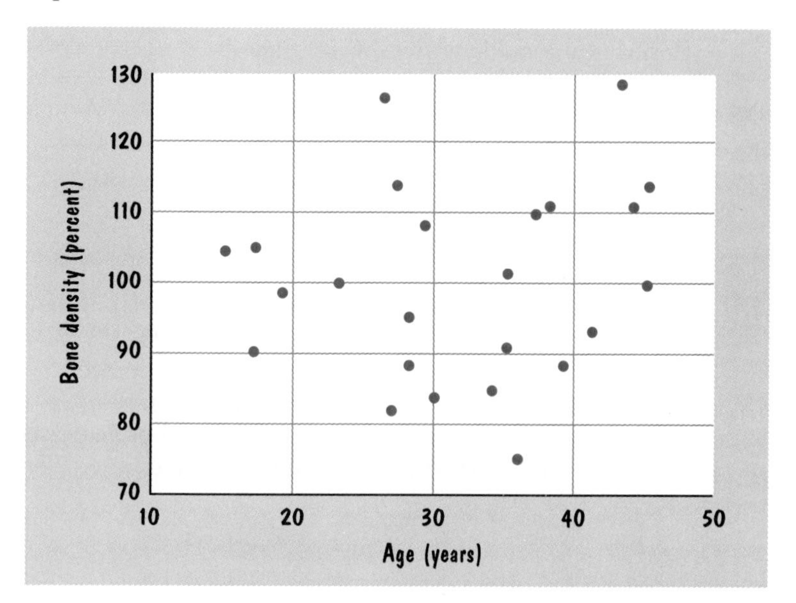

2.45 What is potentially misleading about this graphic, which was used to show the performance of four "historically high-performing mutual funds" from 1100 mutual funds offered by Merrill Lynch?[54] (Each graph shows the return from a $10,000 investment in September 1985; the initial values are somewhat less than $10,000 because of load fees charged for investing in these funds.)

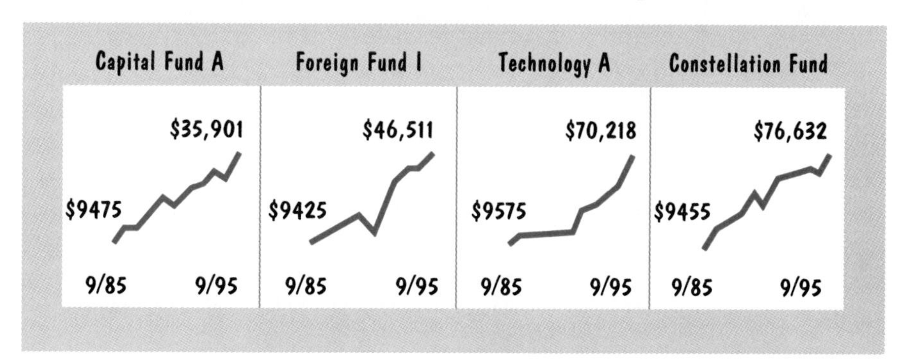

2.46 Shown below are the percentages of the Jewish population of various ages in Germany in 1928 and in 1959 and of a theoretical stationary population which is neither growing nor contracting.[55] Make three histograms, one for each population. (Assume that the interval of 60 and over is from 60 to 79.) What conclusions can you draw from these graphs?

AGE	1928	1959	STATIONARY MODEL
0–19	25.9	14.2	33
20–39	31.3	19.8	30
40–59	27.7	28.2	25
60 and over	14.3	28.2	12

2.47 Below is a graph of the winning times (in seconds) of the men's 1500-meter run in all the Olympic games from 1896 to 1996.

a. In what year was the fastest winning time recorded?
b. What was this winning time (approximately)?
c. Have the winning times generally increased or decreased over this period?
d. Was this increase or decrease larger before or after World War II?

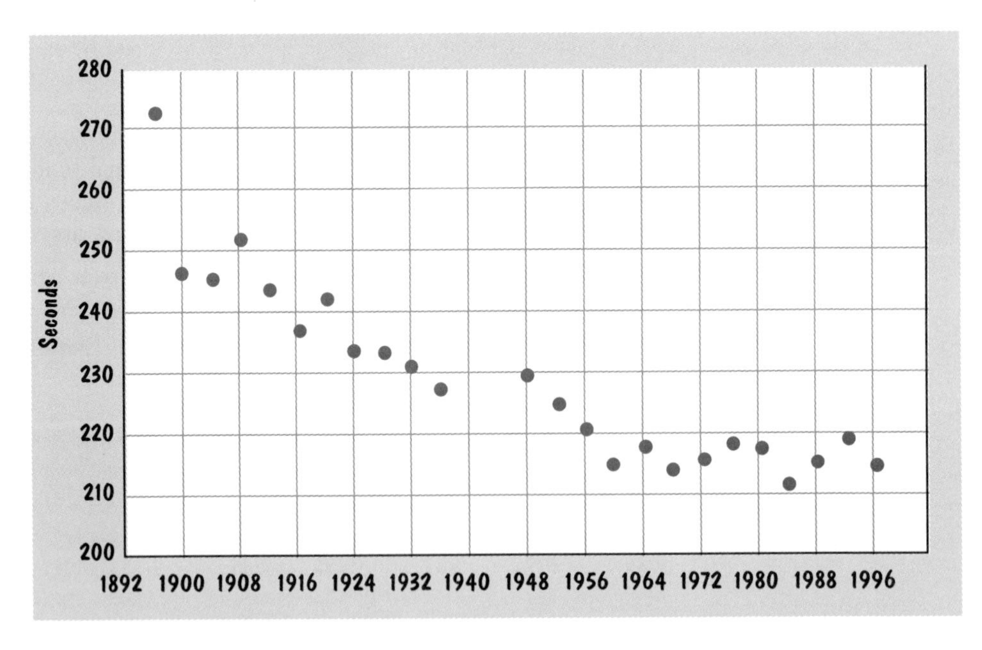

2.48 An article in *Scientific American* illustrated the distribution of water on earth (among oceans, glaciers and polar ice, underground aquifers, lakes and rivers, atmosphere, and biosphere) by showing six circles "in which the amount of water present in various natural reservoirs is represented in terms of comparative spherical volumes."[56] Here, for example, is what the circles for glaciers and underground aquifers looked like:

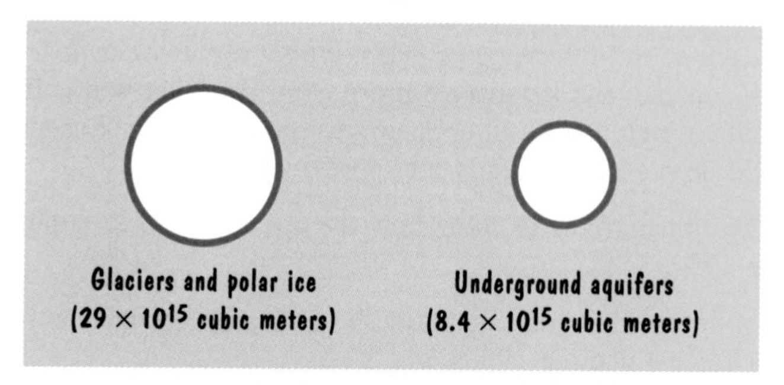

What is misleading about these figures?

2.49 Answer this letter to Ann Landers:

> I've read your column for ages and almost always agree with you. One subject on which we do not see eye-to-eye, however, is senior citizens driving. According to the *Memphis Commercial Appeal,* recent statistics by the National Highway Traffic Safety Administration indicate that drivers over 70 were involved in 4,431 fatal crashes in 1993. That is far fewer than any other age group. The 16- to 20-year-olds were involved in 7,711 fatal crashes.
>
> Elderly drivers also had the lowest incidence of drunk driving accidents. Now, will you give seniors the praise they deserve?[57]

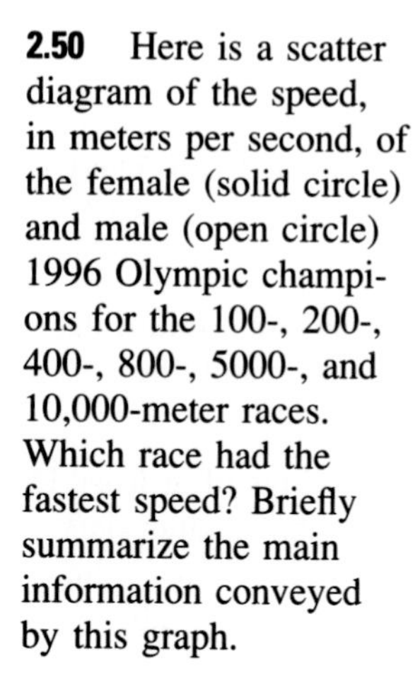
2.50 Here is a scatter diagram of the speed, in meters per second, of the female (solid circle) and male (open circle) 1996 Olympic champions for the 100-, 200-, 400-, 800-, 5000-, and 10,000-meter races. Which race had the fastest speed? Briefly summarize the main information conveyed by this graph.

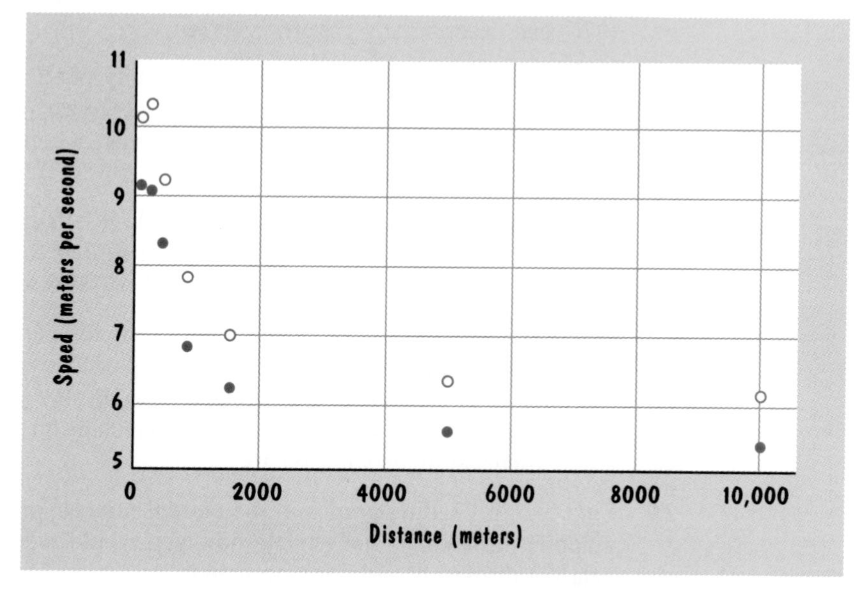

projects

For each of the following projects, type a report in ordinary English, using clear, concise, and persuasive prose. Any data that you collect for this project should be included as an appendix to the report. Data used in your report should be presented clearly and effectively.

2.1 Use two histograms to show how the age of the U.S. population has changed since 1950.

2.2 Use a time-series graph to show the changing gender composition of the U.S. labor force since the 1950s.

2.3 Has there been any apparent trend in voter participation in U.S. presidential elections since the 1930s?

2.4 *Vital Statistics of the United States* contains a great deal of birth and mortality data. Find the most recent edition and the oldest edition at your college library, and use two histograms to show how the ages of women giving birth have changed over time.

2.5 Use the U.S. Department of Energy's monthly publication *Monthly Energy Review* to find annual data back to 1976 on the U.S. retail price of unleaded regular gasoline (cents per gallon, including taxes). *Economic Report of the President* or *Survey of Current Business* will provide data on the overall consumer price index over this same period. Graph the real price of gasoline since 1976, and describe any patterns you see.

2.6 Use the U.S. Department of Energy's monthly publication *Monthly Energy Review* to obtain monthly data over a recent 3-year period on total U.S. energy consumption (quadrillion British thermal units). Graph these data and describe any patterns you see.

2.7 Use the source given at the bottom of Table 2.1 to extend the stock return data back to 1926. Now draw a scatter diagram similar to Fig. 2.36, using the annual stock returns over the entire period of 1926 to 1995.

2.8 Determine the amounts of calcium and fat in comparable servings of at least 15 different kinds of ice cream. Use one box plot to summarize the calcium data and another to summarize the fat data. Also display your data in a scatter diagram with fat on the horizontal axis and calcium on the vertical axis, labeling any particularly interesting points.

2.9 Find two misleading graphs in the current popular press. Explain why each graph is misleading, and explain how you would redraw each graph so that the data are displayed fairly and accurately.

2.10 Make a scatter diagram of the area and population of the 50 states of the United States. Does there seem to be a statistical relationship? Are there any outliers? Calculate the population density (population divided by area) for each state and summarize your data with a histogram.

2.11 For each of at least 25 different carbonated soda drinks (such as Coke, Pepsi, and 7-Up), record the grams of sugar per serving shown on the nutritional label. Do not include any diet sodas. Use a histogram to summarize your data.

2.12 Go to a local store that sells CDs and record the prices of at least 50 recent CDs. Also record the total playing time for those CDs that display this information. Use a histogram to summarize each of these data sets. For those CDs that show the playing times, make a scatter diagram with the playing time on the horizontal axis and the price on the vertical axis.

HAVE YOU EVER WONDERED ?

Mutual funds manage people's investments for them. Money market funds buy Treasury bills and other very short-term bonds. Other mutual funds specialize in long-term bonds, tax-exempt bonds, or precious metals. The most popular mutual funds are those that invest in corporate stock. At the end of 1996, stock mutual funds held more than a trillion dollars' worth of securities.

A mutual fund pools the savings of thousands of investors in order to purchase a large, diversified portfolio of securities, which has considerable appeal for those investors who lack the wealth to amass a diversified portfolio on their own and feel that they lack the time or expertise to select securities wisely. Mutual fund managers argue that they enable investors with limited resources to have their money managed by professionals. Just as you entrust the care of your teeth to a dentist and the care of your eyes to an ophthalmologist, so you should entrust the care of your investments to a mutual fund manager. Some are skeptical. In his book *Confessions of a Wall Street Insider,* C. C. Hazard wrote, "A mutual fund has been defined as an outfit which, for a fee, spares investors the trouble of losing their own money. The fund takes over the task for them."

What does the historical record tell us about mutual fund performance? Example 2.7 suggests that there is little or no consistency in returns. Funds that have excelled in the past are no more likely to do relatively well in the future than are funds that did poorly in the past. Still, this inconsistency would not be so troubling if mutual funds overall were able to beat the stock market. If mutual funds generally beat the market, then investors might be better off picking a random mutual fund than picking a random portfolio of stocks on their own. So, what does the historical record show? Do mutual funds generally do better or worse than the overall stock market? In this chapter, we see how these questions might be answered.

chapter 3

summarizing data

Nothing shocks me anymore, especially when I know that 50 percent of the doctors who practice medicine graduated in the bottom half of their class.
—Ann Landers

IN THE PRECEDING chapter, we saw how histograms could be used to give a simple, organized picture of a set of quantitative data. In this chapter, we will see how one or two descriptive statistics, such as the mean and the standard deviation, often can be used to summarize data effectively. In addition to their value in describing data simply, these statistics can be used to draw inferences about the population from which the data were taken. For example, by comparing the average number of heart attacks in a group of people who took aspirin every other day with the average in a group who did not, we may be able to draw conclusions about the value of aspirin for the entire population.

In this chapter, we will look at how some descriptive statistics are calculated and how they should be interpreted. In later chapters, we will see how these statistics can be used to make statistical inferences.

3.1 The Center of the Data

The most familiar descriptive statistics are *measures of location* that describe the center of the data, without telling us whether the data are tightly clustered in this center or greatly dispersed. To illustrate three different measures of the center of the data, we use first the hypothetical income data shown in Table 3.1.

Table 3.1

Income in Smalland

1996 INCOME (DOLLARS)	NUMBER OF PERSONS
5,000	2
10,000	1
30,000	1
40,000	1
50,000	1
60,000	1
1,000,000	1

The Median

Our first measure is quite literally the center of the data:

> The **median** is the middle value when the data are arranged in numerical order from the smallest value to the largest value.

We can find the median by arranging the data from lowest value to highest value and then, starting at each end of the data, counting inward. When our two counts meet in the middle, we have located the median. For example, the Smalland income data in Table 3.1 can be arranged as follows:

5000 5000 10,000 **30,000** **40,000** 50,000 60,000 1,000,000

Because there are an even number of observations, (8), we split the difference between the two middle observations, 30,000 and 40,000, getting a median of 35,000. This is the center of the income data in Smalland, in that one-half of the people earned less than $35,000 and one-half earned more than $35,000.

When there are an odd number of observations, there is a single middle observation. For example, with these seven values, the median is 30,000:

5000 5000 10,000 **30,000** 30,000 50,000 60,000

This example also shows that some of the observations to the right or left of the median may turn out to be equal to the median. Thus, strictly speaking, we should say that *at least* one-half of the values are smaller than or equal to the median and *at least* one-half are larger than or equal to the median. In popular usage, we generally just say that one-half of the values are smaller than the median and one-half are larger than the median.

We will now look at a different set of data. Each February, *Money* magazine shows the performance records of thousands of mutual funds during the preceding calendar year, with the results listed alphabetically within each fund category. In the February 1995 issue, the 1994 results for stock mutual funds covered 17 pages. Because there is no reason why a fund's performance should have anything to do with the alphabetical placement of its name, I wrote down the name and the 1994 return for each fund at the bottom of these 17 pages. I then looked in the February 1996 issue of *Money* to see how these 17 funds did in 1995. These data are shown in Table 3.2. To find the median return for

Table 3.2

Annual percentage returns for 17 mutual funds

	1994	**1995**
American Leaders A	.0	37.0
Capstone Growth	−7.8	29.2
Dreyfus Appreciation	3.6	37.9
Fidelity Diversified International	1.1	18.0
Flag Inv. Telephone Income A	−6.3	33.8
Goldman Sachs G&I	5.9	33.5
IAI Midcap Growth	5.7	26.1
Keystone American Hart Em. Gro. A	−1.0	22.4
Matrix Growth	−4.8	22.7
New England Balanced A	−2.6	26.3
Pasadena Nifty Fifty A	1.1	28.2
Prudential Equity-Income B	−.8	20.7
Scudder Global Small Co.	−7.7	17.8
STI Classic Cap Gro. Inv.	−8.0	30.3
United Continental Income	−.4	24.8
Vanguard U.S. Growth	3.9	38.4
Zweig Strategy A	1.1	25.1

these 17 funds in 1995, we can arrange the returns in numerical order and then, counting in nine observations from either end, find the median value to be 26.3:

17.8 18.0 20.7 22.4 22.7 24.8 25.1 26.1 **26.3** 28.2 29.2
30.3 33.5 33.8 37.0 37.9 38.4

A similar analysis of the 1994 data shows the median return that year to be −.4 percent:

−8.0 −7.8 −7.7 −6.3 −4.8 −2.6 −1.0 −.8 **−.4** .0 1.1
1.1 1.1 3.6 3.9 5.7 5.9

When there are so many data that it is cumbersome to count in from both ends simultaneously, we can use the general rules given in the how-to-do-it box, depending on whether there are an even or odd number of observations. Either way, at least one-half of the values are smaller than or equal to the median, and at least one-half are larger than or equal to the median.

HOW TO DO IT

Finding the Median

Suppose that we want to find the median value of the 17 annual mutual fund returns for 1995 shown in Table 3.2.

1. We arrange the data in numerical order:

   17.8 18.0 20.7 22.4 22.7 24.8 25.1 26.1 26.3 28.2
   29.2 30.3 33.5 33.8 37.0 37.9 38.4

2. If the number of observations n is even, count in $n/2$ observations from either end; the median is the average of this observation and the next observation. For example, if there are $n = 18$ observations, then $n/2 = 9$, and the median is the average of the 9th and 10th observations when we count in from either end.
3. If the number of observations n is odd, then add 1 so that $(n + 1)/2$ is a whole number, and count in $(n + 1)/2$ observations from either end; the value of this observation is the median. For example, with $n = 17$ observations, $(n + 1)/2 = 9$, and the median is ninth observation in from either end, or 26.3.

The Mean

The **mean** is the simple arithmetic average value of the data.

> The mean of n observations $x_1, x_2, \ldots, x_n$ is the sum of these n values, divided by n:
>
> $$\text{Mean} = \frac{x_1 + x_2 + \cdots + x_n}{n} \tag{3.1}$$

The mean is often written as $\bar{x}$, which is just x with a bar over it (which is pronounced "x bar"), and we can use the shorthand notation

$$\bar{x} = \frac{\Sigma x_i}{n}$$

The Greek letter Σ (uppercase sigma) indicates that the values of x_i should be added.*

For the hypothetical Smalland incomes in Table 3.1, we add these eight values and divide by 8:

$$\begin{aligned}\bar{x} &= \frac{5000 + 5000 + 10{,}000 + 30{,}000 + 40{,}000 + 50{,}000 + 60{,}000 + 1{,}000{,}000}{8} \\ &= \frac{1{,}200{,}000}{8} \\ &= 150{,}000\end{aligned}$$

For the 17 stock mutual fund returns for 1995 shown in Table 3.2, we add the 17 values and divide by 17:

$$\bar{x} = \frac{37.0 + 29.2 + \cdots + 25.1}{17} = 27.8$$

An average return of 27.8 percent sounds pretty impressive, but it is put into perspective by the fact that overall the average return for all stocks in 1995 was 37.4 percent! In 1994, the average return for all stocks was 1.3 percent, while the average return for these 17 mutual funds was a 1.0 percent loss:

*To be mathematically precise, statisticians often use uppercase and lowercase notation to distinguish between a random variable, which can take on different values, and the actual values that happen to occur. We use lowercase notation throughout this book for simplicity and convenience.

$$\bar{x} = \frac{.0 - 7.8 + \cdots + 1.1}{17} = -1.0$$

Is this disappointing record indicative of the performance of the mutual fund industry as a whole? How much confidence can we have in the average returns of 17 mutual funds as a measure of the average returns of all mutual funds? In later chapters, we will see how to draw statistical inferences from such evidence. (The answer turns out to be that even though this sample seems small, these data provide very persuasive evidence that mutual funds, on average, do not do as well as the overall stock market; either they are not very good at selecting stocks, or what talent they do have is more than offset by their management fees and other expenses.)

The mean is the *balance point* of the data in that the cumulative distance from the mean of those observations above the mean is equal to the cumulative distance from the mean of those observations below the mean. Figure 3.1 gives three examples, each with a mean of 50. In Fig. 3.1*a*, there are two observations, 0 and 100; the mean of 50 is the balance point in that one observation's being 50 above the mean is balanced by the other observation's being 50 below the mean. In Fig. 3.1*b*, the observation that is 50 above the mean is balanced by one observation that is 10 below the mean and another observation that is

HOW TO DO IT

Calculating the Mean

Suppose that we want to find the mean of the eight incomes in Table 3.1:

5000 5000 10,000 30,000 40,000 50,000 60,000 1,000,000

1. Determine the number of observations; here $n = 8$.
2. Add the values of the n observations, and divide by n:

$$\bar{x} = \frac{5000 + 5000 + 10{,}000 + 30{,}000 + 40{,}000 + 50{,}000 + 60{,}000 + 1{,}000{,}000}{8}$$

$$= \frac{1{,}200{,}000}{8}$$

$$= 150{,}000$$

If we have lots of data, it is easier to enter the numbers in a pocket calculator or computer software package that will calculate the mean (and other statistics) for us.

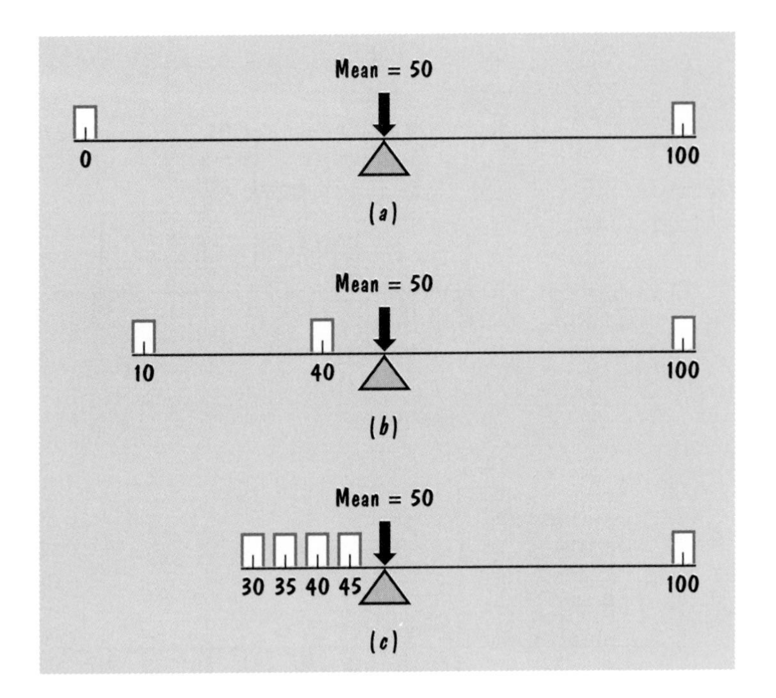

Figure 3.1 The mean is the balance point of the data.

40 below the mean. In Fig. 3.1*c*, the observation that is 50 above the mean is balanced by observations that are 5, 10, 15, and 20 below the mean.

Note in Fig. 3.1*c* that the one observation at 100 pulls the mean above the other four observations so that there are enough observations slightly below the mean to balance out the one observation that is far above the mean. In the same way, the person in Table 3.1 who earns $1 million pulls the mean up to $150,000, which is above the income levels of all seven other Smallanders. You should not automatically interpret the mean as the typical value.

For our sample of mutual fund returns, the mean and median are quite similar: −1.0 and −.4 percent in 1994 and 27.8 and 26.3 percent in 1995. Similarly, the histograms of these returns in Fig. 3.2 reveal no apparent anomalies in these data. For a quite different situation, consider the report by a national magazine[1] that a group of Colorado teachers had failed a history test, with an average score of 67. It turned out that only four teachers had taken the test and one received a score of 20. The other three averaged 83. The one very low score pulled the mean below all the other scores and misled a magazine that interpreted the average score as the typical value. Another example was provided by the director of the University of Virginia's office of career planning.

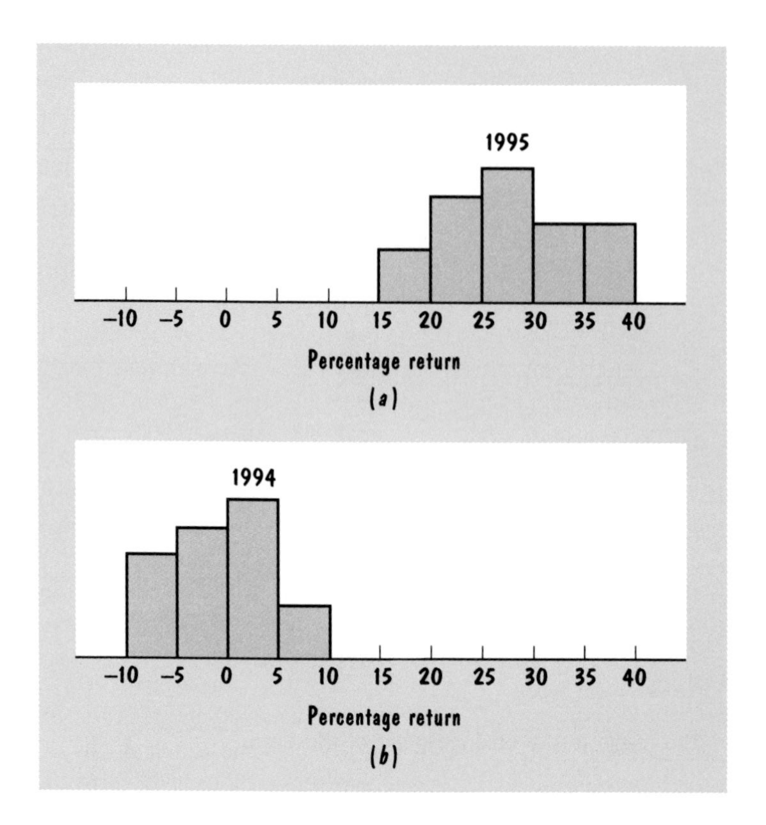

Figure 3.2 Histograms of a sample of mutual fund returns.

Reporting on the jobs taken by the school's 1983 graduates, he noted, "Our highest salaries were for graduates of the department of Rhetoric and Communications Studies, where the beginning average pay was $55,000 a year. Of course, the average height was 6′5″. Thanks Ralph."[2] Ralph Sampson, the first player picked in that year's National Basketball Association draft, pulled up the average salary and the average height.

The mean income tells us how much each person would earn if the total income were equally divided among the people. If income is not equally divided, there may be few people, or even no one, earning the "average" income. It need not be the case that one-half the people earn more than the average income and one-half earn less. As in Smalland and at the University of Virginia in 1983, the mean can be affected greatly by a few extreme observations. It is for this reason that the U.S. Census Bureau reports both the mean and median household income. In 1993 the mean household income in the United States was $41,428 while the median was $31,241. Apparently the mean was pulled above the median by a relatively small number of people with relatively high incomes.

The millionaire in Smalland is an *outlier,* in that this value is very different from the other observations. Ralph Sampson's height was an outlier at the University of Virginia (but not in the National Basketball Association). Outliers can make the mean a misleading measure of the center of the data. Recognizing this potential distortion, some researchers discard data that they consider to be outliers before calculating the mean. Thus, the office of career planning might report, "The average starting salaries of Rhetoric and Communications graduates, other than Ralph Sampson, were . . . "

A more formal procedure is to compute a *trimmed mean,* by discarding a specified percentage of the observations at the two extremes. For example, a 10 percent trimmed mean is arrived at by discarding the highest 10 percent and the lowest 10 percent of the observations and calculating the average of the remaining 80 percent. This procedure is commonly used in many international sports competitions (such as gymnastics) to protect the performers from judges who are excessively generous with their favorites and harsh with other competitors. With ordinary data, a renowned statistician, Francis Y. Edgeworth, argued more than one hundred years ago that the mean is improved "when we have thrown overboard a certain portion of our data—a sort of sacrifice which has often to be made by those who sail upon the stormy seas of Probability."[3]

The $32,500 value calculated for the 10 percent trimmed mean in the how-to-do-it box is a reasonable measure of the center of Smalland's income data. Unlike the mean, the trimmed mean and the median are *robust* or *resistant* to outliers, in that their values are little affected by extreme observations. Look again at the Smalland income data in Table 3.1. Regardless of whether the highest paid earns $100,000 or $100 million, the trimmed mean is still $32,500 and the median is still $35,000. The mean, in contrast, is $37,500 if the highest income is $100,000 and is $12,525,000 if it is $100 million.

An embarrassing illustration of this principle sometimes occurs when an error is made in transcribing data. In July 1986, for example, the Joint Economic Committee of Congress issued a report based on data compiled by researchers at the University of Michigan for the Federal Reserve Board. The report estimated that the share of the nation's wealth owned by the richest .5 percent of U.S. families had increased from 25 percent in 1963 to 35 percent in 1983. Politicians made speeches, and newspapers nationwide reported the story with "Rich Get Richer" headlines. Some skeptics in Washington started poking through the numbers, rechecking the calculations, and discovered that the reported increase was due almost entirely to the erroneous recording of one family's wealth as $200 million rather than $2 million, an error that raised the mean wealth of the rich people who had been surveyed by nearly 50 percent. Somewhere along the line, someone typed two extra zeros and temporarily misled the nation.

HOW TO DO IT

Calculating the Trimmed Mean

Let's find the 10 percent trimmed mean of the eight incomes in Table 3.1.

1. Arrange the data in numerical order:

   5000 5000 10,000 30,000 40,000 50,000 60,000 1,000,000

2. Determine the number of observations; here $n = 8$.
3. Determine the number of observations to be discarded from each end by calculating 10 percent of the number of observations and rounding this up to the next larger whole number—in this case, $(.10)(8) = .8$, rounded to 1.
4. Discard the appropriate number of observations from each end of the data. Here, the remaining data are

   5000 10,000 30,000 40,000 50,000 60,000

5. Compute the mean of the remaining data:

$$\text{Trimmed mean} = \frac{5000 + 10{,}000 + 30{,}000 + 40{,}000 + 50{,}000 + 60{,}000}{6}$$

$$= \frac{195{,}000}{6}$$

$$= 32{,}500$$

For some questions, the mean clearly provides the most appropriate answer. Whether a household will save money over the course of a year depends on its average monthly income and average expenses. A farm's total crop depends on the average yield per acre. The total amount of cereal a company needs to fill 1 million boxes depends on the average net weight. For other questions, other descriptive statistics may be more appropriate.

The Mode

The third measure of the center of a data set is very useful in some situations, but useless in others. The **mode** is the most commonplace value. In Table 3.1, the modal income is \$5000. This is the "typical" income in that there are more people with this income than with any other income. However, the mode need not be common. In our example, 75 percent of the people earn more than $5000, and many earn considerably more.

The mode is seldom used, except when we are working with categorical data that have been grouped into classes or categories that do not have natural numerical values. Table 3.3, for example, shows the number of U.S. white males and black males who died in a recent year because of accidents or violence. We cannot calculate the mean or median cause of death, because the four categories are not numbers that can be added or arranged in numerical order. We can look at the mode, however, and say that the most common cause of accidental or violent death was motor vehicle accidents for white males and homicides for black males.

Similarly, the mode is well suited to answer these questions: What is the most popular television show? What is the best-selling automobile? Which political party has the most members of Congress? For other questions, the mode may be of little or no interest.

Symmetric Data

The nature of the mean, median, and mode and a comparison among them can be brought out by an inspection of the symmetric income data shown in Table 3.4. Because there are 9 people, we find the median by counting in 5

Table 3.3

U.S. deaths due to accidents or violence, 1990

CAUSE OF DEATH	WHITE MALES	BLACK MALES
Motor vehicle accidents	25,441	3,982
All other accidents	23,004	4,634
Suicide	21,445	1,700
Homicide	8,773	9,806

Source: U.S. Department of Commerce, *Statistical Abstract of the United States,* Washington: Government Printing Office, 1996, p. 85.

Table 3.4

Symmetric income data

1996 INCOME (DOLLARS)	NUMBER OF PERSONS
10,000	1
20,000	2
30,000	3
40,000	2
50,000	1

from either end; the median income is $30,000. To find the mean, we add the 9 incomes and divide by 9. This, too, turns out to be $30,000, because the data's symmetry implies that each income above $30,000 is balanced by a corresponding income equally far below $30,000. Finally, the mode is also $30,000. A histogram of these data is shown in Fig. 3.3.

This example illustrates the general principle that if a set of data has a single peak and is symmetric about that peak, then the median, mean, and mode coincide. Conversely, if a set of data has more than one peak or is asymmetric, then the mean, median, and mode will generally not coincide. If the mean, median, and mode differ, then we need to bear in mind what each measures and to use the statistic that is most appropriate for our purposes. The median is the *middle value,* the mean is the *arithmetic average,* and the mode is the *most common value.*

Creating a Spurious Illusion of Precision

When we make a series of calculations, it is important not to round the intermediate results, as the compounding of several small rounding errors may cause a large error in the final result. However, when we have our final result, we should avoid reporting the value of the mean (or other statistics) to a number of digits far beyond that justified by the accuracy of the underlying data. It is

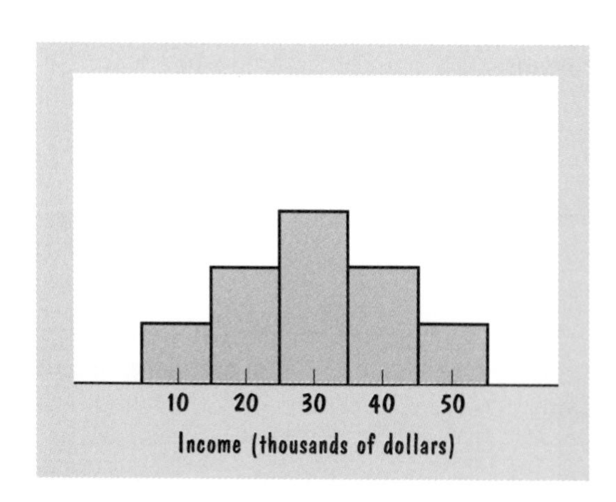

Figure 3.3 Symmetric income data.

tempting to report a great many digits simply because we use a calculator or computer program that does so. Sometimes, the intent is to persuade readers that an extraordinarily impressive scientific study must have been done in order to obtain such precise numbers.

Suppose, for example, that the three starting defensive linemen on a college football team tell us their weights: 260, 280, and 320 pounds. My pocket calculator reports the average of these three numbers as (260 + 280 + 320)/3 = 286.66667. If we report that the average weight of this team's starting defensive linemen is 286.66667 pounds, we are creating a spurious illusion of precision—perhaps hoping that the reader will think that our results are based on an exacting analysis of a substantial amount of carefully collected data (rather than a casual questioning of three people). The weights of football players vary throughout the season and even throughout the day (a player might lose 10 pounds during a game or practice). Each of these players no doubt gave a rough estimate, rounded to the nearest 10 pounds, and we should probably do the same, reporting that the average weight of this team's starting defensive linemen is approximately 290 pounds.

Similarly, the federal government estimated the U.S. population to be 247,343,000 in 1989, 249,924,000 in 1990, and 253,688,000 in 1991. These numbers are given to the nearest thousand because population counts are inexact—even in the decennial census. It would be misleading to report the population in 1991 to be 253,688,123. Similarly, we would be suggesting unwarranted precision in the data if we reported the average U.S. population during these 3 years as

$$\bar{x} = \frac{247{,}343{,}000 + 249{,}924{,}000 + 253{,}688{,}000}{3} = 250{,}318{,}333.33$$

A rounded figure of 250,318,300 or 250,318,000 is a more realistic reflection of the precision of the data.

Simpson's Paradox

Averages have the paradoxical property that if we divide the data into two or more categories and then divide the data in each category into two or more subsets that are not of equal size, then one of the categories can have the highest mean in each of the subsets and yet not have the highest overall mean. This phenomenon is known as *Simpson's paradox.* Consider, for example, these June 1991 data on the percentage of flights that arrived on time in the five major airports served by both Alaska Airlines and Northwest Airlines.[4]

	ALASKA AIRLINES			NORTHWEST AIRLINES		
	NUMBER OF ARRIVALS	NUMBER ON TIME	PERCENTAGE ON TIME	NUMBER OF ARRIVALS	NUMBER ON TIME	PERCENTAGE ON TIME
LOS ANGELES	559	497	88.9	811	694	85.6
PHOENIX	233	221	94.8	5255	4840	92.1
SAN DIEGO	232	212	91.4	448	383	85.5
SAN FRANCISCO	605	503	83.1	449	320	71.3
SEATTLE	2146	1841	85.8	262	201	76.7
TOTAL	3775	3274	86.7	7225	6438	89.1

Even though Alaska Airlines had a better on-time performance in every single city, Northwest Airlines had better overall on-time performance because most of its flights were into Phoenix, where flights generally land on time, and most of Alaska Airlines' flights were into Seattle, where weather problems frequently cause delays. Thus Northwest Airlines advertised that it was "the number one on-time airline," and Alaska Airlines hoped that air travelers would examine the data for individual cities.

Here is another example of a case where disaggregated data can show a pattern that is reversed in aggregated data. Suppose that we are analyzing data on the acceptance and rejection of male and female applicants to graduate school, and we separate the data by department in order to identify cases where there are gender disparities. Perhaps 500 females apply for admission in one department and 200 (40 percent) are rejected, while 3500 females apply to another department and 2100 (60 percent) are rejected. The overall rejection rate for female applicants is not 50 percent (a simple average of 40 percent and 60 percent). Because so many of the females apply to the department with the 60 percent rejection rate, the average rejection rate is pulled up to 57.5 percent:

$$\frac{\text{Total rejected}}{\text{Total applicants}} = \frac{200 + 2100}{500 + 3500} = \frac{2300}{4000} = .575$$

Now suppose that 50 percent of 3000 male applicants to the first department are rejected and that 70 percent of 1000 male applicants to the second department are rejected. Even though males are rejected more often than females in both departments, the overall male rejection rate is somewhat lower, because most males apply to the department that is easier to get into:

$$\frac{\text{Total rejected}}{\text{Total applicants}} = \frac{1500 + 700}{3000 + 1000} = \frac{2200}{4000} = .55$$

Thus, paradoxically, a group could have below-average rejection rates in every department but an above-average overall rejection rate if the members of this group apply disproportionately to departments with high rejection rates.

This very paradox was observed at the University of California at Berkeley in the 1970s. For the graduate school as a whole, women were rejected more often than men. Yet a committee that was formed to search for possible discrimination found no department that rejected substantially more women than men; if anything, the reverse was true. Women had a higher overall rejection rate because they applied in disproportionate numbers to those departments that were the most difficult to get into.[5]

3.1 Here are some data on United Airlines' annual profit, as a percentage of net worth, during an 11-year period characterized by two dramatic increases in energy prices:

YEAR	PROFIT	YEAR	PROFIT
1972	3.1	1978	25.6
1973	7.2	1979	−6.3
1974	12.7	1980	1.8
1975	−.7	1981	−6.3
1976	2.4	1982	1.0
1977	10.1		

Calculate the mean, median, and mode. Does each measure give a similar summary of United Airlines' profits over this period? If they differ, which measures do you think are misleading?

3.2 During Ronald Reagan's 1980 campaign for president of the United States, he repeatedly asked voters if they were better off in 1980 than they were 4 years earlier, when the incumbent, Jimmy Carter, had been elected president. Here are some data on the percentage of unemployment rate and inflation rate during Carter's 4 years in office (1977 to 1980) and Reagan's 8 years (1981 to 1988):

	UNEMPLOYMENT	INFLATION		UNEMPLOYMENT	INFLATION
1977	7.1	6.7	1983	9.6	3.8
1978	6.1	9.0	1984	7.5	3.9
1979	5.8	13.3	1985	7.2	3.8
1980	7.1	12.5	1986	7.0	1.1
1981	7.6	8.9	1987	6.2	4.4
1982	9.7	3.8	1988	5.5	4.4

Calculate the average unemployment rate and the average inflation rate during the Carter years and during the Reagan years. Was there any noticeable difference in average unemployment or average inflation between these two administrations?

3.3 A researcher is analyzing data on the financial wealth of 100 professors at a small liberal arts college. The values of their wealth range from $400 to $400,000, with a mean of $40,000 and a median of $25,000. However, when entering these data into a statistics software package, the researcher mistakenly enters $4 million for the person with $400,000 wealth. How much does this error affect the mean and median?

3.4 A 1993 article in the *Los Angeles Times* reported that because the annual rainfall in Los Angeles varies considerably from year to year, the average rainfall

> means more in places where it rains and snows like clockwork, according to Maurice Roos, the state's chief hydrologist. Typically, that is the case in much of Northern California, particularly along the coast. But in semiarid outposts such as Los Angeles, the only certainty about weather is that there isn't much of it—except of course, when there is. . . .
>
> "We have a lot of years that are dry and not quite so many that are wet," said Roos. . . . Ken Turner, a watershed manager for the state Department of Resources said a lot of confusion could be avoided if officials referred to median precipitation instead of averages . . . But median figures are more difficult to compute and more complex to explain, and no one is expecting them to roll off the tongues of [television personalities] Fritz Coleman or Willard Scott anytime soon.[6]

a. Explain how you would compute and interpret a figure for the median annual rainfall, using 115 years of Los Angeles data.

b. Based on the quotation, is the median rainfall in Los Angeles above or below the mean? Explain your reasoning.

3.5 An old joke is that a certain economics professor left Yale to go to Harvard and thereby improved the average quality of both departments. Is this possible?

3.6 The epigram for this chapter quotes Ann Landers: "Nothing shocks me anymore, especially when I know that 50 percent of the doctors who practice medicine graduated in the bottom half of their class."[7] Is she using the mean, median, or mode? Is there any relationship between the competency of the medical profession and the fact that 50 percent of all doctors graduated in the bottom half of their class?

3.7 In 1798, Henry Cavendish made 23 measurements of the density of the earth relative to the density of water:[8]

5.10 5.27 5.29 5.29 5.30 5.34 5.34 5.36 5.39 5.42 5.44
5.46 5.47 5.53 5.57 5.58 5.62 5.63 5.65 5.68 5.75 5.79
5.85

Calculate the mean, median, mode, and 10 percent trimmed mean. Which is closest to the value 5.517 that is now accepted as the density of the earth?

3.8 Answer this letter to Marilyn vos Savant.[9]

> I'm having a problem with the illustration below, which was captioned, "Men consume about 76% of all alcoholic beverages. Percentage consumed:"

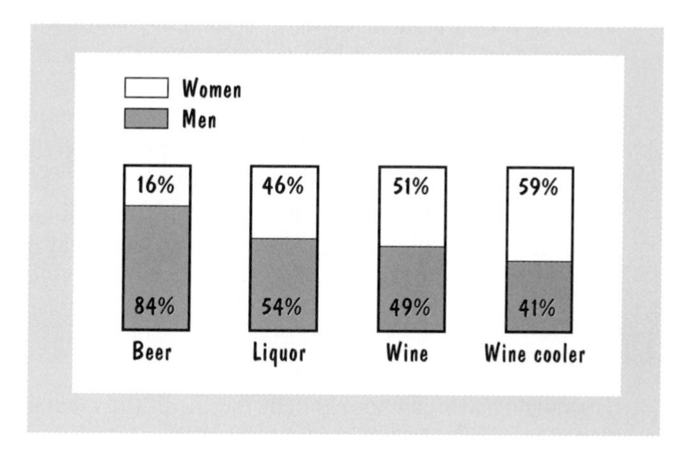

> The illustration seems to suggest: Of alcohol consumed 57% is by men, and 43% is by women. . . . Help!

3.2 The Spread of the Data

The mean is, by far, the most commonly reported summary statistic. However, an average does not tell us the underlying variation in the data, which may be of great interest. Sir Francis Galton once commented:

> It is difficult to understand why statisticians commonly limit their enquiries to Averages, and do not revel in more comprehensive views. Their souls seem as dull to the charm of variety as that of the native of one of our flat English counties, whose retrospect of Switzerland was that, if its mountains could be thrown into its lakes, two nuisances would be got rid of at once.[10]

When your instructor hands back your midterm examinations, you naturally will be interested not only in how you did, but also in how the class as a whole did. If the instructor announces that the mean score is 80, you probably would also like to know something about the spread (or variation) in the data.

Did every student get an 80? Or did one-half get 100 and one-half get 60? Or did most students get 100 and a few get 0? In a small class, the instructor may in fact write every score on the chalkboard. With 300 tests, however, the instructor does not want to write every score, and the students do not want to look at every score. The problem is even more acute with larger data sets, such as the income of 100 million U.S. households.

One way that we can organize and digest massive amounts of data is to sort the values into a small number of intervals and then draw a histogram, as explained in Chap. 2. While a histogram gives us a visual impression of the spread in the data, sometimes we would like to have a simple statistic that gauges the spread numerically, so that we can compare the spread of these data to that of other data. Are the midterm scores more spread out than the homework scores, or than the midterm scores in other sections or in other years? Are the incomes in one country more compact or more dispersed than in other countries, or in this country during other historical periods?

Ranges

Fortunately, we can supplement the mean (or median) with other summary statistics that provide information about the spread or variation in the data. One obvious choice is the **range,** which is simply the difference between the largest and smallest values. For the 1995 mutual fund returns in Table 3.2, the largest value is 38.4 and the smallest value is 17.8, giving a range of 17.8 to 38.4. The problem with using the range to measure the variation in the data is that it looks at only the two extreme values, and extremes can be very atypical. Which of the following two data sets is more compact, and which is more dispersed? Ten values are 17.8, and 10 values are 38.4; or 1 value is 17.8, 1 value is 38.4, and 18 values are 28? Common sense tells us that the second set of data is less dispersed; yet each set has exactly the same range. The problem with the range is that it only looks at the difference between the two extreme values and ignores the dispersion among all the rest of the data. If we are interested in more than the two extremes, we need a summary measure of the overall variation among the data.

Another possibility is to identify a range that gauges the spread but is not overly dependent on the extreme values. For example, Fig. 3.4 illustrates how data can be divided into four groups of equal size. One-fourth of the observations have values between the minimum and the first quartile. The first quartile is also known as the *25th percentile,* since this value is larger than 25 percent of the data. Another one-fourth of the data lie between the first quartile and the median (which is the 50th percentile); one-fourth are between the median and the third quartile (the 75th percentile), and one-fourth are between the third quartile and the maximum. The **interquartile range** is equal to the difference

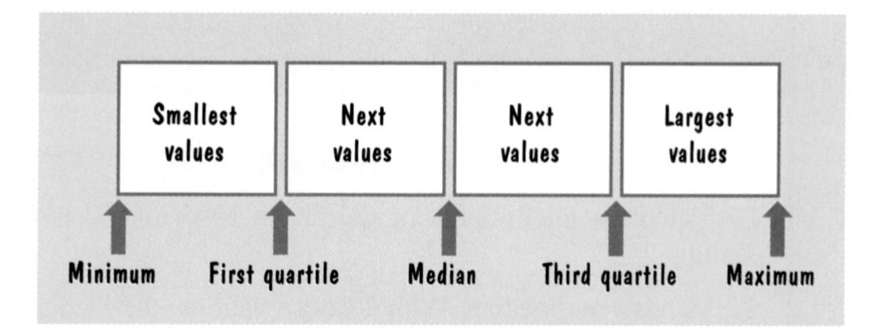

Figure 3.4 The minimum, first quartile, median, third quartile, and maximum divide the data into four groups containing an (approximately) equal number of observations.

between the first and third quartiles, which is the range encompassed by the middle half of the data.

In practice, the data will not usually divide perfectly into four groups of exactly equal size, and some statisticians (and statistics software) use slightly different rules for calculating the first and third quartiles; these differences are small and can be safely ignored. The conventions we will follow are illustrated in Table 3.5, using the 1995 mutual fund returns that were given in Table 3.2, but now arranged in numerical order.

Table 3.5

Finding the quartiles for the 1995 mutual fund returns in Table 3.2

38.4	Maximum (38.4)
37.9	
37.0	
33.8	Third quartile (33.65)
33.5	
30.3	
29.2	
28.2	
26.3	Median (26.3)
26.1	
25.1	
24.8	
22.7	First quartile (22.55)
22.4	
20.7	
18.0	
17.8	Minimum (17.8)

HOW TO DO IT

Determining the Interquartile Range

Let's find the interquartile range for the 1995 mutual fund returns in Table 3.2.

1. Identify the median. With 17 observations, the median is the ninth observation, or 26.3.
2. Look at the observations below the median and find the *first quartile,* which is the median of these observations. Here, because there are eight observations below the median, the first quartile is the average of the fourth and fifth observations: (22.4 + 22.7)/2 = 22.55.
3. Look at the observations above the median, and find the *third quartile,* which is the median of these observations. Here, looking at the eight observations above the median, the third quartile is (33.5 + 33.8)/2 = 33.65.
4. The interquartile range is equal to the difference between the values of the first and third quartiles; here, 33.65 − 22.55 = 11.1.

To determine the interquartile range, first we locate the median. With 17 observations, the median is equal to the ninth observation, or 26.3. The *first quartile* is the median of the observations below the median; with 8 returns below the median, the first quartile is the average of 22.4 and 22.7: (22.4 + 22.7)/2 = 22.55. Analogously, the *third quartile* is the median of the observations above the median: (33.5 + 33.8)/2 = 33.65. The interquartile range is equal to the difference between the values of the first and third quartiles: 33.65 − 22.55 = 11.1.

Thus we can describe the 1995 returns for these 17 funds by observing that the lowest return was 17.8, the highest was 38.4, the median was 26.3, and the interquartile range of 22.55 to 33.65 encompassed the middle half of the returns.

Box Plots

The box plot was created by John Tukey to show the center and spread of a data set in a clear, economical fashion. Consider again the 1995 mutual fund returns in Table 3.2. Figure 3.5*a* shows these 17 data points on a numerical line with labels for the minimum, first quartile, median, third quartile, and

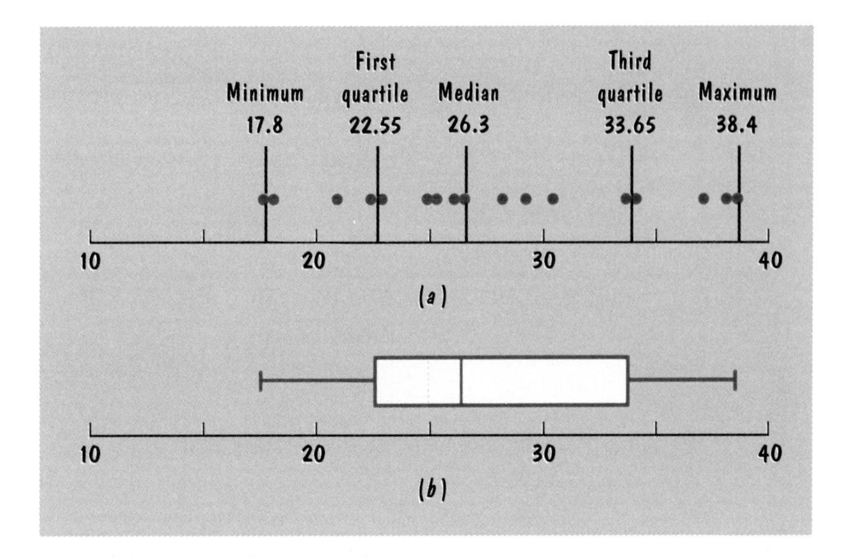

Figure 3.5 A box plot for 17 mutual fund returns in 1995.

maximum. In Fig. 3.5*b*, we use a box to connect the first and third quartiles, use horizontal lines to connect the minimum and maximum values to this box, and remove some of the clutter so that we can focus on the five values that describe this set of data.

This graphical summary is called a **box plot** or, more descriptively, a **box-and-whisker diagram.** The ends of the box are at the first and third quartiles, so that the box has a length equal to the interquartile range and encompasses approximately one-half of the data. The median is denoted by the line inside the box. The ends of the two horizontal lines coming out of the box (the "whiskers") show the minimum and maximum values. This figure shows, in a relatively simple way, the median, the center half of the data, and the extremes. The box plot conveys a considerable amount of information, but is less complicated than a histogram. It is also relatively robust, in that the box itself (but not the whiskers) is resistant to outliers.

Figure 3.6 shows the box plot for the 1995 fund returns together with a box plot of these 17 funds' 1994 returns. In addition to summarizing these data visually, these box plots show vividly that the lowest 1995 return was considerably higher than the highest 1994 return and that, whether measured by the spread between the minimum and maximum returns or by the width of the box (the interquartile range), there was a greater dispersion in the 1995 returns than in the 1994 returns.

For a different example, Table 3.6 shows the quartiles for the populations of

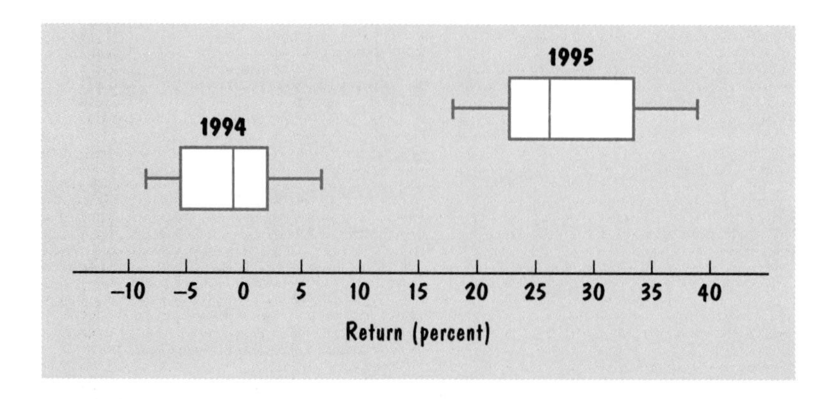

Figure 3.6 Box plots of 17 mutual fund returns in 1994 and 1995.

the 10 largest U.S. cities in 1990. Figure 3.7 shows box plots for these data and for the 10 largest U.S. cities in 1970 and 1980. These are *modified box plots,* which follow Tukey's recommendation for stopping each whisker at the most extreme point that is not an outlier, leaving the outliers as points suspended in space. One widely used definition of an outlier is a value whose distance from the box is more than 1.5 times the length of the box (that is, 1.5 times the interquartile range). For the 1990 data in Table 3.6, the interquartile range is $2.784 - 1.007 = 1.777$, and an observation that is larger than $2.78 + 1.5(1.77) = 5.44$ or smaller than $1.01 - 1.5(1.77) = -1.65$ is considered an outlier. By this definition, New York is the only outlier in 1990, and San Antonio and Los Angeles are the most extreme points that are not outliers.

Table 3.6 Population of the 10 largest U.S. cities in 1990, in millions

City	Population	
New York	7.323	
Los Angeles	3.485	
Chicago	2.784	Third quartile (2.784)
Houston	1.631	
Philadelphia	1.586	Median (1.586 + 1.111)/2 = 1.349
San Diego	1.111	
Detroit	1.028	
Dallas	1.007	First quartile (1.007)
Phoenix	.983	
San Antonio	.936	

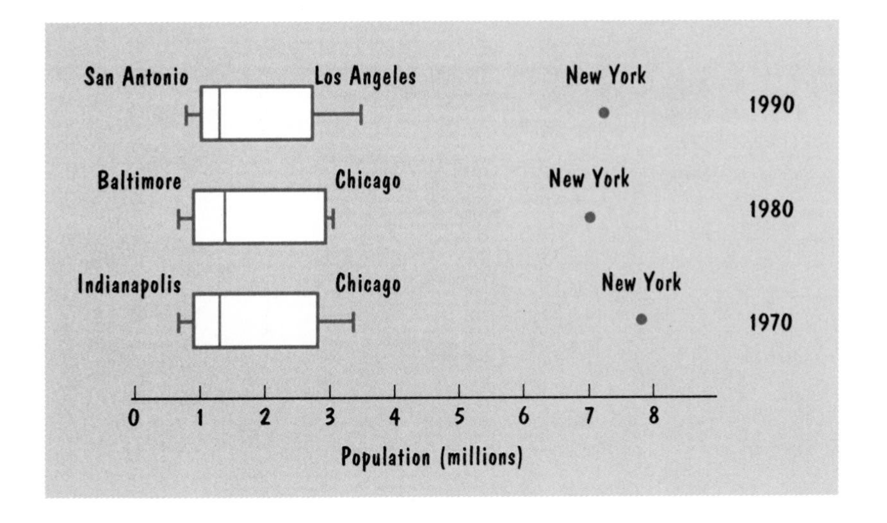

Figure 3.7 Modified box plots of the populations of the 10 largest U.S. cities, in millions.

The Standard Deviation

Another intuitively appealing way to gauge the variation or spread in the data is to compute the average deviation from the mean by subtracting the mean from each observation, adding each of these deviations from the mean, and then dividing by the number of observations. However, using the Smalland income data as an example, Table 3.7 shows that the average deviation from the mean is zero, which is hardly a useful statistic. In fact, because the mean is the balance point of any set of data, *the average deviation from the mean is always zero.*

Because the positive and negative deviations from the mean offset each other, giving an average deviation of zero, we learn nothing at all about the size of the deviations. To eliminate this offsetting of positive and negative deviations, we could take the absolute value of each deviation and then calculate the *average absolute deviation,* which is the sum of the absolute values of the deviations from the mean, divided by the number of observations. Table 3.7 shows this calculation.

The average absolute deviation is easy to calculate and interpret; however, there are two reasons why it is seldom used. First, while it is easily calculated for a specific set of data, it presents formidable problems for the mathematical analysis of general cases. It is very difficult to work with absolute values and

Table 3.7

Three measures of variation in a data set

INCOME (THOUSANDS OF DOLLARS)	DEVIATION FROM MEAN	ABSOLUTE VALUE OF DEVIATION	SQUARED DEVIATION
5	−145	145	21,025
5	−145	145	21,025
10	−140	140	19,600
30	−120	120	14,400
40	−110	110	12,100
50	−100	100	10,000
60	−90	90	8,100
1000	850	850	722,500
Sum	0	1700	828,750
Average	0	212.5	118,392.86

Note: The average squared deviation was calculated by dividing by 8 − 1 = 7.

to see how they change as the data change in mathematically well-defined ways. Second, there is an attractive alternative that, as we will see in later chapters, plays a prominent role in probability and statistics.

Remember that we took absolute values of the deviations to keep the positive and negative deviations from offsetting each other. Another technique that accomplishes this same end is to square each deviation, since the squares of positive and negative deviations are both positive. The average of these squared deviations is the *variance* of a data set, and the square root of the variance is the standard deviation:

The **variance** of n observations $x_1, x_2, \ldots, x_n$ is the average squared deviation of these observations about their mean:

$$\text{Variance} = \frac{(x_1 - \bar{x})^2 + (x_2 - \bar{x})^2 + \cdots + (x_n - \bar{x})^2}{n - 1} \tag{3.2}$$

The standard deviation s is the square root of the variance:

$$s = \sqrt{\text{variance}}$$

Note that the variance of a set of data is calculated by dividing the sum of the squared deviations by $n - 1$, rather than n. It can be shown mathematically that if the variance in a set of data is used to estimate the variance of the population from which these data came, this estimate will, on average, be too low if we divide by n, but will, on average, be correct if we divide by $n - 1$. (This is an example of the kinds of general theorems that mathematicians can prove when they work with the mathematically tractable standard deviation, rather than the average absolute deviation.)

HOW TO DO IT

Calculating the Standard Deviation

Table 3.7 shows the calculation of the variance for the Smalland income data. (To simplify the calculations, the data are in thousands of dollars.)

1. Calculate the mean of the n observations; here

$$\bar{x} = \frac{5 + 5 + 10 + 30 + 40 + 50 + 60 + 1000}{8}$$

$$= \frac{1200}{8}$$

$$= 150$$

2. Calculate the squared value of the deviation of each observation from the mean; here, $(5 - 150)^2 = 21{,}025$, and so on.
3. The variance is equal to the sum of all the squared deviations, divided by $n - 1$. Here,

$$\text{Variance} = \frac{21{,}025 + 21{,}025 + \cdots + 8100 + 722{,}500}{8 - 1}$$

$$= \frac{828{,}750}{7}$$

$$= 118{,}392.86$$

4. The standard deviation s is equal to the square root of the variance; here,

$$s = \sqrt{118{,}392.86} = 344.08$$

5. With lots of data, it is easier to enter the numbers in a pocket calculator or computer software package that calculates the standard deviation.

Because each deviation is squared, the variance has a scale that is much larger than the underlying data. The standard deviation, in contrast, has the same units and scale as the original data. With the exception of this scale difference, the variance and standard deviation are equivalent measures of the dispersion in a set of data. A data set that has a higher variance than another data set also has a higher standard deviation.

In most cases, the standard deviation, the average absolute deviation, and the interquartile range will agree in their rankings of the amount of variation in different data sets. Exceptions can occur when there are outliers that, because of the squaring of deviations, have a big effect on the standard deviation. Just as the median is more resistant to outliers than the mean is, so the average absolute deviation and the interquartile range are more resistant to outliers than is the standard deviation. Nonetheless, because of its mathematical tractability and importance in probability and statistics, the standard deviation is generally used to gauge the variation in a set of data.

Let's now look again at the mutual fund returns in Table 3.2. From just staring at the numbers in the table, it is hard to tell whether there was more dispersion in the 1994 or 1995 returns. The box plots in Fig. 3.6 suggest more dispersion among the 1995 returns, and the standard deviations confirm this visual impression:

1994:

$$\text{Mean} = \frac{.0 + \cdots + 1.1}{17} = -1.00$$

$$\text{Variance} = \frac{[.0 - (-1.00)]^2 + \cdots + [1.1 - (-1.00)]^2}{17-1} = 21.0950$$

$$\text{Standard deviation} = \sqrt{21.0950} = 4.59$$

1995:

$$\text{Mean} = \frac{37.0 + \cdots + 25.1}{17} = 27.78$$

$$\text{Variance} = \frac{(37.0 - 27.78)^2 + \cdots + (25.1 - 27.78)^2}{17 - 1} = 43.6819$$

$$\text{Standard deviation} = \sqrt{43.6819} = 6.61$$

The 6.61 percent standard deviation of the 1995 returns is larger than the 4.59 percent standard deviation of the 1994 returns, indicating that there was more variation among these returns in 1995 than in 1994.

The interpretation of the value of the standard deviation is not as easy as the interpretation of the mean. The -1.00 percent average return on these mutual funds in 1994 implies that if you had invested an equal amount in each of these 17 funds, your average return would have been -1.00 percent. But what does the 4.59 percent standard deviation mean? A famous theorem and two rules of thumb provide guidance. The remarkable theorem called *Chebyshev's inequality* states that in any set of data at least $1 - 1/k^2$ of the data are within k standard deviations of the mean. For $k = 2$, at least $1 - 1/2^2 = 3/4$ of the data are within 2 standard deviations of the mean. The two rules of thumb in those cases where a histogram has the bell shape mentioned in Chap. 2, are (1) roughly two-thirds of the data are within 1 standard deviation of the mean and (2) approximately 95 percent are within 2 standard deviations. Table 3.8 applies these rules to our mutual fund data. Overall, 58 percent (20/34 = .58) of the returns are within 1 standard deviation of the mean, and 100 percent are within 2 standard deviations.

Thus, if we are told (as we are in Example 3.4) that over the period from 1926 through 1995 the annual returns on corporate stock have averaged 12.5 percent with a standard deviation of 20.4 percent, and if we have seen (as we saw in Fig. 2.12) that a histogram of these returns is roughly bell-shaped, then we can conclude that stock returns have been between 12.5 − 20.4 = −7.9 and 12.5 + 20.4 = 32.9 percent in approximately two-thirds of these years and between 12.5 − 2(20.4) = −28.3 and 12.5 + 2(20.4) = 53.3 percent in approximately 95 percent of these years.

The Profitability of Five Companies

Table 3.9 shows the annual profits (as a percentage of net worth) for five different companies over a 10-year period that encompassed two economic recessions. These are a mass of data that is not easy to absorb. The mean and

Table 3.8

The number of mutual fund returns within 1 and 2 standard deviations of the mean

	1994	1995
Mean	−1.00	27.78
Standard deviation	4.59	6.61
Mean plus or minus 1 standard deviation	−5.59 to 3.59	21.17 to 34.39
Fraction of data within 1 standard deviation	$\frac{9}{17} = .53$	$\frac{11}{17} = .65$
Mean plus or minus 2 standard deviations	−10.18 to 8.18	14.56 to 41.00
Fraction of data within 2 standard deviations	$\frac{17}{17} = 1.00$	$\frac{17}{17} = 1.00$

standard deviation can help us summarize these data in a way that facilitates meaningful comparison:

COMPANY	MEAN	STANDARD DEVIATION
American Water Works	7.61	.68
Brown & Sharpe	7.62	7.39
Campbell Soup	13.56	1.05
McDonald's	20.04	1.02
Pan American	−.98	14.18

American Water Works, the largest private water company in the United States, sells water in 20 states at prices set by regulatory commissions to give a "fair" rate of return. Over this 10-year period, its average annual profit rate was 7.61 percent; the very small standard deviation accurately reflects the fact that, because of close regulatory supervision, this company's profit rate varied little from year to year.

Brown & Sharpe earned almost exactly the same average return as American Water Works, but with a much higher standard deviation. As a manufacturer of tools and equipment, Brown & Sharpe's sales and profits are sensitive to the nation's economy. The company lost money in 1975, when the economy was in recession, but made large profits in 1979, when the economy was booming. Over this 10-year period, American Water Works and Brown & Sharpe were, on average, equally profitable, but American Water Works' profits were much more stable.

Campbell Soup is an interesting contrast; its standard deviation is almost as low as that of American Water Works, but Campbell Soup's average profit rate

Table 3.9

Five companies' profits as a percentage of net worth

	AMERICAN WATER WORKS	BROWN & SHARPE	CAMPBELL SOUP	MCDONALD'S	PAN AMERICAN
1972	7.2	2.1	11.0	17.9	−7.0
1973	6.6	5.8	13.6	19.9	−4.7
1974	6.8	6.5	13.7	20.4	−26.7
1975	7.3	−5.4	13.1	21.0	−18.0
1976	7.8	2.3	14.3	21.0	−2.3
1977	8.0	7.7	13.9	21.3	12.5
1978	7.7	15.2	14.2	20.4	18.3
1979	7.5	18.4	14.8	19.8	10.5
1980	8.7	16.8	14.0	19.4	10.0
1981	8.5	6.8	13.0	19.3	−2.4

is nearly double that of the first two companies. Year in and year out, it made a roughly constant 13 to 14 percent profit. McDonald's (the hamburger company) did even better, earning an average profit of 20 percent, with the same low standard deviation as that of Campbell Soup. This is why so many investors liked McDonald's stock and why its price per share doubled over this 10-year period (from \$24 per share at the beginning of 1972 to \$48 per share at the end of 1981). At the other extreme is Pan American, at the time the second-largest U.S. Airline, which lost money in 6 of these 10 years. Its profits ranged from a high of 18.3 percent to a low of −26.7 percent. The average was a tiny −.98 percent, and the standard deviation was a huge 14.18 percent. No wonder the price of this stock fell from a high of \$17.80 per share in 1972 to a low of \$2.40 per share in 1981. This volatile company went bankrupt in 1991, and it reemerged several years later as a reorganized, slimmed-down shell of its former self.

Beef or Chicken?

In 1994 *Consumer Reports* reported the results of a study of the amount of fat per ounce and cholesterol per ounce in 9 hamburgers (basic and with the "works") and 13 chicken sandwiches, pieces, and nuggets served by national fast-food restaurant chains.[11] The results are summarized in the box plots below. The chicken dishes tended to have somewhat more fat per ounce, although there was slightly more variability in the burgers. The cholesterol differences are much more pronounced, with the chicken dishes having substantially higher levels and much greater variation. Although the dish with lowest cholesterol is some kind of chicken, the cholesterol per ounce in every burger is less than or equal to that in the median chicken dish.

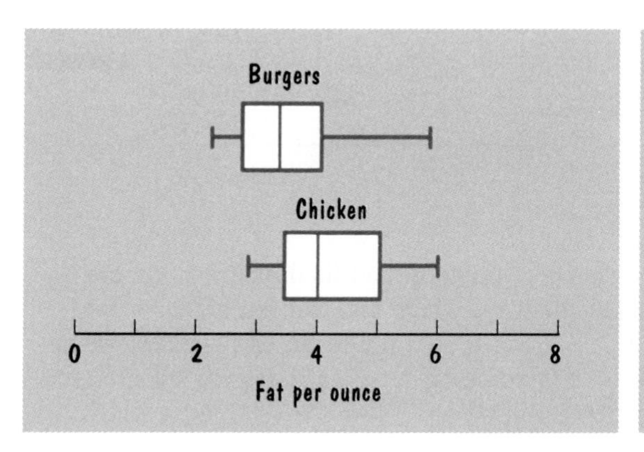

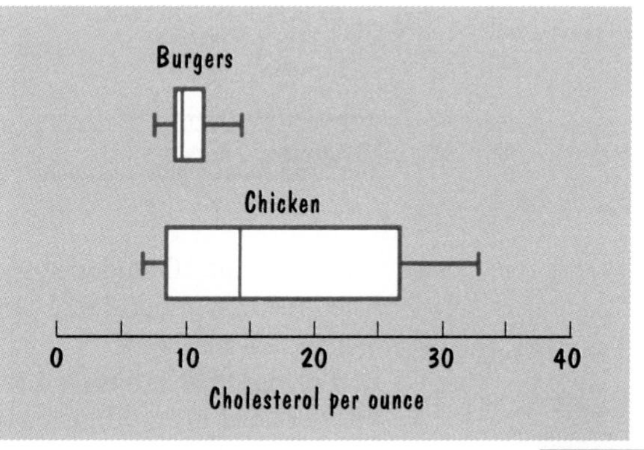

Risk Has Its Rewards

The standard deviation is a commonly used yardstick for comparing the risks of alternative investments. An investment that sometimes has large positive returns and sometimes has large negative returns has a high standard deviation. An investment whose return varies little from year to year has a low standard deviation.

If investors are risk-averse, they will shun risky investments unless the price is right—that is, unless they are compensated for taking these risks with relatively high anticipated returns. A high return cannot, of course, be guaranteed for a risky investment because, by definition, a risky investment has a very uncertain return. However, an investment with a very uncertain return can be priced to give a relatively high average return. If investors are predominantly risk-averse, we should find that investments with high standard deviations tend to do better on average than investments with low standard deviations.

Table 2.1 shows the annual percentage returns from four types of securities from 1971 through 1995. These annual returns have been computed back to 1926, and the means and standard deviations for this longer historical period are shown below. These are the returns that were experienced—how people did, rather than how they thought they would do. Nonetheless, the results are an intriguing confirmation of our common sense. The safest asset, Treasury bills, has been the least rewarding, while the riskiest, corporate stock, has done the best on average. The annual return on stocks was 54 percent one year and −43 percent another year. During these 70 years, stocks have been a risky, but also a rewarding, investment.

	AVERAGE RETURN	STANDARD DEVIATION
TREASURY BILLS	3.77	3.27
TREASURY BONDS	5.54	9.25
CORPORATE BONDS	6.03	8.83
CORPORATE STOCK	12.52	20.42

3.9 The scores of 10 students on a statistics midterm and final examination are listed next. Without doing any calculations, which test had the higher mean? Which had the higher standard deviation? Would the mean and standard deviation of the final examination scores increase or decrease if Jessica's score on the final had been 90 instead of 99?

	JUAN	JESSICA	LAURA	BETH	ED	MIKE	TERRI	WEI	PAUL	CHRIS
MIDTERM	98	96	95	92	85	85	80	76	72	65
FINAL	92	99	84	94	92	86	75	88	85	74

3.10 Calculate the mean and standard deviation and use a histogram and modified box plot to summarize these data on annual inches of precipitation from 1961 to 1990 at the Los Angeles Civic Center. How many of these observations are within 1 standard deviation of the mean? Within 2 standard deviations?

5.83	15.37	12.31	7.98	26.81	12.91	23.66	7.58	26.32	16.54
9.26	6.54	17.45	16.69	10.70	11.01	14.97	30.57	17.00	26.33
10.92	14.41	34.04	8.90	8.92	18.00	9.11	9.98	4.56	6.49

3.11 Hundreds of earthquakes occur each month in Yellowstone National Park. The histogram below shows the magnitudes of the 259 earthquakes that were recorded in January 1995.[12] Give a rough estimate of the mean and standard deviation. Explain your reasoning.

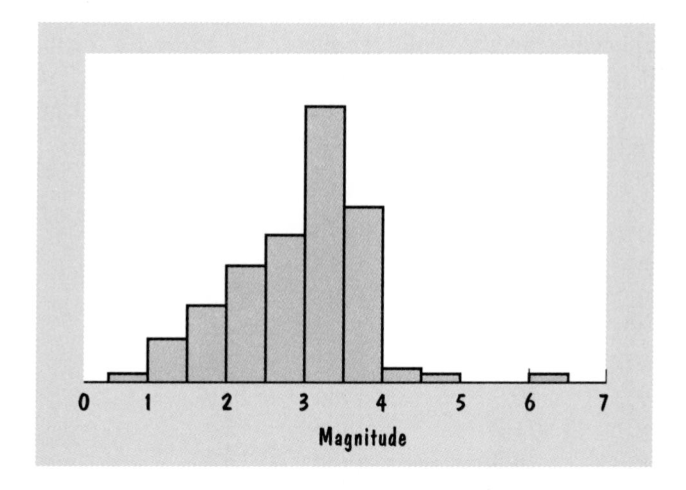

3.12 The average U.S. household has 2.4 persons. Explain why this statistic does not imply that all apartments and houses should be designed to accommodate 2 to 3 persons.

3.13 A 1994 article in the *American Scientist* reported the results of a survey of 17 scientific experts about the chances of the earth's experiencing catastrophic global warming.[13] Here, for example, is a summary of their assessments of the probability of a 3° Celsius (3°C) increase in the global average temperature by 2090. How has this graph been misdrawn?

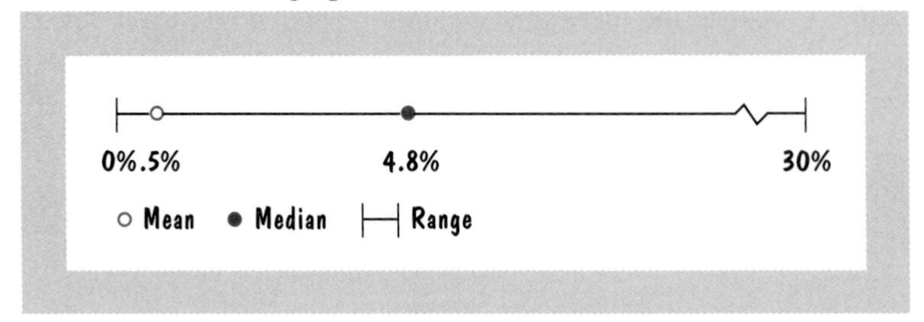

3.14 Using the Vostok data in Exercise 2.13, identify any observations of the atmospheric carbon dioxide concentration *C* that are more than 2 standard deviations from the mean or more than 1.5 times the interquartile range from either the first or the third quartile.

3.15 A stock's price/earnings (*P/E*) ratio is the per-share price of its stock divided by the company's annual profit per share. The *P/E* ratio for the stock market as a whole is used by some analysts as a measure of whether stocks are cheap or expensive, in comparison with other historical periods. Here are some annual *P/E* ratios for the New York Stock Exchange (NYSE):

YEAR	*P/E*	YEAR	*P/E*
1970	15.5	1979	7.4
1971	18.5	1980	7.9
1972	18.2	1981	8.4
1973	14.0	1982	8.6
1974	8.6	1983	12.5
1975	10.9	1984	10.0
1976	11.2	1985	12.3
1977	9.3	1986	16.4
1978	8.3		

Calculate the mean and standard deviation. The stock market reached a peak in August 1987 when the Dow Jones Industrial Average topped 2700. The Dow slipped back to 2500 in October 1987 and then to 2250. Then, on a single day, October 19, 1987, the Dow fell by 508 points. At its August 1987 peak, the market's price/earnings ratio was 23. Was this *P/E* value more than 1 standard deviation above the mean *P/E* for 1970 to 1986? Was it more than 2 standard deviations above the mean? Was it more than 1.5 times the interquartile range above the median? Draw a box-and-whisker diagram, using these *P/E* data for 1970 to 1986.

3.16 Shown below are box plots of the estimated 1996 market values of professional sports teams in major league baseball (MLB), the National Basketball Association (NBA), and the National Football League (NFL).[14]

a. Which league has the highest median team value?

b. Which league has the team with the lowest value?

c. Write a brief summary of the information provided by these box plots.

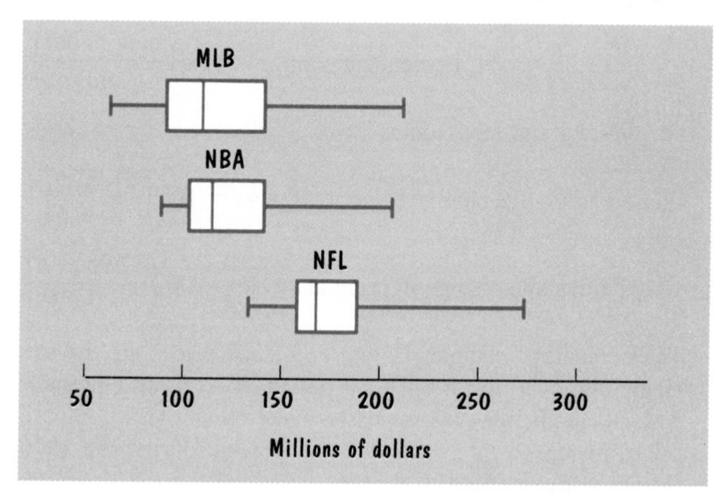

3.3 Working with Percentage Changes

Often we are interested in data that have been converted from levels to percentage rates of change. Suppose, for example, that we are comparing the behavior of egg prices and meat prices during the 1970s and that some of our data are as follows:

	1970	1980
One dozen large eggs	\$.61	\$1.00
1 pound of sirloin steak	\$1.35	\$2.95

The price of a dozen large eggs increased by \$1.00 − \$.61 = \$.39, and the price of 1 pound of sirloin steak increased by \$2.95 − \$1.35 = \$1.60. However, the \$.39 and \$1.60 changes are not directly comparable because they depend on the arbitrary choice of units for these products. If we had recorded the price of a ¼ pound of sirloin steak, instead of 1 pound, the increase in the price of steak would have been only \$.40—virtually the same as the increase in the price of a dozen eggs. If we had monitored the price of a single egg, its price would have increased by only 3 cents.

To make a meaningful comparison, we can calculate the percentage changes in these prices between 1970 and 1980. In general, the percentage change is computed by dividing the change by the value before the change and then multiplying by 100 to convert the fraction to a percentage:

$$\text{Percentage change} = 100\left(\frac{\text{new} - \text{old}}{\text{old}}\right) \tag{3.3}$$

For the egg and steak data:

$$\text{Percentage change in price of eggs} = 100\left(\frac{\$1.00 - \$.61}{\$.61}\right) = 63.9$$

$$\text{Percentage change in price of steak} = 100\left(\frac{\$2.95 - \$1.35}{\$1.35}\right) = 118.5$$

The percentage changes in egg and steak prices are the same if we work with a single egg or a dozen eggs, ¼ pound of steak or 1 pound. Thus the use of percentage changes allows us to make meaningful comparisons. Here, egg prices increased by a little over 60 percent during the 1970s, while steak prices increased by nearly 120 percent.

Unbelievable Percentages Are Probably Using the Wrong Base

Sometimes, people use an unconventional base when calculating a percentage change and come up with an unbelievable number. For example, *Newsweek* once reported that Chinese Premier Mao-Tse Tung had reduced the salaries of some government officials by 300 percent.[15] Now, unless a person's income is reduced to less than zero (which perhaps was possible in Mao's China), it cannot be reduced by more than 100 percent. After readers pointed out that this 300 percent figure was unbelievable, *Newsweek* checked its arithmetic and decided that salaries had actually been reduced by 66.7 percent. How did this arithmetic error happen? Suppose that the salary initially had been 300 and was reduced to 100. *Newsweek* apparently divided the old salary of 300 by the new salary of 100, obtained 3.00, and called this a 300 percent decrease. There are two mistakes here. First, the numerator should have been the change in the salary (100 − 300), not the level of the salary (300). Second, *Newsweek* should have divided by the old salary (300), not the new salary (100). To calculate the percentage change correctly, we divide the change in the salary by the old salary:

$$\text{Percentage change} = 100\left(\frac{\text{new} - \text{old}}{\text{old}}\right) = 100\left(\frac{100 - 300}{300}\right) = -66.7$$

Even when the percentage change is calculated correctly, questionable numbers can result from the use of a peculiar base period or from an unspecified base period. If I report that "The price of orange juice has increased by 50 percent," we need to know what base period I am using in my calculations. Has the price of orange juice increased by 50 percent since yesterday or since 1900? If I do not specify the base period (or I give a peculiar base period), I may be trying to magnify or minimize the percentage change to support a position, rather than present the facts honestly. To make the price increase seem large, I may compare it to that in a period when orange juice was cheap, perhaps decades ago. Someone else who wants to report a small price increase may choose a base period when orange juice was expensive, perhaps after bad weather had ruined most of the crop. An honest analysis will compare the price to a natural base period and identify that base period; for example, the price of orange juice has increased by 2 percent during the past year and by 35 percent over the past 10 years. If orange juice prices fluctuate considerably, we can use a time-series graph, as explained in Chap. 2, to show these ups and downs.

Because there usually is no room to specify the base period in a brief newspaper headline, we sometimes find two accurate, but seemingly contradictory headlines appearing on the very same day. On January 24, 1995, *The New York Times* reported, "American Express Net Climbs 16%," while *The Wall Street Journal* reported, "Net at American Express Fell 16% in 4th Quarter." These were not typographical errors. Both stories were accurate, and each explained the base period that it used. The *Times* compared American Express's earnings in the fourth quarter of 1994 to its earnings in the third quarter of 1994; the *Journal* compared American Express's earnings in the fourth quarter of 1994 to its earnings in the fourth quarter of 1993.

Similarly, these three headlines appeared on January 30, 1996:[16]

- "Survey Finds Health Costs rose in '95," *The New York Times*
- "Health Care Costs Slowing for Employers," *USA Today*
- "Employee Health-Care Costs Were Steady," *The Wall Street Journal*

The headline in *The New York Times* was based on the fact that health care costs had risen 2.1 percent in 1995, after falling by 1.1 percent in 1994. The *USA Today* headline was based on the fact that the 2.1 percent increase in 1995 and 1.1 percent decline in 1994 were below the "five years of double-digit gains in the late 80s and early 90s." The headline in *The Wall Street Journal* was based on the fact that health care costs for working employees were up a

negligible 0.1 percent, while a 10 percent increase for retirees raised the overall increase to 2.1 percent.

Do Not Use Percentages to Gauge Quantities

We cannot assume that because two percentages are equal, the underlying quantities must be equal. For example, if a household's income and expenses both increase by 10 percent, is the additional income equal to the additional expenses? If a business's revenues and costs both increase by 10 percent, is its profit unchanged? If a government's tax revenue and spending both increase by 10 percent, is its budget deficit unaffected? In all three cases, the answer is not necessarily. Suppose that a small business's annual revenue is initially $100,000 and that its annual expenses are initially $90,000, so that it makes a profit of $10,000. If its revenue increases by 10 percent to $110,000 and its expenses increase by 10 percent to $99,000, then the increase in revenue is larger than the increase in expenses and its profit will increase by 10 percent to $11,000. This calculation illustrates the mathematical fact that whenever two variables y and x both increase by the same percentage, the difference $y - x$ also increases by that same percentage. The more general principle is that because percentages may use different bases, we cannot gauge the relative magnitudes of quantities just by looking at percentages.

Another, seemingly paradoxical example of this principle is that if something increases by a certain percentage and then decreases by the same percentage (or vice versa), it does not end up where it started. For instance, if the price of a stock is $10 per share, a 50 percent increase will raise the price to $15 and a subsequent 50 percent decrease will reduce the price to $7.50, which is 25 percent less than the initial price. This paradox occurs because we are not taking 50 percent of the same base. A 50 percent increase from $10 is indeed equal to a 50 percent decrease from $10, but we are comparing a 50 percent increase from $10 with a 50 percent decrease from $15. We cannot use percentages to compare quantities unless the bases are equal.

Here is another example. Mary Ann Mason begins her book *The Equality Trap* by stating, "Something has gone very wrong with the lives of women. Women are working much harder than they have worked in recent history."[17] As evidence, she reports that between 1959 and 1983, the average number of hours that women between the ages of 25 and 64 worked outside the home increased by nearly 100 percent while the average number of hours they worked inside the home fell by only 14 percent. Readers are evidently supposed to conclude that the total number of working hours must have increased—and maybe by $100 - 14 = 86$ percent.

But these percentages do not necessarily apply to comparable bases. A 14 percent decrease in a large number of hours could well be larger than a 100 percent increase in a small number of hours. Suppose, for example, that the average number of annual working hours for women in 1959 was 2000 hours inside the home and 200 hours outside the home. Fourteen percent of 2000 hours is 280 hours, and 100 percent of 200 hours is only 200 hours. With these hypothetical numbers, the sum of the hours worked inside and outside the home actually decreased. Readers of *The Equality Trap* cannot tell whether the 100 percent increase is larger than the 14 percent decrease unless they put down that book, find the source of these data, and see what the bases were.

These data were originally reported by Victor Fuchs in a 1986 article in *Science* that combined data from several different sources.[18] The estimates of the number of hours worked inside the home were based on a 1975–1976 survey by the University of Michigan's Institute for Social Research. How, you might ask, could a 1975–1976 survey provide data for 1959 and 1983? Fuchs used the 1975–1976 data to see how working hours inside the home were related to a variety of socioeconomic factors (including race, age, and marital status) and then applied these estimates to socioeconomic data for the population in 1959 and 1983. No one was actually surveyed about working hours inside the home in either 1959 or 1983! Fuchs cautions that these estimates "are subject to a variety of possible errors—reporting, sampling, coding, and estimating—and differences of only a few percent should not be regarded as . . . significant."

Here are Fuchs' estimates of the average number of hours worked during the year:

	FEMALES		MALES	
	1959	1983	1959	1983
INSIDE THE HOME	1689	1453	618	619
OUTSIDE THE HOME	572	929	1875	1667
TOTAL	2261	2382	2493	2286

If we look at the total number of hours, females worked 5 percent more in 1983 than in 1959, and males worked 8 percent less. Males worked 10 percent more than females in 1959 and 4 percent less than females in 1983. The drop in male work outside the home may have been partly due to the choice of years; the unemployment rate was 5.5 percent in 1959 and 9.6 percent in 1983.

These estimates may well reflect profound social changes, or they may be a flawed use of a 1975–1976 survey to draw inferences about 1959 and 1983 behavior. What we can be certain of is that the "100 percent more" and "14 percent less" figures cited in *The Equality Trap* are not sufficient to prove the author's conclusion that "Women are working much harder than they have worked in recent history." The percentages, by themselves, are consistent with women working many more hours or far fewer hours. We simply cannot tell from the percentages alone. In this case, the underlying data indicate that women, on average, worked 5 percent more hours in 1983 than in 1959.

A Small Base Can Mean a Large Percentage

Because a percentage is calculated relative to a base, a very small base can give a misleadingly large percentage. On your second birthday, your age increased by 100 percent. When you graduate from college, your income may increase by several thousand percent. On July 25, 1946, the average hourly rainfall between 6 a.m. and noon in Palo Alto, California, was 84,816 times the average hourly rainfall in July during the preceding 36 years, dating back to 1910, when the Palo Alto weather station first opened.[19] In fact, the rainfall during these 6 hours on the morning of July 25, 1946, was 19 times the total cumulative rainfall for every July for the preceding 36 years! How much rain fell on this incredibly wet July 25? Only .19 inch. During the preceding 36 years, there had been only one July day with measurable precipitation in Palo Alto, and on that day the precipitation was .01 inch.

Sometimes, the unscrupulous use large percentages calculated from small bases to mislead the audience. For example, during congressional hearings on a proposed tariff on products imported from England, proponents pointed out ominously that cotton imports from England had increased by 1238 percent.[20] The value of these increased cotton imports turned out to be less than $1000. One way to guard against the small-base problem is to give the level as well as the percentage, for example, to note that on this unusually wet July day in Palo Alto, California, the rainfall was .19 inch.

Is Wellfleet the Murder Capital of Massachusetts?

Wellfleet is a small town on Cape Cod, renowned for its oysters, artists, and tranquillity. There was considerable surprise when the Associated Press reported that Wellfleet had the highest murder rate in Massachusetts in 1993, with 40 murders per 100,000 residents—more than double the murder rate in Boston, which had 17 murders per 100,000 residents. A puzzled newspaper reporter looked into this statistical murder mystery.[21] She found that there had, in fact, been no murders in Wellfleet in 1993 and that no Wellfleet police officer, including one who had lived in Wellfleet for 48 years, could remember a murder's occurring in Wellfleet.

However, a man accused of murdering someone in Barnstable, which is 20 miles from Wellfleet, turned himself in at the Wellfleet police station in 1993, and the Associated Press had erroneously interpreted this Wellfleet arrest as a Wellfleet murder. Because Wellfleet had only 2491 permanent residents, this one misrecorded murder arrest translated into 40 murders per 100,000 residents. Boston, in contrast, had 98 murders, which works out to 17 murders per 100,000 residents.

The solution to this murder mystery shows how a statistical fluke can make a big difference if the base is small. A misrecorded murder in Boston would not noticeably affect its murder rate. In Wellfleet, a misrecorded murder changes the reported murder rate from 0 to 40. If the Barnstable suspect had gone to the police station in Truro, an even smaller town adjacent to Wellfleet, then Wellfleet's murder rate would have gone from 40 to 0 and Truro's reported murder rate would have jumped from 0 to 67.

One way to deal with small bases is to average the data over several years in order to get a bigger base. Wellfleet's average murder rate over the past 50 years is 1 with the misrecorded arrest and 0 without it—either way confirming that it is indeed a peaceful town.

3.17 Calculate the percentage change between 1970 and 1980 in the prices of these products:

	1970	1980
A 1-pound can of coffee	$.91	$2.82
A ½-gallon of ice cream	$.85	$1.92

3.18 A 1989 radio commercial claimed, "If you have a LifeAlert security system, your chances of becoming a victim of a serious crime are 3000 to 4000 percent less than your neighbors'." Is this possible?

3.19 A doctor once reported on the "remarkable" incidence of Kaposi's sarcoma in the Wabendi tribe in Tanzania. Compared to a national average of .1 case per

10,000 people, this tribe of 8000 had 1.3 cases per 10,000—or 13 times the national average.[22] What is misleading about this remarkable statistic?

3.20 Explain how the following two newspaper headlines could both be accurate: "Orders for Machine Tools Increased 45.4% in October" (*The New York Times,* November 28, 1994) and "October Orders for Machine Tools Decreased 29%" (*The Wall Street Journal,* November 28, 1994).

3.21 Explain why the itemized data cited below do not explain the alleged 33 percent increase in book production and material costs:

> The gap between advancing book prices and authors' earnings, it appears, is due to substantially higher production and material costs. Item: plant and manufacturing expenses alone have risen as much as 10 to 12 percent over the last decade, materials are up 6 to 9 percent, selling and advertising expenses have climbed upwards of 10 percent. Combined boosts add up to a minimum of 33 percent.[23]

summary

Descriptive statistics can often be used to summarize a set of data. The median, mean, and mode describe the center of the data, around which the other values are spread. The median is the middle value when the data are arranged in numerical order from the smallest value to the largest value. The arithmetic mean (often written as $\overline{x}$) is the average value, the sum of the values divided by the number of values:

$$\text{Mean} = \frac{x_1 + x_2 + \cdots + x_n}{n}$$

The mode is the most commonplace value. If a set of data has more than one peak or is asymmetric, then the mean, median, and mode generally do not coincide.

The mean is the balance point of the data in that for every observation that is a certain amount above the mean, other observations are a cumulative offsetting amount below the mean. Outliers, values that are very different from the other observations, can pull the mean substantially above or below the median. In comparison to the mean, the median is more robust or resistant to outliers. Another robust measure of the center of a set of data is the trimmed mean, which is computed by discarding a specified percentage of the data at the two extremes and then calculating the average of the remaining data. The mode is best suited for data that have been grouped into categories that do not have natural numerical values, for example, the identification of heart attack victims by gender.

The value of a mean or other statistic should not be reported to a number of digits far beyond that justified by the accuracy of the underlying data. The mathematical fact that disaggregated data can show a pattern that is reversed in aggregated data is known as Simpson's paradox.

To gauge the spread or variation in the data about its center, we can use the interquartile range or the standard deviation. The first quartile is the median of the observations below the median, the third quartile is the median of the observations above the median, and the interquartile range is the difference between the first and third quartiles. The interquartile range spans the middle half of the data. A box plot (or box-and-whisker diagram) shows the median, the center half of the data, and the extremes. In a modified box plot, outliers that are more than 1.5 times the interquartile range from the box are shown as separate points, and each of the two lines coming out of the box stops at the most extreme point that is not an outlier.

The variance is the average squared deviation of the observations about their mean:

$$\text{Variance} = \frac{(x_1 - \bar{x})^2 + (x_2 - \bar{x})^2 + \cdots + (x_n - \bar{x})^2}{n - 1}$$

The standard deviation s is the square foot of the variance. The interquartile range is more resistant to outliers than is the standard deviation, but because of its mathematical tractability and statistical importance, the standard deviation is generally used to gauge the variation in a set of data.

We often work with percentage rates of change, which are computed by dividing each change by the value before the change (and multiplying by 100 to convert a fraction to a percentage):

$$\text{Percentage change} = 100\left(\frac{\text{new} - \text{old}}{\text{old}}\right)$$

Sometimes, people calculate percentages incorrectly by dividing the change by the value after the change rather than before, and consequently they report erroneously that something that cannot be negative declined by more than 100 percent.

We should not choose the base period so as to magnify or minimize the size of a percentage change. Instead we should choose a natural base period and identify it: The price of orange juice increased by 2 percent during the past year. Because percentages may use different bases, we cannot gauge the relative magnitudes of quantities just by looking at percentages; for example, if income and expenses both increase by 10 percent, the dollar increases in income and in expenses are not necessarily equal. Because a percentage is calculated relative to a base, use of a very small base can result in a misleadingly large percentage.

review exercises

3.22 There was a players' strike in the middle of the 1981 baseball season. Each division determined its season winner by having a playoff between the winner of the first part of the season, before the strike, and the winner of the second part of the season. Is it possible for a team to have the best winning percentage for the season as a whole, yet not win either half of the season and consequently not have a chance of qualifying for the World Series? Use some hypothetical numbers to illustrate your reasoning.

3.23 Table 2.1 shows the annual returns for four different securities. Which security looks as if it had the highest average return? Which looks as if it had the highest standard deviation? Calculate the mean and standard deviation for each of these four securities, and see if you are right. If we gauge risk by these standard deviations, which of these four investments was the safest? Which was the riskiest?

3.24 Do not collect any data, but use your general knowledge to make a rough estimate of the mean and standard deviation of the ages of the faculty at your school. Explain your reasoning.

3.25 The average one-family home built in 1990 had three bedrooms, two bathrooms, and central air conditioning. Is this average a mean, median, or mode? Explain your reasoning.

3.26 In a 1983 federal case,[24] Tenneco was accused of discriminating against black laborers in its hiring policies at a plant in the St. Bernard parish, which is part of the New Orleans metropolitan area. The plaintiff presented data showing that 59.4 percent of all laborers in the New Orleans area were black, but that only 48.3 percent of the laborers employed at the St. Bernard Tenneco plant were black. The defendant argued that measures of the available labor force should take into account commuting distances, and that a reasonable way to do this is to examine the relative number of job applicants at the St. Bernard plant from each of the four parishes that comprise the New Orleans metropolitan area:

PARISH	PERCENTAGE OF JOB APPLICANTS FROM THIS PARISH	PERCENTAGE OF PARISH LABORERS WHO ARE BLACK
Jefferson	7.10	49.29
Orleans	43.10	71.10
St. Bernard	47.16	15.21
St. Tammany	2.64	57.54

Suppose that there are a total of 10,000 people in the available labor force, divided by parish in the same proportions as job applicants are divided by parish. Show how the defendant used these data to calculate the overall fraction of the available labor force that are black. (The Court accepted the defendant's calculation and found that this plant had not discriminated against black laborers.)

3.27 In 1965, the federal government obtained a search warrant and seized some gambling records from a New York bookmaker showing that over a 3-day period he had handled $5715.99 in wagers on three-digit numbers and over a 2-day period he had handled $1077.40 in wagers on two-digit numbers.[25] (The records of two-digit wagers on the third day had been destroyed by the bookmaker.) The government alleged that the bookmaker had been handling three-digit and two-digit wagers 6 days per week for nearly 5 years and consequently owed the government a 10 percent excise tax, plus interest, on these wagers. Because the bookmaker produced no other betting records, the judge ruled that the government could use the data it had seized to estimate the amount wagered daily and then apply this estimate to the entire 5-year period. If you were a statistician employed by the court, what would be your estimate of the amount wagered daily?

3.28 Roll a standard pair of six-sided dice 20 times, each time recording the sum of the two numbers rolled. Calculate the mean, median, and mode of the 20 rolls. Repeat this experiment 5 times. Which of these three measures seems to be the least stable?

3.29 The alumni of Clareville's class of 1929 seem to be getting younger every year. In 1994 their average age was 87; in 1995 it was 86. How can this be?

3.30 Fifty randomly selected college students were asked to write down their grade-point average (GPA) and to indicate where they typically sit in large classrooms. Use the data shown below to calculate the average GPA of the 26 students who sit either in the front or toward the front.

	NUMBER OF STUDENTS	AVERAGE GPA (12-POINT SCALE)
IN THE VERY FRONT	5	10.94
TOWARD THE FRONT	21	9.38
IN THE MIDDLE	11	9.38
TOWARD THE BACK	10	8.37
IN THE VERY BACK	3	10.20

3.31 Observing that 40 percent of the 1989 first-year class at Pomona College is nonwhite,

> Dean of Admissions Bruce Poch noted in early October that the increase has not changed the school's high standards. "Combined median SAT

> (Scholastic Aptitude Test) scores," Mr. Poch stated of the 1989 crop of Pomona students, "increased slightly to 1320 at the same time that ethnic diversity increased substantially, clearly laying to rest the myth that academic excellence and ethnic diversity must be incompatible."[26]

Why, even if SAT scores are the appropriate measure of academic excellence, are the data cited of virtually no help in proving or disproving Poch's conclusion regarding Pomona College?

3.32 While driving 60 miles per hour, you notice that the number of cars you are passing is equal to the number passing you. Is 60 miles per hour the mean, median, or modal speed?

3.33 Below are the self-reported heights (in inches) of singers in the New York Choral Society in 1979.[27] The sopranos and altos are female; the tenors and basses are male. Use the means, medians, standard deviations, and box plots to describe these data.

SOPRANO				ALTO				TENOR		BASS			
64	67	66	65	65	62	70	63	69	70	72	66	71	71
62	65	62	61	62	61	65	64	72	65	70	68	70	68
66	62	65	65	68	66	64	67	71	72	72	71	74	70
65	65	63	66	67	64	63	66	66	70	69	73	70	75
60	68	65	65	67	60	65	68	76	68	73	73	75	72
61	65	66	62	63	61	69		74	73	71	70	75	66
65	63	65		67	66	61		71	66	72	68	69	72
66	65	62		66	66	66		66	68	68	70	72	70
65	62	65		63	66	65		68	67	68	75	71	69
63	65	66		72	62	61		67	64	71	68	70	

3.34 When data were grouped into 16 age categories (0 years, 1 to 2 years, 3 to 7 years, 8 to 12 years, and so on), female death rates were higher in Costa Rica than in Sweden in every single age group; yet, the overall female death rate was lower in Costa Rica than in Sweden (8.12 per 1000 women versus 9.29 per 1000 women).[28] How do you explain this paradox?

3.35 Explain the error in this interpretation of inflation data:

> In the 12-month period ending in December of 1980, consumer prices rose by 12.4 percent after a 13.3 percent increase the year before. Similar measures of inflation over the next three years were 8.9 percent, 3.9 percent, and 3.8 percent. . . . We are certainly paying less for what we buy than we were at the end of the Carter years.[29]

3.36 Temperature is very important to Professor Fix because she cannot jog when the weather is either very hot or very cold. When she was offered jobs in San Francisco and Albuquerque in 1972, a check of the *Statistical Abstract* revealed that, over the 30-year period from 1941 to 1970, the average temperature was 56.9 degrees in San Francisco and 56.8 degrees in Albuquerque. Should Professor Fix conclude that the temperatures are quite similar in these two cities?

3.37 Identify the error in this table of consumer prices that is based on a price index equal to 100 in 1967.[30]

YEAR	PERCENTAGE CHANGE IN PRICES FROM PREVIOUS YEAR	PERCENTAGE CHANGE FROM 1967	PRICE INDEX
1970	6	116	116
1971	4	121	121
1972	3	125	125
1973	6	133	133

3.38 Michelson's October 12 to November 14, 1882, measurements of the speed of light in air (in kilometers per second) were as follows:[31]

299,883	299,796	299,611	299,781	299,774	299,696	299,748
299,809	299,816	299,682	299,599	299,578	299,820	299,573
299,797	299,723	299,778	299,711	300,051	299,796	299,772
299,748	299,851					

Calculate the mean, median, mode, and 10 percent trimmed mean. Which is closest to the value 299,710.5 kilometers per second that is now accepted as the speed of light?

3.39 Most countries have only a handful of commercial banks. The United States has more than 10,000 commercial banks, many of which are quite small, as indicated by these 1990 data. (A bank with $50 million in assets can have as little as $2.5 million in net worth.)

ASSETS (DOLLARS)	NUMBER OF BANKS
0 to 24.9 million	3330
25 million to 49.9 million	3145
50 million to 99.9 million	2782
100 million to 499.9 million	2461
500 million to 999.9 million	253
1 billion to 2.9 billion	202
3 billion or more	172

Without doing any calculations, explain why the median asset size is either larger or smaller than the mean.

3.40 Figure 2.13 shows a histogram of the chest measurements of Scottish militiamen. Make rough estimates of the mean, median, and standard deviation, in each case explaining your reasoning.

3.41 Explain the error in the conclusion reached by a security analyst:

> The Dow Jones Industrial Average peaked at 381.17 during September 3, 1929. The so-called Great Crash pushed this index down 48% by November 13. But by April 17, 1930, the index rebounded 48% from the November bottom. In other words, anyone who bought a diversified portfolio of stocks during September 1929 would have experienced no net change in the value of his portfolio by April of 1930.[32]

3.42 Are great discoveries generally made by people who are young and vigorous or by persons who are old and wise? Below is a tabulation of the ages at which 12 scientists made great discoveries.[33] Use these data to calculate the mean and median ages at which these great discoveries were made.

SCIENTIST	DISCOVERY	AGE
Copernicus	Earth revolves around the sun	40
Galileo	Laws of astronomy	34
Newton	Motion, gravitation, calculus	23
Franklin	Nature of electricity	40
Lavoisier	Burning as oxidation	31
Lyell	Earth evolved gradually	33
Darwin	Natural selection in evolution	49
Maxwell	Equations for light	33
Curie	Radioactivity	34
Planck	Quantum theory	43
Einstein	Special relativity	26
Schröedinger	Equations for quantum theory	39

3.43 A recording of the degree of cloudiness at Greenwich each day during July in the years from 1890 to 1904 obtained the data below.[34] Draw a bar chart with the number of days on the vertical axis, and calculate the mean degree of cloudiness. Explain why the mean either is or is not an accurate description of a typical Greenwich day.

DEGREE OF CLOUDINESS	0	1	2	3	4	5	6	7	8	9	10
NUMBER OF DAYS	320	129	74	68	45	45	55	65	90	148	676

3.44 What is your average speed if you drive 20 miles per hour for 1 hour and 40 miles per hour for 1 hour? What if you drive 20 miles per hour for 20 miles and 40 miles per hour for 20 miles?

3.45 A college offered a free book on the history of the college to each of the alumni whose 1995 gift exceeded the median size of the gifts made in 1994 by members of this person's class; for example, $25 for a member of the class of

1994 and $200 for a member of the class of 1993. Did this college increase or decrease the threshold for a free book by using the median rather than the mean?

3.46 An advertising agency is looking for magazines in which to advertise inexpensive imitations of designer clothing. Their target audience is households earning less than $20,000 per year. A survey finds that readers of one magazine have an average household income of $22,000. Is the agency correct in concluding that the people in its target audience do not read this magazine?

3.47 Do the data in Table 3.3 indicate that deaths in each of the four categories vary by race? Use some calculations to support your position. Use two pie charts to compare these four categories by race.

3.48 It has been reported that the typical heterosexual male has six sexual partners during his lifetime, while the typical heterosexual female has two sexual partners during her lifetime.[35] Based solely on statistical reasoning, does *typical* here refer to the mean or the mode?

3.49 A Volvo advertisement said that the average person in the United States drives for 50 years and buys a new car, on average, every 3¼ years—a total of 15.4 cars during one lifetime. Since Volvos last an average of 11 years in Sweden, Volvo owners only need to buy 4.5 cars in 50 years. Carefully explain why these data are very misleading.

3.50 Exercise 2.34 shows the number of households of different sizes in 1990.

a. Looking at the 93.8 million households, ordered by size, what is the size of the median household?

b. Looking at all 264 million individuals, ordered by the size of the household they live in, what is the size of the household in which the median individual lives?

c. Pomona College reports its average class size as 14 students. However, a survey that asked Pomona College students to give the size of the classes they were enrolled in found that the average class size was 38 students and that 70 percent of the respondents' classes had more than 14 students.[36] Use the insights gained in parts **a** and **b** to explain this disparity.

projects

For each of the following projects, type a report in ordinary English, using clear, concise, and persuasive prose. Any data that you collect for this project should be included as an appendix to your report. Data used in your report should be presented clearly and effectively.

3.1 How old have U.S. presidents been when they first took office? Are presidents today younger or older relative to the life expectancy of the general population than in the past?

3.2 Who gets paid more, major league baseball pitchers or batters?

3.3 Choose two major grocery stores in your area, and try to determine which has lower prices on a sample of 30 commonly purchased items.

3.4 Find the most recent February issue of *Money* magazine, which contains detailed data on mutual fund performance. Randomly select a sample of stock mutual funds, and record their total return during the preceding calendar year. Summarize your data with a box plot and histogram, and compare the mean and median return for the funds in your sample to the return for the S&P 500 index that year. Also calculate the fractions of your sample within 1 and 2 standard deviations of the sample mean.

3.5 Collect data from at least 10 private or public day care centers near your college on the hourly cost of day care for a 4-year-old child who will be in day care for at least 10 hours per week. Summarize these data in the form of a report to parents who may want day care for their child.

3.6 What weights are used by *U.S. News & World Report* in its annual college rankings? Do the weights make sense to you? If you were in charge of these rankings, what weights would you use?

3.7 Go to a local grocery store and collect these data for at least 75 breakfast cereals: cereal name; grams of sugar per serving; and the shelf location (bottom, middle, or top). If the store that you select does not have at least 75 breakfast cereals, then collect data from another store too. Group the data by shelf location and use 3 box plots to compare the sugar content by shelf location.

3.8 Follow the directions for the preceding project, but instead of the grams of sugar per serving, record the price per ounce (or per gram) and the shelf location.

3.9 Find the price per ounce for at least 10 domestic brands and 10 imported brands of bottled water. To avoid price variations among stores, try to gather all of your data at a single store, or at 2 comparable stores. Summarize your data with 2 box plots.

3.10 At a local grocery store, identify 2 categories of chips; (for example, potato chips and tortilla chips), and in each category find the grams of fat and the milligrams of sodium per serving on the nutritional labels of at least 10 different varieties. (Do not include any low-fat or low-salt brands.) Use 2 box plots to compare the fat in these 2 categories; do the same with the sodium data.

3.11 Go to a local grocery store and collect data for at least 20 different soups from each of 2 major soup makers; for example, Campbell's and Progresso. For each of these soups, record the per-serving amounts of calories, fat, and sodium. Summarize these data.

3.12 From the National Weather Service, obtain time-series data on the annual rainfall in a nearby city. Summarize these data with a histogram and calculate the fraction of the readings that are within 1 and 2 standard deviations of the sample mean.

HAVE YOU EVER WONDERED?

Public opinion polls have become a staple of modern politics. Newspapers and magazines regularly gauge the mood of the public by surveying them about specific issues and about their overall approval or disapproval of the nation's political leaders. Potential political candidates use their own private polls to decide if they should run for office and to shape the issues they emphasize in their campaign speeches and political advertising. As the election nears, Gallup and other public polling organizations survey voter preferences and predict the winners. After the election, the victorious candidates continue to monitor public opinion and often modify their positions based on their readings of what people want.

Public opinion polls were even used in communist Poland, but to fight the government rather than to help it govern. In 1985, Solidarity, the outlawed Polish trade union, organized a boycott of the parliamentary elections held by the communist government. Afterward, the government reported that the boycott had failed and voter turnout was high—75 percent in Warsaw, for example. Solidarity countered this claim by releasing a voter survey supervised by a mathematician, Konrad Bielinski. This survey showed that only 60 percent of the voters in Warsaw voted. Political polls are one of the few tools available to private citizens to demonstrate election fraud by the government.

In the United States, the most closely watched polls are the voter surveys that precede each presidential election. Are these preelection polls for entertainment purposes only, or are they a legitimate, scientifically sound procedure for predicting election outcomes? How is it logically possible for a survey of a few thousand persons to represent accurately the preferences of tens of millions of voters? And, no matter what the theory says, how well have the polls *really* done in predicting the election results? Do they miss by an inch or a mile? Let's find out.

chapter 4

producing data

"Data! Data! Data!" He cried impatiently. "I can't make bricks without clay!"

—Sherlock Holmes

THE PREVIOUS TWO chapters were concerned with how to display and summarize data. This chapter focuses on the actual production or collection of empirical data. To draw useful statistical inferences, we must exercise care in gathering the data that will be used to make these inferences. If we want to use patient data to assess the potential benefits and side effects of a medication, we need to think seriously about who will participate and how they will be tested. If we want to use a public opinion poll to predict the outcome of a presidential election, we need to think carefully about who will be surveyed and what questions they will be asked.

The data we collect can usually be considered a sample from a much larger population that we would like to make inferences about.

The entire group of items that interests us is called the **population.** The part of this population that we actually observe is called a **sample.** *Statistical inference* involves using the sample to draw conclusions about the characteristics of the population from which the sample came.

In the medical example, the population consists of all persons who might use this medication; the sample is the group of people that we use to test the medication; and a possible statistical inference is that people who take the medication tend, on average, to live longer than people who do not. In the political example, the population consists of all persons who will vote in the election, the sample is the group of people we interview, and a possible statistical inference is a prediction about who will win the election.

In this chapter, we will see how samples can be used to produce useful and reliable data, and we also will identify some pitfalls that can make samples misleading. The remaining chapters of this book show how sample data can be used to draw statistical inferences about the population from which the sample was drawn.

4.1 The Power of Sampling

We use samples to draw inferences about the population because it is often impractical to scrutinize the entire population. If we burn every lightbulb that a manufacturer produces to see how long each bulb lasts, all we will have is a large electricity bill and a lot of burned-out lightbulbs. Many tests are not this destructive, but are simply too expensive to apply to the entire population. Instead, we sample. A lightbulb manufacturer tests a sample of its bulbs. A software company tests a sample of its disks. Editors sample a few pages of unsolicited manuscripts because they lack the time or patience to read every word of every manuscript that crosses their desks purporting to be the Great American Novel. Medications and opinion polls are administered to a sample of people because it is too expensive to test or survey everyone.

In a classic 1955 court case *(Sears Roebuck v. the City of Inglewood),* a judge fell prey to the common misperception that a small sample cannot possibly be reliable.[1] Inglewood had imposed a .5 percent tax on all retail sales to Inglewood residents made within the city limits. A Sears audit found that over an 850-day period Sears had erroneously paid the city of Inglewood a .5 percent tax on all its sales, including sales to nonresidents. Sears examined the sales receipts for a sample of 33 days and found that 36.69 percent of its sales had been to nonresidents. Applying this sample estimate to its revenue for the entire 850-day period, Sears estimated that it had overpaid Inglewood $28,250.

Although Inglewood's sales tax auditors had used sampling procedures for years in their own work, Inglewood's attorney asked only one question at the trial: Did Sears know with certainty that exactly 36.69 percent of its sales were to nonresidents? The judge was persuaded by this logic and ruled that Sears would have to audit the receipts for all 850 days. Sears employees consequently had to work 3384 hours examining 950,000 sales slips. In addition, the slips for two 4-week periods had been lost and could not be tabulated. Even without these slips, their final tally was that Sears had overpaid Inglewood $26,750.22; if the missing days had been comparable, the total would have been $27,527. Sears employees had to work 3384 hours in order to prove that its sample of 33 days gave an estimate that was within $750 of the result of a complete tabulation. (Because of the missing slips and the possibility of clerical errors, there is no way of knowing for certain exactly how much Sears had overpaid.)

Fortunately, the judicial system's attitude toward sampling has changed considerably over the years. For example, in 1975 the Illinois Department of Public Services examined a sample of 353 of 1302 Medicaid claims filed by one doctor and found that the doctor had been overpaid $5018, an average of $14.215 per claim. Applying this figure to all 1302 claims, the department estimated that the doctor had been overpaid $14.215(1302) = $18,508. The doctor and the Illinois Physicians Union filed suit, arguing that the Department of Public Services should have audited all 1302 claims. The federal court for the Seventh Circuit rejected their argument, noting that "the use of statistical samples has been recognized as a valid basis for findings of fact" and that it is impractical and inefficient to audit every single claim.[2]

As the court noted, there is plenty of evidence that sampling works. The Chesapeake and Ohio Railroad Company once conducted a study to see how much difference there would be between a sample and a comprehensive tabulation of its waybills.[3] A waybill is a document accompanying each railroad car, recording its weight and the route traveled. When a shipment goes over several railroad lines, each railroad is entitled to part of the total shipping charge; but the clerical deciphering of the waybill and the arithmetic computations are tedious and time-consuming.

The Chesapeake and Ohio Railroad Company studied 23,000 waybills accumulated over a 6-month period. A sample of 9 percent of the waybills yielded an estimate of $64,568 as the amount due the Chesapeake and Ohio Railroad. To test the validity of the sampling procedure, the railroad also made a complete tabulation of all 23,000 waybills and came up with a figure of $64,651. There was only an $83 difference (.1 percent) between the results of a sample and a complete tabulation! Because it cost less than $1000 for a sample and about $5,000 for a complete tabulation, they were persuaded that sampling made good economic sense.

In another study, Chesapeake and Ohio looked at the allocation of ticket revenue for passengers who traveled on more than one rail line. The company found that a 5 percent sample gave an inexpensive estimate that was within one-half of a percentage point of the result of a complete tabulation. A similar study by three airlines in the 1970s found that a sample of about 12 percent of the interline tickets gave a division of the ticket revenue that was within .1 percent of the result of a full tabulation. They estimated that sampling saved the airline industry about a half a million dollars annually in clerical expenses.

Complete tabulations are subject to error, as we witness every 2 years when the votes in disputed congressional elections are counted and recounted and seldom give the same answer twice. Auditors, too, find that repeated tabulations of financial data often disagree. In fact, special studies by accountants have found that a statistical sample can be more accurate than a full tabulation! A complete count allows lots of chances to make errors, and it is easy to err if the counters are benumbed by their repetitive task. A partial enumeration by the alert may be more successful than a complete tabulation by the bored.

The Nielsen Ratings

Arthur Charles Nielsen was trained as an electrical engineer, earning the highest grade-point average in the history of the University of Wisconsin's engineering school. During the 1930s, he started a business, the A.C. Nielsen Company, that collected and sold data on product sales in drugstores. He initially compiled these data by having his employees interview drugstore customers, but soon found that his data for two soaps, Lux and Lifebuoy, were not consistent with actual sales. Nielsen's son later explained that "Dad realized that a lot of women said they bought Lux because of its image as a prestige soap for movie stars while Lifebuoy was only for people with b.o."[4]

Nielsen learned from this experience that data on what people do are sometimes more reliable than data on what people say they do. He stopped interviewing customers and began paying druggists to allow his employees to examine sales records. Today, the A.C. Nielsen Company makes most of its money from its retail index, which keeps track of store sales of food and drug products.[5] Manufacturers may know how much they are producing, but they lack up-to-the-minute data on whether their products are selling or gathering dust on store shelves. So they pay the Nielsen Company to monitor a sample of stores for them.

Nielsen is best known for monitoring television viewing habits. The primary tool is an electronic gadget, called a *people meter,* that works as an ordinary remote control but keeps a minute-by-minute record of the stations being watched and allows the viewer to record which family members are watching. Nielsen pays 5000 households throughout the country a small fee for using people meters and allowing their televisions to be connected to their telephone lines. Each day, a Nielsen computer telephones these households and automatically processes the viewing habits recorded during the preceding 24 hours. The company then computes a "rating" and "audience share." The Nielsen *rating* is an estimate of the fraction of all homes in the nation that watched a particular program; the *audience share* is an estimate of the fraction of the television sets that are turned on and are tuned to that show. On a typical evening in prime time, about two-thirds of the nation's homes have televisions on. If they are all watching the same program, then that lucky show will get a Nielsen rating of 66.7 percent and an audience share of 100 percent and the other shows will get big fat zeros. In practice, a show on one of the three major networks is usually in trouble if it gets a rating of less than 15 percent and a share of less than 25 percent.

Because the sample of 5000 households is used to estimate the behavior of 100 million homes, each represents 20,000 homes—an awesome responsibility. When 50 of these chosen households switch channels, that will cost a show a rating point and perhaps its life. With all this power, Nielsen's 5000 households must sign an agreement promising not to let anyone connected with the communication industry know that they are among the powerful few.

One of the advantages of Nielsen's people meters is that they record the actual channels that are turned on. With personal interviews or viewing diaries, there is a great temptation for viewers to vote for shows they like, even if they do not always watch them. Of course, the people meter only shows that the television is on, not that people are watching. A *Los Angeles Times* reporter once broke through Nielsen's code of secrecy and located one of the 5000, whom he dubbed "Deep Eyes." Deep Eyes said that he had his television on at least 50 hours a week, winning points for favored shows, even though he did not spend nearly that much time watching television. After the story was printed in the *Times,* other Nielsen households contacted the reporter to say that they were honest watchers and to express outrage at Deep Eyes's deception. Nielsen is working on a way to combat this deception—an electronic gadget that uses sound waves or heat sensors to determine how many people are in the room.

The 5000 people meters are spread too thinly across the country to give reliable ratings of local stations or of local viewing of the network stations. For these data, Nielsen uses viewer diaries. There are four "sweep weeks" each year, when Nielsen mails out millions of viewer diaries to gather detailed data on local viewing habits and the demographic and economic characteristics of viewers. Advertisers want to know not only how many, but also who, since they do not want to advertise arthritis remedies and dentifrice cleansers on shows watched only by children. The television stations treat sweep weeks in a predictable fashion, showing blockbuster movies and other specials during these four important rating weeks. The next time you notice that all three networks are showing great movies or specials at the same time, you can reasonably conclude that it is a sweep week again.

During the May 1987 sweep week, one Los Angeles television station showed something even more effective than a great movie. On its 11 p.m. news show, this station ran a heavily advertised, special eight-part report on the Nielsen surveys. The competitors complained loudly (even threatening lawsuits), arguing that this was a clever scheme to entice the Nielsen households to watch news reports about themselves. In the 12 weekdays preceding the special reports, the station's 11 p.m. newscast averaged an 8.3 rating. Their rating jumped to 18.2 on the first night of the special report and averaged 10.5 during these eight shows.

If there are problems with the Nielsen television surveys, it is not that the samples are too small. We will see later in this chapter that presidential election polls based on samples of a few thousand people are generally quite accurate. Errors that occur in television surveys and election polls are not caused by the small size of the samples, and would not be cured by interviewing more people. The real challenge is getting honest data—here in finding out what people really watch when it is not a sweep week. In the remainder of this chapter, we will look at some of the methods we can use to try to minimize distortions in data.

Statistical Sampling in World War II

Many British and U.S. statisticians used statistical analysis to help their countries during World War II. In the process, they refined their techniques and proved the value of statistical sampling. There are many examples, but we mention only two here.

Many consumer products were rationed during the war, with the U.S. Office of Price Administration (OPA) issuing ration coupons that had to be used to make purchases. To know how many ration coupons to issue for tires, for example, the OPA needed estimates of the inventories held by tire dealers. The OPA first tried to make a complete tabulation by mailing questionnaires to every single dealer in the country, but many dealers did not return the questionnaires. The OPA suspected that the inventories of dealers who failed to respond were quite different from those who did, so it put together a small sample of dealers and used follow-up telephone calls and visits to ensure responses from almost every dealer in the sample. The small but reliable sample proved to be less expensive and more accurate than the large tabulation.

A similar thing happened with estimates of German output. The Germans tried to make a complete tabulation of how much they were producing, but reports from individual factories were often late and not always reliable. British and U.S. statisticians were keenly interested in estimating German war production, too, but they could hardly ask the German factories to send them reports. Instead, they based their estimates on the manufacturing serial numbers of captured equipment. These serial numbers provided a sample that was very small, but reliable. Special studies made after the war discovered that the British and U.S. estimates of German production were more accurate and timely than Germany's own estimates![6]

4.1 Manufacturing processes inevitably fluctuate as there is wear and tear on people and machines. To keep products tolerably close to specifications, companies routinely test production samples. If the output begins to violate specifications, the manufacturing process is stopped and adjustments are made. What are the costs and benefits of this monitoring?

4.2 Some managers drop in unexpectedly on their subordinates to see whether they are working. These surprise visits are intended to be a sample from what population? What are the costs and benefits of using a sample rather than the entire population?

4.3 For the Chesapeake and Ohio study mentioned in this section, the 23,000 waybills were divided into five groups, according to the size of the total shipping charge, as shown below. Why did the company take relatively large samples of the more expensive waybills?

WAYBILL CHARGE	PERCENTAGE SAMPLED
$5.00 or less	1
$5.01 to $10.00	10
$10.01 to $20.00	20
$20.01 to $40.00	50
More than $40.00	100

4.4 Before the Nielsen ratings were introduced, networks gauged the popularity of their shows by weighing sacks of fan mail. Explain why this procedure was unreliable.

4.5 In the 1950s, A.C. Nielsen's researchers invented what came to be called the "Nielsen whoopee cushion." The idea was to monitor who watched television by putting special pressure-sensitive electronic cushions where each family member sat while watching television: a father cushion in the father's chair, a mother cushion in the mother's chair, and children's cushions in the children's chairs. What advantages and disadvantages do these cushions have in comparison with people meters?

4.2 Experimental Design

In chemistry and physics, data can be collected in laboratories with carefully controlled *experiments,* in which the researcher imposes some "treatment" and monitors what happens. In these ideal conditions, several principles guide scientists in designing experiments that will yield useful, reliable data. In the behavioral sciences, data do not generally come from laboratory experiments, but are instead *observational data* from what are euphemistically called "natural experiments." Economists would like to know the effects of income fluctuations on consumer spending, but fortunately they do not have the power to raise and lower the national income to test their theories. Instead, they must be content to record income and consumption data as the economy tosses and turns. The same is true of many interesting social issues, ranging from rent control to capital punishment. Although behavioral scientists may be forced to use observational data, rather than experimental data, we can use lessons from the laboratory to inform us about good experimental design and to help us recognize the limitations of data that are gathered in more natural settings.

The Importance of a Control Group

To analyze some physical or social phenomenon, we generally need to make a comparison. For example, a popular chemistry experiment involves the extraction of caffeine from a cup of coffee by drawing the coffee through a tube filled with silica gel. If we find that the drawing of a cup of espresso coffee through 1000 milligrams of silica gel yields 150 milligrams of caffeine, we will not be able to interpret our results unless we have a point of reference—information about the amount of caffeine in another kind of coffee, a cup of tea, or a chocolate bar and information about the effectiveness of this caffeine extraction process.

Further, our comparison will not be meaningful if we neglect to hold other important, but extraneous, factors constant. A comparison of the caffeine removed from a cup of espresso with the caffeine removed from a chocolate bar will not be meaningful unless the experiment is conducted under otherwise similar conditions. When more than one important factor is present and we cannot untangle which is responsible for the observed effects, the results are said to be **confounded** (mixed up). If 150 milligrams of caffeine are removed from a 4-ounce cup of espresso using 1000 milligrams of silica gel and 80 milligrams of caffeine are removed from an 8-ounce chocolate bar using 500 milligrams of silica gel, the effects are confounded, since we cannot tell whether the observed caffeine differences are due to the different products, the different sizes of the products, or the differing amounts of silica gel used in the experiment.

Here is an example from physics.[7] If an object is attached to a string and swung back and forth as a pendulum, the time it takes to complete a cycle depends, in theory, on the length of the string, but does not depend on the object's mass. Yet in a laboratory experiment with two balls of different weights, the heavier ball consistently took longer to complete a cycle than the lighter ball, when hung from the same string and released from the same angle of swing. The reason? The string was stretched more by the heavier ball, and so the pendulum was longer, which made its cycle time longer. The length of the string was an extraneous factor that was not held constant.

Thus the first principle of good experimental design is to hold constant potentially confounding factors that might influence the results. In physics and chemistry, this is called *conducting an experiment under controlled conditions.* If we hold all relevant factors constant but one, then we are entitled to conclude that it is this factor that causes the changes we observe. If we hold both the length of the string and the angle of swing constant and find that the heavier ball takes longer to complete a cycle, we can conclude that the additional mass is responsible. If we do not, then we cannot. Of course, scientists sometimes overlook factors that turn out to be important and consequently reach erroneous conclusions. The point is not that laboratory experiments are infallible, but rather that in designing experiments we should attempt to control for important extraneous influences.

In medicine and other fields where researchers are studying the effects of a treatment, broadly defined, comparison is generally made to a **control group** that does not receive the treatment. If the control group does not differ systematically from the treatment group in any respect other than the treatment, then the researchers can justifiably conclude that the observed effects are due to the treatment. Again, researchers may inadvertently overlook important factors; the point is that in designing an experiment, they try to keep extraneous influences from distorting their data.

Suppose that we want to study the effects of soil acidity on corn yields. We might add lime to the soil (which reduces soil acidity) and record the corn yield that season. To gauge the effect of soil acidity, we also need data on corn yields from soil of different acidity. It is tempting to compare to corn yields the previous year, before we added lime. If the corn yield was 100 bushels per acre last year and was 90 bushels per acre this year, it appears that the addition of lime reduces corn yields. The problem with these data is that corn yields depend on much more than just soil acidity, and some of these influences—most obviously temperature, sunlight, and rain—vary from year to year. If there was less rain this year than the year before, we do not know whether the observed drop in corn yields was due to the lime or the drought. A better experimental design would be to break up a homogeneous parcel of farmland and observe the corn yields during a single season. The treatment group is the land that receives lime; the control group is the land that does not receive lime, but does not differ systematically from the treatment land in other respects.

If there are systematic differences, the confounding effects may negate the potential benefits of having a control group. For instance, a 1971 study found that people who drink lots of coffee get bladder cancer more often than do people who do not drink coffee. However, a confounding effect was present in that people who drink lots of coffee are also more likely to smoke cigarettes. Is it the coffee or the cigarettes that is responsible for the bladder cancer? In 1993, a rigorous analysis of 35 studies of this question concluded that there is "no evidence of an increase in risk of [lower urinary tract] cancer in men or women after adjustment for the effects of cigarette smoking."[8]

Here is another example of the importance of a control group. A 1984 Associated Press story titled "College Broadens Not Just the Mind, Study Says," reported that students eating in the Penn State dining halls gained an average of 9.1 pounds their first year, 7.3 pounds the second year, 7.8 pounds the third year, and 6.5 pounds the fourth year.[9] Without a control group, one might conclude that the food served in college dining halls is fattening and that health-conscious college students should either eat off-campus or force the dining halls to serve less fattening food. Without in any way defending dining hall food (I went to college, too), we need to take into account the reality that most college-age students are still growing and will presumably gain weight regardless of whether they eat in college dining halls. To see if dining hall food is especially fattening, we need a control group so that we can monitor the weight gains of college-age people who do not eat dining hall food. In fact, the researcher cited in this Associated Press story also interviewed Penn State students who were living off-campus and found their weight gains to be indistinguishable from those of students eating dining hall food.

The Placebo Effect

In medical research, the control group is usually given a *placebo,* which appears identical to the treatment but has no medical value. In studies of vitamin C, for example, those in the control group are given pills that look and taste like vitamin C, but are in fact made of inactive substances that do not affect the body.

Placebos are used so that the subjects do not know whether they are receiving the treatment or are in the control group. If the subjects know which group they are in, they may shade their health reports by saying what they think the researchers want to hear. In observing the subjects, researchers, too, might be influenced by hope that the treatment has the desired effects. To encourage honest reporting by both subjects and researchers, most studies are *double-blind*—neither the subject nor the researcher knows who is in the treatment group and who is in the control group until after all the data have been collected.

A **placebo effect** occurs when people who believe in the power of medicine and are hopeful of a medical cure respond psychologically to treatment, even if the treatment has no medical value. The placebo effect can be so strong that patients given a worthless treatment sometimes show remarkable physical improvement. Because of this placebo effect, it is very important to compare the treatment group with a control group. Without the control group, we do not know how much of the observed improvement is due to the placebo effect and how much is due to the treatment.

For example, medical researchers once tested a vaccine for the common cold, using 548 college students who felt they were especially susceptible to colds.[10] One-half were given the experimental vaccine, and one-half, the control group, were given plain water. The group taking the vaccine reported an astonishing 73 percent decline in colds in comparison with the previous year—astonishing, except for the fact that the control group reported a 63 percent decline! Apparently, this particular season was relatively cold-free, or there were psychological benefits from the presumed vaccine, or the students reported a decline in colds because they thought that this was what the researchers wanted to hear.

If there had been no control group, the researchers would have concluded that the vaccine was a miraculous success. Having a control group for comparison, they concluded that the vaccine had little, if any, effect on colds. The importance of a control group is made even more striking by the researchers' report that they received several letters from doctors similar to this one: "I have a patient who took your cold vaccine and got such splendid results that he wants to continue it. Will you be good enough to tell me what vaccine you are using?" When they checked the records, the researchers often found that the patient with such splendid results had received plain water.

Another example involves a study published in the *Journal of the American Medical Association* showing that gastric freezing, in which a coolant is pumped through a balloon in a patient's stomach, reduced acid secretions and stomach pains in ulcer patients.[11] This experiment had been done without a control group, but because the results were published in such a prestigious journal, gastric freezing was used by doctors for several years. Finally, another study was done, this time using a control group that received a fluid that kept the balloon at body temperature. While 34 percent of those receiving the gastric freezing treatment showed improvement, 38 percent of those receiving the placebo did, too. Evidently, the observed reduction in acid secretion and relief from ulcer pain was due entirely to the placebo effect. Subsequent studies confirmed that gastric freezing had no beneficial effects, and doctors stopped using it.

Control groups are useful not only for clarifying the beneficial effects of a medication, but also for examining potential side effects. For example, tests of the allergy-relief drug Seldane found that of 781 persons who took Seldane for 7 to 14 days, 9 percent reported drowsiness and 6 percent reported headaches—suggesting worrisome side effects.[12] These figures are put into perspective, however, by the fact that of 665 persons who took a placebo in these tests, 8 percent reported drowsiness and 7 percent reported headaches. The presence of a control group alerts us to the fact that many people will report being drowsy or having headaches, regardless of whether they are taking medication. Only by comparing the frequency of these reports in the treatment and control groups can we tell whether the medication is aggravating or alleviating drowsiness and headaches. The procedures developed in later chapters reveal that the small differences reported in these tests do not indicate that Seldane has an effect on either drowsiness or headaches.

The placebo effect and the need for double-blind controls are not limited to medical research. A famous example involves a study over a 5-year period, from 1927 to 1932, of the behavior of workers at Western Electric Company's Hawthorne Works plant.[13] As part of this study, a number of workers who assembled telephone relays were moved into a specially constructed room where they could be watched closely under controlled conditions. In theory, it seemed a good idea to try to replicate the chemist's laboratory, where individual factors could be varied while other factors are held constant. In practice, the observed changes in these workers' productivity were due mostly to the special attention they were receiving. This particular example of how participation in an experiment can affect the behavior of the participants is so well known that this phenomenon is often called the *Hawthorne effect.*

Researchers, too, can be affected by knowledge that an experiment is being conducted and what results might be expected. For example, for the last experiment in a senior-level course in experimental psychology, each of 12 students was given five rats to test on a maze.[14] Six of the students were told that their rats were "maze-bright" because they had been bred from rats that did very well in maze tests; the other six students were told that their rats were "maze-dull." In fact, there had been no special breeding, and these ordinary rats were randomly distributed to the students. Nonetheless, when the students tested the rats in 10 runs each day for 5 days, the rats that had been called maze-bright had, on average, higher scores each day, and their scores also increased more over time—indicating that they were learning faster than the dull rats. Evidently the researchers' expectations were influencing the results. The professors concluded that the students were handling the rats labeled maze-bright differently, perhaps "an extra pat or two for a good performance" or more subtle encouragement transmitted to the rats "via changes in skin moisture, temperature, and the like." The potential effect of the experimenter's expectations on the results is sometimes called *the experimenter effect.*

Selection Bias

The purpose of a control group is to approximate a laboratory experiment in which all influences other than the treatment are held constant. If, in addition to the treatment, there are other important systematic differences between the treatment group and the control group, then we do not know whether our results should be attributed to the treatment or to these other differences. For example, if a study of the effectiveness of a cold vaccine compared the incidence of colds in a group of Florida school children who received the vaccine with a control group of elderly persons in a nursing home in England, we would not know whether the observed differences were due to the vaccine or to differences in ages, weather, and exposure to cold viruses.

More generally, any sample that differs systematically from the population that it is intended to represent is called a *biased sample.* Because a biased sample is unrepresentative of the population, it gives a distorted picture of the population and may lead to unwarranted conclusions. One of the most common causes of biased samples is **selection bias,** which occurs when the selection of the sample systematically excludes or underrepresents certain groups. Selection bias often happens when we use a *convenience sample* consisting of data that are readily available.

If we are trying to estimate how often people get colds and we have a friend who can give us medical records from an elementary school, this is a convenience sample with selection bias. If our intended population is people of all ages, we should not use samples that systematically exclude certain ages. Similarly, the medical records from a prison, military base, or nursing home are convenience samples with selection bias. Military personnel are in better physical health than those living in nursing homes, and both differ systematically from the population as a whole.

Here's a very different example. In the 1980s, Midway Airlines advertised in *The New York Times* and *The Wall Street Journal* that "84 percent of frequent business travelers to Chicago prefer Midway Metrolink to American, United, and TWA."[15] At first glance, it is hard to reconcile this advertisement with the reality that only 8 percent of the passengers flying from New York to Chicago flew on the Midway Metrolink. If 84 percent preferred Midway, a comparable amount should be flying Midway. The solution to this puzzle is that the advertisement was based on a sample with considerable selection bias: a survey of passengers who were flying Midway Metrolink from New York to Chicago. It is not at all surprising that a person who has chosen to fly on an airline prefers that airline. What is truly surprising is that 16 percent of these Midway passengers would have rather been on another airline. But we can hardly expect Midway to advertise that "Sixteen percent of those flying Midway wish they weren't."

Survivor Bias

Retrospective studies look at past data for a contemporaneously selected sample, for example, an examination of the medical records of 65-year-olds. A *prospective* study, in contrast, selects a sample and then tracks the members over time. Retrospective studies are notoriously unreliable, and not just because of faulty memories and lost data. When we choose a sample from a current population in order to draw inferences about a past population, we necessarily exclude members of the past population who are no longer around—an exclusion that causes **survivor bias,** in that we look at only the survivors. If we examine the medical records of 65-year-olds to identify the causes of health problems, we overlook those who died before reaching age 65 and consequently omit data on some fatal health problems. Survivor bias is a form of selection bias in that the use of retrospective data excludes part of the relevant population.

For another example, suppose that we are trying to gauge how well a firm treats its employees. We might look at employees who have been with the firm 20 years and calculate the percentage increase in their wages over this 20-year period. There may be survivor bias here, in that the employees who are still with this firm may not be representative of those hired 20 years ago. Perhaps 99 percent have been fired, and the 1 percent who survived have received pay increases averaging 10 percent per year. It would be misleading to conclude from our analysis of the 1 percent still with the firm that new employees can expect pay increases averaging 10 percent per year.

The bias can go the other way, too. Suppose that this firm is a wonderful training ground for future executives and that most of its employees leave after a few years for high-paying management positions with other firms. Being a benevolent company, the firm does not fire those who cannot find jobs elsewhere, but continues to employ them (at modest salaries). Here, it would be misleading to conclude from our analysis of the "deadwood" still with the firm that a new employee can expect a career with few promotions and small pay raises.

Here is another example. Stock market studies sometimes examine historical data for companies that have been selected randomly from the New York Stock Exchange (NYSE). If we restrict our analysis to companies currently listed on the NYSE, our data will be subject to survivor bias because we will ignore companies that were listed in the past but have subsequently gone bankrupt. If, for example, we want to estimate the average return for an investment in NYSE stocks over the past 50 years and we do not consider the stock of any company that went bankrupt, then we will overestimate the average return.

The 1954 Tests of a Polio Vaccine

Poliomyelitis (polio) is an acute viral disease that causes paralysis, muscular atrophy, permanent deformities, and even death. It is an especially cruel disease in that most of its victims are children. The United States had its first polio epidemic in 1916, and over the next 40 years hundreds of thousands of people were afflicted. In the early 1950s, more than 30,000 cases of acute poliomyelitis were reported each year.

Researchers found that the majority of adults had experienced a polio infection during their lives. In most cases, this infection had been mild and caused the person's body, in trying to ward off the infection, to produce antibodies that made the body immune to another attack. Similarly, they found that polio was rarest in societies with the poorest hygiene. The explanation is that almost all the children in these societies were exposed to the contagious virus while still young enough to be protected by their mother's antibodies, and so they developed their own antibodies without ever suffering from the disease.

Scientists consequently worked to develop a safe vaccine that would provoke the body to develop antibodies, without causing paralysis or worse. A few vaccines had been tried in the 1930s and then abandoned because too often they caused the disease that they had been designed to prevent. By the 1950s, extensive laboratory work had turned up several promising vaccines that seemed to produce antibodies against polio safely.

In 1954, the Public Health Service organized a nationwide test of Jonas Salk's polio vaccine, involving 2 million schoolchildren.[16] Because polio epidemics varied greatly from place to place and from year to year throughout the 1940s and early 1950s, the Public Health Service decided not to offer the vaccine to all children, either nationwide or in a particular city. Otherwise, there would have been no control group and the Public Health Service would not have been able to tell if variations in polio incidence were due to the vaccine or to the vagaries of epidemics. Instead, it was proposed that the vaccine be offered to all second-graders at selected schools. The experience of these children (the treatment group) could then be compared to the school's first- and third-graders (the control group) who were not offered the vaccine. Ideally, the treatment group and the control group would be alike in all respects, but for the fact that the treatment group received the vaccine and the control group did not.

There were two major problems with this proposed experiment. First, participation was voluntary, and those who agreed to be vaccinated tended to have higher incomes and better hygiene and, as explained earlier, to be more susceptible to polio. Thus there was selection bias in the sample. Second, school doctors were instructed to look for both acute and mild cases of polio, and the mild cases were not easily diagnosed. If doctors knew that many second-graders were vaccinated, while first- and third-graders were not, this knowledge might influence the diagnosis. A doctor who hoped that the vaccine would be successful might be more apt to see polio symptoms in the unvaccinated than in the vaccinated.

A second proposal was to run a double-blind test, in which only one-half of the volunteer children would be given the Salk vaccine and neither the children nor the doctors would know whether the child received the vaccine or a placebo solution of salt and water. Each child's injection fluid would be chosen randomly from a box, and the serial number recorded. Only after the incidence of polio had been diagnosed would it be revealed whether the child received the vaccine or the placebo. The primary objection to this proposal was the awkwardness of asking parents to support a program in which there was only a 50 percent chance that their child would be vaccinated.

As it turned out, about one-half of the schools decided to inoculate all second-grade volunteers and use the first- and third-graders as a control group. The remaining half agreed to a double-blind test using the placebo children as a control group. The results are shown in Table 4.1. The Salk vaccine reduced the incidence of polio in both cases, but the gains are more dramatic in the double-blind experiment. The number of diagnosed polio cases fell by about 54 percent in the first approach and by 61 percent with the placebo control group. The effectiveness of the Salk vaccine is partially masked in the first experiment because the second-grade volunteers are different from the first- and third-graders. If the double-blind experiment had not been conducted, the case for

Table 4.1

Results of a 1954 nationwide test of a polio vaccine

	FIRST- AND THIRD-GRADE CONTROL GROUP		DOUBLE-BLIND WITH PLACEBO	
	CHILDREN	POLIO PER 100,000	CHILDREN	POLIO PER 100,000
Treatment group	221,998	25	200,745	28
Control group	725,173	54	201,229	71
No consent	123,605	44	338,778	46

the Salk vaccine would have been less convincing because the reported decline was smaller and because skeptics could have attributed this decline to a subconscious desire by doctors to have the vaccine work.

The 1954 tests were a landmark national public health experiment that provided convincing evidence of the value of a polio vaccine. The Salk vaccine was eventually replaced by the even safer and more effective Sabin preparation. Today, 90 percent of all U.S. children receive three or more oral doses of polio immunization, and the nationwide incidence of polio has been reduced to about a dozen cases per year.

Conquering Cholera

In many important—in fact, life-and-death—situations, it is simply not practical to do controlled laboratory experiments. Instead, researchers must do the best they can with observational data. In the 1800s, London was periodically hit by terrible cholera epidemics. John Snow, a distinguished doctor, suspected that cholera could be caused by drinking water contaminated with sewage, but he could not ethically set up a controlled experiment in which some persons were forced to drink contaminated water. Instead, he used observational data by comparing the cholera mortality rates in an area of London where the water was supplied by two different companies: Southwark & Vauxhall (which pumped water from a nearby section of the Thames river that was contaminated with sewage) and Lambeth (which obtained its water far upstream, from a section of the Thames that had not been contaminated by London sewage). By choosing an area of London in which adjacent houses were served by different water companies, Snow tried to minimize the confounding effects of socioeconomic factors. During the first seven weeks of the 1854 cholera epidemic, there were 1263 cholera deaths in 40,046 houses supplied by Southwark and 98 deaths in 26,107 houses supplied by Lambeth. Assuming that the household sizes were roughly comparable, these data provided convincing evidence of a relationship between drinking contaminated water and the incidence of cholera.[17]

Does Anger Trigger Heart Attacks?

A 1994 Associated Press story in *The New York Times* reported that researchers had found that "People with heart disease more than double their risk of heart attacks when they become angry, and the danger lasts for two hours."[18] This conclusion was based on interviews with 1122 men and 501 women who survived heart attacks. In these interviews, conducted a few days after the attacks, 36 persons reported being angry during the 2 hours preceding the attack, and 9 reported being angry during the day before the attack.

It is certainly a common perception that anger can trigger a heart attack; however, it is also commonly believed that it is healthier to express anger than to "keep it bottled up inside." It certainly would be interesting to know which of these folk myths is closer to the truth. Unfortunately, the data given here tell us nothing at all about the effects of anger on one's risk of having a heart attack. More people were angry 2 hours before the attack than were angry the day before; however, 97 percent of these heart attack victims did not report being angry at all, either on the day of the attack or on the day before!

In addition, this is a retrospective study that involves two kinds of survivor bias. Those who died of heart attacks and those who did not have heart attacks are both excluded from the study. To assess the effect of anger on heart attacks, we need two groups—a test group and a control group—and we need data on both possible outcomes (having a heart attack and not having a heart attack). Perhaps most of the people who were angry did not have heart attacks and were not included in this study. To exaggerate the point, suppose that a complete picture of the data looks like this:

	HEART ATTACK	NO HEART ATTACK	TOTAL
ANGRY	45	1955	2000
NOT ANGRY	1578	0	1578
TOTAL	1623	1955	3578

In these hypothetical data, 2000 people were angry and a control group of 1578 were not angry. As in the actual data cited, 45 of the 1623 people who suffered heart attacks reported being angry. These hypothetical data show that 97 percent of the people who were angry did not have heart attacks and that 100 percent of those who were not angry did have heart attacks. People who were not angry had a much greater risk of heart attack—exactly the reverse of the conclusion drawn by the researchers who examined only people who suffered heart attacks. A scientific study using a control group would surely not yield data as one-sided as this caricature. It might even show that angry people are more likely to experience heart attacks. The data these researchers collected do not really tell us whether anger increases the risk of heart attack.

exercises

4.6 An examination of the records of 850,000 operations performed in 34 hospitals determined the following death rates with four common general anesthetics: 1.7 percent with anesthetic A, 1.7 percent with B, 3.4 percent with C, and 1.9 percent with D. Why should we be cautious in concluding that anesthetic C is twice as dangerous as the other anesthetics?

4.7 A 1991 advertisement for a set of mathematics textbooks called *Ray's Arithmetics* stated that children in the 19th century "used *Ray's Arithmetics*—120 million of them. It was the standard text in an era that produced geniuses like Edison and Ford." Explain why Edison and Ford are not convincing evidence of the value of *Ray's Arithmetics.*

4.8 A vaccine called BCG was tested to see if it could reduce deaths from tuberculosis.[19] A group of doctors gave the vaccine to children from some tubercular families and not to others, apparently depending on whether parental consent could be obtained easily. The results over the subsequent 6-year period (1927 to 1932) were as follows:

	VACCINATED	NOT VACCINATED
NUMBER OF CHILDREN	445	545
DEATHS FROM TUBERCULOSIS	3 (.67 percent)	18 (3.30 percent)

A second study was then conducted in which doctors vaccinated only one-half of the children who had parental consent, using the other half as a control group:

	VACCINATED	NOT VACCINATED
NUMBER OF CHILDREN	556	528
DEATHS FROM TUBERCULOSIS	8 (1.42 percent)	8 (1.52 percent)

Provide a logical explanation of these contradictory findings.

4.9 During World War II, 408,000 U.S. citizens were killed in military duty, and 375,000 U.S. citizens were killed by accidents in the United States. Do these data show that it was almost as dangerous to stay home as to fight in the war?

4.10 The Census Bureau regularly estimates the lifetime earnings of people by age, sex, and education. Using these estimates, the *Cape Cod Times* (July 7, 1983) reported that

> 18-year-old men who receive graduate education will be "worth" twice as much as peers who don't complete high school. On the average, that first 18-year-old can expect to earn just over $1.3 million while the average high-school dropout can expect to earn about $601,000 between the ages of 18 and 64.

Mike O'Donnell was getting poor grades in high school and was thinking of dropping out and going to work for his family's construction firm. But his father read this article and told Mike that dropping out would be a $700,000 mistake. Is Mike's dad right?

4.11 To see whether women are more successful at single-sex or coeducational colleges, a study compared the average college grade-point averages (GPAs) of a sample of women attending a women's college with those of a sample of women attending a coeducational college. Why is this sample design flawed?

4.12 At most colleges and universities, an assistant professor is evaluated during the sixth year and is either terminated or given lifetime tenure. Explain the bias that occurs if we try to measure the job security and the average pay increase of a college's faculty by obtaining data from professors who have been with the college for at least 10 years.

4.3 The Power of Random Selection

Suppose that we want to gauge the physical strength of students at the University of Nebraska. The population is the entire student body; the sample is the students we test. If we use the varsity football team for our sample, perhaps because these data are readily available, our study will have considerable selection bias—the football team is not a representative sample and will give misleading information about the physical strength of the population. The average Nebraska football player may be able to bench-press 400 pounds, but the average student at Nebraska cannot.

We could attempt to put together a representative sample by wandering around the Nebraska campus and carefully selecting students who appear to be "typical," but if we did, we would no doubt end up with a sample with far less variation than the population. We would probably exclude football players completely. We might not include enough women. We would avoid the very small and the very large. Our sample would probably be biased, because those we exclude for being "above average" and those we exclude for being "below average" are unlikely to balance one another out perfectly. Worst of all, these biases would depend, in unknowable ways, on our undoubtedly mistaken perception of the "typical" Nebraska student.

These subjective biases are reminiscent of the story of the Chinese researcher who wanted to estimate the height of the emperor. The researcher was not allowed to look at the emperor. Instead, he estimated the emperor's height by asking the opinion of many people, none of whom had ever seen the emperor. Each person had his or her own, uninformed opinion of how tall the emperor was. It is the same here. If we try to hand pick a representative sample, all we will get is data showing how strong we think Nebraska students are!

We might also be influenced by our preconceived notions of the results we hope to obtain. If we intend to show that Nebraska students are, on average, stronger than the students at another university, this intention may well influence the students we choose to interview.

Simple Random Samples and Randomized Experimental Design

To avoid being influenced by subjective biases, statisticians advise that, paradoxically, the researcher should not personally choose who is selected! A fair hand in a card game is not one in which the dealer turns the deck face up and carefully selects representative cards. A fair hand is whatever results from a blind deal from a well-shuffled deck. What card players call a fair deal, statisticians call a *random sample.*

> In a **simple random sample,** of size n from a given population, each member of the population is equally likely to be included in the sample, and every possible sample of size n from this population has an equal chance of being selected.

For a random sample of five cards, each of the 52 cards in the deck is equally likely to be included in the sample, and every possible five-card hand is equally likely to be dealt.

Some random sampling methods give each member of the population an equal chance of being included in the sample, but do not give every possible sample an equal chance of being selected. For example,

> In a **systematic random sample,** we randomly select one of the first k items in a population list and then select every kth item thereafter.

For instance, a group of students had a team project that involved asking a sample of the 1600 students at their college how many biological children they expected to have during their lifetimes. For a simple random sample of 64 students, they could have placed the 1600 names in a hat, mixed thoroughly, and then drawn out 64 names. Each student is equally likely to be included and all possible 64-person samples are equally likely to be selected. Instead, they obtained a systematic random sample from the alphabetical list of names in the student telephone directory. To obtain 64 names, they needed to select every 25th name, since $64(25) = 1600$. So, they wrote the numbers 1 to 25 on pieces of paper, placed these in a hat, and selected a number, which happened to be 5. The 5th name in the directory became the first selection for their sample. They then selected every 25th name after this: 30, 55, and so on. Each student was equally likely to be included in their sample, but not all possible 64-person samples were equally likely to be selected—there was no chance that students 5 and 6 would both be in their sample. Some of their results are in Exercises 2.2 and 8.31.

A systematic sample is often much easier to construct than a simple random sample, but we must be careful that the list and the sampling interval do not combine to create selection bias. If we are studying restaurant revenue and select every seventh day of the year, all the data will be for the same day of the week. If we are studying housing prices and select every fifth house on a street with five houses on every block, we will have either all corner houses or no corner houses.

With all random sampling methods, even simple random sampling, there is admittedly a possibility that a sample chosen by chance methods will, by the luck of the draw, be unrepresentative. The chance selection of college students might yield nothing but football players. But this slight risk is needed to avoid samples that only give us information about the researcher's opinions.

A subtle advantage of a random sample is that probabilities can be used to make inferences about the population; we will be able to make statements such as, "We are 95 percent confident that the average bench-press strength of the people in our sample will be within 20 pounds of the average bench-press strength of the entire population." If our sample is hand-picked, we have no way of gauging whether the sample provides useful information about the population.

A survey of students at the University of Nebraska involves observational data. The same benefits of random sampling accrue to the production of experimental data under controlled conditions. When the effect of lime on corn yields is investigated, there are pitfalls in hand-picking the parcels of farmland that will be used as the treatment plots and the control plots—pitfalls analogous to those that arise if we hand-pick the Nebraska students we will interview. We may omit land that does not seem "typical," thereby understating the variability in crop yields and biasing our results in unknowable ways. Or, hoping that the treatment will boost corn yields, we may put the best plots in the treatment group. To guard against these biases, we can choose our treatment and control groups randomly.

> A **completely randomized experimental design** assigns items to the treatment and control groups randomly.

We tap the power of random selection by using a simple random sample to gather observational data and by using a completely randomized experimental design to produce experimental data.

In addition to guarding against subjective biases, randomization protects us against the influence of extraneous factors by making it likely that our samples will be representative of the population. If we hand-pick students, and are misinformed about the number of women in the population, we might include far too few women in our sample, or we might include far too many. In a reasonably sized random sample, in contrast, the gender composition of our sample will most likely be close to the gender composition of the population. Similarly, in an experimental setting, the random assignment of items to the treatment and control groups makes it likely that extraneous influences will be present in roughly comparable amounts in each sample. If some pieces of land are more acidic than others, a random assignment that gives each parcel an equal chance of being put in the treatment and control groups will most likely give the two samples similar mixtures of more-acidic and less-acidic soil.

Stratified Sampling and Block Design

Sometimes we can explicitly control for factors that might otherwise influence our results by using modified methods based on subgroups: with observational data, *stratified sampling;* with experimental data, a *block design.* In either case, we try to control for extraneous possible influences by identifying subgroups that differ systematically. In place of a completely random sample, we use random sampling within each preselected subgroup.

Suppose, for instance, that we are surveying people in order to predict the outcome of a presidential election. In a simple random sample, we give each potential voter an equal chance of being included in our sample. However, we might suspect systematic gender differences; perhaps males are more likely to vote Republican than females are. If, by the luck of the draw, a simple random sample ends up with substantially more males in the sample than in the population, it might turn out that the population's preference for the Democratic candidate is masked by the confounding effect of gender: The sample has a preference for the Republican candidate because of the disproportionate number of males in the sample. To control for this gender difference, we can use a *stratified random sample* in which we first divide the population by gender—males in one stratum, females in the other—and then take random samples from each group. By ensuring that our sample has the same female/male proportion as does the population, we have effectively controlled for gender. This same principle can be applied whenever we can identify groups in the population that differ systematically with respect to the matter being studied.

When this principle is applied to experimental data, the subgroups are called *blocks,* rather than strata, but the logic is the same. For instance, suppose that we want to conduct an experiment to investigate the effect of aspirin on the incidence of heart attacks. In a completely randomized design, we randomly separate all the subjects into either the treatment group or the control group. However, we might suspect the presence of systematic gender differences; perhaps males have heart disease more often than females. If, by the luck of the draw, a completely randomized design ends up with substantially more males in the treatment group than in the control group, the benefits of aspirin might be masked by the confounding effect of gender: The positive effects of aspirin on those in the treatment group are offset by the disproportionate number of men in this group. To control for this gender difference, we can use a *randomized block design* in which the subjects are first divided by gender—males in one block, females in the other—and then, within each block, the subjects are randomly assigned to either a treatment group or a control group. By comparing the male treatment group with the male control group (and the female treatment group with the female control group), we have effectively controlled for gender. This principle can, of course, be applied to any factors that might systematically affect our experimental results.

Random Number Generators

How do we actually make random selections? To return to our study of the physical strength of Nebraska students, suppose that we are able to persuade the administration to have every first-year student tested. This will be a large sample, but not a random sample from the Nebraska student body—most students have no chance of being included in this sample. This exclusion causes an obvious selection bias because there are systematic physical differences between first-year students and more mature students. Similarly, graduating seniors, the students in an advanced mathematics class, or the females in a physical education class are not a random sample. In each case, there are easily identifiable reasons why the group is not representative of the student body as a whole.

Ideally, we would like to find a procedure that is equivalent to the following: Put each Nebraska student's name on a slip of paper, drop these slips into a box, mix thoroughly, and pick names out randomly, just as cards are dealt from a well-shuffled deck. Each student, whether weak, strong, or somewhere in between, then has an equal chance of inclusion in our sample. In practice, instead of putting pieces of paper into a box, random sampling is usually done through some sort of numerical identification combined with the random selection of numbers.

The first step is to number each member of the target population. In our study, we might use student identification numbers. In the laboratory, researchers simply assign numbers to each batch, beaker, or bunny. The second step is to select the numbers that will determine the items to be included in the sample. If we use each student whose identification number ends with the number 3, our sample will include about 10 percent of the student body, selected in a manner that is apparently unrelated to physical strength. If a smaller sample is desired, we could use students whose last two digits are, say, 47. If we do not want the responsibility of selecting the digits, we could randomly press the button on a stopwatch that keeps time in hundredths of a second. These crude methods are satisfactory as long as there is no systematic relationship between the selection method and what we are measuring.

Alternatively, we can use slightly more complicated procedures. Suppose that we have 64 patients that we want to divide into two groups, one group to receive a new medication and the other group to receive a placebo. For each patient, we could simply flip a coin, with heads being the new medication and tails the placebo. (A valid test does not require each group to have exactly the same number of patients.) Another procedure is to put the 64 names in alphabetical order, number them 01 to 64, and then select 32 two-digit numbers by repeatedly shuffling and selecting cards from two sets of 10 cards that have been numbered 0 to 9. (Repeated numbers or numbers outside the range of 01 to 64 are ignored.)

A statistician who does not want to fool around with stopwatches, coins, or cards can consult a table of random numbers constructed by someone else using a physical process in which each digit from 0 to 9 has a .1 chance of being selected. The most famous of these tables contains 1 million random digits obtained by the RAND Corporation from an electronic roulette wheel using random-frequency pulses.[20] Some of these digits (which I picked at random) can be found in Table 1 at the end of this book. These random numbers are arranged in 25 columns of 50 two-digit numbers, separated by a space for easier reading.

To use this table to select 32 of the 64 patients that we numbered 01 to 64, first we note that we are interested in two-digit numbers. Next, we need to find a place to begin in Table 1. We could begin at the beginning, with the very first number, 10. But statisticians do not like to be so predictable. The researcher will eventually memorize the numbers at the beginning of the table and thereby destroy their randomness. If I know that 10 is always the first number picked, then later, when I am diagnosing patients, I may remember that patient 10 always gets the medication and consequently I may be more inclined to see an improvement in patient 10's condition.

Statisticians advise randomly selecting a place to begin in a table of random numbers. I used the stopwatch function on my digital wristwatch. I started the stopwatch, stopped it after several seconds, and looked at the reading in the hundredths-of-a-second place; every two-digit number from 00 to 99 is equally likely to be selected by this procedure. First, I went for the row number, and my stopwatch gave 34. Then I went for the column number and got 60; because there are only 25 columns in Table 1, I tried again and got 16. The number in the 34th row and 16th column of Table 1 is 13.

Now I can quickly move up, down, sideways, or even diagonally to find additional numbers. I used my stopwatch again with 00 to 24 being north, 25 to 49 east, 50 to 74 south, and 75 to 99 west, and I got 28 (east). So my first four numbers are 13, 02, 12, and 48. The next two, 92 and 78, are ignored (as are any repetitions) since there are only 64 patients, and I move on to 56, 52, 01, and 06. Having hit the end of this row, I go down to the next row and continue: 46, 05, and so on. Thus the patients numbered 13, 02, 12, 48, 56, . . . receive the new medication.

Random numbers can also be obtained from computer software with built-in procedures for generating a sequence of digits. Sometimes, two large numbers are multiplied and then divided by a third large number. Other programs square a large number and then divide by the product of two prime numbers, using the remainder to repeat the process. In any case, some of the digits in the answer are used as random numbers. Modern computers can make such calculations extremely quickly. Karl Pearson once spent weeks flipping a coin 24,000 times. Now, with a computer program for generating random numbers, 24,000 coin flips can be simulated in a few moments.

Computerized random number generators produce what are called *pseudo-random numbers,* in that the numbers are produced by a deterministic process that gives exactly the same sequence of numbers every time. To randomize the process, we need to select an unpredictable starting position in this deterministic sequence of digits; one popular procedure is to use the value of the tick count in the computer's internal clock. This is the computer equivalent of using a stopwatch to find a starting point in the RAND table of random digits.

Table 4.2 shows 80 random numbers from 1 to 1600 selected by a computer program, Smith's Statistical Package. By entering 1600 and 80 in the boxes, the user specifies that a random sample of 80 numbers will be drawn from the 1600 numbers 1 through 1600. This program then shows the selected numbers, in the order they were selected and in numerical order.

These were the very numbers used by a student doing a term paper on class sizes at her college. There were a total of 1600 students at her college, and their names were listed alphabetically in the student telephone directory. She used the 80 random numbers in Table 4.2 to select a simple random sample of 80 students: the seventh name, twenty-third name, and so on. By making repeated telephone calls, she was able to contact all of the selected students. Some of her results are in Exercise 3.50.

Dr. Spock's Overlooked Women

In 1969, Dr. Benjamin Spock, author of a best-selling book on child care and a vocal opponent of the Vietnam War, was tried in a Massachusetts federal court for conspiracy to violate the Military Service Act of 1967.[21] The court clerk

Table 4.2

Selection of 80 random numbers from 1 to 1600, using Smith's Statistical Package

```
The computer will select a random sample of size k from
a population of size n by selecting k different random
numbers in the range 1 to n.

How many numbers in the population
(1 - 10,000)?                                  n = [1600]

How many numbers in the random sample
(1 - 1000)?                                    k = [  80]

80 random numbers in the range 1 to 1600 in their orig-
inal order:

1295   935   499  1577  1247  1106   802   678  1133   206
 364  1128   538   208  1469    92   927  1136    93  1502
 384   344   736  1438   212   304    34   856   723  1283
 801   300  1346   718    82   461  1586  1311   951   314
 629    23  1457   648   427   596  1568  1501   146   594
 545   430   688  1516  1065  1340  1462  1175    98  1054
 301  1517   142   357   808   574   967  1563   819   252
1378   871   391     7  1281   624    69   509   257   533

80 random numbers in the range 1 to 1600 in numerical
order:

   7    23    34    69    82    92    93    98   142   146
 206   208   212   252   257   300   301   304   314   344
 357   364   384   391   427   430   461   499   509   533
 538   545   574   594   596   624   629   648   678   688
 718   723   736   801   802   808   819   856   871   927
 935   951   967  1054  1065  1106  1128  1133  1136  1175
1247  1281  1283  1295  1311  1340  1346  1378  1438  1457
1462  1469  1501  1502  1516  1517  1563  1568  1577  1587
```

testified that he selected potential jurors by putting his finger randomly on a list of adult residents of the district. Those he selected were sent questionnaires, and after those disqualified by statute were eliminated, batches of 300 were called to a central jury box, from which a panel of 100 (called a *venire*) was selected. As it turned out, even though more than one-half of the residents in the district were female, only 9 of the 100 on Dr. Spock's venire were. Spock was convicted, but appealed.

The jury selection process may have been biased. A finger placed blindly on a list will almost certainly be close to several names, and the court clerk may have favored males when deciding which names to use. The court records do not reveal how the panel of 100 was selected from the central jury box, but this, too, may have been biased by subjective factors.

Several months after the end of the trial, the defense obtained data showing that out of 598 jurors used recently by Dr. Spock's trial judge, only 87 (14.6 percent) were female, while 29 percent of the 2378 jurors used by the other six judges in the district were female. To suggest how this bias may have prejudiced Spock's chances of acquittal, the appeal cited a 1968 Gallup poll in which 50 percent of the males labeled themselves as hawks and 33 percent as doves on Vietnam, compared to 32 percent hawks and 49 percent doves among females. It is also conceivable that women who raised their children "according to Dr. Spock" might have been sympathetic to his antiwar activities. The appeal argued that the probability that a randomly selected jury would be so disproportionately male was small and that, "The conclusion, therefore, is virtually inescapable that the clerk must have drawn the venires for the trial judge from the central jury box in a fashion that somehow systematically reduced the proportion of women jurors." (Techniques beyond the scope of this book show that the probability of 87 or fewer females works out to be about 1 in 28 trillion.)

Spock was eventually acquitted, although on First Amendment grounds rather than because of a flawed jury selection. However, a law passed in 1969 mandated the random selection of juries in federal courts by using statistically accepted techniques, such as random number generators. Subsequently, males and females have been equally represented on venires in federal courts.

What Makes a Book a Best-seller?

The Sunday Book Review section of *The New York Times* contains widely followed lists of best-selling fiction and nonfiction books.[22] Many people consult these lists before they buy books, reasoning that they should read (or at least buy) what other people are reading. (The editor of the *New Republic* once put slips of paper into the middle pages of dozens of best-selling books in Washington, D.C., bookstores, offering a $5 reward for telephoning him; no one called.) Bookstores display *The New York Times* best-seller list prominently; some even have a special section of the store reserved for these best-selling books. Thus success breeds success, in that a book that makes the *Times* best-seller list will sell many more copies, while a book that does not make the list is apt to vanish without a trace.

How does *The New York Times* compile this influential list of best-selling books? It cannot rely on publishers to provide sales data because, recognizing the power of the list, publishers would exaggerate the sales of books they wanted to promote. Instead, the *Times* surveys bookstores—a reasonable course, but one that for decades it pursued in a fairly haphazard manner. Each week,

Times staffers telephoned several bookstore managers and asked them to name, in no particular order, books that were selling well. The managers' answers might be based on actual sales data or on memory, perhaps influenced by a recollection of the titles on the current *Times* best-seller list.

The *Times* now collects actual sales figures, but its data are still flawed. It does not include book clubs and specialty bookstores (which sell lots of Christian, children's, and computer books). They do include sales *to* newsstands, supermarkets, and price clubs—even though these books will be returned if they are not sold.

The *Times'* choice of bookstores provided plenty of opportunities for selection bias and for manipulation. A biography of former Alabama football coach Bear Bryant sold more copies than many books on the *Times* best-seller list, but never made the list because *The New York Times* did not survey enough—if any—bookstores in southern states. A recent biography of Dave Dravecky, a professional baseball pitcher who lost an arm to cancer, sold enough copies to make the *Times'* list, but did not because most sales were through Christian bookstores.

When a novel written by Chuck Barris, a television personality, did not make the best-seller list, he personally purchased thousands of copies from a few bookstores that he knew were surveyed by the *Times,* enough copies to put his book on the list and to generate many more sales to people whose book purchases are influenced by the best-seller list. In 1990 *Barrons'* Alan Abelson reported a similar trick:

> And for those who carp that his nation has lost its fabled business creativity, we ask them to consider the case of Allen Neuharth. Mr. Neuharth is the midwife—or is the proper term midhusband?—of *USA Today,* a novel daily which has demonstrated that with the right graphics a middling high school sophomore newspaper can attain national circulation. Refusing to rest on this laurel, Mr. Neuharth triumphantly departed journalism to take up the autobiographic pen and soon produced an epic entitled *Confessions of an S.O.B.* This sparkling work got the nod from the critics and the public, but alas nod was soon followed by snore. Nothing daunted, Mr. Neuharth devised a way to overcome such shameful apathy toward his literary gem. The foundation he heads bought 2,000 copies in strategically located retail bookstores, thereby boosting *Confessions* from the back of the bin onto the best-seller list.[23]

Randomly Shuffling a Deck of Cards

The construction of a random sample is often compared to the dealing of cards from a well-shuffled deck. Ironically, the randomization of a deck of cards occurs only when a person's physical imperfections cause uncontrollable variations in the way the cards are shuffled. Persi Diaconis, a Harvard statistician and former professional magician, has shown—in both theory and practice—that if a deck of cards is divided into two equal halves and the cards are shuffled perfectly, alternating one card from each half, the deck returns to its original order after eight perfect shuffles. It is our imperfect shuffles that cause the deck to depart in unpredictable ways from its original order. Diaconis showed—again in theory and in practice—that seven shuffles are generally needed to randomize a deck of cards.[24] Six shuffles are usually not adequate, and more than seven shuffles have only a minor effect on the randomness of the deck.

Most card players shuffle only three or four times, which is not sufficient to randomize a deck of cards. In bridge, for example, the cards are played in groups of four, usually four cards of the same suit. If the cards are shuffled insufficiently, cards of the same suit tend to cluster together in the deck and when the cards are then dealt, one after another, to the four players, the players tend to get evenly balanced hands, for example, four cards of one suit, and three cards in each of the other suits. When computerized dealing was introduced at bridge tournaments, players who were accustomed to playing with inadequately shuffled cards complained that the computer program was flawed because it dealt too many hands with six, seven, or eight cards of one suit. In fact, it was the inadequate human shuffling that was flawed.

exercises

4.13 Critically evaluate the following passage:

> [It] is the farmer, not the importuning salesman, who leads in the consumption of alcohol. Results of a survey of drinkers classed by occupation, which was published last week by the Keeley Institute of Dwight, Illinois . . . show that of 13,471 patients treated in this well-known rehabilitation center from 1930 through 1948, a total of 1,553 (11.5%) were farmers. Next in line came salesmen, merchants, mechanics, clerks, lawyers, foremen and managers, railroad men, doctors and manufacturers.
>
> Since Keeley is located in the heart of the farm belt, it might be expected that its proportion of farmer patients would be unusually large. This is not the case, according to James H. Oughton, institute director. The patients in the survey came from all over the world.[25]

4.14 The Educational Testing Service (ETS) wanted a representative sample of college students, so it first divided all schools into groups, such as large public universities, small private colleges, and so on.[26] Then ETS chose a "representative" school from each group. Finally, ETS wrote to the administrators of these selected schools and asked them to choose some students for the study. In what ways might this sample not be representative of the college student population?

4.15 In 1952, the Chicago police department banned the showing of the Italian film "The Miracle." The American Civil Liberties Union (ACLU) subsequently showed this film at several of its private meetings and reported that

> Of those filling out questionnaires after seeing the film, less than 1 percent felt it should be banned. "It thus seems," said Stanford I. Wolff, Chairman of the Chicago Division's Censorship Committee, and Edward H. Meyer, the Chicago ACLU's Executive Director, "that the five members of the [police] Censorship Board do not represent the thinking of the majority of Chicago citizens."[27]

Why is the ACLU's poll not convincing statistical evidence of its conclusion?

4.16 A student tried to see if there is a relationship between a student's social life and academic major. Since all the telephones at his college are hall phones, he did not call individual students but, instead, randomly selected the numbers of the hall phones and surveyed the persons who answered. Why might his results be biased?

4.17 A researcher reported that the number of diet-related articles in the popular press totaled 60 during the entire year 1979, but totaled 50 in the single month of January 1989.[28] As a statistician, what other information would you want to see before concluding that the popular press was much more concerned about dieting in 1989 than in 1979? If you were going to research this topic, what data would you collect?

4.18 Explain why you think that high school seniors who take the Scholastic Aptitude Test (SAT) are not a random sample of all high school seniors. If we were to compare the 50 states, do you think that a state's average SAT score tends to increase or decrease as the fraction of the state's seniors who take the SAT increases?

4.4 Opinion Polls

Public opinion polls present an especially interesting challenge. In presidential elections, for instance, the winner is not determined until tens of millions of votes are cast and counted, but statisticians can use a sample of a few thousand voters to predict the outcome with remarkable accuracy. To make reliable predictions, pollsters need a random sample of the electorate to provide honest answers to fair questions. When they get something else, poll results can be very misleading. One of the most famous examples is the 1936 *Literary Digest* fiasco.

The 1936 *Literary Digest* Poll

In 1932, the Great Depression drove Herbert Hoover out of the White House and brought in Franklin Roosevelt. Roosevelt's efforts to end the depression were largely unsuccessful, but he retained the public's faith that he was trying hard and doing as well as any president could. In 1936, he ran for reelection against Alf Landon, the uncharismatic governor of Kansas. Most veteran political observers thought that Roosevelt would be reelected easily. But the *Literary Digest* attracted considerable attention with its prediction of a landslide victory for Landon, 57 percent to 43 percent. The *Digest* polled 2.4 million people, the largest political poll ever conducted, and claimed that its prediction would be "within a fraction of 1 percent" of the actual vote. The *Digest* had successfully forecast the five previous presidential elections, but this time it was woefully wrong. It was Roosevelt who won the landslide, 62 percent to 38 percent, and with its reputation permanently tarnished, the *Literary Digest* soon folded.

The *Digest* poll was incredibly ambitious. The company mailed questionnaires to 10 million people, a full one-quarter of the voting population. However, in sampling, quality is more important than quantity. Two thousand randomly selected voters are much better than 10 million unrepresentative voters. The *Literary Digest* got the bulk of its names from an examination of every single telephone book in the United States. Nowadays, this sort of compilation would encompass the vast majority of households—everyone but those with unlisted numbers or new numbers and those few households without telephones. Many present-day pollsters conduct polls successfully by having interviewers dial randomly selected telephone numbers.

In 1936, however, telephones were far from universal. Phone service was still relatively new, and the depression had made the telephone a luxury for many households. There were only 11 million residential phones in 1936, and these homes were disproportionately well-to-do and in favor of the Republican candidate Landon. Those people without phones were generally poor and overwhelmingly for Roosevelt. The telephone books in 1936 yielded a biased sample that was unrepresentative of the voter population. The *Literary Digest* also used some names and addresses collected from car registrations, club memberships, and its own subscriber lists, but these sources were even more biased than the telephone directories.

To get a random sample, the *Literary Digest* should have focused on lists of registered voters. The *Digest* was actually one of the first organizations to use voter registration lists, but it used only a few lists and these were swamped by the names from the telephone directories and other unrepresentative sources, giving its 1936 poll a great deal of selection bias.

Nonresponse Bias

The *Literary Digest* poll had another problem that is commonplace among polls conducted by mail. Most people do not spend time carefully reading and responding to junk mail and public opinion surveys. Only those who care a great deal about an issue or an election bother to read a poll's instructions, fill out the form, and mail it in. Those who do take the time and trouble may be atypical of the population as a whole. The systematic refusal of some groups to respond to a poll is called **nonresponse bias.**

For example, the *Literary Digest* mailed voting questionnaires to one-third of the registered voters in Chicago, but only 20 percent of these people bothered to fill out the questionnaires and mail them back to the *Digest.* Landon was favored by more than one-half of the people who responded to this mail poll, but in the election Roosevelt received two-thirds of the Chicago votes. The *Literary Digest* poll apparently suffered from nonresponse bias as well as selection bias.[29]

Statisticians have discovered that one pattern in nonresponses is that mail questionnaires tend to be filled out by middle-income households and discarded by lower-income and upper-income households. Perhaps lower-income households do not trust polls, and upper-income households put a high value on their time. In any case, if a statistician wants all income groups to be fairly represented, a mailed questionnaire is a dubious procedure.

Studies have shown that response rates can be raised by using multiple contacts (including pre-survey notification and post-survey reminders), special postage (for example, certified mail or two-day priority mail), and token financial incentives (the inclusion of a small amount of money with the survey).[30] Although the financial payment is small, it seems to signal the importance of the survey and make recipients feel obligated to respond. Lengthy polls that take hours to complete are naturally the least likely to be returned. My wife and I were once asked to participate in a study of working couples, in which each of us was supposed to fill out a separate 17-page survey with 233 detailed questions. I imagine that few working couples have the time, energy, or patience for such surveys. We did not.

Nonresponses can also be a problem for telephone and in-person surveys. It has been estimated that 30 to 50 percent of those contacted by telephone for preelection surveys refuse to participate.[31] The nonresponse rate, which includes both those who could not be contacted and those who refuse to answer, is even higher. Two of the most respected academic surveys, the General Election Studies and the General Social Survey, try very hard to reduce the nonresponse rate by making multiple attempts to find the elusive and to persuade the reluctant to participate; yet even they have nonresponse rates of 25 to 30 percent. A careful study of the socioeconomic characteristics of those who participate with the population as a whole concluded, "Who is overrepresented in academic surveys? The elderly, blacks, women, the poor, the less-educated. Who is missing? Men, young people, whites, the wealthy."[32]

Because nonrespondents do not participate in surveys, there is no way of knowing for certain if their views differ systematically from those who are willing to be surveyed. However, one distinguishing feature between respondents and nonrespondents is surely that the respondents feel strongly enough about the questions asked to take the time to complete the poll. Who is going to bother to mail in an answer of "I don't know," "No opinion," or "I haven't made up my mind"? It follows that those who do respond will be more sharply divided than is the population as a whole. Suppose that questionnaires are mailed to 1000 people, of whom 200 strongly favor a certain policy, 200 strongly oppose the policy, and 600 are largely indifferent. If the responses come from 100 of those in favor and 100 of those opposed, this survey correctly shows that opinion is evenly divided, but it completely misses the important fact that most people either have not made up their minds or do not really care.

Now let us change the numbers slightly. It is 18 months before a presidential election, and four candidates are beginning their long run for the Democratic nomination. One thousand questionnaires are mailed to 200 people who favor candidate A, 100 who favor B, 50 who favor C, 50 who favor D, and 600 people who do not know any of the candidates, let alone which one they favor. Responses might be from received from 100 people who favor A, 50 who favor B, 25 for C, and 25 for D. This poll shows that A is favored by a 2-to-1 margin over B and that C and D have little support. The fragility of these numbers is completely hidden. The mail poll does not tell us that 60 percent of the people have failed to make up their minds, because these people did not bother to respond. When the undecided do make up their minds, any of these candidates could turn out to be the people's choice.

Let us change the numbers again. Some time has passed, and now there are just two candidates, A and B, left in the race for the Democratic nomination. One thousand questionnaires are mailed out to 200 people who strongly favor A, to 400 people who would vote for B but do not feel very strongly about it, and to 400 people who still have not made up their minds. Perhaps 150 of the A supporters promptly mail in their choice, while only 50 of the B supporters bother to respond. Even though B is favored 2 to 1 by those who have made up their minds, the mail poll shows a 3-to-1 lead for A. Again, the mail poll is distorted by nonresponse bias.

In each case, we cannot tell how many people failed to make up their minds, because the undecided do not bother to express this indecision. In addition, those with strong opinions are more likely to respond than those with more balanced opinions. Unfortunately, the nature of U.S. politics is that candidates who do not do well in the early polls do not get the financial support they need to run a competitive campaign. Thus, special-interest candidates who are strongly favored by a small number of voters have a big advantage over candidates whose appeal is more broadly based, but not as fervent. Mail polls exaggerate this distortion, and this is one reason that impartial political pollsters no longer use mail questionnaires.

Polls that mask the intensity of feelings can be misleading, too. In Massachusetts, for example, pollsters reported that a majority of voters approved of mandatory busing to help integrate the schools, but these polls did not gauge the strength of these feelings. Most of those who favored busing did not feel very strongly about it, certainly not enough for this to be a decisive issue in an election campaign. But those who were opposed to busing were very opposed, enough that they would automatically vote against any candidate favoring busing. Politicians who looked at the polls and thought that they could win votes by endorsing busing soon learned otherwise.

A mail poll is naturally more suspect the fewer the number of people who bother to respond. In the 1940s, the makers of Ipana Tooth Paste boasted that a national survey had found that "Twice as many dentists personally use Ipana Tooth Paste as any other dentifrice preparation. In a recent nationwide survey, more dentists said they recommended Ipana for their patients' daily use than the next two dentifrices combined." The Federal Trade Commission banned this advertisement after it learned that less than 1 percent of the dentists surveyed had named the brand of toothpaste they used and that even fewer had named a brand recommended for their patients.[33]

An even more extreme example occurred in 1983, when the *Winnipeg Sun* asked its readers to vote on whether the city should build a new hockey arena for the Winnipeg Jets.[34] A total of 330 readers mailed in ballots clipped from the newspaper, with 280 in favor of the arena and 150 against it. Despite the very small returns, the *Sun* ran a headline proclaiming, "Winnipeg Wants a Downtown Arena." A year later, a rival newspaper reported that employees of the Winnipeg Jets had mailed in 200 of the 280 yes ballots in a carefully orchestrated plan: buying newspapers in small batches, filling out the ballots with different-color pens and in different handwriting, and mailing the ballots in envelopes of different sizes, with stamps of different denominations, from neighborhood mailboxes throughout the city. Very few people cared enough to share their opinions with the *Sun,* and those who did were not a random sample.

Quota Sampling

The example of a presidential election poll can be used to illustrate other general principles. Ideally, we would like to put every voter's name into a very large box, mix the pieces of paper thoroughly, and then pull out the names for our sample. In practice, we do not know the names of all registered voters, and many registered voters will not vote on election day. The appropriate population is those who will vote, not those who could vote, and there is no way to identify these voters with certainty ahead of time. Furthermore, the names drawn in a genuine random sample will be scattered all across the nation, some in very remote spots. Does it make sense to spend 2 days of a pollster's time finding a recluse in Solitary, North Dakota? Interviewers will spend more time traveling than interviewing. In addition, some people will have moved, and many will not be home when the pollster knocks on the door.

Because of these problems, most pollsters do not attempt to generate a true random sample in which every member of the population is listed and has an equal chance of being selected. One alternative is *quota sampling,* in which a sample is constructed by filling quotas of certain characteristics that are thought to be true of the population as a whole. Based on surveys of past elections, pollsters can estimate the fraction of voters who are in various age, sex, geographic, and economic categories. Interviewers can then poll people who are in these same categories. For example, in the 1948 presidential election, one interviewer for the Gallup poll was told to interview 15 people in St. Louis, of whom seven were to be male and eight female.[35] In addition, the interviewer was told that nine of the people polled should live downtown and six should live in the suburbs. Six of the men were to be white and one black. Three of the men were to be under 40 years of age, and four were to be over 40. Two of the white men should pay less than $18 a month for rent, three should pay between $18 and $44, and one should pay more than $44. There were analogous quotas for the eight women. Beyond these quotas and the requisite mental gymnastics, the interviewer was free to select the persons to be polled.

Quota sampling avoids the drudgery of tracking down a random sample, but gives the interviewer too much discretion. One mechanical problem is the verification of the categories: Where do you draw the line between city and suburb? And people may misstate (or decline to state) their age and rent. To play it safe, the interviewer may avoid people near the dividing points. If you poll only people who are under 25 or over 55 years old, the subjects will definitely be under or over age 40, as required. But playing it safe means that people near 40 years of age will be slighted in the survey, and they may well have different opinions from those under age 25 or over age 55.

In addition, most interviewers will meet quotas in the easiest possible ways, which means that people in risky, unfamiliar, or inconvenient places will be ignored. Interviewers do not want to waste their time or risk their lives. If it is easiest for the pollster to interview people during the day, those with jobs during the day will be slighted. If door-to-door interviews in the evening are easiest for the interviewer, those who go out at night will be missed.

A student who takes a part-time polling job may interview a lot of other students and maybe some professors. Someone who works for a large company will probably interview a lot of coworkers. Some interviewers will misstate ages, addresses, and so on, because that is a lot easier than finding people who fit perfectly into rigid categories. Some will fill out the questionnaires themselves to avoid venturing into unfamiliar parts of town. Even those conscientious interviewers who try to find people to fill every category perfectly will be hesitant to interview those who appear to be disreputable and possibly dangerous. In theory, quotas ensure a more balanced sample; in practice, quota sampling tends to be convenience sampling.

After the 1936 *Literary Digest* fiasco, political pollsters turned from mail questionnaires to quota sampling as an inexpensive way of obtaining representative samples. For the reasons just discussed, quota sampling is likely to be biased, too, although perhaps not as severely as mail polls. Given the freedom to select their poll subjects, interviewers tend to select people who dress well, live in nice neighborhoods, and are disproportionately Republican.

The Gallup poll, for example, was just starting at the time of the 1936 presidential election and, using quota sampling, did better than the *Literary Digest* poll. Still, the Gallup poll overestimated the vote that the Republican candidate would receive in each of the first three elections it covered:

	GALLUP PREDICTION OF REPUBLICAN PERCENTAGE	ACTUAL REPUBLICAN PERCENTAGE	PREDICTION MINUS ACTUAL
1936	44	38	+6
1940	48	45	+3
1944	48	46	+2

The selection bias in quota sampling caused Gallup to overestimate the vote for the Republican candidate; luckily, in each case, the bias was not large enough to cause an erroneous prediction of Republican victory. In the next election, though, the vote was close, and Gallup was one of several polling organizations that incorrectly predicted that the Republican candidate would win.

Franklin Roosevelt died in 1945, and the presidency passed to his vice-president, Harry Truman, a scrappy, but largely unknown, small-time politician from Kansas City. In 1948, Truman had to seek election on his own against Thomas Dewey, who had achieved national fame as a crime-fighting New York City district attorney and then gained prominence as the governor of New York. Gallup predicted that Dewey would get 50 percent of the vote and Truman only 44 percent, with 6 percent going to two minor-party candidates. The other major pollsters, using quota sampling, too, agreed that Dewey would win easily. The October issue of *Fortune* stated:

> Barring a major political miracle, Governor Thomas E. Dewey will be elected the thirty-fourth president of the United States in November. So decisive are the figures given here this month that *Fortune,* and Mr.

> Roper, plan no further detailed reports. . . . Dewey will win a popular majority only slightly less than that accorded Mr. Roosevelt in 1936 when he swept by the boards against Alf Landon.[36]

The editors of the *Chicago Tribune* were so certain of the outcome that they printed up the Dewey victory headlines before the votes were counted. But there was a Republican bias in the polls' quota sampling. Truman got 50 percent of the vote and Dewey only 45 percent. Truman went to bed thinking that he had lost and woke up learning that he had won. He triumphantly posed for photographs, waving a newspaper that had prematurely reported Dewey's victory. Truman went on to become a successful president, while the pollsters looked for an alternative to quota sampling.

Cluster Sampling

The mistaken pollsters were an easy target. One often repeated quip was that, "Everyone believes in public opinion polls—from the man in the street right on up to President Dewey." Even the former editor of the *Literary Digest* managed a few digs: "Nothing malicious, mind you, but I get a very good chuckle out of this."[37] The embarrassed political pollsters concluded that they had made two big mistakes. The first was that they had stopped polling too early, as a lot of people did not decide to vote for Truman until the last 2 weeks before the election. The second was that their samples had not been representative of the voter population.

They wanted to find a way to make their surveys more like genuine random samples, without incurring the staggering expenses associated with a true random sample. What they settled on is a sequential sampling procedure called *cluster sampling,* in which they first sample a map and then sample people. The nation is divided into geographic regions and then into states, counties, and cities of different sizes, with a random number generator used to pick specific cities (with each city's chance of being selected proportionate to its size). In each of the selected cities, a random number generator is used to choose voting precincts. Using voter registration lists, a random number generator chooses households that will be interviewed. (In public opinion polls that are not intended to be restricted to registered voters, interviewers walk down specified streets, knocking on the doors of occupied housing units.) To maintain randomness, interviewers are told to interview a specific person, not necessarily the person who answers the door. Otherwise, a systematic bias may be introduced by the fact that age, gender, and work habits may be related to both who opens the door and the person's political preferences.

The sequential nature of the sampling process makes it unnecessary to list every person in the country. Yet the random selection at each stage means that, in theory, every person has an equal chance of being included in the sample.

As in a pure random sample, inconvenient, out-of-the-way locations may well be chosen. Yet clustering makes it more economical to sample out-of-the-way places because several persons are interviewed within each city. Cluster sampling is more convenient for interviewers, without giving them the discretion to choose their own sample.

Telephone interviews have become increasingly common because it is much more economical for pollsters to "let their fingers do the walking." A typical cluster design involves dividing the country into geographic areas and, within each area, making random selections from listed telephone numbers. The last two digits of each selected number are replaced by two digits chosen by a computerized random number generator, so that unlisted numbers may be dialed.

The final results are adjusted in various ways to offset biases that might be introduced by cluster sampling, for instance, that some homes have more than one telephone line and are consequently more likely to be chosen, or that some demographic groups are more likely to be home when the interviewer knocks on the door or calls on the telephone. Thus, pollsters' practices are a common-sense blend of quota and random sampling, a hybrid with most of the economic advantages of quota sampling and most of the power of random sampling. It is a logical and economically feasible approximation of a true random sample. And it works. Table 4.3 shows Gallup's accuracy in predicting the outcomes of presidential elections since 1948.

The Republican bias has been eliminated by cluster sampling, in that the Republican vote is now underestimated as often as it is overestimated. In

Table 4.3

Gallup's presidential election forecasts, 1952–1996

	GALLUP PREDICTION OF REPUBLICAN PERCENTAGE	ACTUAL REPUBLICAN PERCENTAGE	PREDICTION MINUS ACTUAL
1952	51	55.4	−4.4
1956	59.5	57.8	+1.7
1960	49	49.9	−.9
1964	36	38.7	−2.7
1968	43	43.5	−.5
1972	62	61.8	+.2
1976	49	47.9	+1.1
1980	47	50.8	−3.8
1984	59	59.1	−.1
1988	56	53.9	+2.1
1992	37	38.1	−1.1
1996	41	41.4	−.4

addition, the size of the prediction error has dropped dramatically. With quota sampling, Gallup's average prediction error in the 1936, 1940, 1944, and 1948 elections was 4 percent. With the use of cluster sampling, the average prediction error has been less than 2 percent. The largest error in the last 40 years, 3.8 percent in 1980, was less than the average error with quota sampling. (A *New York Times*-CBS postelection poll found that 20 percent of the voters changed their minds during the last 4 days of the 1980 campaign and 60 percent of this group decided not to vote for the Democrat, Jimmy Carter.)

Gallup and other political pollsters have become more accurate while sharply reducing the size of their samples. Gallup's embarrassing 1948 prediction was based on a poll of 50,000 registered voters. Nowadays, Gallup obtains more accurate predictions from interviews with approximately 2000 people. To be reliable, a sample needs to be random much more than it needs to be large.

Getting Honest Answers

Some pitfalls in public opinion polling cannot be avoided just by sampling methods. Many registered voters do not vote, and the pollsters' objective is to predict the preferences of those who vote, not the preferences of those who could vote. Good pollsters try to identify nonvoters by asking such questions as "Did you vote in the last election?" and "Do you feel strongly about the upcoming election?"

Another response problem is that some people in a sample may not be home at the right time, even if the interviewer tries again and again. And even if people are home, they may have things to do that seem more important than answering nosy questions. Pollsters often piggyback surveys, having their interviewers ask hundreds of questions for many different clients. After an hour or two of this grilling, many bored or irritated subjects will understandably become more careless in their answers or simply stop answering additional questions.

One reason that some people refuse to participate in telephone surveys is that they have learned from experience that telemarketers often try to disguise a sales pitch as a public opinion poll. Legitimate market researchers call this deception *selling under the guise* ("sugging," pronounced "shugging"). I experienced an example in the spring of 1991, when I received a bulk-rate letter from what purported to be a marketing firm located in Pennsylvania. The letter began,

> We are conducting a research study on radio listening preferences in the Los Angeles area, and we are requesting your participation. Your opinions will help shape the kind of music presented to the Los Angeles audience.
>
> You have been carefully selected to represent a specific demographic segment of the general population. In order to obtain better information, each participant has been assigned a specific radio station. We would

> greatly appreciate your cooperation by listening to your assigned station for at least one hour and completing the brief survey card enclosed. The station you have been assigned is . . .

Three of these letters were mailed to my home, and not so coincidentally, all three asked me to listen to the same radio station.

Another problem is that, being human, some interviewers may be a bit lazy or even dishonest. Two weeks before the 1968 presidential election, *U.S. News & World Report* reported that a poll by a Wisconsin editor had found that George Wallace was favored over Richard Nixon by a 5-to-3 margin.[38] The Wisconsin editor conducted this poll by talking to eight men in a bar! It is easier to give a questionnaire to your friends or fill it out yourself than to track down the people who are supposed to be interviewed, particularly if they live in an unfamiliar or seemingly dangerous neighborhood. Good pollsters discourage such dishonesty by making random follow-up calls to see if the people who were supposed to be interviewed really did get interviewed.

Another problem is that the answers may not be candid. Many of us suppress feelings that might be provocative or embarrassing and, instead, say what we think people want us to say. Langbourne Rust, a prominent market researcher, once asked a group of second-graders to tell him what they thought about a certain brand of toothpaste. They said that they had used it and liked it a lot because it tasted good and looked good. There was, in fact, no such toothpaste. Rust had asked the students about a nonexistent brand in order to demonstrate that, from an early age, people tend to tell interviewers what they think the pollsters want to hear.

A political pollster once asked people what they thought of the Pepper-Johnson Bill being debated by Congress and found that most people expressed an opinion, for or against, even though there was no such bill. Evidently, expressing an uninformed opinion was less embarrassing than admitting that they had never heard of this bill. Faulty memories can also cloud answers. A telephone survey the day after the 1993 Super Bowl found that 6 percent of those polled recalled seeing advertisements for AT&T during the game, while only 2 percent recalled seeing advertisements for Gillette.[39] Ironically, Gillette had spent $3.4 million on four 30-second commercials that introduced a new line of men's toiletries, while AT&T had not advertised at all!

A related postelection phenomenon is that more people claim they voted for an election winner than actually did. An extreme example is John F. Kennedy, who received barely one-half of the votes cast in 1960. After his assassination and elevation to national hero, pollsters found it almost impossible to find anyone who would admit to voting against Kennedy in 1960. Kennedy's halo effect is undoubtedly exaggerated by the fact that his 1960 opponent was Richard Nixon, who was later disgraced. Nowadays, most people say they did not vote for Nixon in 1972 either, even though he won that election by a landslide over George McGovern.

People are also apt to misstate their feelings when asked controversial questions. For instance, the National Opinion Research Center used two teams of interviewers, one white and one black, to poll 500 southern blacks during World War II.[40] One of the questions asked was whether blacks would be treated better or worse if the Japanese conquered the United States. The black interviewers reported that 9 percent of those polled thought blacks would be treated better and 25 percent thought blacks would be treated worse. The white interviewers, in contrast, found that only 3 percent thought blacks would be treated better, while 45 percent thought blacks would be treated worse. Blacks were evidently less candid with white interviewers, or else white interviewers were more reluctant to poll blacks who did not look friendly.

In the 1990 Nicaraguan presidential election, Daniel Ortega of the ruling Sandinista National Liberation Front ran against an outsider, Violeta Barrios de Chamorro.[41] Three independent opinion polls conducted during the month preceding the election showed Ortega winning a landslide victory over Chamorro by 16 to 24 percentage points. Two days before the election, Ortega told a reporter that defeat was "inconceivable." Yet, Chamorro won the election with a decisive 55 percent of the vote. One pollster later said that voter surveys had underestimated the citizens' fears of the ruling Sandinistas. Many voters apparently thought that the people conducting the polls were Sandinistas. They publicly voiced support for Ortega but, in the privacy of their secret ballots, voted against him.

Asking the Right Question

Sometimes, it is the wording of the question, rather than the demeanor of the questioner, that biases the answer. For example, in January 1994, the U.S. Bureau of Labor Statistics changed the wording of its monthly unemployment survey. The most important change was in the wording of the very first question, which had three versions, depending on the age and gender of the person being interviewed. If an adult male answered the door, he was asked "What were you doing most last week, working or something else?" If an adult female answered the door, she was asked, "What were you doing most last week, keeping house or something else?" Children of either gender were asked, "What were you doing most last week, going to school or something else?"

The new questionnaire continues by asking whether anyone in the household owns a business or a farm. If so, the person is asked, "Last week, did you do any work for pay or profit?" If not, the person is asked, "Last week, did you do any work for pay?" In trial tests during 1993, 60,000 households were surveyed using the old questions, and an additional 12,000 households were interviewed using the new questionnaire. During these trials, the estimated unemployment rate averaged 7.1 percent with the old questions and 7.6 percent with the new questions. The difference was particularly pronounced among adult women, whose unemployment rate averaged .8 percentage point higher by using the new wording.

Sometimes, seemingly small differences in wording can have a big effect on the answers. When college students who had watched a film of an automobile accident were asked, "About how fast were the cars going when they contacted each other?" The average answer was 31.8 miles per hour. When another group of college students was shown the same film and asked, "About how fast were the cars going when they collided with each other?" the average answer was 40.8 miles per hour.[42] Wording can also be so confusing that we do not know what to make of the results. In 1994, the Roper polling organization asked this question: "Does it seem possible or does it seem impossible to you that the Nazi extermination of the Jews never happened?"[43] To say that you believe the Holocaust happened, you would have to give a double-negative answer: It is impossible that it never happened.

Sometimes, the wording is deliberately chosen to influence the answers. A 1964 issue of *Fact* magazine had this attention-grabbing headline: "1,189 Psychologists Say Goldwater Psychologically Unfit to Be President." The story was based on a poll of 12,000 doctors, asking such loaded questions as

- Does he seem prone to aggressive behavior and destructiveness?
- Can you offer any explanation for his public temper tantrums?
- Do you believe that Goldwater is psychologically fit? No or yes?

About 20 percent of the doctors responded, with the provocative headline based on the fact that about one-half of these answered no to the vague question of Goldwater's psychological fitness (with the wording reversing the usual order of yes and no). Goldwater sued for libel and was eventually awarded $50,000 in damages.

For a less extreme example, consider this question that the Gallup poll once asked: "The U.S. Supreme Court has ruled that a woman may go to a doctor to end pregnancy at any time during the first three months of pregnancy. Do you favor or oppose this ruling?" Gallup found that 47 percent were in favor, while 44 percent were opposed. These answers were undoubtedly influenced by the careful avoidance of the word *abortion* and by the inclusion of the Supreme Court ruling. Another pollster substituted the phrase *for an abortion* for *to end pregnancy* and found that only 41 percent approved while 48 percent were opposed. They then omitted the Supreme Court reference by asking this question: "As far as you yourself are concerned, would you say that you are for or against abortion, or what do you think?" Now they found that only 36 percent were in favor and 59 percent were opposed.[44]

The wording of a question sometimes has a decisive effect on the answers. Unfortunately, it is difficult to word complex questions fairly in a way that will still yield simple yes or no answers. As in all fields, there are some unscrupulous pollsters whose primary objective is to get answers that their clients want to hear. The founder of the Roper poll once lamented, "While we are probably the newest profession in existence, we have managed in a few short years to take on many of the characteristics of the world's oldest profession."[45]

For example, some polls are biased by the fact that they are mailed out with material arguing one side of a particular issue. I used to get letters from my congressman boasting about how hard he was fighting to control wasteful government spending. Included with these letters were questionnaires that asked whether we thought government spending should be reduced or increased. Another Congressman used the question shown below, purportedly to gauge the opinion of voters in his district regarding military spending:[46]

> This year's defense budget represents the smallest portion of our national budget devoted to defense since Pearl Harbor. Any substantial cuts . . . will mean the U.S.A. is no longer number one in military strength.
> Which one of the following do you believe:
> *A.* We must maintain our number-one status.
> *B.* I don't mind if we become number two behind Russia.

Here's one last example that illustrates many of these pitfalls. *The New York Times* once ran a story headlined "Doctors Reported for Security Plan" that began, "A recent poll in New Jersey indicates that most of the country's doctors would like to come under the Social Security old age and survivors insurance program."[47] It turns out that this story was based on a postcard questionnaire that had been included in an issue of *Bulletin of the Essex County Medical Society* that contained an article arguing that doctors should be included in the social security program. Can you spot the major reasons to give little credence to this poll?

First, there is a potential selection bias because Essex County doctors may not be representative of "most of the country's doctors." Second, there is the nonresponse bias inherent in a mail questionnaire (which 80 percent of the readers did not answer). Third, those who responded may have been influenced by having just read arguments on one side of the issue. Fourth, they may have been influenced by the fact that the publication of the article together with the questionnaire made it clear what answers the editors wanted to hear. Once again, for an opinion poll to be reliable, we need to have a random sample that provides honest answers to fair questions.

New Coke versus Old Coke

In 1985, after 99 years with essentially the same taste, Coca-Cola decided to switch to a new, high-fructose corn syrup, to make Coke taste sweeter and smoother—more like its arch rival, Pepsi. This historic decision was preceded by a top-secret $4 million survey of 190,000 people, in which the new formula beat the old by 55 percent to 45 percent. What Coca-Cola apparently neglected to take into account was that many of the 45 percent who preferred old Coke did so passionately. The 55 percent who voted for new Coke might have been able to live with the old formula, but many on the other side swore that they could not stomach new Coke.

Coca-Cola's announced change provoked outraged protests and panic stockpiling by old-Coke fans. Soon, Coca-Cola backed down and brought back old Coke as "Coke Classic." A few cynics suggested that Coca-Cola had planned the whole scenario as a clever way of getting some free publicity and causing, in the words of a Coca-Cola senior vice-president for marketing, "a tremendous bonding with our public." For 1985, new Coke captured 15.0 percent of the entire soft drink market and Coke Classic 5.9 percent with Pepsi at 18.6 percent. In 1986, new Coke collapsed to 2.3 percent, Coke Classic surged to 18.9 percent, and Pepsi held firm at 18.5 percent.

In 1987 *The Wall Street Journal* commissioned an interesting survey of 100 randomly selected cola drinkers, of whom 52 declared themselves beforehand to be Pepsi partisans, 46 Coke Classic loyalists, and 2 new-Coke drinkers.[48] In the *Journal*'s blind taste test, new Coke was the winner with 41 votes, followed by Pepsi with 39 and Coke Classic with 20. Seventy of the 100 people who participated mistakenly thought they had chosen their favorite brand; some were very indignant. A Coke Classic drinker who chose Pepsi said, "I won't lower myself to drink Pepsi. It is too preppy. Too yup. The New Generation—it sounds like Nazi breeding. Coke is more laid back." A Pepsi enthusiast who chose Coke said, "I relate Coke with people who just go along with the status quo. I think Pepsi is a little more rebellious, and I have a little bit of rebellion in me."

In 1990, Coca-Cola relaunched new Coke with a new name—Coke II—and a new can with some blue color, Pepsi's traditional color. Coca-Cola executives and many others in the soft drink industry remain convinced that cola drinkers prefer the taste of new Coke,[49] even while they remain fiercely loyal to old Coke and Pepsi—a loyalty due perhaps more to advertising campaigns than to taste. Given the billions of dollars that cola companies spend persuading consumers that the cola's image is an important part of the taste experience, blind taste tests may simply be irrelevant.

Telling Bill Clinton the Truth

Two weeks before the 1994 congressional elections, the *New Yorker* published an article about the Democratic party's chances of keeping control of Congress.[50] President Bill Clinton's top advisers were confident that the Republicans were unpopular with voters—a confidence based on extensive surveys showing strong disapproval. But the article's author observed that

> What Stephanopoulos did not mention was that the Democratic pollsters had framed key questions in ways bound to produce answers the President presumably wanted to hear. When the pollsters asked in simple unadorned, neutral language about the essential ideas in the Republican agenda—lower taxes, a balanced budget, term limits, stronger defense, etc.—respondents approved in large numbers.

Knowing this, one of the president's pollsters explained how she had to "frame the question very powerfully." Thus one of the questions asked by the president's pollsters was as follows:

> [Republican candidate X] signed on to the Republican contract with their national leadership in Washington saying that he would support more tax cuts for the wealthy and increase defense spending, paid for by cuts in Social Security and Medicare. That's the type of failed policies we saw for twelve years, helping the wealthy and special interests at the expense of the middle class. Do you approve?

When the *New Yorker* writer suggested to a top Clinton adviser that such loaded questions give the president a distorted picture of voter opinion, the adviser responded, "That's what polling is all about, isn't it?" The *New Yorker* also quoted an unidentified "top political strategist" for the Democrats who was not part of Clinton's inner circle:

> The President and his political people do not understand what has happened here. Not one of them ever comes out of that compound. They get in there at 7 a.m. and leave at 10 p.m., and never get out. They live in a cocoon, in their own private Disney World. They walk around the place, all pale and haggard, clutching their papers, running from meeting to meeting, and they don't have a clue what's going on out here. I mean, not a clue.

The election confirmed this strategist's insight and refuted the biased polls. The Republicans gained 8 seats in the U.S. Senate and 56 seats in the House of Representatives, winning control of Congress for the first time in 40 years. The Republicans also gained 12 governorships, giving them a total of 30 (including seven of the eight largest states), and gained more than 400 seats in the state legislatures, giving them majorities in 17 states formerly controlled by Democrats.

When the President held a press conference the day after the election, a *Washington Post* columnist wrote that Clinton was "pretty much in the Ancient Mariner mode, haunted and babbling." The *New Yorker* reported, "It was a painful thing to watch ... [The] protestations of amity and apology were undercut by the President's overall tone of uncomprehending disbelief."[51] Contrary to the opinion of Clinton's advisers, polling is not all about asking loaded questions. Clinton would have been much better served by fairly worded surveys that had given him and his advisers a clear idea of what voters wanted. After the 1994 debacle, Clinton replaced most of his pollsters and, once he had better information about voters' concerns, won reelection easily in 1996.

4.19 A Chinese census for famine relief showed a population of 105 million. Yet a census taken 5 years earlier for tax purposes showed a population of only 28 million. Why do you suppose these census figures were so far apart?

4.20 Ann Landers once asked her readers, "If you had it to do over again, would you have children?" She received 10,000 responses, of which 70 percent said no, they would not. Why should we be cautious in interpreting this poll?

4.21 Congresswoman Louise Day Hicks mailed 200,000 questionnaires to people in her district, asking whether they favored using mandatory busing to integrate public schools. The *Boston Globe* reported that 80 percent of the 23,000 people who responded were opposed to busing, which "proved decisively that the 'silent majority' does care about the country and its problems and, given the opportunity, will speak out on the issues."[52] Why should we be cautious in interpreting these poll results? Who was the real silent majority here?

4.22 In July 1983, Congressmen Gerry Studds (a Massachusetts Democrat) and Daniel Crane (an Illinois Republican) were censured by the House of Representatives for having had sex with teenage pages. A questionnaire in the Sunday (July 17) *Cape Cod Times* asked readers whether Studds should "resign immediately," "serve out his present term, but not run for reelection," or "serve out his present term, and run for reelection." Out of a Sunday circulation of 51,000 people, 2770 returned this questionnaire. Of these, 45.5 percent (1259) wanted Studds to resign immediately, while 46 percent (1273) wanted him to serve out his term and run for reelection. Another 7.5 percent (211) wanted him to serve out his term but not seek reelection, and 1 percent (27) were undecided. Why might these survey results be misleading?

4.23 A poll by *The Wall Street Journal* and NBC on President Ronald Reagan's performance included this question "Do you think Mr. Reagan will change his management style, or do you think he is too set in his ways to change?"[53] Nationwide, 56 percent of those sampled answered that Reagan will not change his management style, leading to an article in *The Wall Street Journal* entitled "President Is Viewed as Too Set in His Ways to Change His Governing Style, Poll Shows." If you were a spokesperson for Reagan, how might you object to the wording of the survey question?

4.24 Shere Hite sent detailed questionnaires (which, according to Hite, took an average of 4.4 hours to fill out) to 100,000 women and received 4,500 replies, 98 percent saying that they were unhappy in their relationships with men. A *Washington Post*-ABC News poll telephoned 1505 men and women and found that 93 percent of the women considered their relationships with men to be good or excellent. How would you explain this difference?

summary

A population is the entire group of items that interests us; a sample is the part of the population that we actually observe. Samples are used in a variety of situations because a complete enumeration is too expensive or difficult. We sample food when we take a small taste. College courses are a sample of a discipline, to help you make career decisions. Your dates can be a sample of married life. Employers use school grades, letters of recommendation, and job interviews as a sample of a job applicant's intelligence, skills, and personality. Statistical inference is used to make inferences about the population from which the sample came.

A reliable measure of the effects of some physical or social phenomenon generally requires either an experiment conducted under controlled conditions (in which other important factors are held constant) or comparison to a control group that does not receive the treatment. In medical research, the control group is usually given a placebo, which appears identical to the treatment being investigated, but is known to have no effects. In double-blind studies, neither the subject nor the researcher knows who is in the treatment group and who is in the control group until after the data have been collected. A placebo effect occurs when people respond positively to treatment that has no medical value. Without a control group, we cannot tell whether observed improvements are due to the placebo effect or to the treatment.

If samples are representative of a population, they can provide reliable information about the population's characteristics. A selection bias occurs when some members of the population are systematically excluded or underrepresented in the group from which the sample is taken. Retrospective studies, using past data for a contemporaneously selected sample, are notoriously unreliable because of faulty memories, lost data, and survivor bias (the exclusion of population members who are no longer around).

Deliberate attempts to construct representative samples are unwise, because representativeness is in the eye of the beholder. If a researcher hand-picks what is intended to be a representative sample, we will learn a great deal about this researcher's opinion, but little more. Instead, statisticians recommend that observational data be based on a simple random sample, in which each member of the population is equally likely to be selected—analogous to a fair deal from a well-shuffled deck of cards. Similarly, experimental data can be produced by using a completely randomized experimental design to allocate items to the treatment and control groups. Sometimes we can control for factors that might otherwise influence our results by using subgroups (stratified sampling with observational data or a block design with experimental data) and then using randomization within each subgroup. The actual process of random selection often relies on a random number generator.

Polls are used to gauge public opinion on all sorts of things, including the political candidates, controversial legislation, new products, and movie plots. Opinion polls can be very interesting and useful, but they require considerable care in construction and interpretation. Nonresponse bias involves the refusal of some groups to participate. In addition, it is sometimes difficult to phrase questions in a meaningful, balanced way that invites simple answers. The statistics of opinion polling are used to select a random sample; the art lies in getting truthful answers to honest questions.

review exercises

4.25 Males who have never smoked average 14.8 days of restricted activity per year, while current smokers average 22.5 days and former smokers average 23.5 days.[54] Do these data show that it is healthiest to never smoke, but once you start, it is better not to stop?

4.26 A variety of data were collected on boys at a summer camp. One camp leader noticed an interesting relationship between foot size and speed:

FOOT LENGTH	AVERAGE TIME FOR 50-YARD DASH
Less than 9 inches	9.5 seconds
9 inches or longer	8.6 seconds

Do big feet help people run faster? Why do you suppose that the boys with big feet were faster than the boys with small feet?

4.27 Critically evaluate:[55]

> The more children you have, the less likelihood there is that you will ever be divorced. This is an idea that many have suspected to be true. But it remained for the Metropolitan Life Insurance Company to prove it with the statistics printed below. The probability that a childless couple will be divorced is about twice as great as it would be if they had two or more dependent children....

NUMBER OF CHILDREN	DIVORCES PER 1000 MARRIAGES
0	15.3
1	11.6
2	7.6
3	6.5
4 or more	4.6

4.28 *Dial* magazine published a long article on how different people are using home computers, based on a survey of its readers.[56] One of its conclusions was that "the huge majority (82 percent) feel 'very favorable' about the computers in their homes. Another 12 percent feel 'somewhat favorable,' and not one of the 1000 in the random sample tabulated for the survey indicated feeling 'very unfavorable.'" Comment on the following description of their random sample of their readers:

> Dial asked its readers who use home computers or who have experience with computers at work to respond to a two-page questionnaire in the magazine. We received nearly 5200 responses; more than 600 of them included letters of up to 15 pages in length. Most of these letters were written on computers.
>
> A random sample of 1000 responses was pulled from the group. The statistics in the article are based on a tabulation of that sample.

4.29 A Harvard study of incoming first-year students found that students who had taken SAT preparation courses scored an average of 63 points lower on the SAT than did first-year students who had not taken such courses (1271 versus 1334). Harvard's admissions director presented these results at a regional meeting of the College Board, suggesting that such courses are ineffective and "the coaching industry is playing on parental uncertainty."[57] Why is this study unpersuasive?

4.30 A razor company used to run a television commercial showing a number of barbers shaving, one after the other, with the same blade. The 12th, 13th, 15th, and 17th barbers were interviewed, and all said the shave was satisfactory—the implication being that this blade was good for at least 17 shaves. Why might this conclusion be unjustified?

4.31 A market research firm in Burbank, California, tested movies and television shows at a movie theater that it ran, called the Preview House. Although it was in a disreputable part of town, no admission was charged at the Preview House, making it a popular inexpensive date for some people. In what ways was the Preview House sample not a scientific random sample?

4.32 *Time* magazine once surveyed college graduates by obtaining lists from colleges throughout the country and mailing questionnaires to all graduates whose last names began with the letters Fa. What advantage is there to this procedure? Can you see any perils?

4.33 Pepsi-Cola had a "blind taste test" in which people tasted Pepsi from a glass marked M and Coca-Cola from a glass marked Q.[58] More than one-half of the people preferred Pepsi. Coca-Cola then ran its own test, letting people drink Coke from a glass marked M and Coke from a glass marked Q. They found that most people preferred Coke from the glass labeled M! This prompted their advertising headline: "The Day Coca-Cola Beat Coca-Cola." Provide an explanation for these results. As an impartial statistician, how would you have designed a blind taste test?

4.34 Explain any possible flaws in this conclusion:

> A drinker consumes more than twice as much beer if it comes in a pitcher than in a glass or bottle, and banning pitchers in bars could make a dent in the drunken-driving problem, a researcher said yesterday.
>
> E. Scott Geller, a psychology professor at Virginia Polytechnic Institute and State University in Blacksburg, Va., studied drinking in three bars near campus. . . . Observers found that, on average, bar patrons drank 35 ounces of beer per person when it came from a pitcher, but only 15 ounces from a bottle and 12 ounces from a glass.[59]

4.35 Why is an estimate of the average annual income of a college's 1980 graduates likely to be biased if this estimate is based on a survey conducted by the college?

4.36 The Berkeley Summer Mathematics Institute encourages minority students to pursue careers in mathematics. Of the 90 students who participated from 1988 to 1991, a 1992 survey found that 61 percent attended or applied to graduate school in mathematics and 29 percent were pursuing undergraduate mathematics degrees.[60] According to the National Bureau of Labor Statistics, only 9.2 percent of all minority employees are working in jobs that require mathematics or computer science. Why are the Institute's survey data inadequate to show that it is successful in interesting minorities in mathematics careers?

4.37 *Consumer Reports* invites readers to rate movies on a scale of 1 to 5: "If you attend movies and would like your votes recorded here, write to. . . ." In its May 1990 issue, it printed ratings for 90 movies, only three of which had been rated by more than 300 readers: *Parenthood* (343 votes and an average rating of 3), *Lethal Weapon II* (325 votes and an average rating of 3), and *Driving Miss Daisy* (304 votes and an average rating of 5). Twenty-seven of the movies rated had fewer than 25 votes. Explain why this poll is not a random sample of reader opinion, and identify some biases that might be expected.

4.38 In *Boykin vs. Georgia Power,* the plaintiff charged that Georgia Power's salary policy discriminated against black employees.[61] Part of the evidence involved a comparison of the 1976 salaries of black and white employees who had been hired to similar entry-level positions in 1970. Why might such data not present an accurate picture of the treatment of black and white employees who had been hired to similar entry-level positions in 1970? Give a specific hypothetical example.

4.39 Chrysler sponsored a study that found that most people in the United States preferred driving a Chrysler to driving a Toyota.[62] Why is this survey probably biased by the fact that the sample consisted of 100 people, none of whom owned foreign cars?

4.40 Since 1985, students at the University of Delaware who are not in the honors program or in remedial sections can choose a first-year composition course that uses IBM computers or one that uses Macintosh computers. In 1990, the assistant director of the university's writing program wrote an article arguing that "there is some sort of effect on students' writing when they use a Mac, different from when they use an IBM."[63] For example, essays written on the Mac tended to be on frivolous topics. How, as a statistics student and Mac enthusiast, could you criticize her student samples as possibly biased?

4.41 *U.S. News & World Report*'s college rankings use the average SAT scores of admitted students who submit SAT scores as part of their admission application. Bowdein College is unusual in that it does not require applicants for admission to submit SAT scores. Explain why this practice is likely to make Bowdein appear either more selective or less selective, relative to colleges that require all applicants to submit SAT scores.

4.42 A Los Angeles television station includes a poll as part of its nightly local news show. A provocative question of the day flashes on the screen, and viewers are asked to call in their yes or no votes. Explain why these poll results may not be representative of community opinion.

4.43 "The most dangerous place to be is in bed, since more people die in bed than anywhere else." Do you agree?

4.44 Television advertisers are concerned about viewers who record programs on video cassette recorders (VCRs) and watch them later, fast-forwarding through the commercials—a practice called *zapping*. A network-sponsored study of zapping asked randomly selected people to watch a pilot television show on a cable station. These people were told that they would be telephoned the next day and asked their opinion of the pilot. When the researchers called, they asked if the volunteers had recorded the show and zapped commercials—and found that very few had. If you worked for an advertising firm, what fault would you find with this study?

4.45 The 7-Eleven stores in New York City let their customers answer a weekly question by choosing to have their coffee in a cup labeled yes or no. A January 1990 survey asked if Congress deserved a pay raise, illustrating the question with a picture of a fat congressman lighting a cigar with a $100 bill; 71 percent of the 682,285 persons served chose a no cup. Why would a statistician be skeptical of these survey results?

4.46 Explain why these data do not prove the claim made in this 1955 newspaper article:

> U.S. Statistics Prove it . . . Marry and Live longer: There's no question about it: Married people live longer than single people; and people who were once married live longer than people who were never married.
>
> The Public Health Service's National Office of Vital Statistics has proved this to be a fact. . . . deaths among bachelors were almost two-thirds greater than among husbands. Among divorced men and widowers, the rate was half again more.[64]

4.47 American Express and the French tourist office sponsored a survey that found that most visitors to France do not consider the French to be especially unfriendly.[65] The sample consisted of "1000 Americans who have visited France more than once for pleasure over the past 2 years." Why is this survey biased?

4.48 In the 1990s, some financially strapped universities offered generous severance packages to any tenured professor who agreed to leave the university. In what way are those who accepted this offer not a random sample of the university's professors? Do you think that this policy raised or lowered the average quality of the university's professors?

4.49 A well-known automobile manufacturer claimed in a 1986 television commercial that 70 percent of its cars that had been registered in the United States since 1974 were still on the road. Does this statistic imply that there is a 70 percent chance that one of this company's cars will last at least 12 years?

4.50 Red Lion Hotels & Inns ran full-page advertisements in 1989 claiming, "For every 50 business travelers who try Red Lion, a certain number don't come back. But forty-nine out of 50 do." The basis for this claim was a survey of people staying at Red Lion, 98 percent of whom said that "they would usually stay in a Red Lion Hotel or Inn when they travel to one of our cities." Carefully explain why the survey results do not prove the advertising claim. You may find a numerical example helpful.

projects

For each of the following projects, type a report in ordinary English, using clear, concise, and persuasive prose. Any data that you collect for this project should be included as an appendix to your report. Data used in your report should be presented clearly and effectively.

4.1 Go to the library and browse through a recent Sunday edition of *The New York Times*, looking for an article that reports the results of a statistical study. (Enclose a duplicated copy of the newspaper article with your report.) Are you able to tell from the newspaper article whether this study used experimental or observational data? What questions do you have about how this study was conducted? Did the newspaper article tell you where to find the complete results from this study?

4.2 Use a table of random numbers or a computer random number generator to select 30 female students and 30 male students at your college. Explain exactly how these names were selected.

4.3 Do more expensive chocolate chip cookies taste better than less expensive ones? Design an experiment to investigate this question. Do not actually conduct a study, but be very specific about how you would do so.

4.4 It has been reported that college students are pessimistic about the future of the social security system. How would you survey students at your college to learn more about how confident they are that they will receive substantial social security checks when they retire? Develop a short questionnaire, and explain exactly how you would select the students to be questioned. (Do not use the survey, just plan it.)

4.5 Suppose that you wanted to compare the political beliefs of students and professors at your school. How would you do so? Develop a short questionnaire, and explain exactly how you would select the students and professors to be questioned. (Do not use the survey, just plan it.)

4.6 Conduct a taste test of either Coke versus Pepsi or Diet Coke versus Diet Pepsi. Survey at least 50 students who identify themselves beforehand as cola drinkers with a preference for one of the brands you are testing. Describe your taste test and summarize your results.

4.7 Dial randomly selected, local off-campus telephone numbers and begin with a greeting similar to this one: "Hello, I'm a college student doing a statistics paper on voting in the 1956, 1960, and 1964 elections, and I would appreciate your help greatly. Did you vote in any of these three elections?" If the person answers no, say thank you and try another number. If the person answers yes, ask whether she or he voted for Eisenhower or Stevenson in 1956, for Kennedy or Nixon in 1960, and for Johnson or Goldwater in 1964. Continue in this fashion until you have 50 people who reveal their votes in each of these elections. Compare these memories with the actual votes received by the candidates in these elections. Explain any substantial discrepancies.

4.8 Dial randomly selected, local off-campus telephone numbers, and begin with a greeting similar to this one: "Hello, I'm a college student doing a statistics paper on voting in the 1972, 1976, and 1980 elections, and I would appreciate your help greatly. Did you vote in any of these three elections?" If the person answers no, say thank you and try another number. If the person answers yes, ask whether he or she voted for McGovern or Nixon in 1972, for Carter or Ford in 1976, and for Carter or Reagan in 1980. Continue in this fashion until you have 50 people who reveal their votes in each of these elections. Compare these memories with the actual votes received by the candidates in these elections. Explain any substantial discrepancies.

4.9 In the spring of 1994, Pomona College changed its general education system from three courses in each of the college's three divisions to one course in each of 10 intellectual skills, one of which is the ability to understand and analyze data. The class of 1998 was the first to use this new general education system. In 1997, the college received a grant to assess whether the new system did in fact enable students to understand and analyze data. Specify a reasonable plan for making this assessment.

4.10 Ask 25 randomly selected people this question: "Karl Marx said, 'Whenever a form of government becomes destructive of these ends, it is the right of the people to alter or to abolish it.' Do you agree?" Now ask 25 randomly selected people this question: "The U.S. Declaration of Independence says: 'Whenever a form of government becomes destructive of these ends, it is the right of the people to alter or to abolish it.' Do you agree?" Compare your responses.

4.11 Have a woman ask 100 randomly selected persons this question: "Have you had a nightmare anytime during the past 12 months?" Record both the response and the subject's gender. Now have a man ask 100 randomly selected persons this same question. Compare your responses.

4.12 If you do not conduct this survey on a Monday, replace "Monday" with the correct day of the week. Give 50 randomly selected students this question and ask that they drop their unsigned answer in a box or bag:

> If your birthdate is an even number, please answer question (a); if it is an odd number, please answer question (b).
> a. Is today Monday?
> b. Have you handed in a school paper during the past two years that was mostly written by someone else?

Assuming that even and odd birthdates are equally likely, estimate the percentage of the population that has handed in papers written by others.

PART 2

Statistical Inference

HAVE YOU EVER WONDERED?

When you donate blood to the Red Cross, ELISA (Enzyme-Linked Immunosorbent Assay) tests are used to detect the presence of antibodies produced by the human immunodeficiency virus type 1 (HIV-1). These tests are very good, but they are not perfect. How can probabilities be used to quantify the accuracy of ELISA tests? More generally, how can probabilities be used in medical diagnoses to help doctors and patients understand the implications of test results and to communicate with one another?

What, exactly, does it mean when a test for a virus is reported to be 95 percent accurate? Does this mean that if you have the virus, then 95 times out of 100 this test will indicate that you have the virus? Or does it mean that if the test says you have the virus, then 95 times out 100 you do indeed have the virus? Or does it mean both? In this chapter, we see how probabilities can be used to answer life-and-death questions and how these probabilities should be interpreted.

chapter 5

explaining data

I know of scarcely anything so apt to impress the imagination as the wonderful form of cosmic order expressed by the "Law of Frequency of Error." The Law would have been personified by the Greeks and deified, if they had known of it. . . . Whenever a large sample of chaotic elements are taken in hand and marshaled in the order of their magnitude, an unsuspected and most beautiful form of regularity proves to have been latent all along. The tops of the marshaled row form a flowing curve of invariable proportions; and each element, as it is sorted into place, finds, as it were, a preordained niche.

—Sir Francis Galton

THE GREAT FRENCH mathematician Pierre Simon Laplace (1749–1827) observed that probabilities are only "common sense reduced to calculation." Explicit calculations have two very big advantages over common sense. Without the rigor of

probability calculations, common sense can be easily led astray. Without the unambiguous language of probability calculations, common sense can be easily misunderstood. For example, medical diagnoses are sometimes clear-cut (the X rays reveal a broken bone) and at other times ambiguous (the patient may or may not have HIV). Probabilities provide a convenient method for determining and communicating an uncertain diagnosis.

Suppose that a college student with no prior evidence of HIV gives a blood sample that is tested for the presence of HIV antibodies. The test comes back positive, but the test is not perfect and the results are not always correct. When told of the test results, the student asks a straightforward question: Do I have HIV? A definitive answer of yes or no is not warranted because the test is imperfect; yet the student should be told, in some honest way, that the results are worrisome.

Words alone are inadequate because doctor and patient may interpret words very differently. When 16 doctors were asked to assign a numerical probability corresponding to the diagnosis that disease "cannot be excluded," the answers ranged from a 5 percent probability to a 95 percent probability, with an average of 47 percent.[1] When they were asked to interpret the term *likely,* the probabilities ranged from 20 to 95 percent, with an average of 75 percent. Even the phrase *low probability* elicited answers ranging from 0 to 80 percent, with an average of 18 percent. If by low probability one doctor means 80 percent and another means no chance at all, then it is better for the doctor to state the probability than to risk a disastrous misinterpretation of ambiguous words.

Doctors can calculate these probabilities with computer software and national databases that combine published medical data with specific information about the individual patient. Later in this chapter, we will see how such data can be used to determine the probability that HIV antibodies are present in a blood sample with worrisome test results.

Probabilities can be used not only for life-threatening diseases, but also for all the daily uncertainties that make life interesting and challenging. To make valid statistical inferences from data, we can use probability models to describe where the data come from, for example, a medical test that is correct or incorrect, a surveyed person who supports or opposes a particular candidate, or a Scholastic Aptitude Test (SAT) score that is above or below 700. In this chapter, first we see how probabilities can be used to quantify uncertainty. Then we look at how two probability models can help us explain and interpret empirical data.

5.1 Using Probabilities to Quantify Uncertainty

When we say that a flipped coin has a .5 probability of landing with its heads side up, we mean that if this coin is flipped an interminable number of times, we anticipate that it will land heads about one-half the time. More generally,

> If an event has a probability P of occurring, then the fraction of the times that it occurs in the very long run will be very close to P.

Obviously, a probability cannot be negative or larger than 1. How do we determine the value within this range? There are three different approaches.

Equally Likely Reasoning

The **equally likely** method reasons that if there are n equally likely possible outcomes, the probability that any one of m outcomes will occur is m/n. When a symmetric coin is flipped fairly, there are two equally likely outcomes: heads or tails. Because these outcomes are equally likely, each has a probability of occurring of $1/2$. When a symmetric six-sided die is rolled fairly, there are six equally likely outcomes, and each therefore has a $1/6$ probability of occurring. What about the probability that the die will be an even number? Because three of the six equally likely outcomes are even numbers, the probability of an even number is $3/6 = 1/2$.

Long-Run Frequencies

The equally likely approach works well with physical objects, such as coins, dice, and cards, where we can reasonably assume the outcomes to be equally likely. In many other situations, there may be compelling evidence that the possible outcomes are not equally likely. A company selling a 1-year life insurance policy to a 20-year-old woman can hardly assume that life and death are equally likely outcomes. To handle such cases, a second approach to probabilities has been developed. If a coin has a $1/2$ probability of landing heads, we can infer that, in a large number of coin flips, heads will come up approximately one-half the time. To reverse this reasoning, if, in a large number of trials, an event occurs one-half the time, we can conclude that its probability of occurring is approximately $1/2$. Thus the **long-run frequency** interpretation of probabilities is that if a certain event has occurred *m* times in *n* identical situations (where *n* is a very large number), then its probability is approximately *m/n*.

I once assigned a homework exercise in which each of 30 students flipped a coin 100 times. Using the students' last names to arrange the outcomes in alphabetical order, I obtained the sequence of 3000 coin flips shown in Figs. 5.1 and 5.2. The proportion that were heads varied considerably during the first 100 flips, but by the end of 3000 flips it was very close to the anticipated .50. The exact heads proportion turned out to be 1511/3000 = .5037. If we did not have our equally likely logic to rely on, we could use these data to estimate the probability of heads to be roughly .5.

An insurance company that cannot use equally likely logic can estimate the probability of a healthy 20-year-old woman's dying within 1 year by looking at the recent experiences of millions of other similar women. If the company has data on 10 million 20-year-old women, of whom 12,000 died within 1 year, the company can estimate this probability as 12,000/10,000,000 = .0012, or slightly larger than 1 in 1000.

The equally likely and long-run frequency approaches to probability are two entirely consistent ways of thinking about probabilities. They differ only in how they go about determining probabilities. In the equally likely method, we count the possible incomes and think about whether they are equally likely. In the long-fun frequency method, we go out and collect data from repeated trials. The equally likely method is more appropriate for games of chance or similar

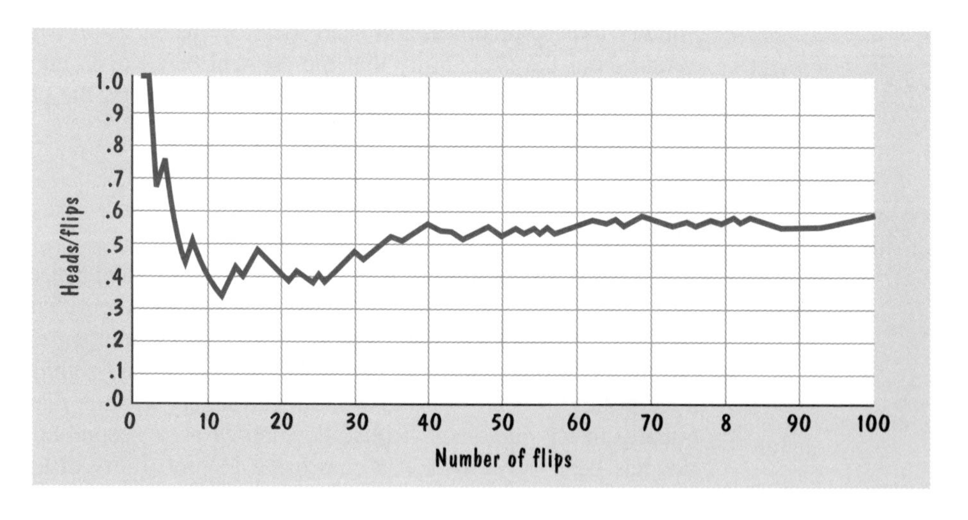

Figure 5.1 The first 100 coin flips.

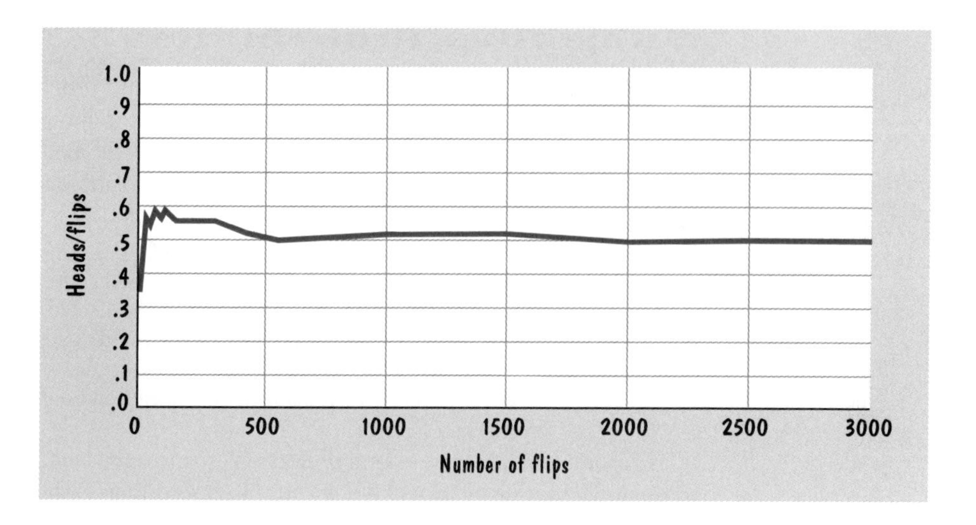

Figure 5.2 All 3000 coin flips.

situations in which the physical apparatus entitles us to make persuasive assumptions about the outcome's relative likelihood. The long-run frequency method is more appropriate when we suspect that the outcomes are not equally likely and we have access to data that can confirm or deny our suspicions. The long-run frequency approach is obviously needed for insurance rates and similar situations in which it is apparent that the equally likely approach cannot be used. The long-run frequency method may also be useful for fine-tuning the probabilities of events that are almost, but not quite, equally likely.

Consider, for instance, the probability that a newborn baby will be a boy or a girl. We might plausibly assume boy and girl babies to be equally likely and consequently assign a ½ probability to each. However, an examination of the data shows that boy babies slightly outnumber girl babies (perhaps nature's way of compensating for the higher male fatality rate). During the years from 1975 to 1984, there were 34,696,000 recorded births in the United States, of which 17,787,000 were boys and 16,909,000 were girls. For these data, the long-run frequency estimate of the probability of a boy is

$$P[\text{boy}] = \frac{17{,}787{,}000}{34{,}696{,}000} = .513$$

Because 51.3 percent of these babies were male (and 48.7 percent female), we estimate that a baby has a .513 probability of being male and a .487 probability of being female.

Subjective Probabilities

The long-run frequency approach allows us to extend probabilities to events that are not equally likely. However, its application is limited to situations in which we have lots of repetitive data, and much of the uncertainty we confront involves virtually unique situations. In 1946, the self-proclaimed Wizard of Odds calculated presidential probabilities as follows:

> Miss Deanne Skinner of Monrovia, California, asks: Can the Wizard tell me what the odds are of the next President of the United States being a Democrat? . . . Without considering the candidates, the odds would be 2 to 1 in favor of a Republican because since 1861 when that party was founded, there have been 12 Republican Presidents and only 7 Democrats.[2]

The outcome of a presidential election is uncertain, and it would be useful to quantify that uncertainty. We would like more than a shrug of the shoulders and a sheepish "Who knows?" However, we cannot reasonably predict U.S. presidential election outcomes by counting the frequencies with which Republicans and Democrats have won in the past. The choice of president is not determined by the flip of a coin every 4 years. The odds change from election to election, depending on the candidates and the mood of the electorate.

In the 18th century, Reverend Thomas Bayes wrestled with an even more challenging problem—the probability that God exists. The equally likely and long-run frequency approaches are useless, and yet this uncertainty is of great interest to many people, including Reverend Bayes. Such a probability is necessarily subjective. The best that anyone can do is to weigh the available evidence and logical arguments and to come up with a personal probability of God's existence. This idea of personal probabilities has been extended and refined by other *Bayesians,* who argue that many uncertain situations can only be analyzed by means of subjective probability assessments. Bayesians are willing to assign probabilities to everything from presidential elections to God's existence.

These personal probabilities can be elicited by offering the person a choice between a gamble that depends either on the occurrence of the specified event or on a game of chance in which the probability of winning can be calculated easily. For example, in January 1995, a former student, Juan Guerra, telephoned me to ask whether I thought U.S. stocks were cheap or expensive. In exchange for this valuable advice, he agreed to repeat a mental experiment that he had participated in during one of my statistics classes. I asked Juan to choose one of the following gambles:

- Receiving $10 if Bill Clinton is reelected president in 1996
- Receiving $10 if, on election day, a red card is drawn from a deck containing 5 red cards and 5 black cards

Juan emphatically chose the card draw, thereby showing that he believed the probability of Clinton's being reelected to be less than .5. I then offered him this choice:

- Receiving $10 if Bill Clinton is reelected president in 1996
- Receiving $10 if, on election day, a red card is drawn from a deck containing 3 red cards and 7 black cards

This time, Juan chose Clinton, showing that he believed the probability of Clinton's being reelected to be larger than $3/10 = .3$. With further questioning, I pinned him down to a .35 probability that Clinton would be reelected, by finding that he was indifferent between these gambles:

- Receiving $10 if Bill Clinton is reelected president in 1996
- Receiving $10 if, on election day, a red card is drawn from a deck containing 35 red cards and 65 black cards

Juan knew a lot about politics. The two presidents preceding Clinton had been Republicans (Ronald Reagan and George Bush), and in 1994 the Republican party won control of both houses of Congress for the first time in 40 years. Many Democrats blamed Clinton for the party's disastrous showing in the 1994 midterm elections, and there was considerable speculation about Democratic candidates who might run against Clinton in the party's primary elections. The economy was doing well, but the Federal Reserve seemed fanatical about inflation and might cause an economic recession in 1996, which voters would blame on Clinton. None of this could be translated directly to a probability, but Juan subjectively weighed all this information, and more, and came up with a personal probability of .35 that Clinton would be reelected. A year later, in January 1996, Juan telephoned me again and said that Clinton's revived popularity had persuaded him to revise his personal probability of Clinton's reelection from .35 to .80.

The **Bayesian** interpretation of probabilities is that an event has a subjective probability P of occurring if you are indifferent between a gamble hinging on this event and one based on the selection of a red card from a deck in which a fraction P of the cards are red.

Many probability theorists are not comfortable with the Bayesian approach. They prefer equally likely or long-run frequency probabilities on which we can all agree and from which we draw common conclusions. How can we have a scientific discourse if a probability is .35 one year and .80 the next, or if one person believes a probability to be .7 and another thinks it is .4? Bayesians respond that these subjective disagreements are all around us and should not be ignored. As Mark Twain remarked, "It is difference of opinion that makes horse races." If we refuse to use probabilities when the equally likely or long-run frequency approach cannot be used, we will be forced to ignore not only horse races, but also many very interesting and far more important questions. Who will win the next presidential election? Will interest rates go up or down? Will this medication work? Is Iraq preparing to invade Kuwait? These are all important uncertainties about which informed and well-intentioned people disagree. Instead of ignoring these disagreements, Bayesians argue that it is better to quantify them by specifying subjective probabilities.

A Scientific Study of Roulette

Outside the United States, roulette wheels generally have 37 slots, numbered 0 to 36. Wheels used in the United States have an additional 00 slot, making 38 slots in all. If the wheel is perfectly balanced, clean, and fair, we can assume that each of the slots is equally likely to catch the ball. In reality, imperfections in the wheels cause some numbers to win slightly more often than others.

In the late 1800s, an English engineer, William Jaggers, took dramatic advantage of these imperfections. He paid six assistants to spend every day for an entire month observing the roulette wheels at Monte Carlo and recording the winning numbers. Jaggers found that certain numbers came up slightly more often than others. He then bet heavily on these numbers and, in 4 days, won 1.5 million francs, nearly $200,000—a fortune in the late 1800s. More recently, in the 1960s, while their fellow students were studying or protesting, a group of Berkeley students were reported to have pulled off a similar feat in Las Vegas. Nowadays, unfortunately, casinos examine and rotate their roulette wheels frequently in order to frustrate the long-run frequency bettors.

Success Probabilities at Sandoz

Since 1970, Sandoz, a large Swiss pharmaceutical company, has been using subjective probabilities to gauge the potential success of its research and development projects.[3] After a potentially useful chemical has been identified by exploratory research, it becomes a "project" and is tested for a minimum of 5 years for efficacy and safety. If it passes these tests (on average, 1 in 10 do), then it is judged technically successful and is registered and marketed.

The firm's decision to approve or reject a project for testing is crucial because the new products that will be marketed 5 to 10 years from now depend on which projects are approved for testing today. Even for those projects that are initially approved, the firm must decide as testing proceeds whether to continue spending resources that might be better used elsewhere. A primary consideration for these important management decisions is each project's chance of eventually being judged technically successful. An equally likely assumption is of little use since not all projects are equally likely to be successful. And because each product is different, it is not reasonable to estimate success probabilities from long-run frequencies.

Instead, Sandoz uses subjective probabilities, reflecting the consensus of a panel of its research and development experts. Once a project is launched, its success probability is updated semiannually. For some projects, the probability of success is revised upward, and as it approaches 1.0, the project is judged to be a technical success. For other projects, the success probability declines as testing proceeds, and management ends further testing.

Sandoz's subjective probabilities are no simple matter. These are the chances that an untested or partially tested compound will eventually be successful. But as one elderly man said about his health, "I've got problems, but life sure beats the alternative." Sandoz reasons that subjective probabilities are better than none at all. And it has been pleased to report that its panels' initial subjective probability assessments have turned out to be a reliable guide to project success: Of those given an x percent chance of succeeding, about x percent do succeed.

exercises

5.1 You are playing draw poker and are dealt four spades and a heart. If you discard the heart and draw a new card, what is the probability that this new card will be a spade, giving you a flush? (Assume that there are no other players, since it can be shown that your chances do not depend on whether there are other players, as long as you do not know what cards they have been dealt.)

5.2 A very successful football coach once explained why he preferred running the ball to passing it: "When you pass, three things can happen [completion, incompletion, or interception], and two of them are bad." Can we infer that there is a ⅔ probability that something bad will happen when a football team passes the ball?

5.3 The most important skating event in the Netherlands is the Elfstedentocht, a race over 124 miles of canals through 11 Dutch cities.[4] This race is only held if the entire course is covered by ice at least 8 inches thick. During the 96 years from 1900 to 1995, the race was held 14 times. Based on these data, what is your estimate of the probability that the race will be held next year?

5.4 The U.S. Postal Service handled 56 billion pieces of mail in 1978, of which 4.5 billion were initially misdelivered and 211,013 were reported lost completely. Based on these data, what is the probability that a randomly selected letter will be misdelivered? Will be lost completely? Why might these calculated probabilities understate or overstate the probability that a letter you send will be misdelivered?

5.5 In early 1987, it was very unclear whether the U.S. economy would boom, bust, or muddle in between. Edward Yardeni, director of economics and fixed income research at Prudential-Bache, explained his position as follows:

> We've been slumpers for the past few months. A slump is still possible, but now we feel more comfortable with a muddling scenario. Until recently, we assigned the slump scenario a probability of 50%, with muddling at 30% and with a boom given a 20% chance. Now, muddling gets 50%, bust gets 30%, and boom stays at 20%.[5]

Would you characterize his probabilities as being equally likely, long-run frequency, or subjective? Why are the other two kinds of probabilities inappropriate here?

5.2 Contingency Tables

We began this chapter with the example of worrisome results from an ELISA test for the presence of HIV antibodies in blood donated to the Red Cross. To analyze ELISA test results with probabilities, we use a **contingency table** in which data are classified in one way by the rows and in another way by the columns. (Because of this two-way classification, a contingency table is sometimes called a *two-way table.*) In our example, the blood sample has two possible conditions, (HIV or no HIV) which we show as two rows in this table, and the test has two possible outcomes (positive and negative), which we show as two columns:

	TEST POSITIVE	TEST NEGATIVE
HIV		
NO HIV		

The entries in this table show the four possible combinations of conditions, for example, having HIV and testing positive. To determine these entries, we need to assume an arbitrary size for the population being tested, say, 1,000,000 blood samples, and we need to specify the fraction of this population that has HIV. We do not know whether college students who donate blood to the Red Cross are more or less likely than other students to have HIV. For illustrative purposes, we use the estimate that roughly 2 in every 1000 U.S. college students have HIV; and therefore we assume that of these 1,000,000 blood samples,

2000 (.2 percent) have HIV and 998,000 (99.8 percent) do not. We show these figures in a total column:

	TEST POSITIVE	TEST NEGATIVE	TOTAL
HIV			2,000
NO HIV			998,000
TOTAL			1,000,000

To fill in the rest of table, we need to know the accuracy of the ELISA tests. When HIV antibodies are present in a blood sample, ELISA tests have a .997 probability of detecting the antibodies and a .003 probability of not detecting them; when the antibodies are not present, ELISA tests have a .985 probability of correctly indicating their absence and a .015 probability of incorrectly signaling the presence of the antibodies.[6] Therefore, for those 2000 cases in which the antibodies are present, we use the estimate that ELISA tests have a .997 probability of detecting these antibodies: .997(2000) = 1994 will show positive test results and .003(2000) = 6 will show negative results. For those 998,000 cases where the antibodies are not present, we use the estimate that ELISA tests have a .985 probability of correctly indicating their absence: .985(998,000) = 983,030 will show negative test results, and .015(998,00) = 14,970 will show positive results. Entering these numbers gives

	TEST POSITIVE	TEST NEGATIVE	TOTAL
HIV	1,994	6	2,000
NO HIV	14,970	983,030	998,000
TOTAL			1,000,000

Adding the columns, we can fill in the rest of the contingency table, as shown in Table 5.1.

Misinterpreting Conditional Probabilities

To use this table, we need to define the **conditional probability** $P[B \text{ if } A]$ as the probability that B will occur if A occurs. (The conditional probability $P[B \text{ if } A]$ is often written as $P[B \text{ given } A]$ or $P[B \mid A]$.) For example,

Table 5.1

ELISA tests for HIV antibodies

	TEST POSITIVE	TEST NEGATIVE	TOTAL
HIV	1,994	6	2,000
NO HIV	14,970	983,030	998,000
TOTAL	16,964	983,036	1,000,000

P[test positive if HIV] signifies the probability that a blood sample with HIV antibodies will test positive. This is very different from the reverse conditional probability P[HIV if test positive], which means the probability that a blood sample that tests positive actually has HIV antibodies. The data in Table 5.1 illustrate this distinction. Of the 2000 samples with the antibodies, 1994 test positive; therefore,

$$P[\text{test positive if HIV}] = \frac{1994}{2000} = .997$$

Of the 16,964 samples that test positive, 1994 have HIV antibodies:

$$P[\text{HIV if test positive}] = \frac{1994}{16{,}964} = .118$$

These numbers demonstrate clearly that we must exercise care in interpreting conditional probabilities. If we look at the first numerical row of Table 5.1, we see that 99.7 percent of the blood samples containing HIV antibodies are correctly identified. Yet, if we look at the first numerical column, only 11.8 percent of the samples that test positive actually contain HIV antibodies.

Before the ELISA test is used, the probability that a randomly selected blood sample has HIV antibodies is .002. After the positive ELISA reading, the revised probability of HIV antibodies is .118. The probability of HIV has increased by a factor of nearly 60, from .2 to 11.8 percent, but is still far from certain. It is much more likely than not that this blood sample does not contain HIV antibodies. The most effective and unambiguous way of communicating this diagnosis is with an 11.8 percent probability, not with words (such as *cannot be excluded, likely,* or *low probability*) that can be easily misinterpreted.

Here is another example of how reverse conditional probabilities can be quite different. At a large university, only 1 out of every 1000 female students might play on the women's basketball team, but 100 percent of the players on the women's basketball team are female:

$$P[\text{on women's basketball team if female}] = .001$$
$$P[\text{female if on women's basketball team}] = 1.0$$

Independent Events and Winning Streaks

The probability expression $P[B]$ asks, Considering all possible outcomes, what is the probability that B will occur? The expression $P[B \text{ if } A]$ asks, Considering only those cases where A occurs, what is the probability that B will also occur? If the probability that B occurs does not depend on whether A occurs, then it is natural to describe A and B as independent: Two events A and B are **independent** if $P[B \text{ if } A]$ equals $P[B]$. (Logically, it must also be true that $P[A \text{ if } B]$ equals $P[A]$.)

In games of chance involving coins, dice, roulette wheels, and other physical objects, each outcome is generally independent of other outcomes—past, present, or future. In any fair game, a player will win some and lose some (or, it often seems, win some and lose many). The wins at times will be scattered and at other times will be bunched together. Some gamblers mistakenly attach a great deal of significance to these coincidental runs of luck. They apparently believe that luck is some sort of infectious disease that a player catches and then takes awhile to get over. For example, Clement McQuaid, "author, vintner, home gardener, and keen student of gambling games, most of which he has played profitably," offers this advice:

> There is only one way to show a profit. Bet light on your losses and heavy on your wins. Many good gamblers follow a specific procedure: a. *Bet minimums when you're losing...* b. *Bet heavy when you're winning...* c. *Quit on a losing streak, not a winning streak.* While the law of mathematic probability averages out, it doesn't operate on a set pattern. Wins and losses go in streaks more often than they alternate. If you've had a good winning streak and a loss follows it, bet minimums long enough to see whether or not another winning streak is coming up. If it isn't, quit while you're still ahead.[7]

You will indeed show a profit if you win your large bets and lose only your small ones, But how are you to know in advance whether you are going to win or lose your next bet? Suppose you are playing a dice game and have won 3 times in a row. You know that you have been winning, and you are probably excited about it, but dice have no memories or emotions. Games were invented by people. Dice do not know the difference between a winning number and a losing number. Dice do not know what happened on the last roll and do not care what happens on the next roll. The outcomes are independent in that the probabilities are constant, roll after roll.

Sometimes you will win 4 times in a row, but often you will win 3 times in a row and then lose the fourth time. Sometimes you will win 5 times in a row, but often you will win four in a row and then lose the fifth. You may fondly recall those occasions when you won 4, 5, or more times in a row and regret afterward that you did not bet more heavily. Unfortunately, there is no way to know in advance when a winning streak will begin or end. Games of chance are the classic example of independent events.

Interpreting Mammogram Results

One hundred doctors were asked this hypothetical question: In a routine examination, you find a lump in a female patient's breast. In your experience, only 1 out of 100 such lumps turns out to be malignant; but to be safe, you order a mammogram X ray. If the lump is malignant, there is a .80 probability that the mammogram will identify it as malignant; if the lump is benign, there is a .90 probability that the mammogram will identify it as benign. In this particular case, the mammogram identifies the lump as malignant. In light of these mammogram results, what is your estimate of the probability that this lump is malignant?

Of the 100 doctors surveyed, 95 gave probabilities of around 75 percent. However, the correct probability is only 7.5 percent, as shown by the following two-way classification of 1000 patients:

	TEST POSITIVE	TEST NEGATIVE	TOTAL
LUMP IS MALIGNANT	8	2	10
LUMP IS BENIGN	99	891	990
TOTAL	107	893	1000

In 10 of these cases (1 percent) the lump is malignant, and in 990 cases it is benign. Looking across the numerical rows, we see that the test gives the correct diagnosis in 80 percent of malignant cases and 90 percent of benign cases. Yet, looking down the first numerical column, we see that of the 107 patients with positive test results, only 7.5 percent actually have malignant tumors: $8/107 = .075$.

The data given here imply that when there is a malignant tumor, there is a .80 probability that it will be detected by a mammogram; however, in those cases in which the mammogram indicates the presence of a malignant tumor, there is only a .075 probability that it will actually turn out to be malignant. It is very easy to misinterpret conditional probabilities, and these doctors evidently misinterpreted them. According to the researcher who conducted this survey,

> The erring physicians usually report that they assumed that the probability of cancer given that the patient has a positive X-ray . . . was approximately equal to the probability of a positive X-ray in a patient with cancer. . . . The latter probability is the one measured in clinical research programs and is very familiar, but it is the former probability that is needed for clinical decision making. It seems that many if not most physicians confuse the two.[8]

The solution, no doubt, is not for doctors to refrain from using probabilities, but to become better informed about their meaning and interpretation.

5.6 One in 800 U.S. women are infected with HIV. Thirteen percent of all U.S. females are black, but 53 percent of U.S. female HIV carriers are black. What is the probability that a randomly selected black U.S. woman has HIV? What is the probability that a randomly selected nonblack U.S. woman has HIV? To answer these questions, construct a contingency table with a hypothetical population of 10 million women, with black and nonblack for the columns and HIV and no HIV for the rows.

5.7 In *Craig vs. Boren* (1976), the U.S. Supreme Court considered whether important government objectives were served by the gender distinction in an Oklahoma statute that prohibited the sale of 3.2 percent beer to males under the age of 21 and to females under the age of 18. Among the evidence considered in this case were the following data on persons arrested in Oklahoma for driving under the influence (DUI) during the last 4 months of 1973: 92 percent of those arrested were male, 8 percent of the males arrested were under the age of 21, and 5 percent of the females arrested were under the age of 21. Assume a hypothetical population of 10,000 DUI arrests, and construct a contingency table with the columns showing gender (female or male) and the rows showing age (under 21 or 21 and older). Of these DUI arrests, what is the probability that a randomly selected person under the age of 21 is male? Do these data indicate that the gender and age of DUI arrests are dependent or independent?

5.8 A 1968 story in a Denver newspaper argued that women are better drivers than men.[9] Among the evidence cited: "Of 101 drivers involved in an accident while passing on a curve, 15 were women." In addition to automobile data, the newspaper cited these data: "3000 men were injured on bicycles in the state in 1967 and 34 were killed, compared with 662 females injured and 11 killed." Explain why these data are not sufficient to show that women drive more safely than men around curves and while riding bicycles.

5.9 Use some specific hypothetical numbers to explain why these data do not necessarily justify this conclusion by the magazine *California Highways:*

> A large metropolitan police department made a check of the clothing worn by pedestrians killed in traffic at night. About four-fifths of the victims were wearing dark clothes and one-fifth light-colored garments. This study points up the rule that pedestrians are less likely to encounter traffic mishaps at night if they wear or carry something white after dark so that drivers can see them more easily.[10]

5.10 Evaluate this advice for winning at craps in Las Vegas:

> [F]ind a hot table. Never remain at a cold one. Always make it a policy to keep looking—move around! A tip-off might be the yelling crowd where a hot roll may be taking place. Another indication is a lot of money spread every which way around the table by numerous players. . . . [I]t is far more lucrative to tag along on the tail end of a hot roll than to go in fresh on a cold one. And no one moving from table to table will actually catch a hot roll from the beginning. If 65 percent of a streak is caught, it's enough![11]

5.3 The Binomial Model

We turn now to an explanation of how probabilities can be used to explain data, so that we will be able to draw statistical inferences. Remember that there are two kinds of data: *quantitative* data, such as heights and weights, which have natural numerical values, and *categorical* data, such as whether a person is female or male, which do not have natural numerical values but can be grouped into nonnumerical categories. This section describes a simple, yet powerful model for analyzing categorical data. Then we do the same for quantitative data.

The outcome of a coin flip—heads or tails—is an example of categorical data. We can use a simple probability model to describe the "experiment" of flipping a coin and recording the result. When an evenly balanced coin is flipped fairly, there is a .5 probability of heads and a .5 probability of tails. What about for 10 flips? If 10 coins are tossed fairly (or a single coin is tossed 10 times), what is the probability of exactly 5 heads? What are the chances of 6, 7, or even 10 heads? And what about that hypothetical long run? Are we guaranteed 50 percent heads, or is there some chance of 51 percent, or 60 percent, or 100 percent? In this section we see how to answer these questions and many more.

Coin flips are but one example of a great many phenomena that were analyzed by James Bernoulli (1654–1705), one of an eminent family of Swiss mathematicians, and can be described as follows:

In the **binomial model,** there are n trials, or experiments, and

1. Each trial has two possible categories of outcomes, not necessarily equally likely, that are labeled generically *success* and *failure*.
2. The probability of success p is the same for each trial.
3. The trials are independent, in that the outcome of any trial is not affected by the outcomes of other trials

The tossing of many coins or the repeated tossing of a single coin satisfies these binomial-model conditions. So does the roll of dice or the spin of a roulette wheel, if we label some of the outcomes successes, for example, odd numbers or the number 5. So do all the following situations. Each item in a group of electronic components has the same probability of failure (and failures are independent). Each person who makes an airline, hotel, or rental car reservation has a constant, independent probability of not showing up. A person is given independent medical tests, each with the same probability of a particular diagnosis. We conduct a political poll and, for each independent interview, have a constant probability of selecting someone who favors a particular position or candidate. The draw for a jury has a constant probability of picking someone who will find a particular criminal defendant guilty. Each day, the probability that stock prices will increase is constant, no matter what happened to stock prices on other days.

Throughout this book, you will encounter an enormous variety of situations that can be described by the binomial model. There are several examples discussed in this chapter and many more in the exercises. Other examples will appear in later chapters and on your exams. By the time you finish this course, you may be seeing the binomial model all around you, and indeed it is.

Sampling without Replacement

If we draw cards, one after another, from a standard deck of 52 cards, the binomial model is not strictly appropriate because the probabilities change as cards are removed from the deck. Similarly, if we interview 1000 randomly selected households, each selection slightly changes the pool of remaining households and so alters the probability of choosing households with certain characteristics. When a selected item is removed and cannot be selected again, this is called **sampling without replacement.** The binomial model would be appropriate if each selection of cards (or households) were replaced and the "deck" thoroughly reshuffled before the next selection was made. If each selected item is replaced and can be selected again, this procedure is called **sampling with replacement.**

More complicated models can handle the messy case of sampling without replacement. Sometimes, however, the "deck" is so large that the binomial model is a reasonable and convenient approximation. Imagine that we have 1 million decks of cards thoroughly shuffled to form one enormous deck with 52 million cards. Perhaps success is drawing a heart. Since there are 13 million hearts scattered among the 52 million cards, the initial probability of a heart is exactly .25. If the first card drawn is a heart and it is not replaced, the probability of a heart on the second draw falls to

$$\frac{12{,}999{,}999}{51{,}999{,}999} = .2499999856$$

If the first card drawn is not a heart, the probability of a heart on the second draw rises to

$$\frac{13{,}000{,}000}{51{,}999{,}999} = .2500000048$$

For most purposes, it would be perfectly acceptable to assume that the probability of a heart is still .25 and that the binomial model is applicable. There is undoubtedly more error introduced by inadequately shuffling 52 million cards than by neglecting to replace a few cards taken from the deck. Of course, if 26 million cards were drawn without replacement from a deck of 52 million cards, the probability of a heart could change considerably. A rule of thumb is that the binomial model is a good approximation when the sample is not more than 10 percent of the size of the population from which the sample is drawn.

In this same spirit, the binomial model is commonly applied to public opinion surveys, in which a few thousand people are selected from a population of millions, even though the selection is done without replacement (we do not want to interview the same person twice). The slight variation in the probabilities is trivial compared to the more worrisome danger that the deck is not adequately shuffled, giving every person the same probability of being selected and interviewed.

Binomial Probabilities

A **random variable** x is a variable whose numerical value is determined by chance, the outcome of a random phenomenon.* In the binomial model, we can let x be the number of successes in n trials. The probability of exactly x successes is given by the following probability formula:

The **binomial distribution** states that in n binomial trials with probability p of success on each trial, the probability of exactly x successes is

$$P[x \text{ successes}] = \binom{n}{x} p^x (1-p)^{n-x} \tag{5.1}$$

where

$$\binom{n}{x} = \frac{n!}{x!(n-x)!} = \frac{n \cdot (n-1) \cdots (n-x+1)}{x(x-1) \cdots 1}$$

In practice, whenever possible, we avoid using Eq. (5.1) to calculate binomial probabilities by hand. It is easier, faster, and more accurate to use statistics software or a convenient table, such as Table 2 at the end of this book. The How to Do It box compares a calculation by hand with these easy alternatives.

*For simplicity, we do not use upper- and lowercase letters to distinguish between a random variable, which can take on different values, and the actual values that happen to occur.

HOW TO DO IT

Binomial Probabilities

Consider the probability of 3 heads when 5 coins are tossed fairly.

1. Is the binomial model appropriate? Yes. Each observation has two possible outcomes, heads and tails, and the observations are independent with a constant probability of heads or tails.
2. What is a *success,* and what are the values of p, n, and x? If we call heads a success, then the success probability is $p = .5$, the number of trials is $n = 5$, and the number of successes is $x = 3$.
3. Equation (5.1) shows the binomial probability to be

$$
\begin{aligned}
P[3 \text{ successes}] &= \binom{n}{x} p^x (1-p)^{n-x} \\
&= \binom{5}{3} (.5^3)(.5^2) \\
&= \frac{5!}{3!\,2!} (.5^3)(.5^2) \\
&= \frac{5(4)(3)}{3(2)(1)} (.125)(.25) \\
&= \frac{60}{6} (.125)(.25) \\
&= .3125
\end{aligned}
$$

4. An easy alternative is computer software. Table 5.2 shows the output from Smith's Statistical Package. After the user enters the values of p, n, and x in the three boxes, the binomial probabilities are shown in the bottom half of the computer screen.
5. If software is not easily accessible, we can consult Table 2 at the end of this book. Table 5.3 shows how to do this. Look down the left-hand column for $n = 5$, and then look in the top row for $p = .5$; the six probabilities for $x = 0, 1, 2, 3, 4,$ and 5 are shown in the column under $p = .5$, across from $n = 5$. The probability we want is .3125 for $x = 3$.

Table 5.2

Binomial probabilities from Smith's Statistical Package for $p = .5$, $n = 6$, and $x = 3$

```
If the probability of success is p, then the probability
of exactly x successes in n trials is given by the
binomial distribution.

    What is this success probability?        p = [ .5 ]

    How many trials?                         n = [ 5 ]

    How many successes?                      x = [ 3 ]

In 5 independent trials, each with success probability
.5, the probability of exactly 3 successes is .312500.

The probability that the number of successes will be

    Fewer than 3 is .500000

    More than 3 is .187500
```

Table 5.3

Probability of 3 heads in 5 flips of a fair coin ($n = 5$, $p = .5$, $x = 3$)

		p										
n	*x*	.01	.05	.10	.15	.20	.25	.30	.35	.40	.45	.50
5	0	.9510	.7738	.5905	.4437	.3277	.2373	.1681	.1160	.0778	.0503	.0313
	1	.0480	.2036	.3280	.3915	.4096	.3955	.3602	.3124	.2592	.2059	.1563
	2	.0010	.0214	.0729	.1382	.2048	.2637	.3087	.3364	.3456	.3369	.3125
	3	.0000	.0011	.0081	.0244	.0512	.0879	.1323	.1811	.2304	.2757	.3125
	4	.0000	.0000	.0004	.0022	.0064	.0146	.0284	.0448	.0768	.1128	.1563
	5	.0000	.0000	.0000	.0001	.0003	.0010	.0024	.0053	.0102	.0185	.0313

It is important to recognize that the binomial formula gives the probability of *exactly* x successes, for example, the probability of exactly 3 heads in 5 coin tosses. If we want to know the probability of 3 or more heads, then we must add the probability of 3 heads, the probability of 4 heads, and the probability of 5 heads.

The right-hand column of Table 5.3 shows the binomial probabilities for the six possible outcomes ($x = 0, 1, 2, 3, 4$, and 5) when 5 coins are tossed fairly. A graph of the complete set of probabilities, as shown in Fig. 5.3, is called a **probability histogram,** or a **population histogram,** or a **probability distribution.** A probability histogram is scaled so that the total area inside the rectangles is equal to 1. Here, with the bar widths equal to 1, the height of each bar is equal to the probability of that number occurring.

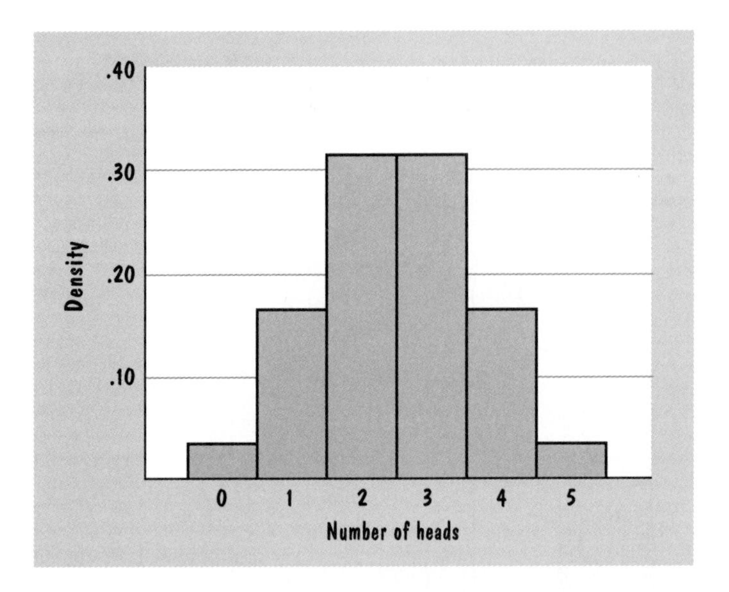

Figure 5.3 Binomial probability histogram for five coin flips (n = 5, p = .5).

A probability histogram looks like the data histograms introduced in Chap. 2, but is conceptually different since it refers to the probabilities of various outcomes occurring, rather than the actual frequencies with which the outcomes happen to occur in a particular set of experiments. These two histograms—data and probability—are related by the fact that in the long run, the observed frequencies will be very close to the theoretical probabilities.

To illustrate this relationship, I tossed 5 pennies and got 2 heads. I tossed the pennies again and got 1 head. I repeated this experiment 8 more times and obtained these results for my 10 experiments:

2 1 2 2 5 2 0 4 3 1

Figure 5.4 shows a data histogram for these data in part *a* and for 100 and 1000 such experiments in parts *b* and *c*. As the number of experiments increases, the data histograms in Fig. 5.4 look more and more like the probability histogram in Fig. 5.3.

We can think of the probability histogram as describing what our data would look like if the experiment were repeated an innumerable number of times; this is why a probability histogram is also called a population histogram. If we toss 5 coins once, there is a .3125 probability that we will get 3 heads. If we repeat this experiment many times, we expect to get 3 heads about 31.25 percent of the time. Similarly, we anticipate, in the long run of many such experiments, getting 0 heads about 3.12 percent of the time, 1 head approximately

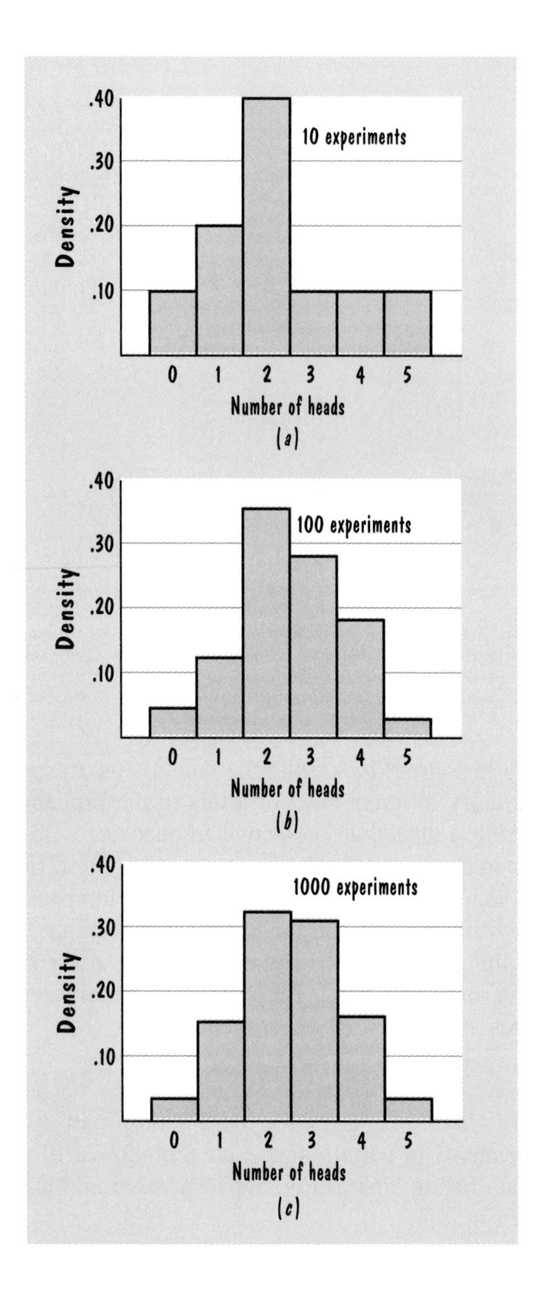

Figure 5.4 Data histograms for (*a*) 10, (*b*) 100, and (*c*) 1000 experiments of tossing five coins.

15.62 percent of the time, and so on. A probability histogram can be thought of as the population from which a single experiment is a random draw.

Note that because heads and tails are equally likely, this particular probability histogram is symmetric: 3 tails are as likely as 3 heads, 4 tails are as likely as 4 heads, and 5 tails are as likely as 5 heads. If success and failure are not equally likely, the binomial probability histogram is not symmetric. If we have a four-sided die, with the sides numbered 1 through 4, and consider success to be the number 1, then the probability of success on each roll is $p = .25$ and the probability of failure is $1 - p = .75$. the roll of 5 such dice can be described by the binomial model, and Table 2 shows the probabilities of 0 to 5 successes. Look on the first page of Table 2, under $p = .25$ and across from $n = 5$: The probability of 0 successes is .2373, the probability of 1 success is .3955, and so on. Figure 5.5 uses these five probabilities to show an asymmetric probability histogram. The most likely outcome is one success. Two successes are slightly more likely than none, and there are rapidly declining probabilities of three, four, or five successes.

The Population Mean and Standard Deviation

It can be shown (with some difficulty) that the mean and standard deviation of the number of successes x in the binomial model's population are as follows.

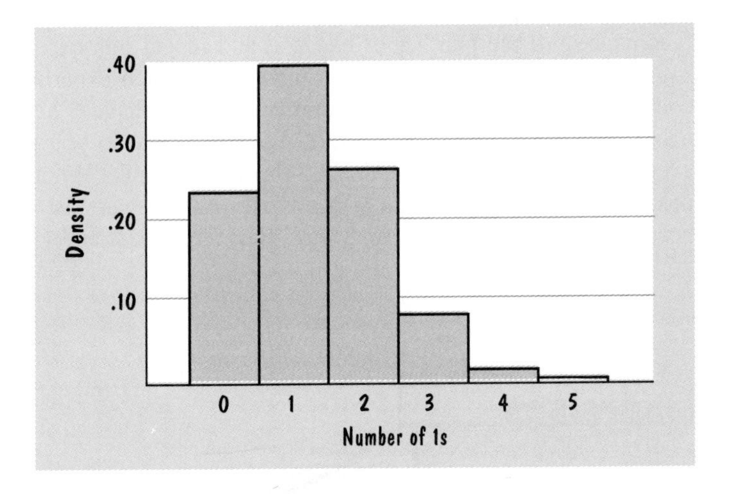

Figure 5.5 Binomial probability histogram for five four-sided dice ($n = 5$, $p = .25$).

In the binomial model, the population mean and standard deviation of the number of successes x are as follows:

Mean of x: $$\mu = np \tag{5.2}$$

Standard deviation of x: $$\sigma = \sqrt{np(1-p)} \tag{5.3}$$

We use the Greek symbols μ (pronounced "mew") and σ (pronounced "sigma") to distinguish the population mean and standard deviation from the sample mean $\bar{x}$ and standard deviation s discussed in Chap. 3, which refer to the data obtained in a particular sample.

We also have formulas for the population mean and standard deviation for the success proportion x/n, the fraction of the trials that are successes:

In the binomial model, the population mean and standard deviation of the success proportion x/n are as follows:

Mean of x/n: $$\mu = p \tag{5.4}$$

Standard deviation of x/n: $$\sigma = \sqrt{\frac{p(1-p)}{n}} \tag{5.5}$$

The formulas for the standard deviation of x and x/n are difficult to explain intuitively, so we just accept them as correct. However, the formulas for the population means of x and x/n agree with common sense. Suppose that we flip 5 coins, record the number of heads x, and repeat this experiment over and over until we have 1 billion experiments, with each experiment consisting of 5 coin flips. In some of these experiments, we will have 5 heads; in others 0 heads; and in still others 1, 2, 3, or 4 heads. What do you predict will be the average number of heads in these 5-flip experiments? What will be the average proportion of the 5 flips that is heads? If your answers are an average of 2.5 heads, which is 50 percent of 5 flips, then Eqs. (5.2) and (5.4) confirm that your intuition is correct.

exercises

5.11 Identify which of the following can be described by the binomial model, and explain your reasoning.

a. Head or tail when a bent coin is flipped
b. Safe or out when a baseball player bats
c. Black or not when a bird lands on a telephone wire
d. Republican or not when a governor is elected in Oregon

5.12 After an Iraqi missile hit the U.S. frigate *Stark,* novelist tom Clancy wrote that

> It's quite possible that the *Stark's* radar did not see the weapons separate from the launching aircraft. I have seen radar miss an aircraft carrier sitting in plain view on the horizon. The same radar will "paint" the blip nine times out of 10, but the laws of statistics mean that occasionally it will miss on two of the three consecutive passes.[12]

Assuming that the binomial model applies with $p = .1$, what is the probability of 2 or more misses in 3 trials?

5.13 A traditional Cayuga game uses six peach pits that have been rubbed smooth and blackened on one side with fire. The six peach pits are placed in a cup, shaken thoroughly, and then inspected. The player is credited with 5 points if all six pits have a similar side face up (either all blackened or all not blackened) and is credited with 1 point if five of the six pits have a similar side face up, with one pit different. Otherwise, the player receives no points. Assuming that each pit is equally likely to land blackened or not-blackened side up, what is the probability of receiving 5 points? What is the probability of 1 point?

5.14 In 1984, a sports columnist for the *Dallas Morning News* had a particularly bad week picking the winners of National Football League (NFL) games—he got 1 right and 12 wrong, with 1 tie. Afterward, he wrote, "Theoretically, a baboon at the Dallas Zoo can look at a schedule of 14 NFL games, point to 1 team for each game and come out with at least 7 winners." The next week, Kanda the Great, a gorilla at the Dallas Zoo, made his predictions by selecting pieces of paper from his trainer. Kanda got 9 right and 4 wrong, better than all six *Dallas Morning News* sportswriters. Assuming no ties, what is the probability that a baboon picking the winners of 13 games will select at least 9? Will select fewer than 2 winners?

5.15 In 1980, the United States tried a daring rescue of U.S. hostages being held by Iranians in the U.S. Embassy in Teheran. The U.S. plan was to bring in soldiers in 8 helicopters traveling 800 miles across the desert at low altitudes during the night. The military commanders believed that at least 6 of the 8 helicopters would have to reach Teheran for the mission to have a chance of succeeding. As it turned out, the rescue attempt was canceled when 3 of the helicopters became disabled. Use the binomial distribution to calculate the probability that at least 6 of 8 helicopters reach Teheran if each has a .75 probability of success, and if each has a .90 probability of success.

5.4 The Fallacious Law of Averages

The binomial distribution was used by James Bernoulli and other mathematicians to show that, in the long run, the success proportion x/n is almost certain to be close to the probability of success p. Thus in a large number of trials, the fraction of coin flips that land heads almost certainly will be close to .5. Unfortunately, far too many have misinterpreted this argument as stating that in the long run, the numbers of heads and tails must be exactly equal and, therefore, any short-run deficit of heads or tails must soon be balanced out by an offsetting surplus. In this section we look at the correct logic and see where intuition has so often been led astray.

The Law of Large Numbers

Figure 5.6*a* to *c* shows the probability histogram for 10, 100, and 1000 coin flips. Notice that each histogram is centered at $x/n = .5$ and that the histogram narrows around $x/n = .5$ as n increases. This is Bernoulli's **law of large numbers:** As n increases, the probability that x/n will be close to the success probability p approaches 1. Equivalently, for large values of n, the probability is small that x/n will be far from p. In our coin flip example, the probability of more than 51 percent heads is .382 with 100 flips, .253 with 1000 flips, and .022 with 10,000 flips.

A Common Fallacy

The law of large numbers is mathematically correct and intuitively obvious, but it is frequently misinterpreted as a fallacious *law of averages,* which states that, in the long run, x/n must be exactly equal to p. For example, some people think that in 1000 coin flips there must be exactly 500 heads and 500 tails; thus if tails come up more often than heads in the first 10, 50, or 100 flips, heads must now come up more frequently than tails in order to "average" or "balance" things out. This erroneous belief is also known as the *maturity of chances,* the *law of equilibrium,* and (most correctly) the *gambler's fallacy.* This belief is wrong, but widespread. For example, one gambler wrote:

> Flip a coin 1000 times and it'll come up heads 500 times, or mighty close to it. However, during that 1000 flips of the coin there will be frequent periods when heads will fail to show and other periods when you'll flip nothing but heads. Mathematical probability is going to give you roughly 500 heads in 1000 flips, so that if you get ten tails in a row,

> there's going to be a heavy preponderance of heads somewhere along the line.
>
> Regardless of mathematics and theory of probability, if I'm in a craps game and the shooter makes ten consecutive passes [wins], I'm going to bet against him on his eleventh throw. And if he wins, I'll bet against him again on the twelfth. Maybe it's true that the mathematical odds on every throw of the dice remain the same, regardless of what's gone before—but how often do you see a crapshooter make 13 or 14 straight passes?[13]

Thirteen consecutive heads are a rare event, but it is a very different question to ask the probability of 13 straight heads, *given that you have already*

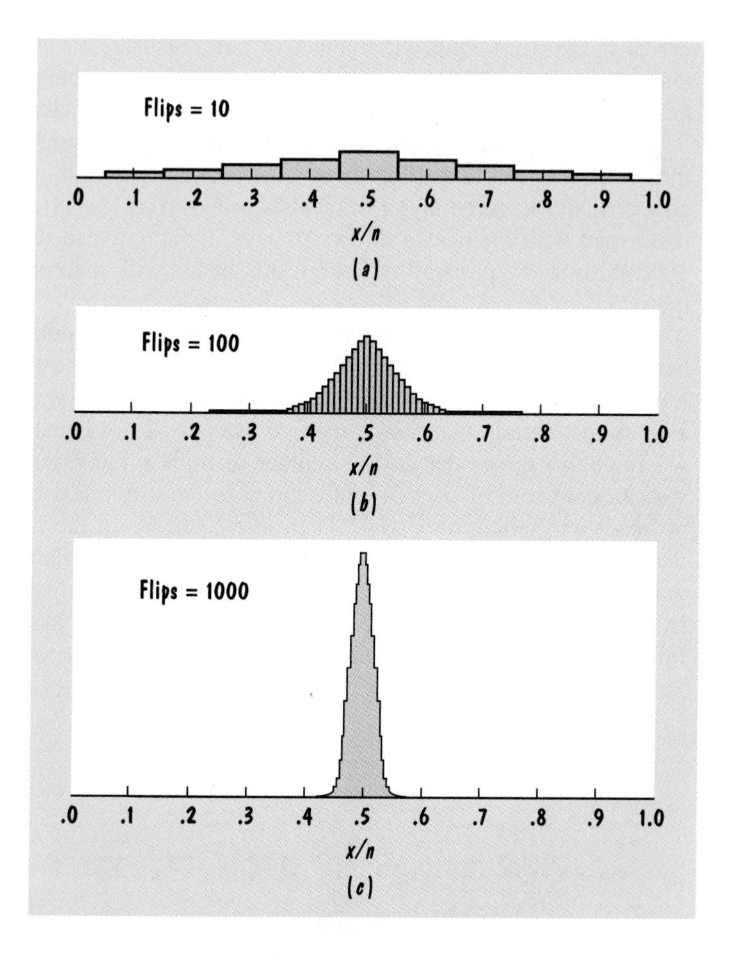

Figure 5.6 Binomial probability histogram for fair-coin flips.

tossed 12 heads in a row. The coin does not remember what happened on the last 12 flips, and it is not interested in whether someone wins or loses a wager. A coin is simply an inanimate object that is repeatedly tossed in the air and examined by curious humans. If the coin is unbent and fairly tossed, then heads and tails are equally likely to appear, no matter what happened on the last flip or the last 999 flips.

The Mathematical Error

The basis for the fallacious law of averages is the mistaken belief that heads and tails must come up equally often; a run of tails must consequently be balanced by a run of heads. The law of large numbers does not say that the *number* of heads must equal the number of tails. Instead, it says that if we begin making a large number of coin flips, the probability is close to 1 that the *fraction* of these forthcoming throws that are heads will be close to .5.

First, notice that the law of large numbers says nothing about the last 12 throws or the last 12 million throws—which have already happened, cannot be changed, and have no effect on future throws. The law of large numbers is concerned with the next n throws. It says, looking ahead to predict the next 1000 throws, the probability is high that heads will come up close to one-half the time.

Second, a *fraction* that is close to .5 is not inconsistent with the *numbers* of heads and tails being unequal. In fact, while we are almost certain that in the long run, the fraction of the flips that are heads will be very close to .5, we are also almost certain that the number of heads will not equal the number of tails!

Table 5.4 shows that as the number of trials n increases, the heads fraction x/n can converge to .5 even while the absolute difference between the numbers of heads and tails grows large. There are 57 heads in the first 100 flips, 7 more than an exact 50 percent; yet we do not need 7 extra tails to average things out. The extra 7 heads will not be offset by 7 tails, but instead will be diluted by additional tosses, in that 7 extra heads are unimportant in a long run of a million or trillion flips. In fact, as shown in Table 5.4, heads can continue com-

Table 5.4

The number of heads need not equal the number of tails

TOSSES n	HEADS x	FRACTION x/n	HEADS MINUS TAILS
100	57	.5700	14
1,000	520	.5200	40
10,000	5,040	.5040	80
100,000	50,060	.5006	120
1,000,000	500,100	.5001	200
1,000,000,000	500,001,000	.500001	2,000

ing up more often than tails even while the relative frequency of heads converges to .5.

We can make an even stronger statement. Not only do we not need the number of heads to equal the number of tails in the long run, but we can, in fact, be almost certain they will *not* be equal. Look at the binomial probabilities in Table 2. With two flips, the probability of 1 head and 1 tail is .5. With 4 flips, the probability of 2 heads and 2 tails drops to .375. The probability that the numbers of heads and tails will be equal keeps falling as n increases: .3125 for $n = 6$, .274 for $n = 8$, and .2462 for $n = 10$. For $n = 50$, the largest value given in Table 2, the probability of 25 heads and 25 tails is .1123. Computer software shows that the probability of an equal number of heads and tails continues to decline as n increases. For $n = 1000$, for example, there is only a .025 chance of exactly 500 heads.

While the probability is small that there will be exactly 500 heads in 1000 flips, the probability is large that there will be close to 500 heads. To illustrate this crucial distinction, Table 5.5 shows, for different values of n, the probability that x/n will differ from .5 by less than .001, and it also shows the probability that the number of heads will differ from the number of tails by more than 100. As n increases, we become more and more certain that the fraction of the flips that are heads will be very close to .50. At the same time, we are increasingly certain that there will be at least 100 more heads than tails or at least 100 more tails than heads. The law of large numbers does not require the numbers of heads and tails to be exactly equal!

Some Examples

The fallacious law of averages is all around us. It is commonly appealed to by gamblers, hoping somehow to find a profitable pattern in the chaos created by random chance. When a roulette wheel turns up a red number several times in

Table 5.5

We are almost certain that for large n, x/n will be very close to .5 and that there will be a large difference between the numbers of heads and tails

NUMBER OF TOSSES n	PROBABILITY THAT x/n WILL BE BETWEEN .499 AND .501	PROBABILITY THAT THERE WILL BE AT LEAST 100 MORE HEADS THAN TAILS OR AT LEAST 100 MORE TAILS THAN HEADS
1,000	.025	.002
10,000	.151	.322
100,000	.529	.754
1,000,000	.955	.919
1,000,000,000	1.000	.998

a row, there are always people eager to bet on black, counting on the law of averages. Others will rush to bet on red, trying to catch a "hot streak." The casino cheerfully accepts wagers from both, confident that future spins do not depend on the past. One of the most dramatic roulette runs occurred at Monte Carlo on August 18, 1913. At one of the tables, black suddenly started coming up again and again. After about 10 blacks in a row, the table was surrounded by excited people betting on red and counting on the law of averages to reward them. But black came up again and again. After 15 straight blacks, there was near panic as people tried to reach the table so that they could place even heavier bets on red. Still black came up. By the 20th black, desperate bettors were wagering every chip they had left on red, hoping to recover a fraction of their losses. When this memorable run ended, black had come up 26 times in a row and the casino had won millions of francs.

In honest games, at Monte Carlo and elsewhere, "balancing systems" based on the law of averages do not work, except by chance. In fact, the few gambling systems that have paid off have been based on the opposite principle—that physical defects in a roulette wheel or other apparatus cause some events to occur more often than others. If the number 6 comes up an unusually large number of times (not twice in a row, but maybe 35 times in 1000 spins), then perhaps some mechanical irregularity is responsible. Instead of betting against 6, expecting its success to be balanced out, we might bet on 6, hoping that some physical imperfection will cause 6 to continue to come up 35 times in every 1000 spins.

The sports pages are another fertile source of law-of-averages fallacies. Penn State's football team beat West Virginia in 1956 and 1957, tied in 1958, and then won the next 13 games in a row. On the eve of the 1972 game, newspapers reported that West Virginia had the law of averages on its side. Penn State won that game and 11 more after that—a long wait for those counting on the law of averages to give West Virginia a victory. On October 27, 1984, West Virginia ended 28 years of frustration by finally beating Penn State, 17 to 14. It was not the law of averages that won the game. Going into the 1984 game, West Virginia had six wins and one loss and was ranked 14th and 18th in the AP and UPI polls; Penn State was 5 to 2 and ranked 19th and 20th. After the game, the West Virginia coach said, "This is the greatest win I've been associated with since I've been here. Beating Oklahoma and Florida and Pitt twice is great, but to beat a team with the class and talent of Penn State ranks above them all." This win was earned by the players, not some fallacious law of averages.

Admittedly, sports do differ from games of chance in at least two ways. First, we may not have much information about the relative strengths and weaknesses of players or teams. If Penn State and West Virginia had never played each other or common opponents, we might be very unsure about the outcome; but when Penn State beats West Virginia year after year, we become convinced that Penn State has the stronger football program. As the victories pile up, it does not become more probable that Penn State will lose the next game. If anything, we become more convinced that Penn State will win yet again.

Second, players do have memories and emotions. They do care about wins and losses, and they do remember the last game. Sometimes a team will be beaten badly and play harder the next time, seeking revenge. At other times, teams that lose a lot of games also seem to lose their self-confidence and grow accustomed to losing. If West Virginia loses to Penn State year after year, West Virginia players may get fired up and Penn State may be nonchalant; but it is also possible that Penn State will play with confidence and West Virginia will be anxious and doubtful of their chances. Destructive self-doubts are seen in almost all sports, and athletes have turned to sports psychologists and even hypnotists to relax and forget past failures.

Similarly, when a baseball player goes hitless in 12 times at bat, a sports commentator often announces that he "has the law of averages on his side" or that he "is due for a hit." The probability of a base hit does not increase just because a player has not had one lately. The 12 outs in a row may have been bad luck, 12 line drives hit right at fielders. If so, this bad luck has not made the player any more likely to have good luck the next time at bat. If it has not been bad luck, then 12 outs in a row suggest that a physical or mental problem may be causing the player to do poorly. The manager of a baseball player who is 0 for 12 should be concerned—not confident that the player is due for a hit. Similarly, a player who has had 4 hits in last 4 at-bats should not be benched

THE BORN LOSER by Art Sansom

because he is "due for an out." Yet that is exactly what happened when a manager once had a pinch hitter bat for a player who already had four hits that day, explaining that players seldom get five hits in a row. (They never will if they do not have the opportunity!)

I once watched a Penn State kicker miss three field goals and an extra point in an early-season football game against the Air Force. The television commentator said that Joe Paterno, the Penn State coach, should be happy about those misses, as he looked forward to some tough games in coming weeks. The commentator explained that every kicker is going to miss some over the course of the season and that it is good to get these misses "out of the way" early in the year. This unjustified optimism is the law-of-averages fallacy again. Those misses do not have to be balanced by successes. If anything, Joe Paterno should have been very concerned by this poor performance, as it suggests that his kicker is not very good. Even if these misses were just bad luck, the kicker is certainly going to be very nervous on his next attempts.

Mary Sue Younger, a statistics professor at the University of Tennessee, has written for another misuse of the law of averages: "I was told by my automobile insurance salesman that some insurance companies will raise your premiums after so many years without a claim, figuring you are coming due for an accident!" If any conclusion can be drawn from many years without an accident, it is probably the opposite: You are a safe driver and should have your premiums reduced.

Hot Streaks in Sports

Sports fans often observe "hot" or "cold" streaks by athletes—the basketball player who makes (or misses) five shots in a row, the football quarterback who completes five passes in a row or throws five straight incomplete passes, the baseball player who gets four hits in a row or makes four outs in a row. What most fans do not appreciate is that even if each outcome is independent of previous outcomes, we will still observe streaks.

To illustrate this point, I flipped a coin 20 times, which we can interpret as 20 shots by a basketball player or 20 passes by a quarterback, with a head counting as a successful shot or completed pass and a tail counting as a missed shot or an incomplete pass. A string of consecutive heads is a hot streak, and a string of tails is a cold streak. Here are my results:

H H T T T T H H T H
H H H T H H T T H H

Beginning with the third flip, a streak of four tails in a row occurs. Beginning with the 10th flip, there is a streak of four heads in a row. I had a cold streak and then, a little later, a hot streak, even though every coin flip was independent!

I repeated this experiment 9 more times. In 7 of my 10 experiments, I had a hot or cold streak of four or longer. I had two streaks of 4, 4 streaks of 5, and one streak of 10 in a row! The table that follows shows the theoretical probability of observing streaks of various lengths in a sequence of 20 coin flips.[14] There is a .768 probability of a streak of four or more, and 7 of my 10 experiments yielded such streaks. My streak of 10 in a row is unusual, but it has about an 11 percent chance if, as here, the experiment is repeated 10 times.

Such experiments with coins do not prove that human hot or cold streaks are not due to physical or emotional factors. However, these experiments do show that the occurrence of a hot or cold streak does not, by itself, prove that unusual physical or emotional forces are at work. Streaks occur even with completely independent coin tosses. The important implication for players, coaches, and fans is that a hot streak does not guarantee continued success and a cold streak does not guarantee continued failure, any more than four heads in a row guarantee a heads on the next coin flip.

LENGTH OF STREAK	PROBABILITY
Fewer than 3	.021
3	.211
4	.310
5	.222
6	.121
7	.061
8	.029
9	.013
More than 9	.012

5.16 At the midpoint of the 1991 Cape Cod League baseball season, Chatham was in first place with a record of 18 wins, 10 losses, and 1 tie. (Ties are possible because games are sometimes stopped due to darkness or fog.) The Brewster coach, whose team had a record of 14 wins and 14 losses, said that his team was in a better position than Chatham: "If you're winning right now, you should be worried. Every team goes through slumps and streaks. It's good that we're getting [our slump] out of the way right now."[15] Explain why you agree or disagree with his reasoning.

5.17 Explaining why he was driving to a judicial conference in South Dakota, the chief justice of the West Virginia state supreme court said, "I've flown a lot in my life. I've used my statistical miles. I don't fly except when there is no viable alternative."[16] What do you suppose the phrase *used my statistical miles* means? Explain why you agree or disagree with the judge's reasoning.

5.18 Answer this question that a reader asked Marilyn vos Savant, who is listed in the *Guinness Book of World Records Hall of Fame* for highest IQ:[17]

> At a lecture on fire safety that I attended, the speaker said: "One in 10 Americans will experience a destructive fire this year. Now, I know that some of you can say you have lived in your homes for 25 years and never had any type of fire. To that I would respond that you have been lucky. But it only means that you are moving not farther away from a fire, but closer to one."

Is this last statement correct?

5.19 Explain why you either agree or disagree with Edgar Allan Poe:

> Nothing . . . is more difficult than to convince the merely general reader that the fact of sixes having been thrown twice in succession by a player at dice, is sufficient cause for betting the largest odds that sixes will not be thrown in the third attempt. . . . It does not appear that the two throws which have been completed, and which now lie absolutely in the Past, can have influence upon the throw which exists only in the Future. . . . [T]his is a reflection which appears so exceedingly obvious that attempts to controvert it are received more frequently with a derisive smile than anything like respectful attention. The error here involved—a gross error redolent of mischief—I cannot pretend to expose within the limits assigned me at present.[18]

5.20 Explain why you either agree or disagree with the following:

> This man has kept records of play on Blackjack, craps, and Roulette, and the house percentage on all three games works out inexorably, within a fraction of a percentage point—but there are times shown by his records when there are lengthy streaks of steady house wins along with other streaks of house losses.
>
> His records show that such streaks exist. But no matter how long he studies his figures, he hasn't been able to arrive at any pattern in them. He can see that the streaks are there, but he hasn't anything even resembling an explanation of why they occur when they do or why they last for a certain period.
>
> "Nonetheless," he insists, "if I'm scoring red and black in a Roulette game and red has come up 500 times in 900 spins of the wheel, I'm going to put my money on black the next 100 spins. What's more, I'll bet you that I come out ahead of the game.[19]

5.5 The Normal Distribution

In Chap. 2, we saw that histograms are often shaped like a bell: There is a single peak with the bars declining symmetrically on either side of the peak—gradually at first, then more steeply, then gradually again. W. J. Youdon, a distinguished statistician, described this familiar shape this way:

> T h e
> n o r m a l
> law of error stands
> out in the experience of
> mankind as one of the
> broadest generalizations of
> natural philosophy. It serves as
> the guiding instrument in researches in
> the physical and social sciences and in
> medicine, agriculture, and engineering. It is an
> indispensable tool for the analysis and the interpretation
> of the basic data obtained by observation and experiment.

The German 10-mark note is decorated with a bell-shaped curve, along with a portrait of Karl Gauss, who investigated the widespread prevalence of this curve. In the classroom, some professors are said to "grade by the curve." On the best-seller lists, a controversial book was called *The Bell Curve.*

What is this bell-shaped curve, and why do we see it in so many contexts? Why does it describe so well data on such disparate phenomena as the heights of humans, the weights of tomatoes, scores on the SAT test, baseball batting averages, and the location of molecules and stars? We will answer these questions in this section.

Probability Density Curves

The binomial model involves *discrete* random variables, where we can count the number of possible outcomes. The coin can be heads or tails; the die can be 1, 2, 3, 4, 5, or 6; there can be from 0 to 100 defective items. Other random variables can take on a continuum of values. For these *continuous* variables, the outcome can be any value in a given interval. For example, Fig. 5.7 shows a spinner for randomly selecting a point on a circle. We can imagine that this is a clean, well-balanced device in which each point on the circle is equally likely to be picked. But how many possible outcomes are there? How many points are there on the circle? In theory, there are an uncountable infinity of points in that between any two points on the circle, there are still more points.

Weight, height, and time are other examples of continuous variables. Even though we might say that Sarah Cunningham is 19 years old, a person's age can, in theory, be specified with infinite precision. Instead of saying that she is 19 or 20, we could say that she is 19½; or 19 years 7 months; or 19 years, 220 days, and 10 hours. With continuous variables, we can specify finer and finer gradations within any interval.

How can we specify probabilities when there are an uncountable number of possible outcomes? In Fig. 5.7, each point on the circle is equally likely, and a point surely will be selected; but if we give each point a positive probability p, the sum of this uncountable number of probabilities will be infinity, not 1. Mathematicians handle this vexing situation of an uncountable number of possible outcomes by assigning probabilities to *intervals* of outcomes, rather than to individual outcomes. For example, the probability that the spinner will stop between .25 and .50 is ¼.

We can display these interval probabilities by using a continuous **probability density curve,** as in Fig. 5.8, in which the probability that the outcome is

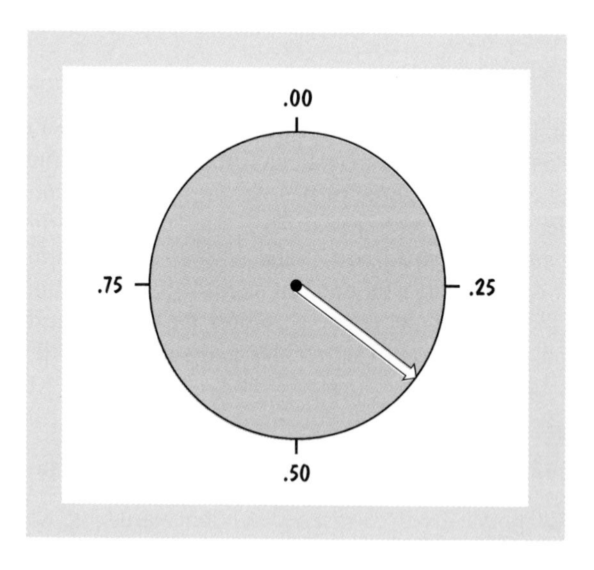

Figure 5.7 Pick a number, any number.

in a specified interval is given by the corresponding area under the curve. The shaded area shows the probability that the spinner will stop between .25 and .50. This rectangular area is (base)(height) = (.25)(1.0) = ¼. What is the probability that the spinner will stop between 0 and 1? This probability is the entire area under the curve: (base)(height) = (1)(1.0) = 1. In fact, the height of the probability density curve, 1.0, was derived from the requirement that the total probability must be 1. If the numbers on our spinner went from 0 to 12, as in a clock, the height of the probability density curve would have to be $1/12$ for the total area to be (base)(height) = (12)($1/12$) = 1.

We can summarize our discussion as follows. A *continuous random variable* x can have any value in an interval of real numbers. The probability that the value of x will be in any specified interval is shown by the corresponding area under the graph of a probability density curve. The density cannot be negative, and the total area under the curve must be 1.

The smooth density curve for a continuous random variable is analogous to the jagged probability histogram for a discrete random variable. The population mean and the standard deviation consequently have the same interpretation. The population mean is the anticipated long-run average value of the outcomes; the standard deviation measures the extent to which the outcomes are likely to differ from the mean. The population mean is at the center of a symmetric density function; in Fig. 5.8, for example, the mean is .5. More generally, however, the mean and standard deviation of a continuous random variable cannot be calculated without using advanced mathematics.

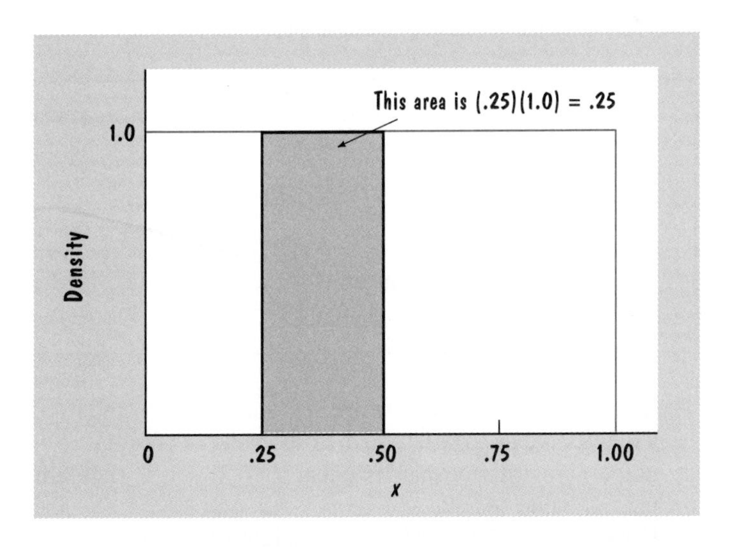

Figure 5.8 A continuous probability distribution for the spinner in Fig. 5.7.

Standardized Variables

Many random variables are the cumulative result of a sequence of random events. For instance, a random variable giving the sum of the numbers when eight dice are rolled can be viewed as the cumulative result of eight separate random events. The percentage change in a stock's price over a 6-month period is the cumulative result of a large number of random events during that interval. A person's height at 11 years of age is the cumulative result of a great many random events, some hereditary and others having to do with diet, health, and exercise.

These three different examples—dice rolls, stock price changes, and height—involve very different units of measurement—number, percent, and inches. In addition, each variable depends on the specific horizon, whether we are looking at 8 dice or 800 dice, stock price changes during 6 months or 6 years, and the heights of 11-year-olds or of 21-year-olds. In each of these examples, as we lengthen the horizon, both the mean and the standard deviation of the variable change. If we were to draw histograms for varying horizons for these three quite different variables, the histograms would look very dissimilar. However, in the 18th and 19th centuries, researchers discovered that when variables are *standardized,* in a particular way that will soon be explained, their histograms are often virtually identical! This remarkable similarity is perhaps the most important discovery in the long history of probability and statistics.

We have seen that the mean and standard deviation are two important tools for describing probability distributions. One appealing way to standardize variables is to transform them so that they have the same mean and the same standard deviation. This reshaping is easily done in the statistical beauty parlor.

To standardize a random variable x, we subtract its mean μ and then divide by its standard deviation σ:

$$z = \frac{x - \mu}{\sigma} \tag{5.6}$$

No matter what the initial units of x, the **standardized random variable** z has a mean of 0 and a standard deviation of 1.

The standardized variable z measures how many standard deviations x is above or below its mean. If x is equal to its mean, z is equal to 0. If x is 1 standard deviation above its mean, z is equal to 1. If x is 2 standard deviations below its mean, z is equal to -2.

For example, if we look at the height of a randomly selected U.S. woman between the ages of 25 and 34, we can consider this height to be a random variable x drawn from a population with a mean of 66 inches and a standard deviation of 2.5 inches. Here are the standardized z values corresponding to five different values of x:

x, INCHES	$z = (x - 66)/2.5$, STANDARD DEVIATIONS
61.0	−2
63.5	−1
66.0	0
68.5	+1
71.0	+2

Instead of saying that a woman is 71 inches tall (which is useful for some purposes, such as clothing sizes), we can say that her height is 2 standard deviations above the mean (which is useful for other purposes, such as comparing her height with the heights of other women).

Another reason for standardizing variables is that it is difficult to compare the shapes of distributions when they have different means and/or standard deviations. Figure 5.6 showed the probability histograms for the heads proportion x/n for 10, 100, and 1000 coin flips. All three histograms are centered at $p = .5$ (50 percent heads), but the standard deviation declines as n increases. If the graphs had been drawn with x, the number of heads, on the horizontal axis, then the centers of the three distributions would have been at 5, 50, and 500 and the standard deviation would increase as n increased.

By converting these three variables to standardized z values which have the same mean (0) and the same standard deviation (1), we can focus our attention on the shapes of these probability histograms without being distracted by their location and spread. The results of this standardization are seen in Fig. 5.9, which shows that as the number of flips increases, the probability distribution becomes increasingly shaped like a bell.

Now suppose that we want to examine the three probability distributions for a random variable x equal to the sum of the numbers obtained upon rolling 2, 10, and 100 standard six-sided dice. If we work with the nonstandardized variable x, we again have the problem that each probability histogram has a different mean and standard deviation. In addition, we find it difficult to compare the probability histograms for dice rolls with those for coin flips. With 10 dice rolls, the mean is 35 and the standard deviation is 5.40; with 1000 coin flips, the mean is 500 and the standard deviation is 15.81. Instead of trying to compare distributions with very different centers and spreads, we can convert the dice rolls to standardized z values, just as we did with the coin flips.

Figure 5.10 shows the results of this standardization for 2, 10, and 100 dice rolls. As with the coin flips, the probability histogram becomes increasingly bell-shaped as the number of trials increases. (Because the number of equally likely outcomes is larger with a die than with a coin, fewer trials are needed

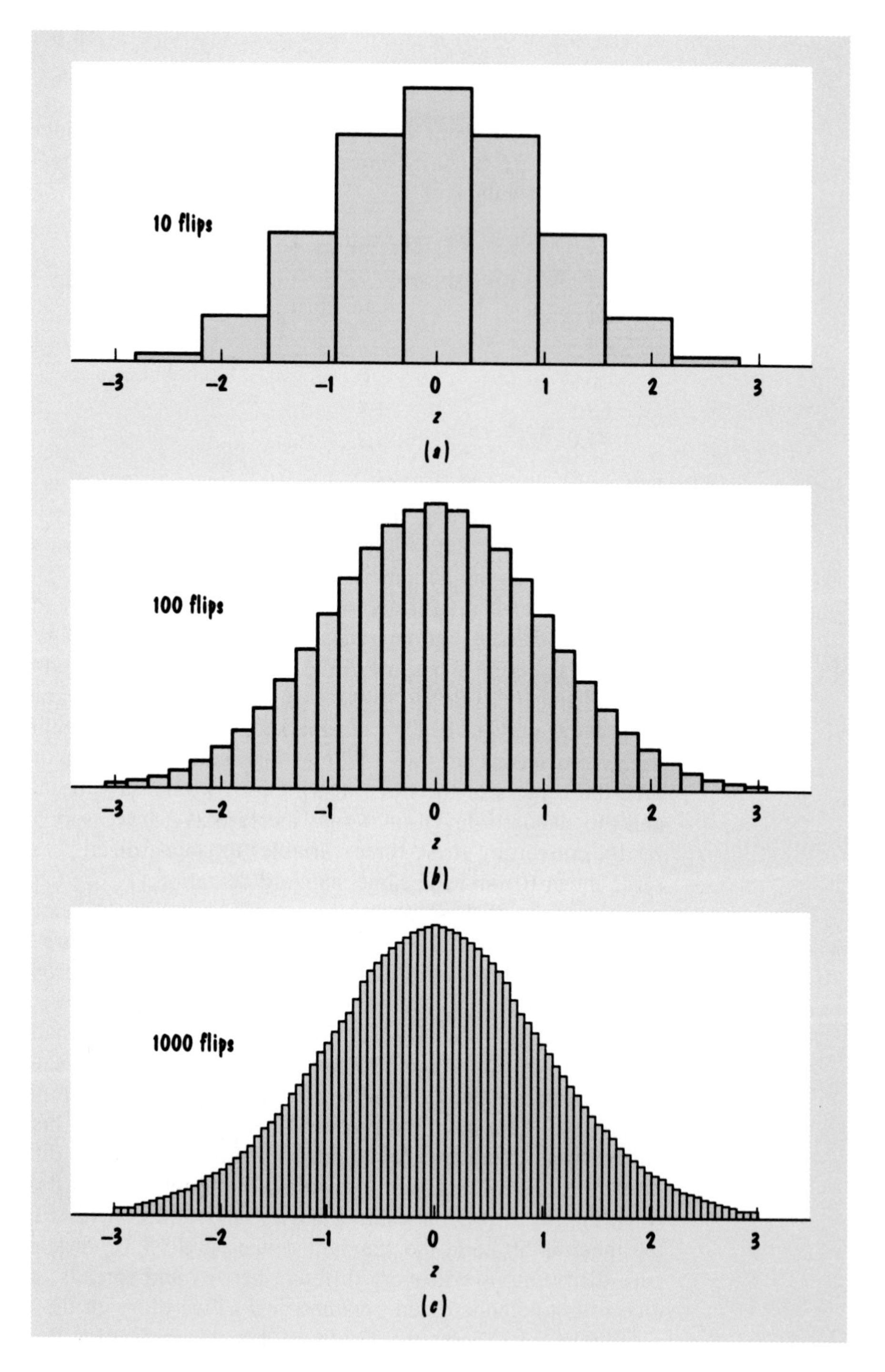

Figure 5.9 Probability histogram for fair-coin flips, using standardized z.

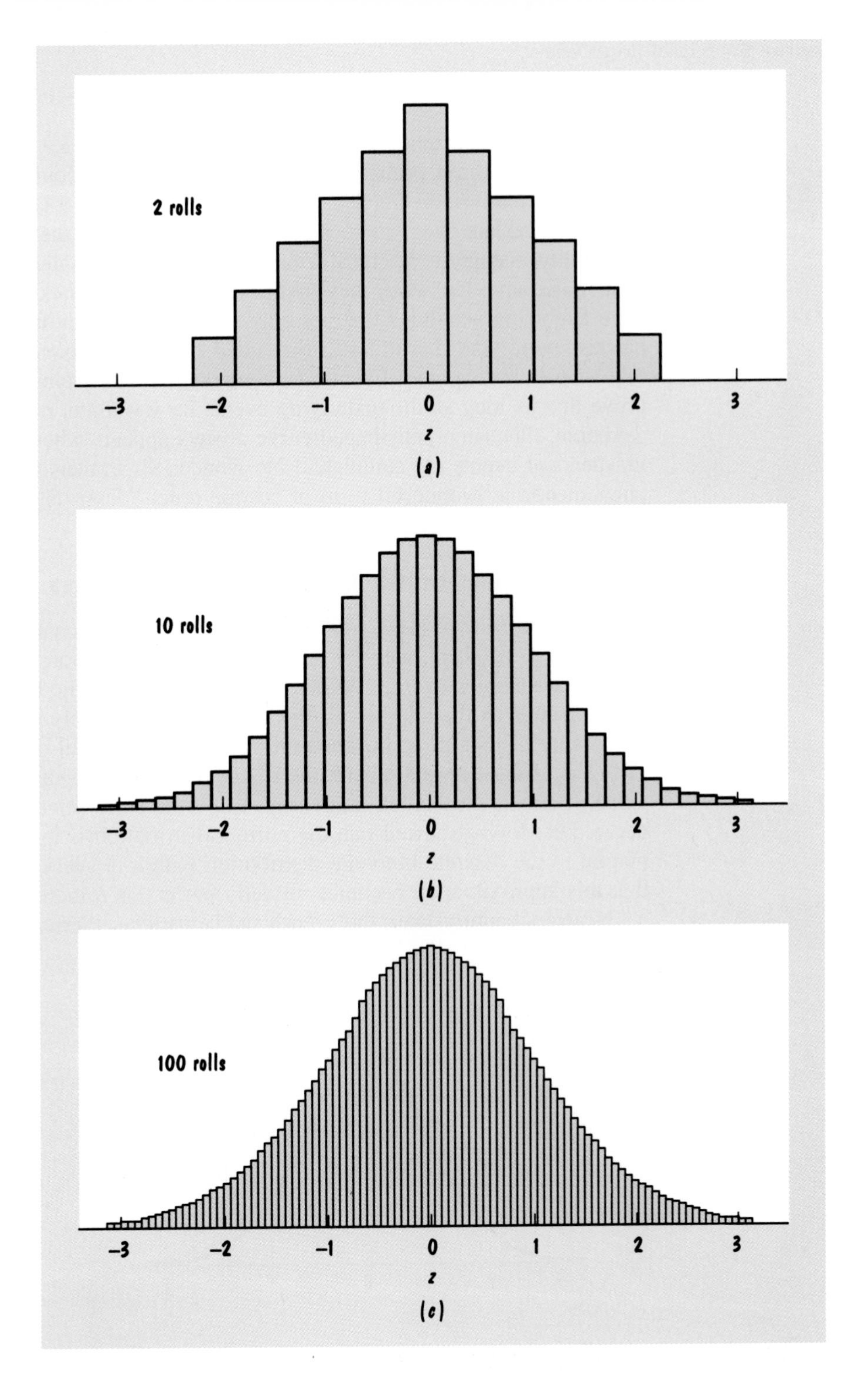

Figure 5.10 Probability histogram for six-sided dice, using standardized z.

for dice rolls to become bell-shaped.) Comparing Figs. 5.9 and 5.10, we see that the standardized probability distributions for 1000 coin flips and 100 dice rolls are virtually indistinguishable. When we cumulate a large number of independent uncertain events, either coin flips or dice rolls, the same bell-shaped probability histogram emerges! You can imagine the excitement that mathematicians must have felt when they first discovered this remarkable regularity. They were analyzing situations that not only were governed by unpredictable chance but also were very dissimilar (a two-sided coin and a six-sided die), and yet a regular pattern emerged. Even more remarkably, they were eventually able to prove that as long as the underlying events have a finite, nonzero standard deviation, this same bell-shaped curve always appears when a large number of independent events are cumulated! No wonder Sir Francis Galton called this phenomenon a "wonderful form of cosmic order" deserving deification.

The Central Limit Theorem

Abraham De Moivre (1667–1754), a French-born mathematician who emigrated to England, was the first to deduce the equation for the bell-shaped curve that emerges in Figs. 5.9 and 5.10. He analyzed the binomial model for fair coin flips as the number of flips increased indefinitely and found that the probability histogram approached the bell-shaped **normal distribution** shown in Fig. 5.11. The normal distribution is continuous, with the probability that the value of z is in a specified interval given by the corresponding area under the curve. De Moivre showed that the normal distribution is a continuous approximation to the discrete binomial distribution with a .5 success probability and that this approximation becomes virtually perfect as n increases.

Nearly a century later, the French mathematician Pierre Simon Laplace

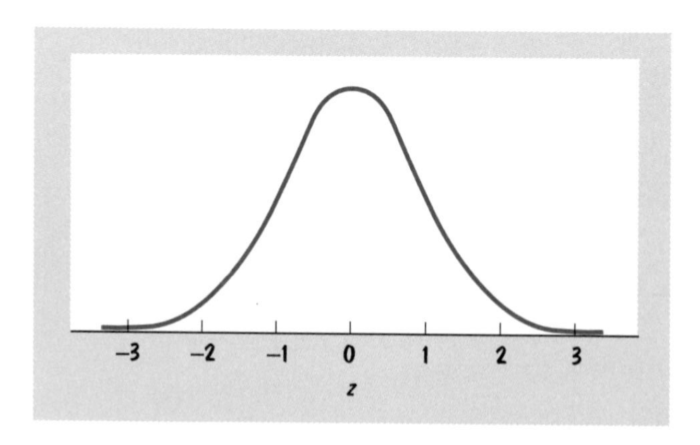

Figure 5.11 The normal distribution.

(1749–1827) proved that this convergence is true for all binomial distributions, with any success probability (except 0 and 1, of course), and for several other distributions, too. Laplace believed that his remarkable theorem could be used to explain most of the uncertainties that fill our lives, and he consequently applied the normal distribution to a wide variety of phenomena in both the natural and social sciences. Karl Gauss (1777–1855), the German "prince of mathematicians," applied the normal distribution to measurements of the shape of the earth and the movements of planets. His work was so extensive and influential that the normal distribution is often called the *Gaussian distribution.*

Others, following in the footsteps of Laplace and Gauss, applied the normal distribution to all sorts of physical and social data. They found that empirical data often conform to a normal distribution and proved that many specific probability distributions converge to a normal distribution when cumulated. In the 1930s, mathematicians proved that this convergence is true for a very broad range of probability distributions. This theorem, the culmination of 200 years of investigation, is one of the most famous mathematical theorems:

> The **central limit theorem** states that if z is a standardized sum of n independent, identically distributed (discrete or continuous) random variables with a finite, nonzero standard deviation, then the probability distribution of z approaches the normal distribution as n increases.

As remarkable as it is, the central limit theorem would have little practical value if the normal curve emerged only when n was extremely large. The normal distribution is important because it so often appears even when n is quite small. Look again at the case of $n = 10$ coin flips in Fig. 5.9; for most purposes, a normal curve would be a completely satisfactory approximation to that probability distribution. If the underlying distribution is reasonably smooth and symmetric (as with dice rolls and coin flips), the approach to a normal curve is very rapid and values of n larger than 20 or 30 are sufficient for the normal distribution to provide an acceptable approximation. A very asymmetric distribution, such as a .99 probability of success and .01 probability of failure, requires a larger number of trials (approximately 500).

Furthermore, the central limit theorem would be just a mathematical curiosity if the assumption that the cumulated variables are independent and identically distributed had to be satisfied strictly. This assumption is appropriate for dice rolls and other repetitive games of chance, but in practical affairs, it is seldom if ever exactly true. Probabilities vary from one trial to the next as conditions change and, in many cases, because of the outcomes of earlier trials. For example, height and weight depend on heredity and a lifetime of diet and exercise—factors that are not independent and do not have identical probability distributions. Yet histograms of height and weight are bell-shaped. A baseball player's chances of getting a base hit depend on the player's health, the opposing pitcher, the ballpark, the month, and whether it is a day game or a night game. Yet histograms for the total number of base hits over a season are bell-shaped. A student's ability to answer a particular test question depends on the question's difficulty and on the student's ability and alertness; but histograms of test scores are bell-shaped.

The central limit theorem is remarkably robust in that even if its assumptions are not exactly true, the normal distribution is still a pretty good approximation. To illustrate its power, consider the following sequence of events. For the first event, there is a .5 probability of success and a .5 probability of failure. Thereafter, whenever a success occurs, the probability is .8 that the next event will be a success and .2 that it will be a failure. Whenever a failure occurs, the probability is .8 that the next event will be a failure and .2 that it will be a success. This is a stylized representation of a situation in which events tend to repeat themselves. It violates the central limit theorem's assumptions because the probabilities are not constant and outcomes are not independent. And yet Fig. 5.12 shows that the normal distribution is apparent after just 20 trials. For $n = 1$, the probability distribution is rectangular; for $n = 2$, it is U-shaped. But as the number of trials increases, the probability distribution is increasingly bell-shaped.

When we cumulate a large number of outcomes, it is most likely that some outcomes will be relatively large and others small, giving a sum that is near the population mean for the sum. For the sum to be far above or far below its mean, the individual outcomes must be persistently either above or below their means, which is unlikely in a large number of trials. Thus the damping effect of summing many outcomes gives a symmetric bell shape.

The example in Fig. 5.12 illustrates the robustness of the central limit

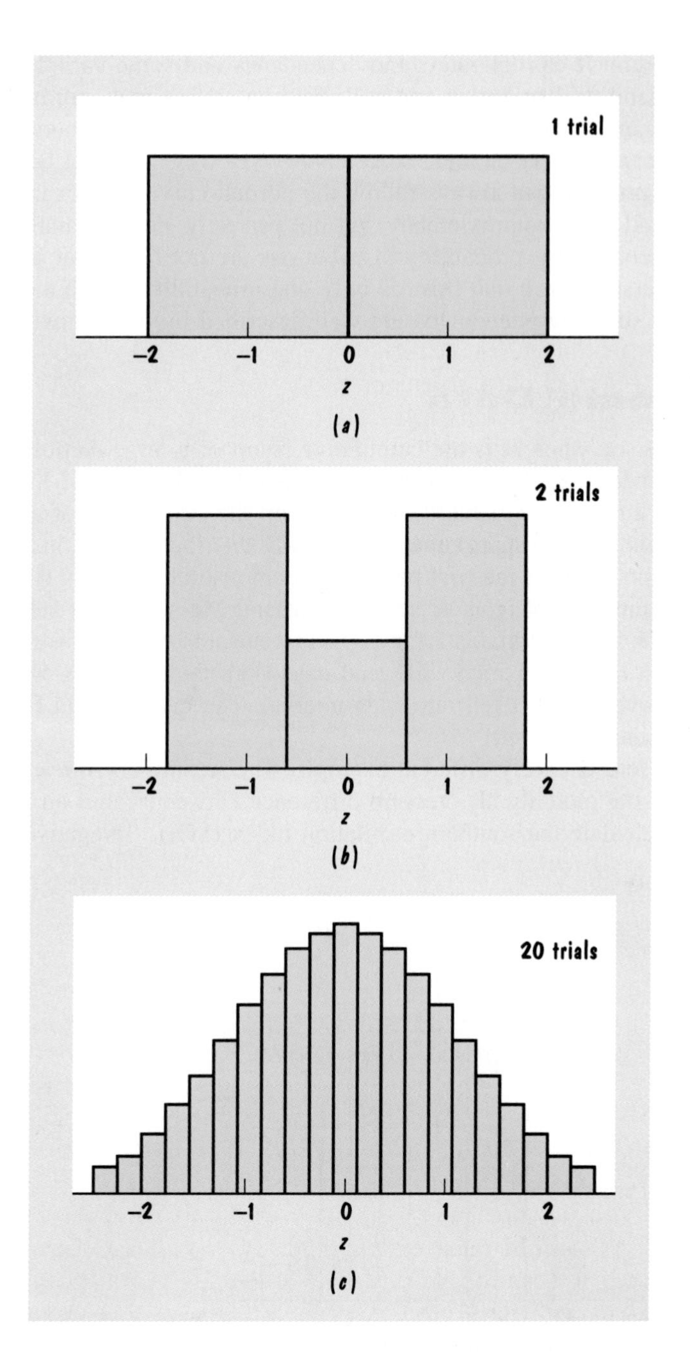

Figure 5.12 A sequence of dependent events.

theorem: It can tolerate some dependence and some variation in each trial's probability distribution and still yield an approximate normal distribution for the sum of the outcomes. This is why the normal distribution is so popular and the central limit theorem so celebrated. However, do not be lulled into thinking that probabilities always follow the normal curve. The examples that we have looked at are approximately, but not perfectly, normal, and there are many phenomena whose probability distributions are not normal at all. My purpose is not to persuade you that there is only one probability distribution, but to explain why so many phenomena are well described by the normal distribution.

Some Data

A person's height is the cumulative result of a large number of uncertain events involving heredity, diet, health, and exercise. The central limit theorem suggests that a random variable that depends on the cumulation of a large number of events may be approximately normally distributed, and this is, in fact, true of heights. One of the first persons to demonstrate this was H. P. Bowditch, who measured the heights of several thousand Massachusetts school children in the 1870s.[20] Using his data for 1293 11-year-old boys, I constructed 15 height intervals, each 1 inch wide, and calculated the frequency with which the observed heights fell into each interval. The histogram in Fig. 5.13 has the now familiar bell shape.

Here is a very different example. The Australian Bureau of Meteorology uses the monthly air pressure difference between Tahiti and Darwin, Australia, to calculate the southern oscillation index (SOI).[21] Negative values of the SOI

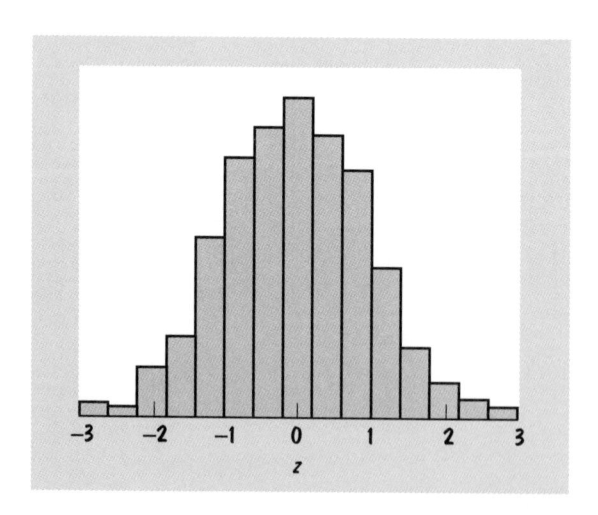

Figure 5.13 The heights of 11-year-old boys.

indicate an El Niño episode, which is usually accompanied by less-than-usual rainfall over eastern and northern Australia; positive values of the SOI indicate a La Niña episode, which is usually accompanied by more-than-usual rainfall over eastern and northern Australia. The histogram in Fig. 5.14 of the annual averages from 1876 through 1995 is roughly bell-shaped.

Other researchers have examined all sorts of data, and they found over and over that the histograms are bell-shaped. A normal distribution appears when we examine the weights of humans, dogs, and tomatoes. The lengths of thumbs, widths of shoulders, and breadths of skulls are all normally distributed. Scores on IQ, SAT, and Graduate Record Examination (GRE) tests are normally distributed. So are the number of kernels on ears of corn, ridges on scallop shells, hairs on cats, and leaves on trees. If some phenomenon is the cumulative result of a great many separate influences, then the normal distribution may be a very useful approximation.

Baseball Batting Averages

Baseball is a wonderful source of statistical data. Perhaps the most well-known baseball statistic is a player's *batting average,* which is equal to the number of base hits divided by the number of official times at bat (walks do not count as official times at bat). A player with a .300 batting average has gotten a base hit 30 percent of the time.

Each player's probability of a base hit depends on the mental and physical conditions of the player and of the opposing pitcher, the shape of the ballpark,

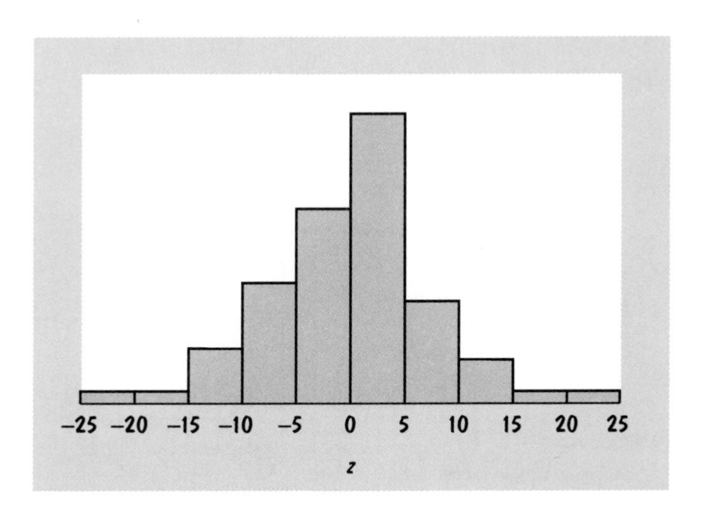

Figure 5.14 Annual values of the southern oscillation index, 1876 through 1995.

and even whether the game is played during the day or night. Dedicated students of the game collect all sorts of data on a player's performance under varying conditions, for example, how well a certain player bats in August with runners on base against a certain pitcher during night games at Fenway Park.

Although batting is not exactly a sequence of independent trials with constant success probability, if we look at the cumulative result of a large number of times at bat, we may find an approximate normal distribution. To see if this is so, I looked at the annual batting averages of nine players with lifetime batting averages of approximately .300: Hank Aaron, Lou Brock, Al Kaline, Mickey Mantle, Willie Mays, Minnie Minoso, Joe Morgan, Duke Snider, and Carl Yastrzemski. I considered only years in which the player had at least 400 times at bat. In each of these years, the season's batting average is the cumulative result of at least 400 times at bat, although admittedly not with constant and independent probabilities.

The table below shows the data grouped into 14 intervals, and the bar graph reveals its approximate bell shape. Most years, the season average is near .300, but there are some years in which a player's batting average is much higher or lower than .300. The popular press attributes these good and bad years to fluctuating skills. An alternative explanation is good luck and bad.

BATTING AVERAGE	NUMBER OF YEARS	BATTING AVERAGE	NUMBER OF YEARS
.225–.234	1	.295–.304	23
.235–.244	3	.305–.314	21
.245–.254	2	.315–.324	13
.255–.264	7	.325–.334	12
.265–.274	11	.335–.344	3
.275–.284	14	.345–.354	3
.285–.294	20	.355–.364	2

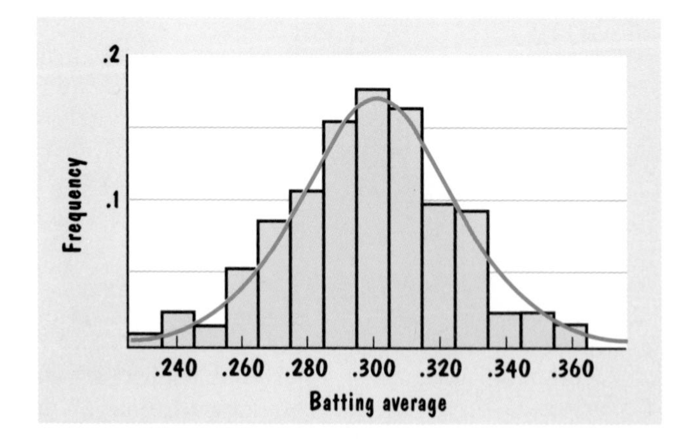

5.21 Which of the following variables are discrete and which are continuous?

a. The number of females born in Denver on January 1, 2000
b. The time at which the first female is born in Denver on January 1, 2000
c. The weight of the first female born in Denver on January 1, 2000
d. The temperature at noon in Denver on January 1, 2000
e. The number of cars sold in Denver in the year 2000
f. The number of Denver citizens who vote in the presidential election in the year 2000

5.22 Assume that the high temperature on the next January 1 in the city where you are attending school is a random variable that can be described by the normal distribution. Do not collect any data, but use your general knowledge to draw a rough sketch of this distribution. You do not need to label the vertical axis, but you must label the horizontal axis carefully.

5.23 A nationwide test has a mean of 75 and a standard deviation of 10. Convert the following raw scores to standardized z values: $x = 94, 89, 83, 81, 78, 76, 74, 71$, and 67. What raw score corresponds to a z value of 1.5?

5.24 In the 1800s, a Belgian statistician, Adolphe Jacques Quetelet, published data on the heights of men who had been drafted by, but not necessarily inducted into, the French army. The heights generally followed a normal distribution, but had a noticeable dip at 1.57 meters, which was the minimum height for draftees to be inducted into the army. Provide a plausible explanation for this dip.

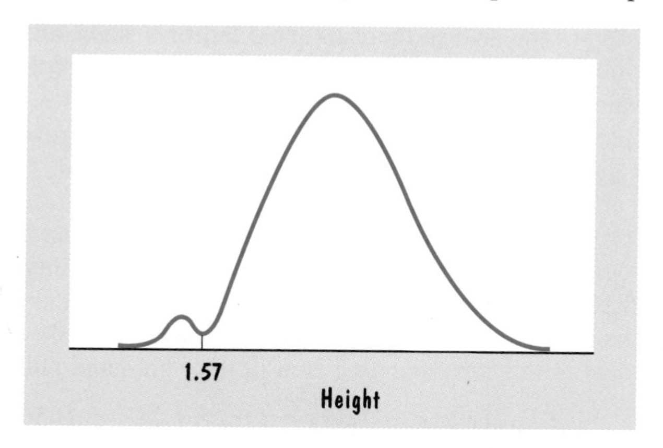

5.25 Karl Pearson and Alice Lee recorded the heights (in inches) of 1052 mothers in 1903.[22] Draw a histogram and then draw a freehand smooth curve that fit these data. Does your curve look like a normal distribution?

HEIGHT	FREQUENCY	HEIGHT	FREQUENCY
56–57	.006	63–64	.155
57–58	.017	64–65	.109
58–59	.033	65–66	.075
59–60	.076	66–67	.039
60–61	.129	67–68	.015
61–62	.155	68–69	.007
62–63	.174	69–70	.004

5.6 Finding Normal Probabilities

The density curve for the normal distribution is graphed in Fig. 5.11. The probability that the value of z will be in a specified interval is given by the corresponding area under this curve. However, there is no simple formula for computing areas under a normal curve. These areas can be determined from complex numerical procedures, but nobody—not even a hard-core mathematician—wants to do these computations every time a normal probability is needed. Instead, they consult statistical software or a table, such as Table 3 at the end of this book, that shows the normal probabilities for hundreds of values of z. We will, too.

The Normal Probability Table

Table 3 shows the area in the right-hand tail of a standardized normal distribution; its use is demonstrated in Fig. 5.15. The left-hand column of Table 3 shows values of z in .1 intervals, such as 1.1, 1.2, and 1.3. For finer gradations, the top row shows .01 intervals for z. By matching the row and column, we can, for instance, find the probability that z is larger than 1.25. The logic of this calculation is shown in Fig. 5.15.

For probabilities other than the right-hand tail, we can use the fact that the normal curve is symmetric about $z = 0$, with .5 probability in each half. Figure 5.16 shows three typical calculations. To determine the probability that z is less than -1.25, we note that, because of symmetry, the area in a left-hand tail is equal to the corresponding area in the right-hand tail:

$$P[z < -1.25] = P[z > 1.25] = .1056$$

To determine the probability that z is between 0 and 1.25, we use the fact that the total area in the right-hand tail is .5:

$$P[0 < z < 1.25] = .5 - P[z > 1.25] = .5 - .1056 = .3944$$

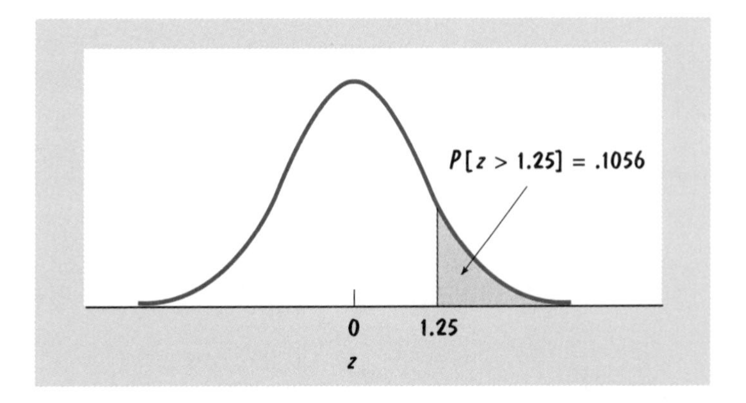

An excerpt from Table 3:

z	.00	.01	.02	.03	.04	.05	.06	.07	.08	.09
1.1	.1357	.1335	.1314	.1292	.1271	.1251	.1230	.1210	.1190	.1170
1.2	.1151	.1131	.1112	.1093	.1075	.1056	.1038	.1020	.1003	.0985
1.3	.0968	.0951	.0934	.0918	.0901	.0885	.0869	.0853	.0838	.0823

Figure 5.15 Right-hand tail probabilities.

To determine the probability that z is between .60 and 1.25, we subtract the area to the right of 1.25 from the area to the right of .6:

$$P[.60 < z < 1.25] = P[z > .60] - P[z > 1.25] = .2743 - .1056 = .1687$$

Thus Table 3 gives all the information we need to determine the probability that the value of a standardized normal variable z is in any specified interval. For areas more complicated than the right-hand or left-hand tail, it is usually safest to make a quick sketch, as in Fig. 5.16, so that the probabilities that must be added or subtracted are readily apparent. We do some more examples in this chapter, and there are, of course, exercises for practice.

Nonstandardized Variables

So far, we have calculated probabilities only for a normally distributed variable z that has been standardized to have a mean of 0 and a standard deviation of 1. Nonstandardized variables can have positive or negative means and standard deviations that are larger or smaller than 1 (but not negative). The mean tells us where the center of the distribution is, and the standard deviation gauges the spread of the distribution about its mean. Together, the mean and standard deviation describe a normal distribution.

Figure 5.17 shows two distributions with the same standard deviation, but

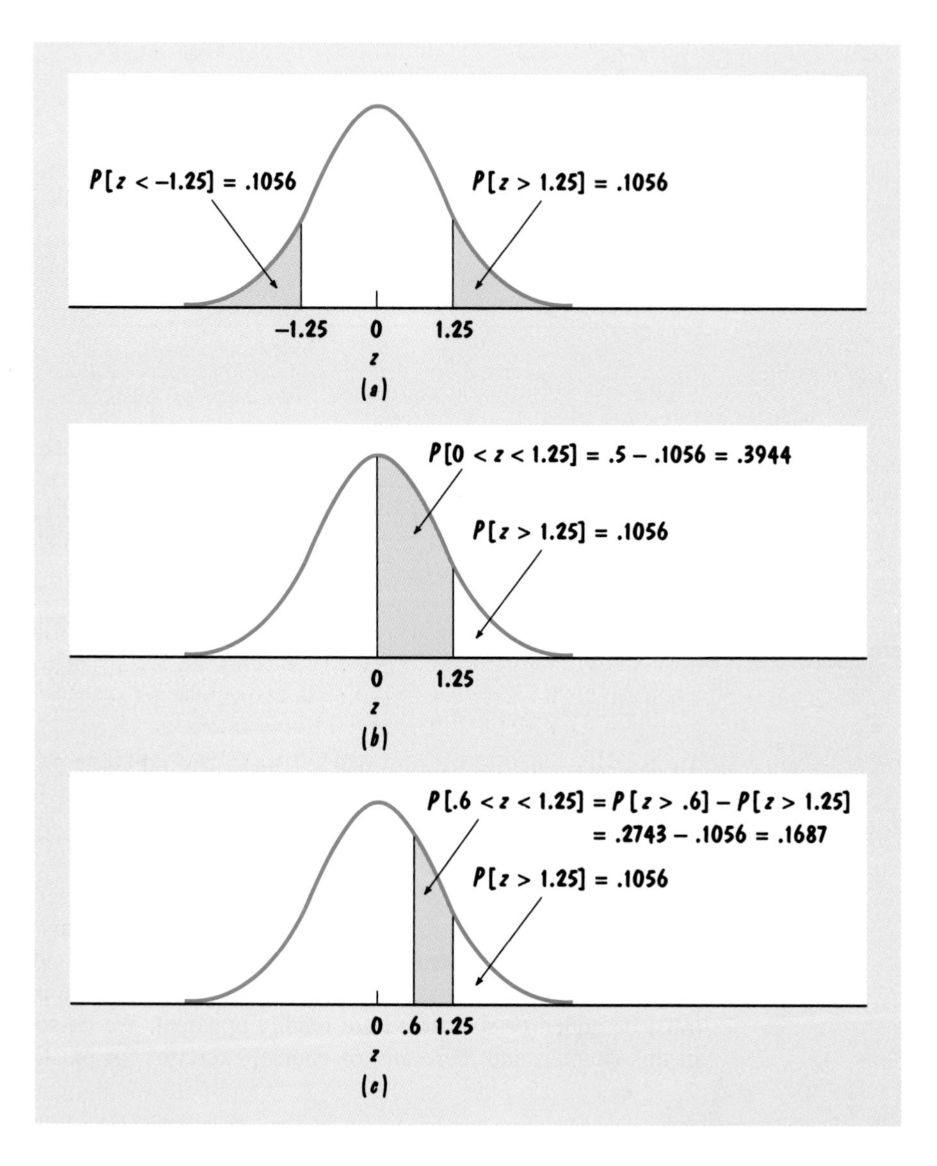

Figure 5.16 Three normal probability calculations.

different means. Perhaps these random variables are the percentage returns on two investments, one with a 10 percent mean and the other with a 20 percent mean. In the long run, it is almost certain that the second investment will have the higher average return. In the short run, the equal standard deviations indicate that these two investments are equally risky in that their returns are equally uncertain. For instance, the probability that the return will turn out to

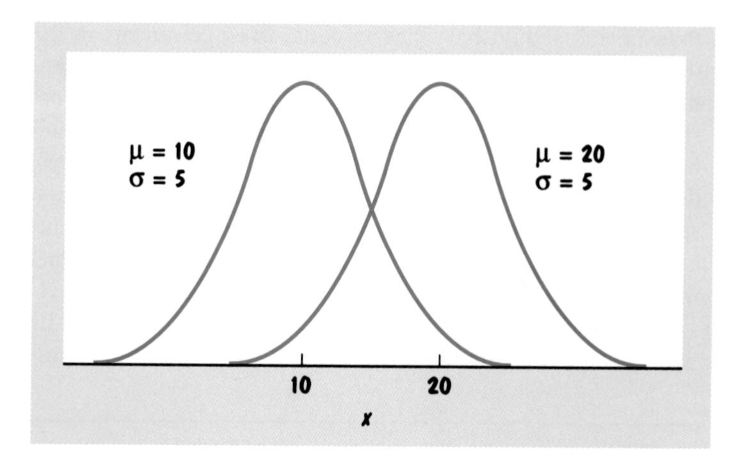

Figure 5.17 The mean locates the center of a normal distribution.

be more than 10 percentage points below the mean is the same for both investments. It is natural to describe these two investments by saying that they have mean returns of 10 and 20 percent and that we are equally confident of how close the actual return will be to the respective means. If offered a choice, investors would choose the second investment.

Figure 5.18 shows a different situation. Here, the means are the same, but the standard deviations differ. If these are percentage returns, then these two

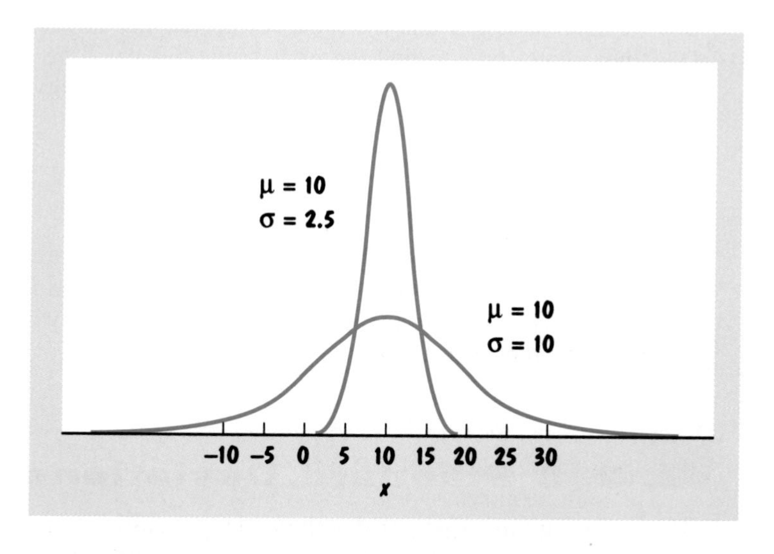

Figure 5.18 The standard deviation describes the spread of a normal distribution.

investments should have nearly equal average returns in the long run. But for the short run of a single outcome, the unequal standard deviations imply that these investments are not equally risky. For the investment with a 2.5 percent standard deviation, we are almost certain that the return will turn out to be between 5 and 15 percent. For the investment with a 10 percent standard deviation, there is a substantial probability that the return will be outside this range. Risk-averse investors will prefer the first investment because it is expected to do as well, on average, as the second investment and its return is more certain.

How do we determine probabilities for a normally distributed variable x with mean μ and standard deviation σ? We can use this mathematical theorem:

> If a random variable x has the normal distribution $N[\mu, \sigma]$, then the standardized random variable $z = (x - \mu)/\sigma$ has the normal distribution $N[0, 1]$.

The letter N signifies a normal distribution, with the first number inside the brackets being the mean and the second number being the standard deviation. Thus, the statement that "x has the normal distribution $N[10, 5]$" (or, more simply, "x is $N[10, 5]$") means that x is normally distributed with a mean of 10 and a standard deviation of 5.

To determine a probability for a nonstandardized, normally distributed random variable x, we simply determine the corresponding value of the standardized random variable z and then find the probability in Table 3. The How-to-Do-It box shows how to find the probability that x will be greater than 15 if the distribution of x is $N[10, 5]$. What about the probability that the value of x will be greater than 20? This z value is

$$z = \frac{x - \mu}{\sigma} = \frac{20 - 10}{5} = 2.0$$

and Table 3 shows the probability to be .0228. For a normally distributed variable with a mean of 10 and a standard deviation of 5, the value $x = 20$ is 2 standard deviations above the mean; there is consequently a .0228 probability that x will be greater than 20.

1, 2, 3 Standard Deviations

The following three rules of thumb can help us estimate probabilities for normally distributed random variables without consulting Table 3:

HOW TO DO IT

Determining a Normal Probability

Suppose that x is $N[10, 5]$ and we want to find the probability that the value of x will be greater than 15.

1. Determine the appropriate value of z by subtracting the mean of x from the value of x that we are interested in and then dividing by the standard deviation of x:

$$z = \frac{x - \mu}{\sigma} = \frac{15 - 10}{5} = 1.0$$

2. We can interpret this z value by remembering that standardization transforms variables from their natural units (inches, pounds, percentage) to standard deviations. Here, $x = 15$ corresponds to $z = 1$ because 15 is 1 standard deviation above the mean of x.
3. Locate the z value in Table 3, and find the probability. For $z = 1.00$, we look in the leftmost column (labeled *Cutoff*) and read down this column until we find 1.0. Then we look to the right until we find the second digit after the decimal point. For $z = 1.00$, this is the column labeled **.00,** and we see that there is a .1587 probability that z will be greater than 1.00. Thus there is a .1587 probability that x will be greater than 15.
4. Alternatively, we can use computer software. Table 5.6 shows the computer output from Smith's Statistical Package. After the user enters the values of μ, σ, and x in the three boxes, the normal probabilities are shown in the bottom half of the computer screen.

$$P[-1 < z < 1] = .6826$$
$$P[-2 < z < 2] = .9544$$
$$P[-3 < z < 3] = .9973$$

These useful benchmarks are displayed in Fig. 5.19. A normally distributed random variable has about a 68 percent (roughly two-thirds) chance of being within 1 standard deviation of its mean, a 95 percent chance of being within 2 standard deviations of its mean, and better than a 99.7 percent chance of being within 3 standard deviations of its mean. Turning these around, we can say that a normally distributed random variable has less than a .3 percent chance of being more than 3 standard deviations from its mean, roughly a 5 percent

Table 5.6

Normal probabilities from Smith's Statistical Package for $\mu = 10$, $\sigma = 5$, and $x = 15$

```
This program gives the probability that the value of a
normally distributed random variable will be larger than
a specified value.

    Mean of X:                                   [ 10 ]
    Standard deviation of X                      [  5 ]
    Which particular value of X?                 [ 15 ]

For a normal distribution with
    mean 10 and standard deviation 5,
the probability that the value of X will be
    larger than 15 is .15865
    smaller than 15 is .84135
```

chance of being more than 2 standard deviations from its mean, and a 32 percent chance of being more than 1 standard deviation from its mean.

Consider, for example, the baseball batting averages in Example 5.5. The graph shown there is an empirical histogram, rather than a density curve, but nonetheless we can gauge how closely that histogram can be approximated by the normal distribution by seeing what fractions of the observations are within 1, 2, and 3 standard deviations of the mean. For these 135 batting averages, the mean is 297.78 and the standard deviation is 25.52. The range encompassed by the mean ± 1 standard deviation is from $297.78 - 25.52 = 272.26$ to $297.78 + 25.52 = 323.30$. For a normal density curve, there is a .6826 probability of being within 1 standard deviation of the mean. For these empirical data, 92 of 135 observations (68.15 percent) are within 1 standard deviation of the mean. Similarly, 127 observations (94.07 percent) are within 2 standard deviations of the mean, and 100 percent are within 3 standard deviations. This histogram is well approximated by the normal distribution.

Coin Weights

The graph that follows shows the distribution of the weights of quarters when minted and after 5 and 20 years of circulation.[23] Each distribution is approximately normal, for reasons explained by the central limit theorem. The initial weight is the cumulative result of many steps in the minting process, and as time passes, each coin's weight is altered by the cumulative effects of repeated handling.

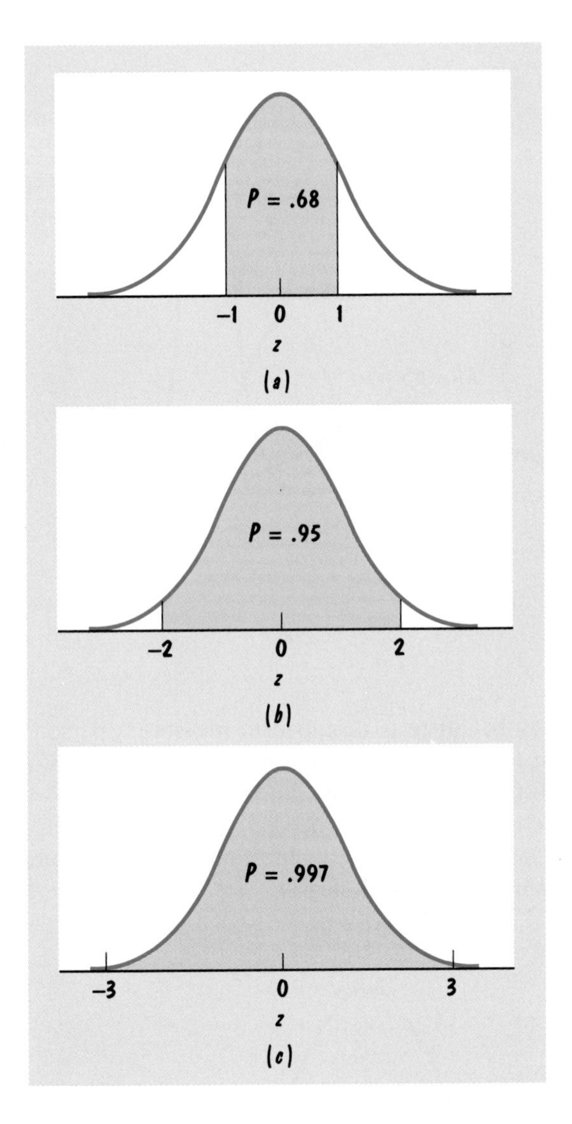

Figure 5.19 Three normal probability benchmarks.

Notice that the average weight declines over time, causing the distribution to slide to the left, as wear and tear take their toll. The standard deviation also increases, widening the distribution as the experiences of coins diverge. Some are handled very gently and retain almost all their initial weight, while others are nicked and bruised almost beyond recognition.

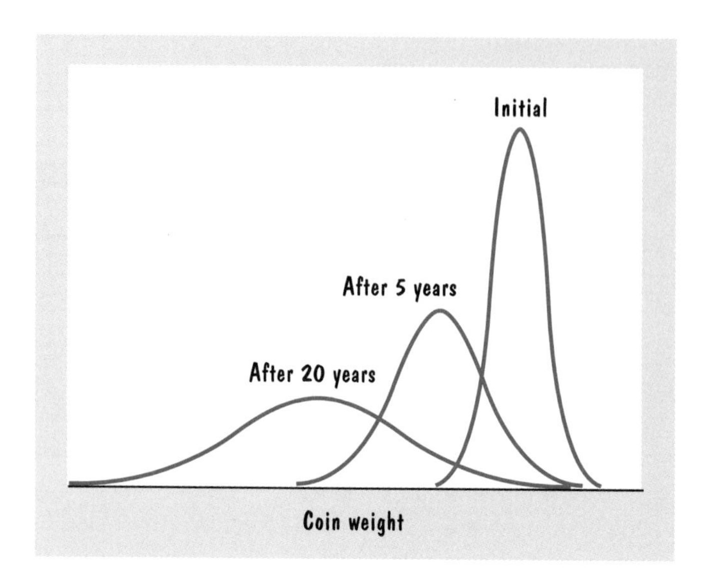

EXAMPLE 5.7

IQ Tests

There are a number of tests designed to measure a person's intelligence quotient (IQ). The first IQ test consisted of 54 mental "stunts" that two psychologists, Alfred Binet and Theodore Simon, devised in 1904 to identify students in the Paris school system who needed special assistance. IQ tests today are designed to measure general intelligence, including an accurate memory and the ability to reason logically and clearly.

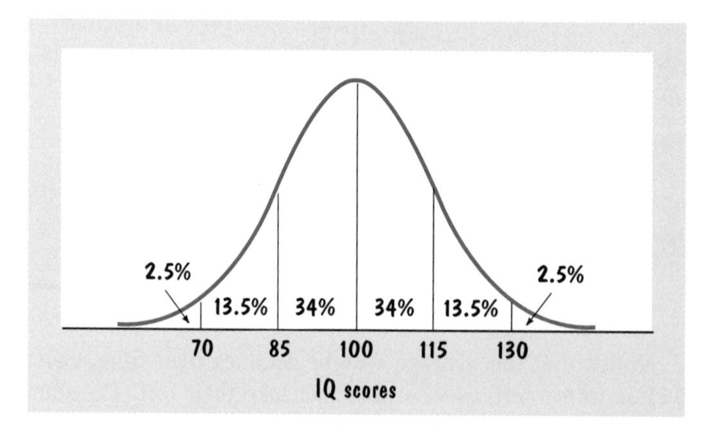

Because an individual's score on an IQ test depends on a very large number of hereditary and environmental factors, the central limit theorem explains why IQ scores are approximately normally distributed. One of the most widely used tests today is the Wechsler Adult Intelligence Scale, which has a mean IQ of 100 and a standard deviation of 15. Thus a score of 100 indicates that a person's intelligence is average. About one-half of the people tested score above 100, while one-half score below 100. Our 1-standard-deviation rule of thumb implies that about 32 percent of the population will score more than 15 points away from 100—16 percent above 115 and 16 percent below 85. Our 2-standard-deviations rule implies that about 5 percent of the population will score more than 30 points away from 100—2.5 percent above 130 and 2.5 percent below 70.

5.26 Use Table 3 to determine these four probabilities for the standardized random variable z that is normally distributed with a mean of 0 and a standard deviation of 1. In each case, sketch the probability distribution of z, and shade in the area showing the probability.

a. The probability that the value of z will be positive
b. The probability that the value of z will be greater than 1.75
c. The probability that the value of z will be less than -1.50
d. The probability that the value of z will be between $-.50$ and 1.30

5.27 This figure shows the probability distribution for a random variable z that is normally distributed with a mean of 0 and a standard deviation of 1. Each of the three numbers under the curve is the probability that the value of z will be in the corresponding interval. Use Table 3 to determine the z values a, b, and c.

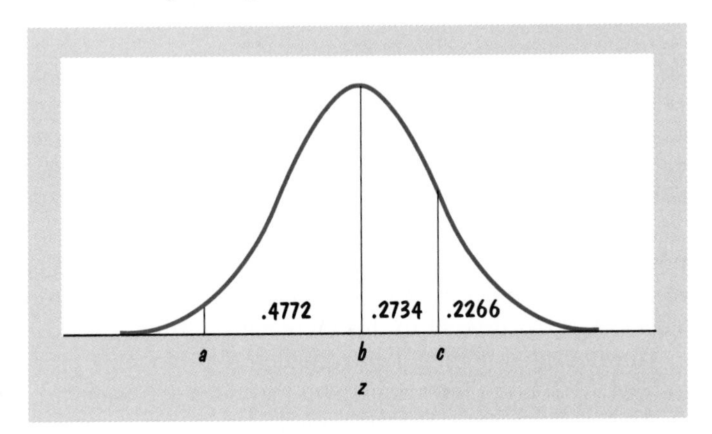

5.28 If IQ scores are normally distributed with a mean of 100 and a standard deviation of 15, what is the probability that a randomly selected person has an IQ above 150? A psychologist estimated that Galileo had an IQ of 185 and that Goethe had an IQ of 210. What is the probability that a randomly selected person has such a high IQ?

5.29 A woman wrote to Dear Abby, saying that she had been pregnant for 310 days before giving birth.[24] Completed pregnancies are normally distributed with a mean of 266 days and a standard deviation of 16 days. What is the probability that a completed pregnancy lasts at least 270 days? At least 300 days? At least 310 days?

5.30 The *Los Angeles Times* reported that

> In 1977, 14.97 inches fell at the downtown collection post, close enough to make it the most normal year on record [i.e., the closest to the long-run average] . . . A review of downtown rain charts dating back to the late 1800s reveals a remarkable absence of normality: Not only has the rainfall never been precisely normal, but we've been in striking distance only a handful of years in this century.[25]

During the 115 years from 1878 through 1992, annual rainfall at the Los Angeles Civic Center averaged 14.89 inches with a standard deviation of 6.75 inches. If a year's annual rainfall x is normally distributed with a mean of 14.89 inches and a standard deviation of 6.75 inches, what is the probability that the value of x will be within .08 inch of 14.89 inches (that is, between 14.81 and 14.97 inches)?

5.7 Approximating the Binomial Distribution

We have seen that a probability histogram for the binomial distribution converges to a normal distribution as the number of trials increases. We can consequently use the normal distribution to calculate approximate values of binomial probabilities. Figure 5.20 shows the familiar bell shape for a binomial probability histogram with success probability $p = .50$ and $n = 100$ trials. To find the exact binomial probability that x is greater than or equal to 60, we have to compute and add 41 separate (and complicated) binomial probabilities. Instead, we can make a quick approximation by finding the area to the right of $x = 60$ under a normal curve fit to this binomial probability histogram.

To calculate a normal approximation to a binomial probability, we need to use Eqs. (5.2) and (5.3) which give the mean and standard deviation of the number of successes x for a binomial distribution:

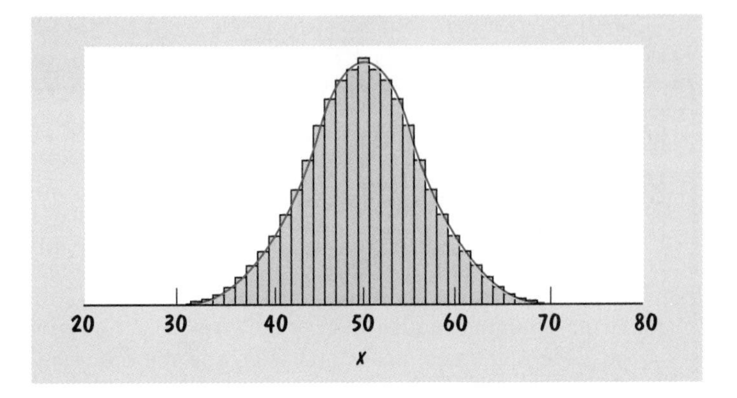

Figure 5.20 Binomial distribution for $p = .5$ and $n = 100$.

$$\text{Mean of } x: \quad \mu = np$$
$$\text{Standard deviation of } x: \quad \sigma = \sqrt{np(1 - p)}$$

By using these formulas for the mean and standard deviation of x, we can convert to a standardized z value and use Table 3 to find the relevant probability, as explained in the How-to-Do-It box. The .0228 value found by the normal approximation in the How-to-Do-It box is close to the exact binomial probability .0284 (which I found by using statistical software). For a larger number of trials, a normal approximation will be even closer to the exact binomial probability. In general, a satisfactory normal approximation requires a large enough value of n to put the mean np safely away from the extreme values $x = 0$ and $x = n$. To ensure that the mean is at least 5 units away from these extremes, we can use the rule of thumb that a normal approximation is reasonably accurate if np and $n(1 - p)$ are both larger than 5.

A more fundamental question is, Why would we use a normal approximation at all? Decades ago, when binomial probabilities had to be calculated by hand, normal approximations were a welcome alternative to excruciatingly tedious computations. Today, researchers have an even better alternative—statistical software that will calculate exact binomial probabilities. This, too, is what you should use.

The purposes of this section are somewhat different. First, we provided hard evidence of the power of the central limit theorem by showing that the normal distribution is a reasonable approximation to the binomial distribution. Second, when we gauge the accuracy of political polls and other empirical data in later chapters, we will rely heavily on the fact that the binomial distribution is approximately normal with a mean and standard deviation given by Eqs. (5.2) and (5.3). The calculations in this section justify that reliance.

HOW TO DO IT

Making a Normal Approximation to a Binomial Probability

If a fair coin is flipped 100 times, what is the probability of 60 or more heads?

1. Is the binomial model appropriate? Yes. Each observation has two possible outcomes, heads and tails, and the observations are independent with a constant probability of heads or tails.
2. What is a *success,* and what are the values of p, n, and x? If we call heads a success, then the success probability is $p = .50$, the number of trials is $n = 100$, and the number of successes is $x \geq 60$.
3. Are np and $n(1 - p)$ both larger than 5, so that a normal approximation is justified? Here $np = 50$ and $n(1 - p) = 50$.
4. Use Eqs. (5.2) and (5.3) to calculate the mean and standard deviation of x:

   Mean of x: $\mu = np = 100(.5) = 50$

   Standard deviation of x: $\sigma = \sqrt{np(1 - p)}$

   $= \sqrt{100(.5)(1 - .5)} = \sqrt{25} = 5$

5. Determine the appropriate value of z by subtracting the mean of x from the value of x that we are interested in and then dividing by the standard deviation of x:

$$z = \frac{x - \mu}{\sigma} = \frac{60 - 50}{5} = 2.0$$

6. Locate the z value in Table 3, and find the probability. For $z = 2$, Table 3 shows that there is a .0228 probability that z will be greater than 2; thus there is approximately a .0228 probability of 60 or more heads in 100 coin flips.

exercises

5.31 In 1989, *Sports Illustrated* reported that Kansas City's Pat Tabler, a career .290 hitter, had 38 hits in 66 times at bat with the bases loaded, an astonishing .576 batting average in these key situations.[26] Use a normal approximation to the binomial distribution to calculate the probability that an event with a .290 probability of success would happen 38 or more times in 66 trials.

5.32 A sportswriter counted the total number of words spoken by Dick Vitale (7706) and Tim Brando (5360) during their ESPN coverage of a 1989 basketball game between Georgetown and North Carolina.[27] If each word is considered to be an independent observation, with a .5 probability of being spoken by Vitale, what is the probability that such a high proportion of these words will be spoken by Vitale? (Use a normal approximation.) Why is each word not an independent observation?

5.33 The U.S. Postal Service (USPS) has on-time standards for its mail delivery; for instance, letters from Washington, D.C., to Manhattan should arrive within 2 days. The USPS claims that it delivers 90 percent of the mail on time in Manhattan. In the summer of 1988, *New York* magazine tested this claim by mailing 144 letters from around the country to Manhattan Zip codes and found that 50 percent of these letters did not arrive on time. If the USPS's claims are correct, what is the probability that 72 or more of these 144 letters will not arrive on time? Assume that the binomial model applies and use a normal approximation.

5.34 Mendel postulated that the self-fertilization of hybrid yellow-seeded sweet peas would yield offspring with a .75 probability of being yellow-seeded and a .25 probability of being green-seeded. In 1865, he reported that 8023 such experiments yielded 6021/8023 = .7505 yellow-seeded plants and 2002/8023 = .2495 green-seeded plants. It has been suggested that these results are too good to be true, in that Mendel must have reported what he wanted rather than what he observed. If these were honest trials that could be described by the binomial model, what is the probability that the number of yellow-seeded plants would be between 6013 and 6021 (that is, no more than 4 plants from the mean, 6017.25)? Use a normal approximation.

5.35 Karl Pearson reported flipping a coin 24,000 times and obtaining 12,012 heads and 11,988 tails. If heads and tails are equally likely, what is the probability that the number of heads would be between 11,988 and 12,012 (that is, no more than 12 from the mean)? Use a normal approximation.

summary

Probabilities are used to quantify life's uncertainties. There are three basic approaches: the enumeration of equally likely outcomes, the computation of observed long-run frequencies, and a subjective blending of fact and opinion. The classic equally likely approach is most appropriate for games of chance—

the roll of dice, spin of a roulette wheel, and deal of cards. The long-run frequency approach is designed for cases in which there are lots of repetitive data and the outcomes are apparently not equally likely. The subjective approach is closely associated with Reverend Thomas Bayes and is based on personal opinion.

The conditional probability of A's occurring if B occurs is not the same as the conditional probability of B's occurring if A occurs. For example, the probability that a female will be on the women's basketball team is not the same as the probability that a person on the women's basketball team is female. Events A and B are independent if the probability of A's occurring does not depend on whether B has occurred.

The binomial model can be used to describe the generation of data that fall into two categories—success and failure—if the probability of success is constant and the outcomes are independent. This model can be applied to a wide variety of situations, including games of chance, manufacturing processes, and public opinion polls. Binomial probabilities are determined most easily by statistical software; some can be found in Table 2 at the end of this book.

A graph of binomial probabilities is called a probability histogram, or population histogram, or probability distribution. The population mean and standard deviation for the number of successes x are as follows:

$$\text{Mean of } x: \quad \mu = np$$

$$\text{Standard deviation of } x: \quad \sigma = \sqrt{np(1-p)}$$

The population mean and standard deviation of x/n, the fraction of the trials that are successes, are

$$\text{Mean of } x/n: \quad \mu = p$$

$$\text{Standard deviation of } x/n: \quad \sigma = \sqrt{\frac{p(1-p)}{n}}$$

The fallacious law of averages says that an unusual run of successes must be balanced out by a run of failures so that x will equal np exactly.

A continuous random variable x can have any value in an interval of real numbers. The probability that the value of x will be in a specified interval is shown by the corresponding area under a probability density curve. A (discrete or continuous) random variable x is standardized by subtracting its mean μ and then dividing by the standard deviation σ: $z = (x - \mu)/\sigma$. The standardized random variable z has a mean of 0 and a standard deviation of 1. The central limit theorem explains why so many random variables are approximately normally distributed.

Table 3 at the back of this book shows the probabilities in the right-hand tail of a standardized normal distribution and can be used to find the probabilities associated with nonstandardized variables. For example, if x is normally distributed with a mean $\mu = 10$ and a standard deviation $\sigma = 5$, then the probability that the value of x will be larger than 20 is equal to the probability that the value of $z = (x - \mu)/\sigma$ will be larger than $(20 - 10)/5 = 2.0$, which Table 3 tells us is .0228.

A rough rule of thumb is that a normally distributed random variable has about a .68 (roughly two-thirds) probability of being within 1 standard deviation of its mean, a .95 probability of being within 2 standard deviations, and better than a .997 probability of being within 3 standard deviations.

The normal distribution is a reasonably accurate approximation to a binomial distribution with success probability p if the mean np is at least 5 units away from both 0 and n; equivalently, if np and $n(1 - p)$ are both larger than 5. To make this approximation, we convert to a standardized z value by using the equations for the mean and standard deviation of the number of successes x for a binomial distribution.

review exercises

5.36 Explain why you either agree or disagree with this reasoning: "Females make up half the human race, Cathy. If you answer the phone, you've got a fifty-fifty chance of hearing a woman's voice on the line."[28]

5.37 A book on probabilities advises

> [N]ext time you pat your little nephew on the head and ask him what he wants to be when he grows up, don't expect a quick and intelligent answer. The poor kid has the odds stacked 17,452 against him. That's how many specified occupations there are in the *Dictionary of Occupational Trades*.[29]

Explain why you either would or would not use 1/17,452 as the probability of your nephew's correctly selecting his future occupation.

5.38 In three careful studies, lie detector experts examined several persons, some known to be truthful and others known to be lying, to see if the experts could tell which were which.[30] Overall, 83 percent of the liars were pronounced "deceptive," and 57 percent of the truthful people were judged "honest." If you use these data and assume that 80 percent of the people tested are truthful and 20 percent are lying, what is the probability that a person pronounced deceptive is in fact truthful? What is the probability that a person judged honest is in fact lying? (Use a contingency table with 100,000 people tested.)

5.39 Approximately 1.5 percent of people in the United States are schizophrenic. Computerized axial tomography (CAT) scans show brain atrophy in 30 percent of the people diagnosed as schizophrenic and in only 2 percent of the persons diagnosed as not schizophrenic.[31] In the 1982 trial of John Hinckley for the attempted assassination of President Ronald Reagan, the defense attorney tried to present evidence that a CAT scan of Hinckley had shown brain atrophy, thereby indicating that Hinckley was schizophrenic. Consider the hypothetical case of 10,000 randomly selected people in the United States who are given CAT scans, and use a contingency table to calculate the fraction of those persons with brain atrophy who are schizophrenic. Is a CAT scan showing brain atrophy persuasive evidence that the person is schizophrenic?

5.40 The U.S. Supreme Court once compared 6-person and 12-person juries and declared that "there is no discernible difference between the results reached by the two different sized juries." Since jurors are not allowed to discuss the case with anyone until all the testimony has been heard, we will look at an initial secret ballot conducted by the jury immediately after the testimony has been completed. Assume that jurors are a random sample from a population in which a fraction $p = .9$ would vote for conviction in this initial secret ballot and a fraction $1 - p$ would vote for acquittal. Compare the probabilities that all members of a 6- and 12-person jury will vote for conviction. What if $p = .8$?

5.41 During the January 1984 NFC championship game between Washington and San Francisco, it appeared that the game might be decided by a field goal. One of the CBS commentators, Jack Buck, pointed out ominously that Mark Mosely had already missed four field goals in the game. Hank Stram, the other CBS commentator, responded, "The percentages are in his favor." Explain why you either agree or disagree.

5.42 In a 1979 court case, a military supplier argued that the U.S. government had tested too small a sample when it rejected a shipment of 20,000 nose fuse adapters, a component of artillery shells.[32] Military procurement standards specified that a shipment of 10,001 to 35,000 adapters would be tested by sampling 315 pieces, yet the government rejected this shipment of 20,000 pieces after testing 20 adapters and finding them all to be defective. What is the probability that every item in a random sample of 20 items from an infinite population will be defective if the probability of selecting a defective item is $p = .10$?

5.43 A basketball article in the January 16, 1983, *Los Angeles Times* was titled, "Wright Defies the Percentages, and USC Holds Off Oregon, 62–54." The article explained that "Oregon's strategy was obvious in the closing minutes—foul Gerry Wright, a 37.5 percent free throw shooter. But Wright, a reserve center, didn't fold under the pressure. He made 3 of his 6 foul shots in the final two minutes."

a. Assuming the binomial model to be appropriate, what is the probability that Wright would make more than 2 of 6 free throws?

b. Is the fact that he did so an unlikely defiance of the percentages?
c. What assumptions are needed for the binomial model to be appropriate?

5.44 Answer this question that a reader asked Marilyn vos Savant, who is listed in the *Guinness Book of World Records Hall of Fame* for highest IQ: "During the last year, I've gone on a lot of job interviews, but I haven't been able to get a decent job. . . . Doesn't the law of averages work here?"[33]

5.45 In his final newspaper column, Melvin Durslag reminisced about some of the advice he had received in his 51 years as a sports columnist, including this suggestion from a famous gambler: "Nick the Greek tipped his secret. He trained himself so that he could stand at the table eight hours at a time without going to the washroom. It was Nick's theory that one in action shouldn't lose the continuity of the dice."[34] Explain why you either agree or disagree with Nick's secret.

5.46 "Five out of ten people who have this disease die from it. It is lucky you came to me; my last five patients all died." Would you be comforted by such report? If there is a .5 probability of dying from this disease, what is the probability that 5 out of 5 patients will die from it?

5.47 During the years from 1900 to 1989, the annual precipitation in the contiguous United States has averaged 730.95 millimeters with a standard deviation of 57.11 millimeters. If annual precipitation is normally distributed with a mean of 730.95 millimeters and a standard deviation of 57.11 millimeters, what is the probability that next year's precipitation will be greater than 865.6 millimeters (the amount of precipitation that occurred in 1983, a year with extreme El Niño conditions)? What is the probability that next year's precipitation will be less than 602.2 millimeters (the amount that occurred in 1910)?

5.48 In the early 1960s, the U.S. National Center for Health Statistics estimated the heights of U.S. males between the ages of 25 and 34 to be normally distributed with a mean of 5 feet 9 inches and a standard deviation of 2.65 inches. Now it estimates that the heights of U.S. males between the ages of 25 and 34 are normally distributed with a mean of 5 feet 10 inches and a standard deviation of 2.65 inches. If, in each era, there were 30 million males between the ages of 25 and 34, calculate the effect of this change on the number of males in this age bracket who are 6 feet 3 inches or taller.

5.49 The heights of U.S. females between the age of 25 and 34 are approximately normally distributed with a mean of 66 inches and a standard deviation of 2.5 inches. What fraction of the U.S. female population in this age bracket is taller than 70 inches, the height of the average adult U.S. male of this age?

5.50 A geology professor mapped certain geological formations on the surface of Venus, as illustrated in the following two-dimensional rectangle.[35]

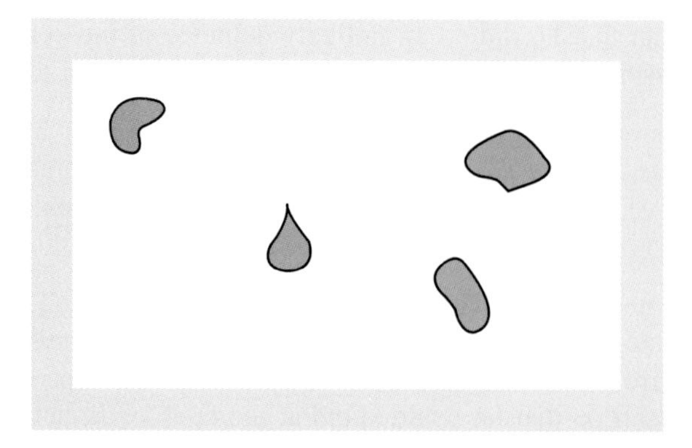

In all, these formations covered 4.9 percent of the surface of Venus. This professor also determined that a total of 932 meteorites had left visible craters on the surface of Venus, of which 42 were inside these geological formations. If every square inch of Venus is equally likely to be hit by a meteorite, so that the location of each crater is completely random, what is the probability that exactly 42 craters would be inside these formations? (Ignore the possibility that a crater might be partly inside a formation and partly outside.)

projects

For each of the following projects, type a report in ordinary English, using clear, concise, and persuasive prose. Any data that you collect for this project should be included as an appendix to your report. Data used in your report should be presented clearly and effectively.

5.1 Use computer software to simulate 1000 flips of a fair coin. Record the fraction of flips that were heads after 10, 100, and 1000 flips. Repeat this experiment 100 times, and then use three histograms to summarize your results.

5.2 From 10 people, elicit subjective probabilities of a Democrat's winning the next presidential election. How easy or difficult was it to determine each of these personal probabilities? How compact or dispersed are the probabilities?

5.3 The first paragraph of Example 5.3 poses a hypothetical question about the probability that a lump is malignant in light of worrisome mammogram results. Type this hypothetical scenario (all the first paragraph, beginning with "In a routine examination") on a sheet of paper, and ask 25 people to read over the description and then give an estimate of the probability that the lump is malignant. Use a histogram to summarize your results.

5.4 Flip a coin 200 times (use a real coin, not a computer simulation). Draw a graph showing the fraction of the flips that were heads as the flips progressed. Draw a second graph showing the total number of heads minus the total number of tails as the flips progressed.

5.5 Flip a coin 200 times (use a real coin, not a computer simulation). Divide your results into 20 groups of 10 flips, and count the number of heads in each of these 20 groups. How often did you get 10 heads in 10 flips? Nine heads? And so on. Use a histogram to summarize your results. Now divide your results into 10 groups of 20 flips, and identify the longest streak of consecutive heads or consecutive tails in each of these 20 groups. How often did you get a streak of 20? A streak of 19? And so on. Use a histogram to summarize your results.

5.6 Ask 25 persons this question: Suppose you have a chance to win $1000 by calling the next flip of a coin correctly. The last four flips have been heads. Do you call heads or tails? After they answer, ask them to explain their choices.

5.7 Ask 50 randomly selected people this question and then summarize your results:

> After a wonderful life on earth, you find yourself standing at the Pearly Gates in front of St. Peter. He tells you that, to get inside, you must predict the outcome of the next spin of a black/red roulette wheel. The last 20 spins have ended in black. Although I know you're correct when you say the odds are still 50/50 on each spin, I'd side with probability here and choose red. Which would you choose?[36]

5.8 One way to compare outstanding baseball pitchers from different historical periods is to see how many standard deviations each pitcher's earned-run average (ERA) was from the average ERA that year. Considering only the ERAs of pitchers who pitched at least 100 innings, compare Greg Maddux's performance in 1995 with that of Sandy Koufax in 1966.

5.9 One way to compare outstanding baseball batters from different historical periods is to see how many standard deviations each player's batting average (BA) was from the average BA that year. Look at the most recently completed major league baseball season and consider only the BAs of players with at least 400 official times at bat. In each league (American and National), identify the player with the highest batting average and compare his performance with that of Ted Williams in 1941.

5.10 Table 2.1 shows the annual percentage returns from four investments during the period from 1971 through 1995. Use the reference given with this table to extend these data back to 1926. Now, for each of these four securities, calculate the mean and standard deviation of the annual returns. Then see what fraction of the actual returns is within 1 standard deviation of the mean and what fraction is within 2 standard deviations of the mean.

5.11 Ask a random sample of either 100 male or 100 female students at your school to tell you their heights. Try to make your sample truly random; do not just wander through a dining hall, stopping at tables. Otherwise, you may survey a table of basketball players eating together, or you may avoid tables that do not seem typical. When you have your data, display them in a histogram. Calculate the mean and standard deviation and the fractions of the heights that are more than 1, 2, and 3 standard deviations from the mean.

5.12 Find a listing of all major league batting averages for a recently completed season. For either the American or National league, record the batting averages of every nonpitcher who had at least 400 official times at bat. Display your data in a histogram and calculate the mean and standard deviation. What fractions of the batting averages are within 1, 2, and 3 standard deviations of the mean?

5.13 In the most recent issue of *The Wall Street Journal,* find the listing of the New York Stock Exchange composite transactions. For each stock, the last two columns in this table give the closing price that day and the net change in the closing price from the day before. These figures can be used to calculate the percentage price change from the day before. For example, if the closing price is 26 and the net change is +1, then the closing price the day before was 25 and the percentage change is $100(1/25) = 4.00\%$. Randomly select one of the first 10 stocks listed, and then look at every 10th stock thereafter. For example, if your randomly selected number is 4, look at stocks 4, 14, 24, and so on. For each of these stocks, calculate the percentage change in the closing price. Display these percentage changes in a histogram, and calculate the mean and standard deviation and the fractions of the changes that are within 1, 2, and 3 standard deviations of the mean.

5.14 Look at the most recently completed major league baseball season, and identify the two players in the American League and the two players in the National League who hit the most home runs that season. For each of these players, use the data for this season and the two preceding seasons to estimate the probability p that this player will hit a home run during a time at bat. Also use your data for each of these four players over all three seasons to estimate the maximum number of times at bat n during a season. Now, assuming the binomial model is appropriate, use your estimates of p and n to estimate each player's chances of hitting more than Roger Maris's record of 61 home runs in a season.

5.15 By placing your fingers on your neck near your Adams apple, you can measure your pulse rate by counting the rhythmical throbs of your arteries for one minute. Have each team member measure his or her pulse rate at least 50 times at intervals at least one hour apart. Try to avoid taking readings in unusual circumstances, for example, after exercising. For each team member, summarize the data with a box plot and histogram, and calculate the fraction of the readings that are within one and two standard deviations of the sample mean.

HAVE YOU EVER WONDERED?

Food products sold in the United States must be clearly marked to show the net weight and nutritional information for a standard serving, such as the number of calories and the percentages of the recommended daily amounts of fat, cholesterol, carbohydrates, and various vitamins and minerals. The contents of each and every package are, of course, not exactly identical. Because of imperfections in the manufacturing process, some net weights will be slightly higher or lower than others, and some will have more or less calories, cholesterol, and so on. Producers do not analyze the contents of every box and print different nutritional labels on each individual box, because to do so would be prohibitively expensive. Instead, they use a random sample to estimate the average characteristics of all boxes and then label the boxes with these averages.

Very similar issues arise in other, very different contexts. For instance, in the weeks and months preceding important elections, the candidates and the public are naturally curious about voter preferences. If the election were held today, who would win? But it is prohibitively expensive to interview every voter. Instead, pollsters use a sample of a few thousand people to estimate the preferences of all voters.

How much faith can we place on nutritional estimates that are based on a minuscule fraction of all the food produced, or on political polls that are based on a minuscule fraction of all the people who will vote? If the estimates—nutritional or political—are said to be "plus or minus 3 percent," where does the 3 percent come from and what does it mean? This chapter shows you how to make these kinds of estimates and how to interpret your results.

chapter 6

estimation

In solving a problem of this sort, the grand thing is to be able to reason backward. . . . Most people, if you describe a train of events to them, will tell you what the result would be. . . . There are few people, however, who, if you told them a result, would be able to evolve from their own inner consciousness what the steps were which led up to that result. This power is what I mean when I talk of reasoning backward.

—Sherlock Holmes, A Study in Scarlet

SAMPLING PROVIDES AN economical way to estimate the characteristics of a large population. Samples are used to estimate the amount of cholesterol in a person's body, the average acidity of a farmer's soil, and the number of fish in a lake. Medical samples are used to estimate how people will react to a medication. Production samples are used to estimate the fraction of a company's products that are defective, marketing samples to estimate how many people will buy a new product. The federal government uses samples to estimate the unemployment rate and the rate of inflation. Weather stations use samples at different locations to estimate average

temperature and rainfall. Public opinion polls are used to predict the winners of elections and to estimate the fraction of the population that agree with certain positions.

In each case, sample data are used to estimate a population value. But exactly how should the data be used to make these estimates? And how much confidence can we have in estimates that are based on a small sample from a large population? In this chapter we answer these questions.

To begin, we need to distinguish again between quantitative data (such as heights and weights) and categorical data (such as gender), which do not have natural numerical values. With quantitative data, we generally want to estimate the population mean, for instance, the average cholesterol level of middle-aged smokers or the average acidity of farmland. With categorical data, we usually want to estimate a success probability or a population proportion; for example, the probability that a person with lung cancer who receives a certain treatment will live more than 5 years or the percentage of voters who favor a certain candidate. The underlying logic is very similar in either case, but the details are slightly different. We begin with quantitative data and then show how our procedures can be modified to handle categorical data.

6.1 Estimating a Mean When the Standard Deviation Is Known

A certain brand of cereal is sold in a box marked *net weight 24 ounces.* There may, on average, be 24 ounces of cereal in the boxes sold by this company, but imperfections in the manufacturing and packaging process inevitably cause variations in the net weights of individual boxes. A more accurate label might read: *on average, net weight 24 ounces.* However, no one can know the average net weight of all the cereal boxes sold by this company unless every single box is opened and weighed. This is where statistics comes in. The company, the government, or a consumer research group can use the average weight of a sample of cereal boxes to estimate the average weight of the entire population of cereal boxes.

The population mean μ is a **parameter** whose value is unknown, but can be estimated. A sample statistic, such as the sample mean, that is used to estimate the value of a population parameter is called an **estimator.** The specific value of the estimator obtained in a particular sample is an **estimate.**

A random sample of 10 boxes yielded the following net weights (in ounces):

24.10 24.01 23.82 24.05 23.93 24.05 23.86 23.90 24.05 24.13

The average net weight of these 10 boxes is a natural estimate of the average net weight of all the boxes; that is, we use the sample mean $\bar{x}$ to estimate the population mean μ. With these particular data, the sample mean is

$$\bar{x} = \frac{24.13 + 24.01 + \cdots + 24.10}{10}$$

$$= 23.99$$

But how seriously can we take an estimate of the average weight of millions of cereal boxes when our estimate is based on just 10 boxes? We know that if we were to repeat this experiment, we would almost certainly get a different sample and most likely get a somewhat different value for the sample mean. Because our samples are chosen randomly, **sampling variation** will cause the sample mean to vary from sample to sample, sometimes being larger than the population mean and sometimes lower. How much faith can we place in the mean of one small sample?

Sampling Error

The sample mean is a random variable that depends on which boxes happen to be selected for the random sample. By the luck of the draw, a particular sample may contain mostly boxes with below-average weights or mostly boxes with above-average weights. The difference between the value of one particular sample mean and the average of the means of all possible samples of this size is known as the **sampling error.** Sampling error is not due to a poorly designed experiment or sloppy procedure. It is the inevitable result of the fact that the observations in our sample are chosen by chance. In contrast, **systematic errors** or biases cause the sample means to differ, on average, from the population parameter we are trying to estimate.

Suppose that we are trying to estimate the average net weight of cereal boxes from a population whose average net weight is 24 ounces, but our scale underreports every weight by 1 ounce. If we were to weigh a great many random samples from this population with this defective scale, the average of these sample means would be 23 ounces. The difference between the mean of one particular sample and this 23-ounce average is the sampling error. The difference between the 23-ounce average of our sample means and the true 24-ounce average net weight of all cereal boxes is the systematic error.

Systematic errors can be detected and corrected. We can repair our scale or get a new one. Sampling errors, in contrast, inevitably occur when we take samples from a population. Even with an accurate scale, the sample mean will be sometimes above, and sometimes below, 24 ounces. While we cannot correct for sampling error, we can be aware of its consequences. In practice, we take one sample, calculate the sample mean, and use this value as an estimate of the population mean. It is tempting to regard this number as definitive. That temptation should be resisted. We must always remember that our particular sample is just one of many samples that might have been selected, and that other samples would have yielded somewhat different sample means.

We cannot say whether a particular sample mean, such as 23.99, is above or below, close to or far from, the population mean because we do not know the value of the population mean. We might have obtained a sample mean of 23.99 from a population with a mean of 24.00, 23.99, 10.00, or 100.0, and we do not know for sure which is the case. But the situation is not hopeless. We can use probabilities to deduce how likely it is that a sample will be selected whose mean is within a specified distance from the population mean.

Let's follow Sherlock Holmes' advice and reason backward. Because the net weight of a cereal box depends on a great many imperfections in the preparation and packaging of the cereal, the central limit theorem suggests that net weights may be normally distributed. For the sake of argument, suppose hypothetically that the net weight of each box is, in fact, a random variable from a normal distribution with a mean of $\mu = 24$ ounces and a standard deviation of $\sigma = .1$ ounce. As shown in Fig. 6.1, when a box is selected at random, there is a .5 probability that the net weight will be more than 24 ounces and a .5 probability that the net weight will be less than 24 ounces. Remembering the 2-standard-deviations rule of thumb for normal distributions, there is approximately a .95 probability that the net weight will be within 2 standard deviations

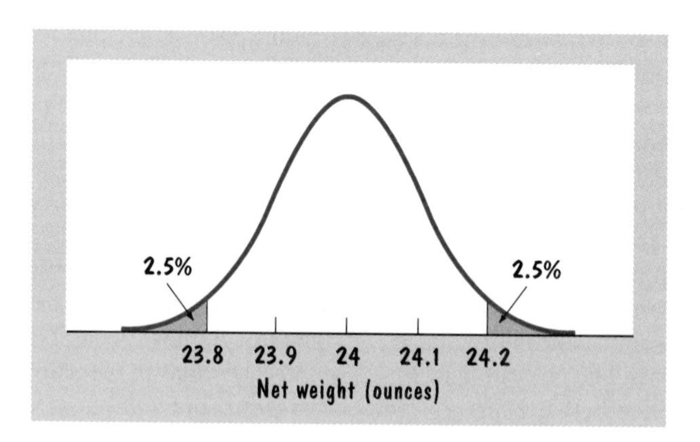

Figure 6.1 A hypothetical distribution of the net weight of cereal boxes.

of the mean—between 23.8 and 24.2 ounces. The fact that all 10 of the observations in our sample are within this interval (although one just barely) is certainly consistent with the assumption that the next weights come from a normal distribution with a mean of 24 ounces and a standard deviation of .1 ounce. Of course, these data are consistent with other, slightly different assumptions, too. However, they are inconsistent with some assumed values; for example, if the net weights came from a normal distribution with a mean of 10 ounces and a standard deviation of .1 ounce, it is highly unlikely that a random sample would give 10 observations that average 23.99 ounces.

The Sampling Distribution of the Sample Mean

To make this insight more concrete, we can investigate the probability distribution of the sample mean.

> The **sampling distribution** of a statistic, such as the sample mean, is the probability histogram or density curve that describes the population of all possible values of this statistic.

First, consider the shape of the sampling distribution for the sample mean. The value of the sample mean is determined by the selection of the n observations in the sample:

$$\bar{x} = \frac{x_1 + x_2 + \cdots + x_n}{n}$$

It can be shown mathematically that if the individual observations are drawn from a normal distribution, then the sampling distribution for the sample mean is also normally distributed, no matter what the size of the sample.

Even if the observations are drawn from a population that does not have a normal distribution, the sampling distribution of the sample mean will approach a normal distribution as the sample size increases. Here's why. Each observation in a random sample is an independent random variable drawn from the same population. The sample mean is the sum of these n outcomes, divided by n. Except for the unimportant division by n, these are the same assumptions as in the central limit theorem! Therefore the sampling distribution for the mean of a random sample from any population approaches a normal distribution as n increases.

The central limit theorem not only explains why data histograms are often bell-shaped, but also tells us that the sampling distribution for the mean of a reasonably sized random sample is bell-shaped. The only caution that we need exercise is to be sure that the sample is large enough for the central limit theorem to work its magic. With something like the weights of cereal boxes, which are themselves approximately normally distributed, a sample of 10 observations is large enough. If the underlying distribution is not normal, but roughly symmetric, a sample of size 20 or 30 is generally sufficient for the normal distribution to be appropriate. If the underlying distribution is known to be very asymmetric, then 50, 100, or more observations may be needed.

Figure 6.2 shows the sampling distribution for the mean of a sample taken from a uniform distribution in which each number from 1 to 9 has the same probability of being selected. As the sample size increases, it is increasingly likely that the sample mean will be close to 5. The only way to get a mean of 1 in a sample of size 10 is to draw 10 straight 1s. In contrast, there are lots of ways to get a sample mean of 5; for example, ten 5s; or five 4s and five 6s in any order; or one 3, two 4s, or four 5s, two 6s, and one 7 in any order. In all, there are 167,729,959 ways to get a sample mean of 5, making this a much more likely outcome than a sample mean of 1. There are slightly fewer ways to get a sample mean of 4.9 or 5.1 and even fewer ways to get a sample mean of 4.8 or 5.2. For samples as small as 10, Fig. 6.2 shows that the sampling distribution for the sample mean is well approximated by the bell-shaped normal curve.

Figure 6.3 shows the sampling distribution for the sample mean when samples of size 2 and 30 are taken from a distribution, labeled sample size = 1, that is asymmetric about the population mean $\mu = 1$. Even for a sample of size 2, the sampling distribution for the sample mean begins its convergence to a bell shape. For a sample of size 30, the sampling distribution is, for all practical purposes, a normal distribution.

It is important that you recognize the difference between the probability distribution for a single value of x and the sampling distribution for the sample mean $\bar{x}$. A good example of this distinction is that the sampling distribution for the mean of a sample of size 30 in Fig. 6.3 does not look at all like the probability distribution for a single outcome ($n = 1$). This because the sample mean is an average of the results of 30 separate draws from the distribution labeled sample size = 1. When we select a random sample of 30 observations, some of the outcomes most likely will be above 1 and others below 1, thereby giving a sample mean that is, in most cases, not far from 1. It is unlikely that the mean of a reasonably large sample will be far from 1 because, for this to occur, the individual outcomes in the sample must repeatedly be on one side of the population mean, which is unlikely in a large sample. The damping effects of averaging many outcomes give the sampling distribution for the sample mean a symmetric bell shape.

In addition to the general shape of the sampling distribution, we need to know how its mean and standard deviation are related to the population mean μ and standard deviation σ of the population from which the individual observations

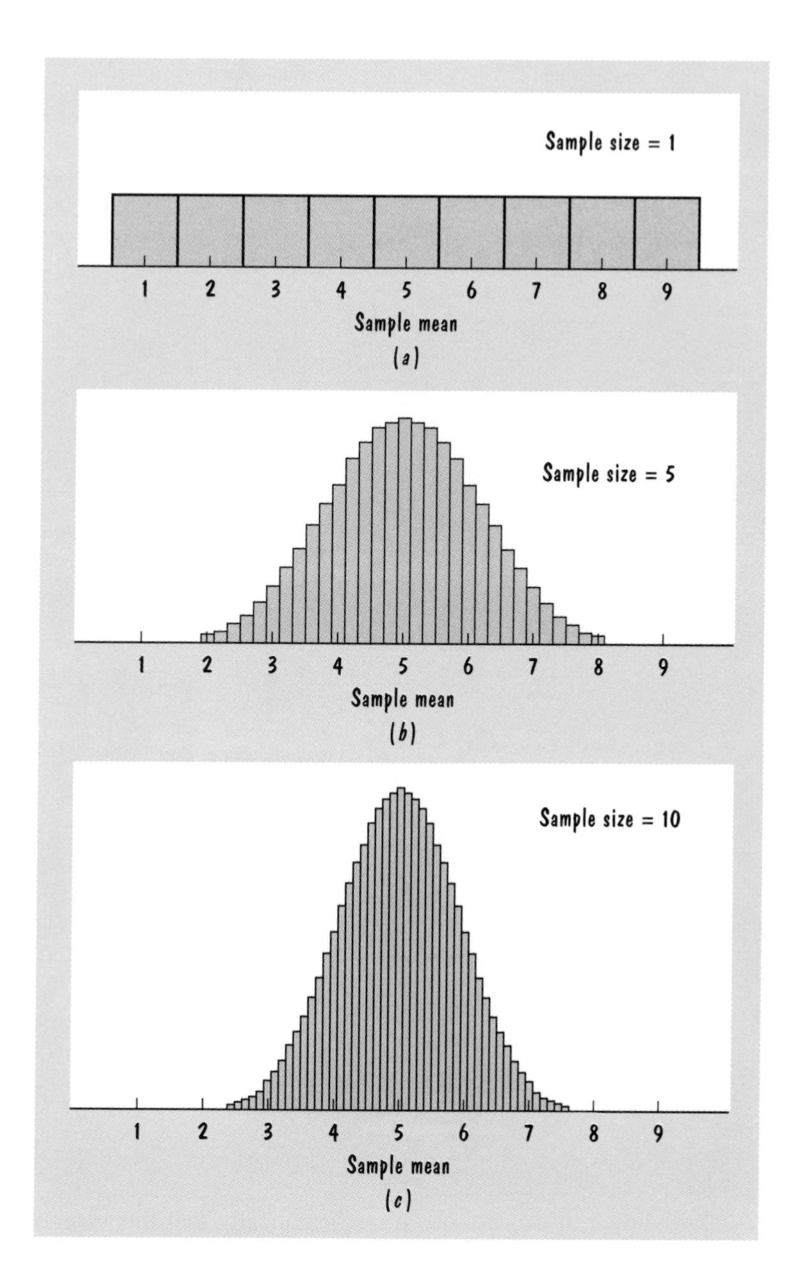

Figure 6.2 The sampling distribution of sample means from a uniform distribution.

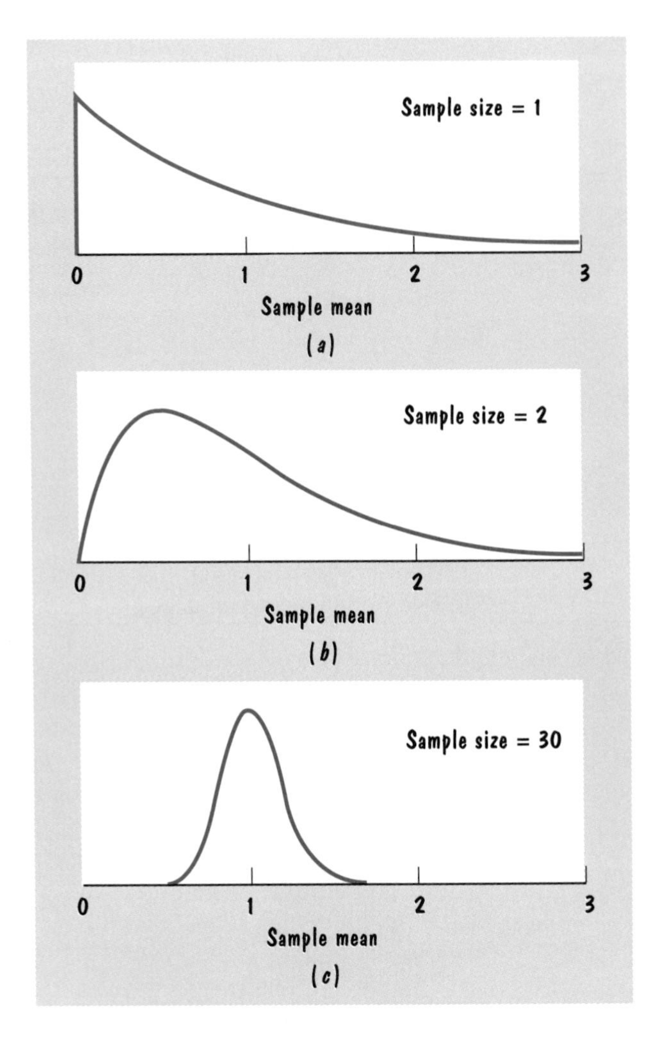

Figure 6.3 As the sample size increases, the sampling distribution for the sample mean approaches the normal distribution.

are drawn. It can be shown mathematically that this sampling distribution has a mean equal to μ and a standard deviation equal to σ, divided by the square root of the size of the sample:

$$\text{Mean of } \bar{x} = \mu$$

$$\text{Standard deviation of } \bar{x} = \frac{\sigma}{\sqrt{n}}$$

We can therefore summarize our discussion of the sampling distribution of the sample mean as follows:

> If each of n observations in a random sample is from a normal distribution with a mean μ and standard deviation σ, then the sampling distribution of the sample mean is a normal distribution with a mean μ and a standard deviation equal to σ, divided by the square root of the sample size:
>
> $$\bar{x} \text{ is } N\left[\mu, \frac{\sigma}{\sqrt{n}}\right] \tag{6.1}$$
>
> Even if the underlying population is not normal, a sufficiently large sample will ensure that the sampling distribution of the sample mean is approximately normal.

This is undoubtedly one of the most important formulas in statistics; it is absolutely crucial that you remember it and understand it.

The Mean of the Sampling Distribution

The sample mean is a random variable whose value is determined by the luck of the draw in selecting the random sample. While we can never know with certainty exactly how close a particular sample mean is to the unknown population mean, we can use the mean and standard deviation of the sampling distribution to gauge the reliability of the sample mean as an estimator of the population mean. A sample statistic is said to be an **unbiased estimator** of a population parameter if the mean of the sampling distribution of this statistic is equal to the value of the population parameter. Because the mean of the sampling distribution of the sample mean is μ, the sample mean is an unbiased estimator of the population mean.

Unbiased estimators have considerable appeal. It would be discomforting to use an estimator that you know to be systematically too high or too low. A statistician who uses unbiased estimators can anticipate estimation errors that, over a lifetime, average close to zero. Of course, average performance is not the only thing that counts. A British Lord Justice once summarized his career by saying, "When I was a young man practicing at the bar, I lost a great many cases I should have won. As I got along, I won a great many cases I ought to have lost; so on the whole justice was done." The conscientious statistician should be concerned with not only how good the estimates are on average, but also how accurate they are in particular cases.

The Standard Deviation of the Sampling Distribution

One way of gauging the accuracy of an estimator is by its standard deviation. If an estimator has a large standard deviation, there is a substantial probability that the estimate will be far from its mean. If an estimator has a small standard deviation, there is a high probability that the estimate will be close to its mean.

Equation (6.1) states that the standard deviation of the sampling distribution for the sample mean is equal to σ, divided by the square root of the sample size n. As the number of observations increases, the standard deviation of the sampling distribution declines. To understand this phenomenon, remember that standard deviation is a measure of the uncertainty of the outcome. With a large sample, it is extremely unlikely that all the observations will be far above the mean and equally improbable that all the observations will be far below the mean. Instead, it is almost certain that some of the observations will be above the mean and others below, and that the average will be close to the mean. For large samples, the standard deviation of the sample mean goes to zero, reflecting our commonsense judgment that when the sample encompasses the entire population, the sample mean is certain to be equal to the population mean. As Sherlock Holmes observed, "While the individual man is an insolvable puzzle, in the aggregate he becomes a mathematical certainty. You can, for example, never foretell what any one man will do, but you can say with precision what an average number will be up to. . . . So says the statistician."[1]

Let us apply Eq. (6.1) to the cereal example, again assuming hypothetically that the manufacturing and packaging process yields boxes with net weights that are normally distributed with a mean $\mu = 24$ ounces and a standard deviation $\sigma = .1$ ounce. If we select 10 boxes randomly, the sampling distribution of this sample mean is a normal distribution with a mean of $\bar{x}$ equal to $\mu = 24$ ounces and a standard deviation of $\bar{x}$ equal to $\sigma/\sqrt{n} = .1/\sqrt{10} = .032$ ounce. Figure 6.4 shows this probability distribution. In accord with our 2-standard-deviations rule of thumb, Table 3 in the back of the book shows that there is a .95 probability that a standardized normal variable z will be between -1.96 and $+1.96$; this implies that there is a .95 probability that a normally distributed random variable will be within 1.96 standard deviations of its mean. Because we are working with the sampling distribution for the sample mean, we have a .95 probability that the sample mean $\bar{x}$ will be in the interval

$$\mu - 1.96(\text{standard deviation of } \bar{x}) \quad \text{to} \quad \mu + 1.96(\text{standard deviation of } \bar{x})$$

In our cereal example with $\sigma = .1$ and $n = 10$, the standard deviation of $\bar{x}$ is $\sigma/\sqrt{n} = .1/\sqrt{10} = .032$. Therefore, there is a .95 probability that the sample mean will be within .062 ounce of the population mean:

$$\mu - 1.96(.032) \quad \text{to} \quad \mu + 1.96(.032) \quad = \quad \mu - .062 \quad \text{to} \quad \mu + .062$$

If we use the sample mean to estimate the population mean μ, there is only a .05 probability that our estimate will be off by more than .062 ounce.

If we use a larger sample, we can be even more confident that our estimate will be close to μ. Figure 6.4 shows how the population distribution for the sample mean narrows as the size of the sample increases. If we base our estimate of the population mean on a sample of 10 boxes, there is a .05 probability that our estimate will miss by more than .062 ounce. With 25 boxes, the standard deviation of $\bar{x}$ is $.1/\sqrt{25} = .02$, and there is a .05 probability that our estimate will be more than $1.96(.02) = .039$ ounce away from the population mean. With 1000 boxes (not shown), there is a .05 probability that we will miss by more than .006 ounce.

Confidence Intervals

We can rephrase these probability computations to show the confidence that we have in using the sample mean to estimate the population mean. If there is a .95 probability that the sample mean will turn out to be within 1.96 standard deviations of the population mean μ, then there is a .95 probability that the interval from

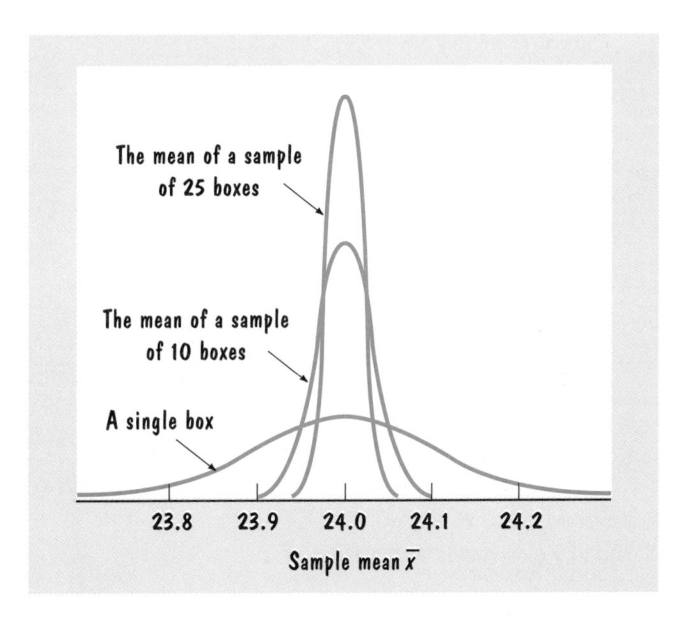

Figure 6.4 The sampling distribution of the mean of a sample of cereal boxes as the sample size increases.

$$\bar{x} - 1.96(\text{standard deviation of } \bar{x}) \quad \text{to} \quad \bar{x} + 1.96(\text{standard deviation of } \bar{x})$$

will include the value of μ. Such an interval is called a **confidence interval,** and the .95 probability is the interval's **confidence level.** The shorthand formula for a 95 percent confidence interval for the population mean μ is

$$\begin{aligned} 95\% \text{ confidence interval for } \mu &= \bar{x} \pm 1.96 \text{ (standard deviation of } \bar{x}) \\ &= \bar{x} \pm 1.96 \frac{\sigma}{\sqrt{n}} \qquad (6.2) \end{aligned}$$

Figure 6.5 shows the logic of confidence intervals. There is a .95 probability that the sample mean $\bar{x}$ will turn out to be within 1.96(standard deviation of $\bar{x}$) of μ, that is, between μ − 1.96(standard deviation of $\bar{x}$) and μ + 1.96(standard deviation of $\bar{x}$). There is consequently a .95 probability that the interval $\bar{x}$ − 1.96(standard deviation of $\bar{x}$) to $\bar{x}$ + 1.96(standard deviation of $\bar{x}$) will encompass μ. There is a .05 probability that the sample mean will, by the luck of the draw, turn out to be more than 1.96(standard deviation of $\bar{x}$) from the population mean μ and that the confidence interval will consequently not include μ.

In our cereal example, Eq. (6.2) gives this 95 percent confidence interval for the average weight of all cereal boxes in the population:

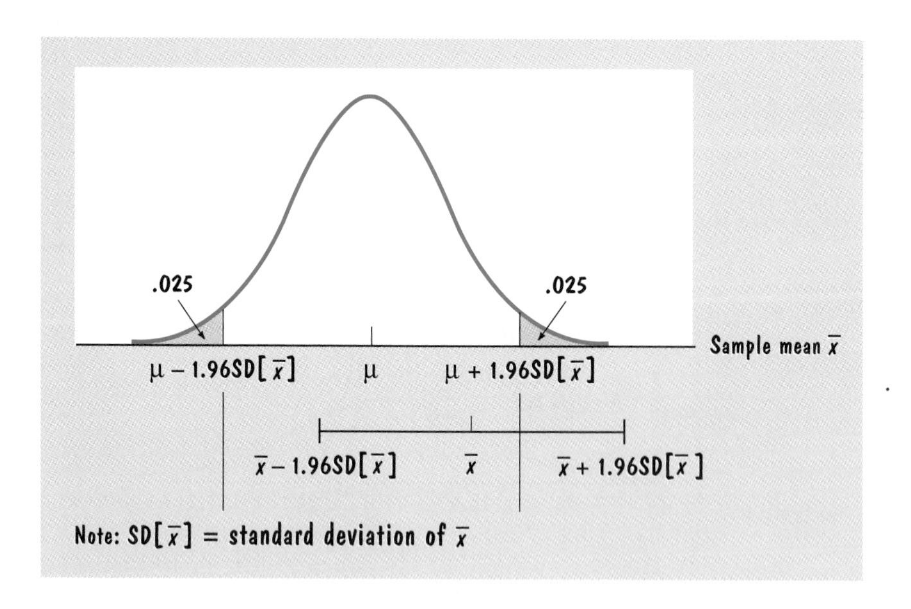

Figure 6.5 If the sample mean $\bar{x}$ turns out to be within 1.96 standard deviations of $\bar{x}$ of the population mean μ, then a 95 percent confidence interval encompasses μ.

$$\begin{aligned}
95\% \text{ confidence interval for } \mu &= \bar{x} \pm 1.96(\text{standard deviation of } \bar{x}) \\
&= 23.99 \pm 1.96\left(\frac{.1}{\sqrt{10}}\right) \\
&= 23.99 \pm .062
\end{aligned}$$

Ninety-five percent confidence levels are standard, but there is no compelling reason why we cannot use others. If we want to be more confident—for example, 99 percent certain—that our interval will contain μ, then we must use a wider interval. If we are satisfied with a 50 percent chance of encompassing μ, then a narrower interval will do.

> Whatever our choice of a confidence level, we can use Table 3 to determine the z value z^* that corresponds to this probability, for example, $z^* = 1.96$ for 95 percent confidence and $z^* = 2.575$ for 99 percent confidence. The confidence interval is then
>
> $$\begin{aligned}
\text{Confidence interval for } \mu &= \bar{x} \pm z^*(\text{standard deviation of } \bar{x}) \\
&= \bar{x} \pm z^* \frac{\sigma}{\sqrt{n}} \qquad (6.3)
\end{aligned}$$

To illustrate, we will determine a 99 percent confidence interval. Table 3 states that there is a .99 probability that a standardized normal variable z will be between -2.575 and $+2.575$; this implies that there is a .99 probability that the sample mean will be within 2.575 standard deviations of μ. Therefore, a 99 percent confidence interval for μ is

$$\bar{x} \pm 2.575(\text{standard deviation of } \bar{x})$$

In our cereal example with 10 boxes having an average weight of 23.99 ounces, a 99 percent confidence interval is

$$\begin{aligned}
99\% \text{ confidence interval for } \mu &= \bar{x} \pm 2.575(\text{standard deviation of } \bar{x}) \\
&= 23.99 \pm 2.575\left(\frac{.1}{\sqrt{10}}\right) \\
&= 23.99 \pm .081
\end{aligned}$$

As anticipated, a 99 percent confidence interval is somewhat wider than a 95 percent confidence interval. Thus there is a tradeoff between the confidence level and the width of the confidence interval. To be more certain that our interval includes μ, we must widen the interval.

The general procedure for determining confidence intervals should be apparent by now, and it is summarized in the How-To-Do-It box.

HOW TO DO IT

Estimating the Population Mean When the Standard Deviation Is Known

We want to use a sample of 10 cereal box weights to estimate the average net weight of all cereal boxes; the standard deviation is known to be .1 ounce.

1. Calculate the sample mean; in our cereal example, this is 23.99 ounces.
2. Calculate the standard deviation of the sample mean by dividing the known standard deviation σ by the square root of the sample size n. In our cereal example, standard deviation of $\bar{x} = \sigma/\sqrt{n} = .1/\sqrt{10}$.
3. Select a confidence level—a popular choice is 95 percent, but we can use 99 percent or another level if we feel this is more appropriate—and consult the normal distribution in Table 3 to determine the z value z^* that corresponds to this probability; for example, $z^* = 1.96$ for a 95 percent confidence interval, and $z^* = 2.575$ for a 99 percent confidence interval.
4. A confidence interval for the population mean is equal to the sample mean plus or minus z^* standard deviations of the sample mean:

$$\text{Confidence interval for } \mu = \bar{x} \pm z^*(\text{standard deviation of } \bar{x})$$
$$= \bar{x} \pm z^* \frac{\sigma}{\sqrt{n}}$$

For the cereal boxes, a 95 percent confidence interval is

$$95\% \text{ confidence interval for } \mu = 23.99 \pm 1.96\left(\frac{.1}{\sqrt{10}}\right)$$
$$= 23.99 \pm .062$$

that is, from 23.928 to 24.052.

The reason for the uncertainty in our estimate is that net weights vary from box to box and the sample mean consequently depends on the luck of the draw. This uncertainty is due to sampling variation, not to a flawed procedure or a biased weighing scale.

The spread in a confidence interval is a **margin for sampling error** (or **margin of error**):

$$\text{Margin for sampling error} = \pm z^*(\text{standard deviation of } \bar{x})$$
$$= \pm z^* \frac{\sigma}{\sqrt{n}}$$

For a 95 percent confidence interval, $z^* = 1.96$ and the margin for sampling error is ± 1.96 standard deviations of the sample mean.

A slip that people sometimes make is to forget that we are concerned with the standard deviation for the sampling distribution of the sample mean. They do not divide σ by the square root of the sample size, and consequently they get the wrong answer. Remember, we are using the sample mean to estimate μ, and the accuracy of our estimate depends critically on the sample size. With a large sample, there is less chance that the sample mean will be far from μ, and we can therefore be more confident that the sample mean is an accurate estimator of μ. This intuition shows up in our computations when we calculate the standard deviation of the sample mean by dividing σ by the square root of the sample size. If we forget to take the sample size into account, we understate the reliability of our estimator.

Notice, too, that it is the sample mean that varies from sample to sample, not the population mean. A 95 percent confidence interval for μ is interpreted as follows: "There is a .95 probability that the sample mean will turn out to be sufficiently close to μ that my confidence interval includes μ. There is a .05 probability that the sample mean will happen to be so far from μ that my confidence interval does not include μ." The .95 probability refers to the chances that random sampling will result in an interval that encompasses the fixed parameter μ, not the probability that random sampling will give a value of μ that is inside a fixed confidence interval. In our cereal box example, the population mean μ is not a random variable and does not have a .95 probability of falling between 23.928 and 24.052. It either is or is not in this interval. A clever analogy can be made to a game of horseshoes. The population mean is the fixed stake, and a confidence interval that results from a random sample is a horseshoe thrown at the stake. There is a .95 probability that we will throw a horseshoe that rings the stake, not a .95 probability that the stake will jump in front of the horseshoe.

Figure 6.6 shows 20 hypothetical 95 percent confidence intervals. We can anticipate that 19 out of 20 times, on average, the sample mean will be within 1.96 standard deviations of the population mean, so that a 95 percent confidence

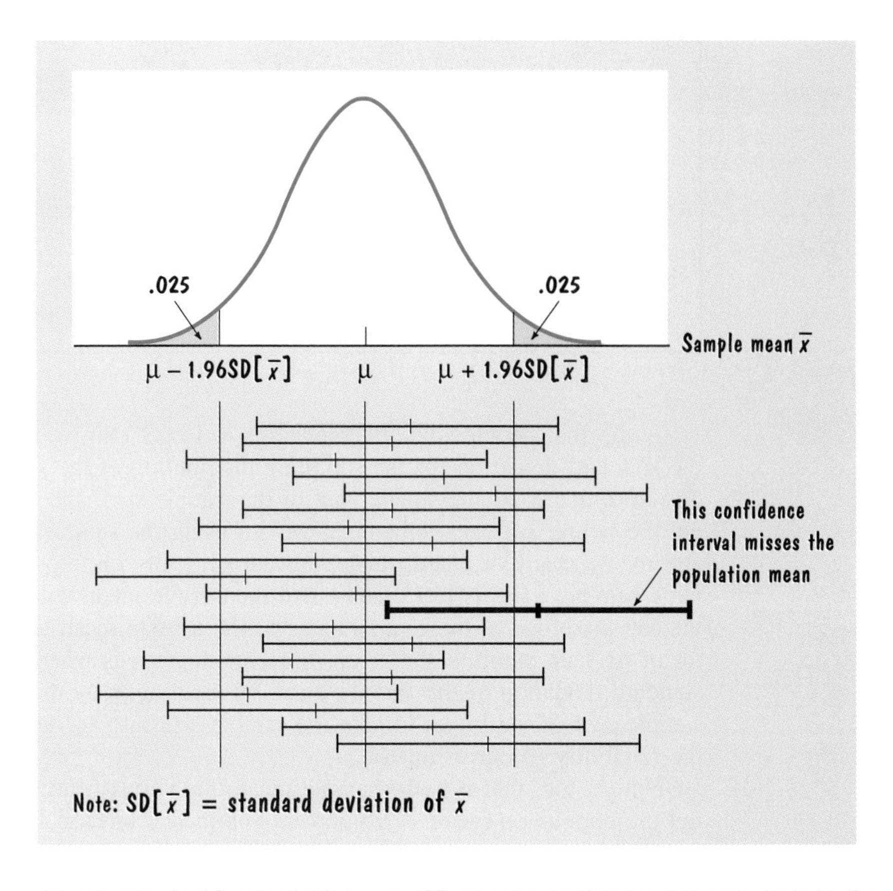

Figure 6.6 On average, in 19 out of 20 cases, 95 percent confidence intervals include the population mean.

interval includes the value of the population mean. One out of 20 times, on average, the sample mean will be more than 1.96 standard deviations from the population mean, and a 95 percent confidence interval will not include the value of the population mean.

When interpreting our confidence interval, we also must remember that it only applies to that population for which our data represent a random sample. If we take a random sample of net weights for one cereal brand from one production plant operated by one cereal company, our confidence interval for the mean net weight doesn't necessarily apply to other brands, other plants, or other companies. If we take a random sample of students at one college, our results may not be applicable to students at other colleges. If we test a medication on male patients, our results may not be applicable to females.

Choosing a Sample Size

We took a random sample of 10 cereal boxes and found the average net weight to be 23.99 ounces. Recognizing that this is just one of many samples that might have been selected, we measured the sampling error by calculating a 95 percent confidence interval, based on the fact that there is a .95 probability that a normally distributed random variable will be within 1.96 standard deviations of its mean. With $\sigma = .1$ and $n = 10$, we report our estimate as $23.99 \pm .062$ ounces.

How could we obtain a more precise estimate, that is, one with a smaller margin for sampling error? Remember that the margin for sampling error is

$$\text{Margin for sampling error} = \pm\, z^*(\text{standard deviation of } \bar{x})$$

$$= \pm\, z^* \frac{\sigma}{\sqrt{n}}$$

The z value z^* comes from the normal distribution table, and the standard deviation σ is determined by the manufacturing and packaging process. The only factor that the statistician can control is the sample size n. A larger sample size will give a smaller margin for sampling error, as shown by these illustrative calculations for a 95 percent confidence interval with $\sigma = .1$:

SAMPLE SIZE	MARGIN FOR SAMPLING ERROR
10	±.0620
40	±.0310
160	±.0155
640	±.0077
2,560	±.0039
10,240	±.0019

Because the formula for the confidence interval uses the square root of the sample size, we must increase the number of observations by a factor of 4 in order to halve the margin for sampling error (and halve the width of the confidence interval).

We can use these kinds of trial calculations to decide how many boxes to sample. If we will settle for a .062-ounce margin for sampling error, then 10 boxes are sufficient. If we need an estimate with only a .0019 margin for sampling error, then 10,240 boxes are needed. The cost of a more precise estimate is the additional time and energy spent on a larger sample, plus the larger number of boxes that will be tested instead of sold.

In making this choice, it is generally best to lay out a table of alternatives, as I have done, so that you can get a feel for how the margin for sampling error varies with the size of the sample. Such a table is not needed if you know the desired margin of error in advance. Then you can simply set this value m equal to the margin for sampling error given in Eq. (6.3)

$$m = z^* \frac{\sigma}{\sqrt{n}}$$

and solve for the requisite sample size n:

$$n = \left(\frac{z^*\sigma}{m}\right)^2$$

In our cereal example, if we want an $m = .01$ margin for sampling error, then

$$n = \left[\frac{1.96(.1)}{.01}\right]^2$$

$$= 384.16 \qquad (6.4)$$

rounded up to a sample of size 385.

Sampling from Finite Populations

Notice that one very interesting characteristic of the margin for sampling error is that it does not depend on the size of the population. Regardless of whether the population consists of 1 million or 1 billion boxes, a sample of 385 boxes has a .01 margin for sampling error. At first glance, this conclusion may seem surprising. If we are trying to estimate a characteristic of a large population, then there is a natural tendency to believe that a large sample is needed. If there are 1 billion boxes in the population, a sample of 385 includes only 1 out of every 2.6 million boxes. How can we possibly obtain a reliable estimate with a sample that only looks at 1 out of every 2.6 million boxes?

A moment's reflection reveals why our margin for sampling error does not depend on whether the population consists of 1 million or 1 billion boxes. The chances that the luck of the draw will yield a sample whose mean differs substantially from the population mean depend on the size of the sample and on the chances of selecting boxes that are far from the population mean, not on how many boxes there are in the population. In fact, our calculations implicitly assume that the population is infinitely large, in that we ignore the fact that sampling without replacement alters the characteristics of a finite population as the sampling proceeds. We do so because the sample has to be fairly large relative to the population (say, more than 10 percent) of our confidence interval to change noticeably. Further, if there is an effect, it is to reduce the width of the confidence interval. Our estimates are more reliable if the population is finite, because then there is less chance of obtaining an unlucky sample of observations that are consistently above or below the population mean. By assuming an infinite population, the confidence interval in Eq. (6.3) understates the reliability of our estimates.

Does the Enzyme Work, or Doesn't It?

Researchers at a pharmaceutical company tested a new enzyme that they hoped would increase the yield from one of their manufacturing processes.[2] The data were calculated as the ratio of the yield with the enzyme to an assumed yield with the old process (based on a great deal of past experience). An observed value of 110.2 meant that, in this particular test, the yield was 10.2 percent higher with the enzyme than with the old process. If the enzyme had no effect on the yield, the measured yields should have averaged about 100.

As it turned out, a sample of 41 observations found an average yield of 125.2 with a standard deviation of 20.1. However, a member of the company's research department was unimpressed:

> There is no good theoretical reason for believing that the mean yield should be higher than 100 with the enzyme. Moreover, in the 41 batches studied, the standard deviation was 20.1. In my opinion, 20 points represents a large fraction of the difference between 125 and 100. There is no real evidence that the enzyme increases yield.

This statistical reasoning is wrong! It is the standard deviation of the sample mean—not the standard deviation of the individual observations—that is of interest. If the standard deviation of an individual observation x is 20.1, then the standard deviation of the sample mean $\bar{x}$ is only

$$\begin{aligned}\text{Standard deviation of } \bar{x} &= \frac{\sigma}{\sqrt{n}} \\ &= \frac{20.1}{\sqrt{41}} \\ &= 3.14\end{aligned}$$

If the enzyme has no effect on the yield, so that each observation is from a distribution with a mean of 100 and a standard deviation of 20.1, then there is indeed a substantial chance that an individual observation might be as high as 125, but there is virtually no chance that the average of 41 observations will be so high. We can calculate this probability by using the mean and standard deviation of the sampling distribution for the sample mean to convert to a standardized z value:

$$
\begin{aligned}
z &= \frac{\bar{x} - \mu}{\sigma/\sqrt{n}} \\
&= \frac{125 - 100}{20.1/\sqrt{41}} \\
&= \frac{25}{3.14} \\
&= 7.96
\end{aligned}
$$

The probability of a z value greater than 7.96 is minuscule. Another way to demonstrate that these data do indeed indicate that the enzyme increases the average yield is to calculate a 95 percent confidence interval for μ:

$$
\begin{aligned}
95\% \text{ confidence interval for } \mu &= \bar{x} \pm 1.96(\text{standard deviation of } \bar{x}) \\
&= 125.2 \pm 1.96(3.14) \\
&= 125.2 \pm 6.2
\end{aligned}
$$

This is far from 100.

Checking the Weight of Gold Coins

The American Coinage Act of 1792 specified that a gold $10 eagle coin must contain 247.5 grains of pure gold. Because of imperfections in the manufacturing process, each coin cannot weigh exactly 247.5 grains. Yet Congress wanted the weights not to deviate substantially from 247.5 grains—neither too heavy, with the Mint being overly generous, nor too light, with the public being shortchanged (and eventually unwilling to accept gold eagles at face value).

In 1837, Congress passed another law, stating that there would be an annual assay, which permitted a .25-grain deviation in the weight of a single eagle and a .048-grain deviation in the average weight of 1000 eagles weighed as a group. Failure of these tests would show that the Mint officers "shall be deemed disqualified to hold their respective positions." By specifying different tolerances for a single coin and for 1000 coins, this law recognized the statistical argument that the standard deviation of a sample mean is smaller than the standard deviation of a single observation. On the other hand, the values .25 and .048 do not correspond exactly to the statistical fact that the standard deviation is inversely related to the square root of the sample size.

Suppose, for example, that the minted coins are a random sample from a normal distribution with a mean of 247.5 grains and an unknown standard deviation σ. Would the Mint be more likely to fail the test of a single coin or the test of the average weight of 1000 coins? Because the distribution of the weight of a single coin x is $N[247.5, \sigma]$, we can convert to a standardized z value for determining the probability that a single randomly selected coin is too heavy:

$$z = \frac{x - \mu}{\sigma} = \frac{(247.5 + .25) - 247.5}{\sigma} = \frac{.25}{\sigma}$$

Using Eq. (6.1), we can find the z value for calculating the probability that the average weight of 1000 randomly selected coins is too heavy:

$$z = \frac{\bar{x} - \mu}{\sigma/\sqrt{n}} = \frac{(247.5 + .048) - 247.5}{\sigma/\sqrt{1000}} = \frac{1.52}{\sigma}$$

(Analogous calculations can be made for the z values for the probabilities that the coins are too light.)

No matter what the value of the standard deviation σ, the probability that z will be larger than $1.52/\sigma$ is smaller than the probability that z will be larger than $0.25/\sigma$. Therefore, the chances of failing the 1000-coin test are less than the chances of failing a single-coin test. For example, if the manufacturing process has a standard deviation of .50 grain (representing one-half of 1 percent of the coin's intended weight), there is a .62 probability of failing a one-coin test and a .002 probability of failing a 1000-coin test. Perhaps Congress did this on purpose, intending to encourage the Mint to make 1000-coin tests. Or perhaps the members of Congress had not studied statistics sufficiently.

exercises

6.1 The life of a Rolling Rock tire is normally distributed with a mean of 30,000 miles and a standard deviation of 5000 miles. What is the probability that a randomly selected tire will last more than 40,000 miles? What is the probability that the average life of four randomly selected tires will be more than 40,000 miles?

6.2 Each package of Cow Country butter says *net weight = 16 ounces*. In fact, the net weight is normally distributed with a mean of 16.05 ounces and a standard deviation of .05 ounce. A government agency has decided to test a sample of Cow Country butter packages to see if the average net weight is at least 16 ounces. What is the probability that the average weight of n randomly selected packages will be less than 16 ounces if $n = 1$? If $n = 4$? If $n = 16$? Why does this probability change as n increases?

6.3 Over the period from 1929 to 1933, scientists measured the speed of light in air 2500 times and obtained an average value of 186,270 miles per second, with a standard deviation of 8.7 miles per second. Assuming these to be independent measurements from a distribution with a standard deviation of 8.7 miles per second, find a 99 percent confidence interval for the speed of light.

6.4 Statisticians sometimes report 50 percent confidence intervals, with the margin for sampling error known as the *probable error*. For example, an estimate $\bar{x}$ of the average useful life of a television picture tube is said to have a probable error of e minutes if there is a .50 probability that the interval $\bar{x} \pm e$ includes the value of the population mean. Calculate the probable error if the standard deviation is known to be 2.5 years and the average useful life of a random sample of 25 picture tubes is found to be 8.15 years.

6.5 In a 1975 lawsuit concerning the discharge of industrial wastes into Lake Superior, the judge observed that a court-appointed witness had found the fiber concentration to be ".0626 fibers per cc., with a 95 percent confidence interval of from .0350 to .900 fibers per cc."[3] Explain the apparent statistical or typographical error in this statement.

6.6 Example 6.2 compares a test of a single coin with a test of the average weight of 1000 coins. If Congress had wanted the Mint to have an equal probability of failing these tests when it produces coins with weights that are normally distributed with a mean of 247.5 and an unknown standard deviation σ, and permitted a .25-grain deviation in the weight of a single coin, how large a deviation should it have permitted in the average weight of 1000 coins?

6.7 Explain why you either agree or disagree with this reasoning: "[T]he household [unemployment] survey is hardly flawless. Its 60,000 families constitute less than .1 percent of the work force."[4]

6.8 A statistics textbook gives this example:

> Suppose a downtown department store questions forty-nine downtown shoppers concerning their age. . . . The sample mean and standard deviation are found to be 40.1 and 8.6, respectively. The store could then estimate μ, the mean age of all downtown shoppers, via a 95% confidence interval as follows:

$$\bar{x} \pm 1.96\left(\frac{\sigma}{\sqrt{n}}\right) = 40.1 \pm 1.96\left(\frac{8.6}{\sqrt{49}}\right)$$
$$= 40.1 \pm 2.4$$

Thus the department store should gear its sales to the segment of consumers with average age between 37.7 and 42.5.[5]

Explain why you either agree or disagree with this interpretation: 95 percent of downtown shoppers are between the ages of 37.7 and 42.5.

6.2 Estimating a Mean When the Standard Deviation Is Not Known

So far, to simplify matters, we have assumed that the standard deviation of net cereal weights is known to be $\sigma = .1$ ounce. In practice, we seldom know the value of the population standard deviation. It, too, is a parameter that must be estimated. The most natural estimator of σ, the standard deviation of the population, is s, the standard deviation of the sample data from this population. The formula for the sample standard deviation is given in Eq. (3.2) and is repeated here. The variance of n observations $x_1, x_2, \ldots, x_n$ is the average squared deviation of these observations about their mean:

$$\text{Variance} = \frac{(x_1 - \bar{x})^2 + (x_2 - \bar{x})^2 + \cdots + (x_n - \bar{x})^2}{n - 1} \tag{6.5}$$

The standard deviation s is the square root of the variance.

When the standard deviation of an estimator, such as the sample mean, is itself estimated from the data, this estimated standard deviation is called the estimator's *standard error.* The standard error of the sample mean is calculated by replacing the unknown parameter σ in Eq. (6.1) with its estimate s:

> The standard error (or estimated standard deviation) of the sample mean is
>
> $$\text{Standard error of } \bar{x} = \frac{s}{\sqrt{n}}$$

The t Distribution

The need to estimate the standard deviation creates another source of uncertainty in gauging the reliability of the sample mean $\bar{x}$ as an estimator of the population mean μ. The calculation of a confidence interval, using Eq. (6.3), requires a value for the standard deviation σ:

$$\text{Confidence interval} = \bar{x} \pm z^* \frac{\sigma}{\sqrt{n}}$$

If we must estimate σ, then we cannot be quite so confident that our estimated interval will encompass μ. Sometimes the confidence interval will miss μ, not because the sample mean is off target, but because our estimate of the standard deviation is too low and our confidence interval is consequently too narrow. To compensate for the fact that our estimate of the standard deviation will sometimes be too low, we need to widen our confidence interval by replacing z^* with a somewhat larger number.

In 1908, W. S. Gosset figured out the exact amount of widening needed when the data are drawn from a normal distribution. Gosset was a statistician employed by the Irish brewery Guinness, which encouraged statistical research but not publication. Because of the importance of his findings, he was able to persuade Guinness to allow his work to be published under the pseudonym *Student* and his calculations became known as the **Student's *t* distribution.**

When the mean of a sample from a normal distribution is standardized by subtracting the mean μ of its sampling distribution and dividing by the standard deviation $\sigma/\sqrt{n}$ of its sampling distribution, the resulting z variable

$$z = \frac{\bar{x} - \mu}{\sigma/\sqrt{n}}$$

has a normal distribution. Gosset determined the sampling distribution of the variable that is created when the mean of a sample from a normal distribution is standardized by subtracting μ and dividing by its standard error (the sample standard deviation s divided by the square root of the sample size):

$$t = \frac{\bar{x} - \mu}{s/\sqrt{n}} \tag{6.6}$$

The exact distribution of t depends on the sample size, because as the sample size increases, we are increasingly confident of the accuracy of the estimated standard deviation. For an infinite sample, the estimate s will equal the actual value σ, and the distribution of t and z coincide. With a small sample, s may be either larger or smaller than σ, and the distribution of t is consequently more dispersed than the distribution of z.

Figure 6.7 compares the t distribution for a sample of size 5 with the normal

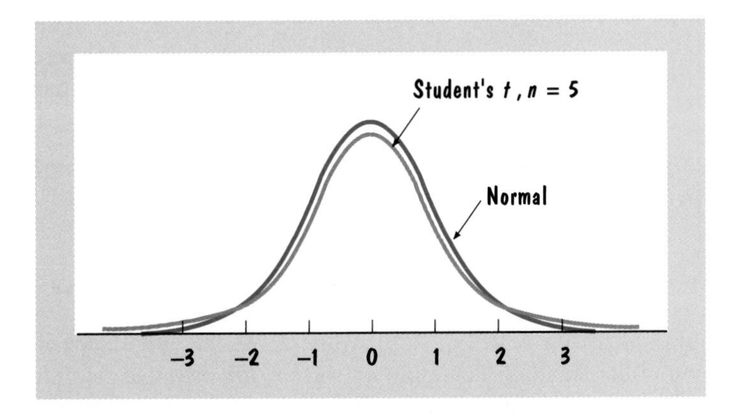

Figure 6.7 Normal and Student's t distributions.

distribution. As shown, there is a somewhat smaller probability of being close to 0 and a somewhat larger probability of being in the tails. With smaller samples, the t distribution is even more dispersed. With larger samples, the t distribution becomes indistinguishable from the normal distribution. The probabilities for various t distributions are shown in Table 4 at the end of this book. Instead of identifying each t distribution by the size of the sample, they are identified by the number of **degrees of freedom:**

$$\text{Degrees of freedom} = \text{Number of observations} - \text{Number of parameters that must be estimated beforehand}$$

Here, we calculate s by using n observations and one estimated parameter (the sample mean); therefore, there are $n - 1$ degrees of freedom.

There is another way to think about degrees of freedom that is more closely related to the name itself. We calculate s from n squared deviations about the sample mean. But we know from Chap. 3 that the sum of the deviations about the sample mean is always 0. Thus if we know the values of $n - 1$ of these deviations, we know the value of the last deviation, too. Only $n - 1$ deviations are freely determined by the sample.

Confidence Intervals

The logic of using a confidence interval to gauge the precision of our estimate is not affected by our use of an estimated standard deviation. There is a slight adjustment in the actual calculation in that we replace the known value of σ with its estimate s and replace the z value with a somewhat larger t value.

After selecting a confidence level, perhaps 95 or 99 percent, we consult the appropriate t distribution in Table 4 to determine the t value t^* that corresponds to this probability. The confidence interval is then

$$\text{Confidence interval for } \mu = \bar{x} \pm t^* \text{ (standard error of } \bar{x}\text{)}$$
$$= \bar{x} \pm t^* \frac{s}{\sqrt{n}} \qquad (6.7)$$

In comparison with Eq. (6.3), while gives the confidence interval when we know the standard deviation σ, we use the estimate s in place of the known value of σ, and we use a t value t^* that depends on the sample size in place of the z value z^*.

One way of gauging the greater dispersion of the t distribution is to compare the widths of, say, 95 percent confidence intervals for several different sample sizes:

SAMPLE SIZE n	DEGREES OF FREEDOM $n - 1$	95 PERCENT CONFIDENCE INTERVAL
2	1	$\bar{x} \pm 12.71$(standard error of $\bar{x}$)
10	9	$\bar{x} \pm 2.26$(standard error of $\bar{x}$)
20	19	$\bar{x} \pm 2.09$(standard error of $\bar{x}$)
30	29	$\bar{x} \pm 2.05$(standard error of $\bar{x}$)
Infinite	Infinite	$\bar{x} \pm 1.96$(standard error of $\bar{x}$)

For a reasonably sized sample, the difference is pretty slight. For 30 observations, it is more important to ensure that the sample is random than to worry about whether the margin for sampling error is precisely 2.05 or 1.96 standard deviations. Nonetheless, t values are generally used in statistical work, and you should use them, too.

Gosset derived the t distribution by assuming that the sample data are taken from a normal distribution. Subsequent research has shown that, because of the power of the central limit theorem, confidence intervals based on the t distribution are remarkably accurate even if the underlying data are not normally distributed, as long as we have at least 15 observations from a roughly symmetric distribution or at least 30 observations from a clearly asymmetric distribution.[6] A histogram can be used for a rough symmetry check.

For our cereal box example, there are only 10 observations, but the central limit theorem suggests that the variation in weights from box to box are due to the cumulation of a great many factors and are therefore likely to be approximately normally distributed. The histogram of our data in Fig. 6.8 does not contradict this assumption. So we proceed to use Eq. (6.5) to calculate the sample variance of our data:

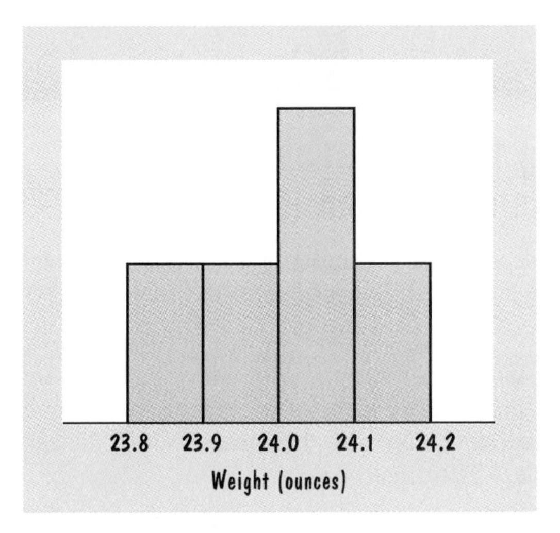

Figure 6.8 A histogram of cereal box weights.

$$\text{Variance} = \frac{(24.10 - 23.99)^2 + \cdots + (24.13 - 23.99)^2}{10 - 1}$$
$$= .0112$$

The standard deviation is the square root of the variance: $s = \sqrt{.0112} = .1056$. The standard error of the sample mean is

$$\text{Standard error of } \bar{x} = \frac{s}{\sqrt{n}}$$
$$= \frac{.1056}{\sqrt{10}}$$
$$= .033$$

As was the case when the standard deviation is known, an easy mistake is to forget to divide s by the square root of n, especially since we already divided by $n - 1$ when calculating s. To avoid making this mistake, remember that s is an estimate of the standard deviation of an individual observation and that our confidence interval depends on the standard deviation of the sampling distribution for the sample mean, which we estimate by dividing s by the square root of the sample size.

To determine our confidence interval, we also need to choose a confidence level and find the appropriate t value. We have $n - 1 = 10 - 1 = 9$ degrees of freedom, and Table 4 shows that a 95 percent confidence interval with .025

HOW TO DO IT

Estimating the Population Mean When the Standard Deviation Is Estimated

We want to use a random sample of 10 cereal box weights to estimate the average weight of all cereal boxes; the standard deviation must be estimated.

1. Calculate the sample mean; in our cereal example, this is 23.99 ounces.
2. Calculate the standard error of the sample mean by dividing the sample standard deviation s by the square root of the sample size n. For our sample of 10 cereal weights

   24.10 24.01 23.82 24.05 23.93 24.05 23.86 23.90 24.05 24.13

   the sample variance is

$$\text{Variance} = \frac{(24.10 - 23.99)^2 + \cdots + (24.13 - 23.99)^2}{10 - 1}$$
$$= .0112$$

   The standard deviation is the square root of the variance: $s = \sqrt{.0112} = .1056$. The standard error of the sample mean is $s/\sqrt{n}$; here, $.1056/\sqrt{10} = .033$.

probability in each tail implies $t^* = 2.262$. Substituting our various pieces of information into Eq. (6.7), we have our 95 percent confidence interval for the average weight of cereal boxes in the population from which this random sample was drawn:

$$\begin{aligned} 95\% \text{ confidence interval for } \mu &= \bar{x} \pm t^*(\text{standard error of } \bar{x}) \\ &= \bar{x} \pm t^* \frac{s}{\sqrt{n}} \\ &= 23.99 \pm 2.262 \left(\frac{.1056}{\sqrt{10}}\right) \\ &= 23.99 \pm .08 \end{aligned}$$

A less tedious alternative is to enter our 10 observations into a statistical software package and let the computer do all the calculations for us. For example, Table 6.1 shows the output from the Minitab program. The command

3. Select a confidence level, such as 95 percent, and look in Table 4 with $n - 1$ degrees of freedom to determine the t value t^* that corresponds to this probability. Here, with $10 - 1 = 9$ degrees of freedom, a 95 percent confidence interval with .025 probability in each tail implies $t^* = 2.262$.
4. A confidence interval for the population mean is equal to the sample mean plus or minus t^* standard errors of the sample mean:

$$\text{Confidence interval for } \mu = \bar{x} \pm t^*(\text{standard error of } \bar{x})$$

$$= \bar{x} \pm t^* \frac{s}{\sqrt{n}}$$

With our cereal data, a 95 percent confidence interval is

$$95\% \text{ confidence interval for } \mu = 23.99 \pm 2.262 \left(\frac{.1056}{\sqrt{10}}\right)$$

$$= 23.99 \pm .08$$

5. Alternatively, we can use computer software, as illustrated by the Minitab output in Table 6.1.

Table 6.1

A Minitab confidence interval for the mean weight of cereal boxes

```
MTB> TINTERVAL 95 C1

         N      MEAN     STDEV   SE MEAN    95.0 PERCENT C.I.
C1      10    23.990      .106      .033    (23.914, 24.066)
```

`TINTERVAL 95 C1` asks the program to use the t distribution to determine a 95 percent confidence interval for the data in column C1 of Minitab's data worksheet. The next two lines are the output: variable name, number of observations, mean of the sample data, standard deviation of the sample data, standard error of the sample mean, and end points for a 95 percent confidence interval.

Early in this chapter, we asked, "How seriously can we take an estimate of the average weight of millions of cereal boxes when our estimate is based on just 10 boxes?" Now we have the answer. If our 10 boxes are a random sample, there is a .95 probability that the 95 percent confidence interval we calculate will include the average weight of all the boxes in the population. With this particular sample, our 95 percent confidence interval is 23.99 ounces plus or minus .08 ounce, that is, between 23.91 and 24.07 ounces.

Using Confidence Intervals to Protect the Environment

The U.S. Environmental Protection Agency (EPA) typically sets industrial effluent discharge standards by measuring the discharge by companies using the *best practical control technology currently available* (BPCTCA). These sample data (on the daily effluent of soap, for instance) are used to calculate a sample mean and standard deviation and, from these, a 99 percent confidence interval for the population mean effluent discharge μ from companies using the BPCTCA. Because the standard deviation must be estimated, a t distribution is used to determine the upper and lower limits for these confidence intervals. For example, with 30 observations, there are $30 - 1 = 29$ degrees of freedom, and Table 4 tells us that a 99 percent confidence interval is

$$\overline{x} \pm 2.756(\text{standard error of } \overline{x})$$

For an exemplary plant using the BPCTCA, there is a .99 probability that the EPA's 99 percent confidence interval will encompass the value of the population mean for this plant's effluent discharge. For a plant using less effective technology with a population mean that is substantially higher than that for a BPCTCA plant, a sample is likely to yield a mean that is outside the BPCTCA confidence interval, causing the plant to be cited by the EPA for violating the effluent discharge standards.

exercises

6.9 A sociologist wants to estimate the average age at which women are widowed in a certain city. A random sample of 25 widows gives a sample mean age of 68.3 years with a sample standard deviation of 12.1 years. Assuming that these data are a random sample from a normal distribution, calculate a 90 percent confidence interval for the average age of the population. Without doing any calculations, would a 99 percent confidence interval be wider or narrower?

6.10 Shown below are the number of kills by Petrea Moyle, a college volleyball player, in each of 12 matches:[7]

14 19 24 22 19 17 28 23 23 23 17 12

Was the number of kills in any of these matches more than 2 standard deviations away from her average number of kills? Assuming these data to be a random sample from a normal distribution, calculate a 99 percent confidence interval for the population mean for the number of kills when this woman plays a college volleyball match. Identify a crucial assumption here that seems especially questionable.

6.11 The first U.S. citizen to win the Nobel Prize in physics was Albert Michelson (1852–1931), who was given the award in 1907 for developing and using optical precision instruments. His October 12 to November 14, 1882, measurements of the speed of light in air (in kilometers per second) were as follows:[8]

299,883	299,796	299,611	299,781	299,774	299,696	299,748
299,809	299,816	299,682	299,599	299,578	299,820	299,573
299,797	299,723	299,778	299,711	300,051	299,796	299,772
299,748	299,851					

Assuming that these measurements were a random sample from a normal distribution, does a 99 percent confidence interval include the value 299,710.5 that is now accepted as the speed of light?

6.12 Henry Cavendish (1731–1810) was an English physicist and chemist who discovered many of the fundamental laws of electricity and conducted a number of important experiments, including a demonstration that water is a compound of hydrogen and oxygen. In 1798, he made 23 measurements of the density of the earth relative to the density of water:[9]

5.10	5.27	5.29	5.29	5.30	5.34	5.34	5.36	5.39	5.42	5.44
5.46	5.47	5.53	5.57	5.58	5.62	5.63	5.65	5.68	5.75	5.79
5.85										

Assuming that these were independent measurements from a normal distribution, does a 99 percent confidence interval include the value 5.517 that is now accepted as the density of earth?

6.13 Table 3.2 gives 2 years of annual returns for 17 randomly selected stock mutual funds. In each of these years, assume that these data come from a normal distribution, and calculate a 95 percent confidence interval for the average return for all mutual funds. Now see if each interval includes that year's return for the Standard & Poor's 500 index: 1.3 percent in 1994 and 37.4 percent in 1995.

6.14 In the 20th century, 12 different players have been the world chess champion:

CHAMPION	YEARS HELD TITLE	AGE WHEN FIRST WON TITLE
Emanuel Lasker (Germany)	1894–1921	26
Jose Capablanca (Cuba)	1921–1927	32
Alexander Alekhine (Soviet Union)	1927–1935, 1937–1947	35
Max Euwe (Holland)	1935–1937	34
Mikhail Botvinnik (Soviet Union)	1948–1957, 1958–1960, 1961–1963	36
Vasily Smyslov (Soviet Union)	1957–1958	36
Mikhail Tal (Soviet Union)	1960–1961	23
Tigran Petrosian (Soviet Union)	1963–1969	33
Boris Spassky (Soviet Union)	1969–1972	32
Robert Fisher (United States)	1972–1975	29
Anatoly Karpov (Soviet Union)	1975–1985	24
Gary Kasparov (Soviet Union)	1985–	22

Treating these 12 ages as a random sample, estimate a 95 percent confidence interval for the mean age at which a world chess champion first becomes champion. Does this seem to you to be a random sample from a population?

6.15 In 1728, Isaac Newton reported the data below on the average reign of kings in 11 different kingdoms.[10] Although he presented no calculations, he concluded that the reign of kings should be reckoned "at about eighteen or twenty years a-piece."

KINGDOM	YEARS	NUMBER OF KINGS	AVERAGE REIGN, YEARS
Judah	390	18	21.67
Israel	259	15	17.27
Babylon	209	18	11.61
Persia	208	10	20.80
Syria	244	16	15.25
Egypt	277	11	25.18
Macedonia	138	8	17.25
England (1066–1714)	648	30	21.60
France (first 24)	458	24	19.08
France (second 24)	451	24	18.79
France (last 15)	315	15	21.00

Treating these data as a random sample of $n = 11$ independent observations of kingdoms, calculate a 95 percent confidence interval for the average reign during a kingdom. Is it reasonable to assume that these data are a random sample from a population?

6.16 In 1993, based on 11 separate studies, the Environmental Protection Agency estimated that nonsmoking women who live with smokers have, on average, a 19 percent higher risk of lung cancer than do similar women living in a smoke-free home. The EPA reported a 90 percent confidence interval for this estimate as 4 percent to 34 percent. After lawyers for tobacco companies argued that the EPA should have used a standard 95 percent confidence interval, an article in *The Wall Street Journal* reported that "Although such a calculation wasn't made, it might show, for instance, that passive smokers' risk of lung cancer ranges from say, 15% *lower* to 160% higher than the risk run by those in a smoke-free environment."[11]

a. Explain why, even without consulting a probability table, we know that a standard 95 percent confidence interval for this estimate does not range from -15 to $+160$ percent.

b. In calculating a 90 percent confidence interval, the EPA used a t distribution with 10 degrees of freedom. Use this same distribution to calculate a 95 percent confidence interval.

6.3 Estimating a Success Probability or Population Proportion

We can now apply the procedures developed so far in this chapter to data that are categorical, in that they do not have natural numerical values. With such data, random samples are used to estimate a success probability or population proportion. For example, samples of manufactured products are used to estimate the fraction that are defective. Responses to medications given to a sample of patients are used to estimate the probability that a patient will benefit from the medication and the probability that there will be serious side effects. Public opinion polls are used to estimate the proportion of the population who favor certain candidates, behave in certain ways, or have certain beliefs.

To structure our discussion, we focus on a political poll. Among the population of all voters, a fraction p would vote for candidate A if the election were held today. If a small random sample of size n is drawn from this large population, each draw has a probability p of picking a person who prefers A and a probability $1 - p$ of picking someone who prefers another candidate. As explained in Chap. 5, the binomial model provides a very useful framework for analyzing categorical data. Here, our random sample can be considered a sequence of n independent observations with the constant probability p of "success."

In probability analysis, our reasoning runs from an assumed probability to possible outcomes: We use the binomial model to tell us the probability of x successes in n observations if the probability of success on each observation is p. For example, if 40 percent of all voters prefer A, the binomial distribution tells us the probability that a random sample of 1000 voters would find that more than 500 (50 percent of those surveyed) prefer A. In statistical analysis, our logic runs in the opposite direction, from outcomes to probability: We observe x successes in n observations and use these data to estimate the unknown probability p:

> In a random sample with categorical data that yield x successes in n observations, the success proportion $\hat{p} = x/n$ is an estimator of p, the probability of success for each observation.

The estimator $\hat{p}$ is pronounced "p hat." Our general approach is the same as when we used the sample mean $\bar{x}$ for quantitative data to estimate the population mean μ. It is easy to see why this should be so. For each observation of categorical data, we could assign a value 1 if there is a success and 0 if there is a failure. The success proportion $\hat{p} = x/n$, giving the total number of successes divided by the number of observations, is really just a sample mean—the average number of successes—and consequently can be analyzed by the very same procedures that we already learned: The sample mean is an estimator, and the mean and standard deviation of its sampling distribution can be used to determine a confidence interval for that estimate.

The Mean and Standard Deviation

The only quirk lies in the relationship between the mean and standard deviation for a binomial distribution. For normally distributed random variables, the mean μ and standard deviation σ are specified separately. For the binomial distribution, Eqs. (5.4) and (5.5) show that the mean and standard deviation of the success proportion x/n depend on the same parameter, the success probability p:

$$\text{Mean of } x/n: \qquad \mu = p \tag{6.8}$$

$$\text{Standard deviation of } x/n: \qquad \sigma = \sqrt{\frac{p(1-p)}{n}} \tag{6.9}$$

Instead of specifying the mean and standard deviation separately, we specify p, which simultaneously determines the mean and standard deviation.

If a fraction $p = .40$ of the population prefers candidate A, then the mean and standard deviation for the sampling distribution for the sample proportion $\hat{p} = x/n$ for a sample of size $n = 25$ are

$$\text{Mean of } x/n: \qquad \mu = p = .4$$

$$\text{Standard deviation of } x/n: \qquad \sigma = \sqrt{\frac{p(1-p)}{n}} = \sqrt{\frac{.40(1-.40)}{25}} = .098$$

Here are the means and standard deviations of the sample proportion for various sample sizes:

SAMPLE SIZE n	MEAN OF x/n	STANDARD DEVIATION OF x/n
25	.40	.098
100	.40	.049
400	.40	.024
1600	.40	.012

Notice that the mean of the sampling distribution for the sample proportion is .40, no matter what the size of the sample, but that the standard deviation declines as the sample increases: The sample proportion $\hat{p} = x/n$ is an unbiased estimator of the population proportion p, and the value of this estimator is less likely to be far from p as the sample gets larger. (Again, because the standard deviation depends on the square root of the sample size, we must quadruple the sample in order to halve the standard deviation.) The above standard deviation calculations and our 2-standard-deviations rule of thumb tell us that there is a .95 probability that the sample proportion will be within $2(.049) = .098$ of $p = .40$ for a sample of size 100 and within $2(.024) = .048$ of $p = .40$ for a sample of size 400.

Since the sample proportion x/n can be interpreted as a sample mean, the central limit theorem tells us that, like the mean of any sufficiently large sample, the sampling distribution for x/n is approximately a normal distribution:

$$\hat{p} = \frac{x}{n} \text{ is } N\left[p, \sqrt{\frac{p(1-p)}{n}}\right] \tag{6.10}$$

A normal approximation to a binomial distribution is generally satisfactory if both np and $n(1-p)$ are larger than 5. For our polling example with $p = .40$, even a sample as small as $n = 25$ easily satisfies these criteria: $np = 25(.40) = 10$ and $n(1-p) = 25(.60) = 15$. Figure 6.9 confirms that the sampling distribution for x/n has the familiar bell shape.

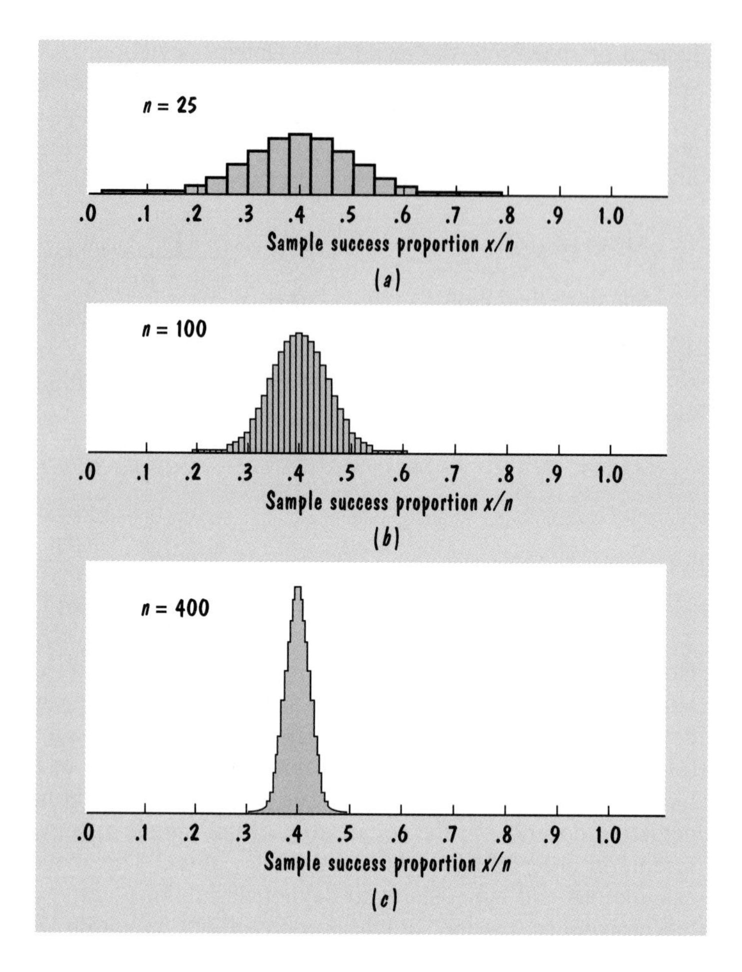

Figure 6.9 The sampling distribution for a sample of voter preferences, $p = .40$.

Confidence Intervals

Because there is a .95 probability that $\hat{p} = x/n$ will be within 1.96 standard deviations of p, a 95 percent confidence interval for p is

$$\begin{aligned} 95\% \text{ confidence interval for } p &= \hat{p} \pm 1.96(\text{standard deviation of } \hat{p}) \\ &= \frac{x}{n} \pm 1.96\sqrt{\frac{p(1-p)}{n}} \end{aligned}$$

More generally, for a given confidence level, such as 95 or 99 percent, Table 3 gives the appropriate z value z^* for the confidence interval:

$$\text{Confidence interval for } p = \hat{p} \pm z^*(\text{standard deviation of } \hat{p})$$
$$= \frac{x}{n} \pm z^* \sqrt{\frac{p(1-p)}{n}} \tag{6.11}$$

Notice, though, that the calculation of a confidence interval requires a value for p, which is the very thing that we are trying to estimate. A logical solution is to use our estimate $\hat{p} = x/n$ and call the estimated standard deviation of $\hat{p}$ the **standard error** of $\hat{p}$.

Using Table 3 to determine the appropriate z value z^* (for example, $z^* = 1.96$ for 95 percent confidence and $z^* = 2.575$ for 99 percent confidence), the confidence interval for p is

$$\text{Confidence interval for } p = \hat{p} \pm z^*(\text{standard error of } \hat{p})$$
$$= \hat{p} \pm z^* \sqrt{\frac{\hat{p}(1-\hat{p})}{n}}$$
$$= \frac{x}{n} \pm z^* \sqrt{\frac{(x/n)(1-x/n)}{n}} \tag{6.12}$$

For our polling example, suppose that a random sample of 1000 voters finds 407 preferring candidate A. A 95 percent confidence interval is

$$95\% \text{ confidence interval for } p = \hat{p} \pm z^*(\text{standard error of } \hat{p})$$
$$= \frac{x}{n} \pm 1.96 \sqrt{\frac{x/n(1-x/n)}{n}}$$
$$= \frac{407}{1000} \pm 1.96 \sqrt{\frac{407/1000(1-407/1000)}{1000}}$$
$$= .407 \pm .030$$

Table 6.2 shows the output from Smith's Statistical Package. After the user enters the values of n and x and the desired confidence level in the three boxes, the confidence interval for p is shown in the bottom half of the computer screen.

To reinforce our logic, consider this very different example. Suppose that historically only 40 percent of the people receiving the standard treatment for a certain type of cancer live more than 3 years. However, of a random sample of

Table 6.2

A confidence interval for p from Smith's Statistical Package

```
A sample success proportion can be used to obtain a
confidence interval for the success probability in the
population that yielded this random sample.

In this random sample,

   How many sample observations?              n =  [1000]

   How many successes in this sample?         x =  [ 407]

   Desired confidence level (for example, 95):     [  95]

The estimate of the success probability is .4070

Using a normal approximation to the binomial distribu-
tion, a 95 percent confidence interval is

   .4070 plus or minus .0304
```

30 patients who were given a new genetically engineered product, 20 lived more than 3 years. We want to use these sample data to estimate the probability p that a cancer patient receiving this treatment will live more than 3 years.

First, we calculate the success proportion in the sample:

$$\begin{aligned}\hat{p} &= \frac{x}{n}\\ &= \frac{20}{30}\\ &= .667\end{aligned}$$

To see if we can use a normal approximation to construct a confidence interval, we check that $n\hat{p}$ and $n(1 - \hat{p})$ are both larger than 5: $n\hat{p} = 30(.667) = 20$ and $n(1 - \hat{p}) = 30(1 - .667) = 10$. Okay. For a 95 percent confidence interval, we use Eq. (6.12) with $z^* = 1.96$:

$$\begin{aligned}\text{95\% confidence interval for } p &= \hat{p} \pm z^*(\text{standard error of } \hat{p})\\ &= \hat{p} \pm z^*\sqrt{\frac{\hat{p}(1-\hat{p})}{n}}\\ &= .667 \pm 1.96\sqrt{\frac{.667(1-.667)}{30}}\\ &= .667 \pm .169\end{aligned}$$

Thus, these data indicate that a 95 percent confidence interval for the probability that a cancer patient who receives this new treatment will live more than three years is .667 ± .169. For a more precise estimate (with a narrower margin of error), we need to test more than 30 patients.

Choosing a Sample Size

One way of deciding the size of our random sample is to consider our desired margin for sampling error, and then to set this desired value m equal to the margin for sampling error given in Eq. (6.11):

$$m = z^* \sqrt{\frac{p(1 - p)}{n}}$$

Rearranging, we see that the requisite sample size is

$$n = \left(\frac{z^*}{m}\right)^2 p(1 - p)$$

We do not know the value of p or what our estimate of p will turn out to be. However, trial and error confirm that the maximum value of $p(1 - p)$ is at $p =$ .5. Therefore, if we set $p = .5$, we can obtain a conservative value for the necessary sample size—conservative in the sense that if our estimate of p does not equal .5, then our margin for error will be somewhat smaller than our target margin.

$$n = \left(\frac{z^*}{2m}\right)^2 \tag{6.13}$$

To illustrate the procedure, suppose that we want a 95 percent confidence interval for our political poll to have a 2 percent margin for sampling error. Equation (6.13) tells us that we need a sample size of 2401:

$$\begin{aligned} n &= \left(\frac{z^*}{2m}\right)^2 \\ &= \left[\frac{1.96}{2(.02)}\right]^2 \\ &= 2401 \end{aligned}$$

By comparing alternatives, we can make an informed choice based on the tradeoff between a smaller margin of error and a larger (and more expensive) sample. These illustrative calculations (which are also graphed in Fig. 6.10) for a 95 percent confidence interval show how a larger sample size gives a smaller margin for sampling error.

HOW TO DO IT

Estimating a Success Probability or Population Proportion

To estimate the fraction p of all voters who prefer candidate A, we select a random sample and find that 407 of 1000 voters surveyed prefer A.

1. Calculate the success proportion in the sample; here,

$$\begin{aligned}\hat{p} &= \frac{x}{n} \\ &= \frac{407}{1000} \\ &= .407\end{aligned}$$

2. Are $n\hat{p}$ and $n(1 - \hat{p})$ both larger than 5, so that a normal approximation is justified? Here, $n\hat{p} = 1000(.407) = 407$ and $n(1 - \hat{p}) = 1000(1 - .407) = 593$.
3. Select a confidence level and consult the normal distribution in Table 3 to determine the z value z^* that corresponds to this probability; for example, $z^* = 1.96$ for a 95 percent confidence interval, and $z^* = 2.575$ for a 99 percent confidence interval.
4. A confidence interval for the success probability (or population proportion) p is equal to the sample success proportion plus or minus z^* standard errors of the sample success proportion:

$$\begin{aligned}\text{Confidence interval for } p &= \hat{p} \pm z^*(\text{standard error of } \hat{p}) \\ &= \hat{p} \pm z^* \sqrt{\frac{\hat{p}(1 - \hat{p})}{n}}\end{aligned}$$

In our polling example, a 95 percent confidence interval is

$$\begin{aligned}95\% \text{ confidence interval for } p &= \hat{p} \pm z^* \sqrt{\frac{\hat{p}(1 - \hat{p})}{n}} \\ &= .407 \pm 1.96 \sqrt{\frac{.407(1 - .407)}{1000}} \\ &= .407 \pm .030\end{aligned}$$

5. Alternatively, we can use computer software, as illustrated by the Smith's Statistical Package output in Table 6.2.

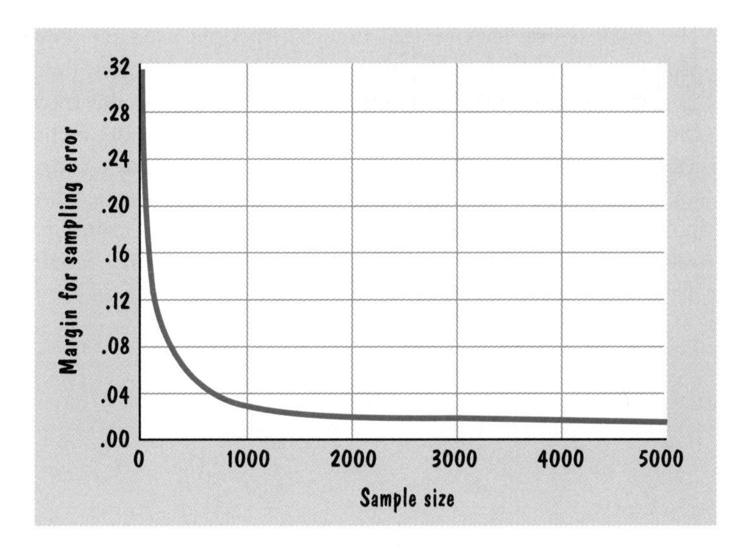

Figure 6.10 Margin for sampling error for a 95 percent confidence interval.

SAMPLE SIZE	MARGIN FOR SAMPLING ERROR
10	$\pm.310$
100	$\pm.098$
500	$\pm.044$
1000	$\pm.031$
2500	$\pm.020$
5000	$\pm.014$

A sample of size 10 would almost always be rendered useless by the 31 percent margin for sampling error. Usually, a better choice is a sample in the range of 1000 to 2500 with a margin of error of 2 to 3 percent.

 A 1996 Gallup Poll

We can use the final Gallup poll in the 1996 presidential election poll to illustrate these procedures further. During the weekend before the election, the Gallup organization conducted telephone interviews with 2400 people, of whom 1573 were identified as likely voters, based on the strength of their feelings and the reported frequency with which they had voted in the past. In addition to weeding out people they did not think would vote, the Gallup organization used cluster sampling, as explained in Chap. 4, rather than genuine random sampling.

Nonetheless, we assume that the 1573 people used for its election prediction were a random sample from the population of election-day voters.

We can interpret the population parameter p as the fraction of the voting population who were going to vote for Bob Dole, the Republican candidate in 1996. If the Gallup poll was a random sample from a large population, then each voter had an equal chance of being selected and, with every selection, the probability of choosing a Dole voter was p. Of the sample of $n = 1573$ likely voters, Gallup identified $x = 629$ who indicated they would vote for Dole. Thus the sample proportion was .40:

$$\begin{aligned}\hat{p} &= \frac{x}{n} \\ &= \frac{629}{1573} \\ &= .400\end{aligned}$$

Recognizing that this is but one of many samples that might have been selected, we want to gauge the reliability of this estimate by calculating a margin for sampling error and a corresponding confidence interval. Using the sample proportion $p = x/n$ as our estimate of p when calculating the standard error, we see that a 95 percent confidence interval is

$$\begin{aligned}\text{95\% confidence interval for } p &= \hat{p} \pm z^*(\text{standard error of } \hat{p}) \\ &= \hat{p} \pm z^* \sqrt{\frac{\hat{p}(1-\hat{p})}{n}} \\ &= .400 \pm 1.96 \sqrt{\frac{.400(.600)}{1573}} \\ &= .400 \pm .024\end{aligned}$$

This calculation indicates that Bob Dole will receive 40.0 percent of the vote, with a 2.4 percent margin of error—meaning we anticipate that, on average, 19 of 20 such confidence intervals will include the actual percentage that the candidate receives in the election. As it turned out, Dole received 40.7 percent of the vote, which is inside this confidence interval.

Table 6.3 shows some data for several Gallup presidential election polls. If we ignore the 1952 and 1956 elections, when Gallup used quota sampling, the actual error has been within the sampling error bounds in 8 of 10 elections. (In the volatile 1980 election, 20 percent of the voters changed their minds during the last 4 days before the election.) Given that these are not simple random samples; that pollsters must predict not only who people favor, but also who will vote; that some voters do not tell the truth; and that some voters change their minds during the time between the poll and the election, this is an

Table 6.3

The Gallup presidential poll's theoretical sampling errors and actual prediction errors

YEAR	NUMBER POLLED	NUMBER OF LIKELY VOTERS	PERCENTAGE OF SAMPLING ERROR IF SIMPLE RANDOM SAMPLE	ACTUAL PERCENTAGE ERROR
1952	5385	3350	±1.7	4.4
1956	8144	4950	±1.4	1.7
1960	8015	5100	±1.4	.9
1964	6625	4100	±1.5	2.7
1968	4414	2700	±1.9	.4
1972	3689	2100	±2.1	.3
1976	3439	2000	±2.2	1.0
1980	3500	1950	±2.2	3.7
1984	3456	1985	±2.1	.2
1988	2626	1970	±2.2	2.1
1992	2019	1562	±2.4	1.1
1996	2400	1573	±2.4	.7

Source: American Institute of Public Opinion, *Gallup Poll Monthly,* various issues, and personal correspondence.

Note: Gallup's estimates of its sampling error are somewhat larger since these are not simple random samples.

impressive record. Surveying roughly 2000 likely voters, Gallup is generally able to predict within a few percentage points the outcome of an election involving nearly 100 million voters.

6.17 The Nielsen television ratings are based on a sample of 5000 homes out of a population of 100 million homes. If it estimates that 50 percent of these homes are watching the Super Bowl, what is its margin for sampling error?

6.18 A 1992 survey of 750 black Americans found that 58 percent of those surveyed favored limiting the benefits paid to single mothers who have additional children while on welfare.[12] Estimate the margin for sampling error.

6.19 A telephone poll by *The Wall Street Journal* conducted on November 30 and December 1, 1987, asked a nationwide random sample of 1599 adults a number of questions about an upcoming summit meeting between U.S. President Reagan and Soviet leader Gorbachev. Explain this observation by the *Journal*:

> Chances are 19 of 20 that if all households with telephones in the U.S. had been surveyed using the same questionnaire, the findings would differ from the poll results of all respondents by no more than three percentage points in either direction.[13]

6.20 The New York-based National Association for Female Executives conducted a poll of 1300 of its 250,000 members during the days surrounding Anita Hill's testimony in the Senate hearings on Clarence Thomas' nomination to the U.S. Supreme Court.[14] The survey found that 77 percent considered sexual harassment to be a problem in the workplace. Assuming this to be a random sample from the population of all female executives, estimate a 95 percent confidence interval for the population proportion who consider workplace sexual harassment to be a problem. Identify two important potential biases in this poll.

6.21 *Consumer Reports* characterizes its typical reader as "well educated, married, suburban, a homeowner."[15] In 1990, this magazine mailed a nutrition survey to a sample of 1400 subscribers and received 928 responses, of which 15 percent were by smokers. Assuming that this is a random sample, use a normal approximation to the binomial distribution to calculate 95 and 99 percent confidence intervals for the fraction of all *Consumer Reports* subscribers who are smokers. Why might this be a biased sample of *Consumer Reports* readers?

6.22 A columnist for *The New York Times* called the 1986 campaign to remove California Supreme Court Justice Rose Bird the "most significant political event in the country."[16] On the eve of the election, a Pitzer College poll of 124 registered voters found that 53 planned to vote for Bird and 71 planned to vote against her. Calculate a 95 percent confidence interval for the fraction of all Californians planning to vote for Bird.

6.23 In the 1940s, the Federal Drug Administration seized a shipment of 22,464 rubber prophylactics when tests of a sample of 408 prophylactics found 30 to be defective, with holes that would allow the transmission of disease.[17] In federal court, the manufacturer argued that these 30 defective prophylactics represented only .1 percent of the shipment and that this was not sufficient grounds for seizure. If you were the judge in this case, would you find this argument persuasive? Estimate a 99 percent confidence interval for the fraction of prophylactics in the shipment that were defective.

6.24 Explain the error in this *Newsweek* explanation of the margin of error in a public opinion poll:

> The margin of error, calculated according to a textbook statistical formula, varies inversely with the sample size. In general, about 500

responses gives a possible error of 5 percent either way; 2500 responses decreases it to 1 percent.[18]

summary

Samples are used to estimate characteristics of populations. With quantitative data, a sample mean is used to estimate a population mean; with categorical data, a sample proportion is used to estimate a success probability or population proportion. Such estimators are subject to sampling error because the luck of the draw determines the particular random sample that will be selected.

If a random variable x is normally distributed with a mean μ and standard deviation σ, then the sampling distribution for the mean of a random sample of n observations is a normal distribution with a mean μ and a standard deviation equal to σ, divided by the square root of the sample size n:

$$\bar{x} \text{ is } N\left[\mu, \frac{\sigma}{\sqrt{n}}\right]$$

Even if the underlying distribution is not normal, a sufficiently large sample will ensure that the sampling distribution of the sample mean is approximately normal. A rule of thumb is that a normal approximation is generally satisfactory for a sample of 20 or 30 observations.

The sample mean is an unbiased estimator of μ, and a confidence interval can be used to gauge the degree of sampling error. Because there is a .95 probability that a normally distributed random variable will be within 1.96 standard deviations of its mean, there is a .95 probability that an interval equal to the sample mean plus or minus 1.96 standard deviations of the sample mean will encompass μ:

$$\begin{aligned} 95\% \text{ confidence interval for } \mu &= \bar{x} \pm 1.96(\text{standard deviation of } \bar{x}) \\ &= \bar{x} \pm 1.96\frac{\sigma}{\sqrt{n}} \end{aligned}$$

In general, we select a confidence level, such as 95 or 99 percent, and then consult the normal distribution in Table 3 to determine the appropriate z value z^* and the corresponding confidence interval:

$$\begin{aligned} \text{Confidence interval for } \mu &= \bar{x} \pm z^*(\text{standard deviation of } \bar{x}) \\ &= \bar{x} \pm z^*\frac{\sigma}{\sqrt{n}} \end{aligned}$$

The margin for sampling error is the spread in a confidence interval:

$$\text{Margin for sampling error} = \pm z^*(\text{standard deviation of } \bar{x})$$
$$= \pm z^* \frac{\sigma}{\sqrt{n}}$$

To halve the margin for sampling error, we must quadruple the size of the sample. The above calculation implicitly assumes an infinite population and overestimates the margin of error when we sample without replacement from a finite population.

If, as is usually the case, the population standard deviation σ is not known, it can be estimated with the sample standard deviation s. If the observations come from a normal distribution (or the sample has at least 15 or 30 observations), a confidence interval for μ is

$$\text{Confidence interval for } \mu = \bar{x} \pm t^*(\text{standard error of } \bar{x}) = \bar{x} \pm t^* \frac{s}{\sqrt{n}}$$

where t^* is given by a t distribution with $n - 1$ degrees of freedom. The effect of using a t distribution rather than a normal distribution is to widen the confidence interval to take into account the possible sampling error in estimating σ.

With categorical data, we can use a sample success proportion $p = x/n$ to estimate a success probability (or population proportion) p. The expected value and standard deviation of the estimator $\hat{p}$ both depend on the success probability p:

$$\text{Mean of } x/n = p$$
$$\text{Standard deviation of } x/n = \sqrt{\frac{p(1-p)}{n}}$$

Because the sample proportion x/n can be interpreted as a sample mean, the central limit theorem implies that the sample proportion is approximately normally distributed. By using the appropriate z value z^*, a confidence interval for p is

$$\text{Confidence interval for } p = \hat{p} \pm z^*(\text{standard error of } \hat{p})$$
$$= \hat{p} \pm z^* \sqrt{\frac{\hat{p}(1-\hat{p})}{n}}$$
$$= \frac{x}{n} \pm z^* \sqrt{\frac{(x/n)(1 - x/n)}{n}}$$

review exercises

6.25 To investigate how infants first comprehend and produce words, six British children were observed in their homes every 2 weeks from age 6 to 18 months.[19] Among the data collected was the age at which each child first comprehended spoken English and first produced a word of English:

	AGE (DAYS) AT FIRST WORD		
	COMPREHENSION	PRODUCTION	DIFFERENCE
Andrew	253	368	115
Ben	238	252	14
George	257	408	151
Katherine	220	327	107
Katy	249	268	19
Sebastian	254	326	72

Assuming these six children to be a random sample of all British children, determine a 95 percent confidence interval for the average difference (in days) between when a child first comprehends and produces spoken English. Does your estimate seem fairly precise? Why is it not more precise? Why might this be a biased sample?

6.26 A 1995 survey for Fodor's, a publisher of travel guidebooks, asked 600 U.S. travelers to name their least desirable vacation destination.[20] The winner, with 9 percent of the votes, was Iraq/Iran; in second place, with 7 percent, was New York City. A spokesperson for New York City's mayor scoffed, "They only interviewed 600 people? That's the size of an apartment building." Assuming this to be a random sample, give a 95 percent confidence interval for the percentage of all U.S. travelers who would choose New York City as their least desirable vacation destination.

6.27 Sally measured the carbon percentage (by weight) in a compound 5 times and found an average value of 15.71635 percent with a standard deviation of .02587 percent.[21] Janet performed the same measurement 7 times, using a different technique, and found an average value of 15.68134 percent with a standard deviation of .03034 percent.

a. Assuming the measurements to be random samples from a normal distribution, calculate 95 percent confidence intervals for each of these estimates of the true carbon percentage. Do these estimates appear to be consistent or inconsistent with each other?

b. Whose estimate of the true carbon percentage is more precise? Explain.

c. If you owned a chemical company and had to choose between these two measurement techniques based solely on the information given here, which would you choose? Why?

6.28 A survey of 10,000 adult Europeans obtained the following data on hours spent per day sleeping and watching television:[22]

	MEAN	STANDARD DEVIATION
Sleeping	8.0	1.3
Watching television	3.5	2.0

Use these data to compute 95 percent confidence intervals for the population means.

6.29 In 1988, the PGA tour made a special study of the putting success of professional golfers.[23] At 15 tournaments, they selected one green with a relatively flat and smooth surface and measured and recorded the result of each putt on that green. Here are some of their results:

LENGTH OF PUTT	NUMBER OF PUTTS	FRACTION MADE
5	353	.589
15	167	.168

For each of these two distances, calculate a 95 percent confidence interval for the probability of making a successful putt, assuming that the binomial model applies and using a normal approximation to the binomial distribution. Why might the binomial model be inappropriate?

6.30 *T voicing* refers to the practice of voicing a t that does not begin or end a word as d when it follows a vowel or r and precedes an unstressed symbol. For example, in U.S. English, *petal* and *pedal* are pronounced the same. After a random sample of 220 U.S. English speakers had studied a Dutch dialect for 8 weeks, 121 showed an absence of t voicing.[24] Give a 90 percent confidence interval for the percentage of all U.S. English speakers who would show an absence of t voicing after 8 weeks of studying this Dutch dialect.

6.31 A manufacturing process is designed to produce machine bolts with a diameter that is normally distributed with a mean of 1.000 centimeter and a standard deviation of .050 centimeter. A quality control engineer proposes sampling four bolts every hour. If the manufacturing process is functioning as intended, what are the mean and standard deviation of the sampling distribution of the mean of a four-bolt sample? What upper and lower control limits should the quality control engineer set for the mean of a four-bolt sample for there to be a .01 probability of violating the control limit when the process is functioning as intended? What are the appropriate upper and lower control limits if nine bolts are sampled instead of four? Explain any differences.

6.32 In the game Roshambo (rock-scissors-paper), two players simultaneously move their fists up and down 3 times and then show a fist (rock), two fingers (scissors), or an open hand (paper). Rock beats scissors, scissors beats paper, and paper beats rock. If there is a tie, the players repeat the game. A student played 50 matches against randomly selected opponents and recorded the frequency of the initial moves.[25] Considering these data to be a random sample, estimate the student's probability of playing rock on the initial move, and use a normal approximation to the binomial distribution to give a 99 percent confidence interval. Do the same for the opponents' probability of playing rock on the initial move.

	STUDENT	OPPONENTS
Rock	26	16
Scissors	9	17
Paper	15	17

6.33 A student used a table of random numbers to select 80 books in her library's card catalog.[26] Of these, only 57 could be found in the library. The other 23 were checked out, stolen, or shelved incorrectly. Assume an infinite population, and use a normal approximation to the binomial distribution to calculate a 95 percent confidence interval for the proportion of all the library's books that cannot be found in the library. Use the lower boundary of this interval and the library's estimate that it has 1.4 million books and 6000 users to estimate a minimum for the average number of missing books per user.

6.34 A 1992 survey of 61 female Navy officers found that 40 said they had been sexually harassed.[27] Find a 99 percent confidence interval for the percentage of all female Navy officers who have been sexually harassed. Would this confidence interval be wider or narrower if we took into account the fact that the Navy had 8000 women officers, about 11 percent of its officer corps? Explain your reasoning.

6.35 Explain how you, as a statistician, would evaluate this advice: "Look for a sample of at least 600 interviews. If fewer than 400 people were polled, you can skip the survey."[28]

6.36 Thirty college students were randomly selected and asked how many more years they expected to live.[29] Their answers had a mean of 56.42 years and a standard deviation of 16.40 years. Calculate a 99 percent confidence interval for the average life expectancy of all college students. Do you need to assume that the number of years that college students expect to live is normally distributed?

6.37 Each of 12 college intramural basketball players was asked to take 30 shots 15 feet from the basket—10 shots from the free-throw line, 10 shots from a 45° angle to the basket, and 10 shots from the baseline.[30] Assuming that the binomial model is appropriate, use the results shown below to give a 95 percent confidence interval for the probability of making each of these three shots.

	ATTEMPTS	SUCCESSES
Free-throw line	120	84
45° angle	120	62
Baseline	120	53

6.38 A researcher tested 72 birthday candles by lighting each and calculating the burning time, in seconds.[31] Use his data to estimate the population mean burning time and to make a 99 percent confidence interval for your estimate.

700	695	660	765	620	650	715	715	660	790	635	645
690	705	695	815	715	615	720	655	690	675	790	725
755	660	630	730	800	620	670	645	740	730	720	630
710	660	650	730	785	690	730	615	580	735	690	765
800	725	840	675	645	680	820	775	745	810	780	780
715	790	765	785	720	760	785	770	755	710	735	740

6.39 A government agency wants to estimate the number of fish in a lake. It catches 100 fish, tags them, and puts them back in the lake. A short while later, the agency catches 100 fish and finds that 10 have been tagged. Think of this second batch as a random sample of all the fish in the lake. Based on this sample, what is your estimate of the fraction of fish in the lake that are tagged? Because you know that 100 fish are tagged, what is your estimate of the total number of fish in the lake?

6.40 Ellen Goodman, a syndicated columnist, claims that between 84 and 93 percent of television remote controls are found in men's hands.[32] (The clicker, she jokes, could be seen as the male's last power-wielding symbol.) Using a normal approximation to a binomial distribution, how many couples would she have had to sample to get a 95 percent confidence interval of .84 to .93?

6.41 A study of a sample of electronic mail messages found 64 typographical errors, of which 7 involved the letter e.[33] Use a normal approximation to the binomial distribution to calculate a 95 percent confidence interval for the probability that a typographical error involves the letter e. In ordinary written English,[34] the letter e is used with a frequency of .13. Is this .13 frequency inside your 95 percent confidence interval for typographical errors? What conclusion would you draw if a letter of the alphabet had many more typographical errors than would be expected from its usage frequency?

6.42 A study of a sample of 286 NBA basketball games during the 1989–1990 season found that 6841 personal fouls were called against visiting teams and 6574 against home teams.[35] Use a normal approximation to the binomial distribution to calculate a 95 percent confidence interval for the probability that a randomly selected personal foul is against the visiting team.

6.43 A poll by *The Wall Street Journal* (July 6, 1987) asked 35 economic forecasters to predict the interest rate on 3-month Treasury bills in June 1988. These 35 forecasts had a mean of 6.19 percent and a variance of .47 percent. Assuming these to be a random sample, give a 95 percent confidence interval for the mean prediction of all economic forecasters, and then explain why each of these interpretations is or is not correct:

a. There is a .95 probability that the actual Treasury bill rate on June 1988 will be in this interval.

b. Approximately 95 percent of the predictions of all economic forecasters are in this interval.

c. If the *Journal* took another random sample, there is a .95 probability that the new confidence interval would include 6.19.

6.44 A 37-car train carrying 2500 kegs of beer was derailed during a snowstorm.[36] The shipper's insurance policy covered the complete cost of this accident if more than 10 percent of the kegs were damaged. The shipper examined a sample of 50 kegs and found that 12 (24 percent) were damaged. The insurance company disputed the use of a sample in place of an examination of all 2500 kegs; however, a court awarded damages based on the calculation of a 95 percent confidence interval. Determine a 95 percent confidence interval for p, the fraction of kegs damaged, using a normal approximation to the binomial distribution and assuming an infinite population. Does this confidence interval include $p = .10$? If 10 percent of an infinite population of kegs are damaged, what is the probability that 12 or more kegs will be found damaged in a random sample of 50 kegs? (Use a normal approximation.)

6.45 In an antitrust case against United Shoe Machinery Company, an important statistical issue concerned the company's share of the market for 30 different kinds of machines used in shoe making.[37] The company had its own internal estimates for 1947, but the federal judge handling the case asked the government to conduct a survey in 1949 of 45 shoe manufacturers selected randomly by the judge. Based on the government sample, a 95 percent confidence interval for the firm's market share generally included the value of United Shoe's estimate. However, there were exceptions. With shoe-cementing machines, for example, the company estimated that it had a 63 percent market share, and the government survey found that, of 234 machines used by the 45 companies surveyed, 40.6 percent were made by United Shoe. Use the government data to calculate two 95 percent confidence intervals for United Shoes' cementing-machine market share, one interval assuming $n = 234$ and the other using $n = 45$. Why is it uncertain that $n = 234$ is appropriate? If you were a statistician employed by United Shoe, how would you explain to the judge the fact that the company's estimate is outside both these confidence intervals?

6.46 A 1986 sample of 118 California voters by Pitzer College students found that 66 (56 percent) planned to vote for Alan Cranston for senator while 52 (44 percent) favored his opponent, Ed Zschau. What is the probability that a random sample of 118 people would find more than 56 percent favoring Cranston if, in fact, 50 percent of all voters favor Cranston? (Use a normal approximation to the binomial distribution.) Calculate a 95 percent confidence interval for the fraction of all California voters planning to vote for Cranston. The early election returns, representing nearly 200,000 voters (3 percent of the total), gave Zschau nearly a 2-to-1 lead, but the final returns showed Cranston to be the winner by 50.8 percent to 49.2 percent. Other than luck, why might a poll of 118 people predict the vote more accurately than the first 200,000 votes counted did?

6.47 Forty randomly selected college students were asked how many hours they sleep each night.[38] Use these data to estimate a 95 percent confidence interval for the mean of the population from which this sample was taken:

5	6	6	8	6	6	6.5	4
4	5	5	7	6	6	5	5
6.5	6.5	6.5	7	5	5.5	5	6
5	5	4	8	5	6.5	6	5
5	8	5	7	6	8	7	6

6.48 A marketing survey will be used to estimate the fraction of the population who prefer a new spaghetti formula to the old formula. The pollsters advertise that there is a .95 probability that their sample estimate will be within 2 percentage points of the actual fraction of the population who prefer the new formula. How large a sample are they using?

6.49 An article about computer forecasting software explained how to gauge the uncertainty in a prediction: "[Calculate] the standard deviation and mean (average) of the data. As a good first guess, your predicted value . . . is the mean value, plus or minus one standard deviation—in other words, the uncertainty is about one standard deviation."[39] If the prediction comes from a normal distribution with a known standard deviation and the mean of the sampling distribution for the prediction is equal to the actual value, what is the probability that this interval will encompass the actual value?

6.50 A student mailed a short questionnaire to the 495 professors at her school.[40] One of the questions was: "Overall, would you consider your views more consistently in agreement with those of Democrats or Republicans?" Of the 248 professors who responded, 193 answered Democrats and 55 answered Republicans. Assuming (unrealistically) these responses to be a random sample from an infinite population, calculate a 95 percent confidence interval for the fraction of all the professors at her school who would answer Democrat. If we were to take into account the fact that the population was finite, would this shrink or widen our confidence interval? Give one good reason why these data may not be a random sample.

projects

For each of the following projects, type a report in ordinary English, using clear, concise, and persuasive prose. Any data that you collect for this project should be included as an appendix to your report. Data used in your report should be presented clearly and effectively.

6.1 In this game, three ordinary pennies are tossed in the air simultaneously. If all three coins land with the same side up, either all heads or all tails, the player gets 20 points. Otherwise, the player gets nothing. Play this game 10 times, keeping track of the number of wins and losses, and then calculate the average number of points received per play. Repeat this experiment 100 times. Calculate the total number of times you won 20 points and the number of times you won nothing, and show these in a histogram. Calculate the number of times that the average number of points in an experiment (10 plays) was 0, 1, and so on, and show these results in a histogram.

6.2 In this game, a penny and dime are flipped. You get 4 points if they both land heads, 2 points if they both land tails, and 0 points otherwise. Play this game 10 times, and calculate the average number of points received per play. Repeat this experiment 100 times, and use a histogram to show the number of times that the average number of points was in the intervals of 0 to .2, .4 to .6, .8 to 1.0, 1.2 to 1.4, 1.6 to 1.8, and so on.

6.3 Roll five standard six-sided dice, and calculate the average of the numbers on these five dice. For example, I rolled five dice and got the numbers 1, 6, 3, 2, 1 for an average of $(1 + 6 + 3 + 2 + 1)/5 = 2.6$. Repeat this experiment 100 times, each time recording the average. (Do not record the individual rolls.) Use a histogram to display your results.

6.4 In this game, when a standard 6-sided die is rolled, the number 6 is a success and all other numbers are failures. Obtain 30 standard six-sided dice, roll these simultaneously and determine the number of sample successes (the number of 6s among these 30 dice). Repeat this experiment 300 times and then display your results in a histogram.

6.5 Remove the jacks, queens, and kings from 2 standard decks of playing cards. Mix the remaining cards thoroughly and deal 5 cards. Letting each card be worth its face value (an ace is 1 point, a deuce 2 points, and so on), calculate the average value of these 5 cards—the total points divided by 5. Repeat this experiment 200 times, shuffling the cards thoroughly before each deal. Use a histogram to summarize your data.

6.6 Use statistical software to simulate drawing 100 random samples of size 2000 from a population of voters, of whom 55 percent prefer candidate A.

6.7 Estimate the percentage of the seniors at your college who regularly read a daily newspaper, the percentage who can name the two U.S. senators from their home state, the percentage who are registered to vote, and the percentage who would almost certainly vote if a presidential election were held today.

6.8 Estimate the average number of hours that students at your school sleep each day, including both nighttime sleep and daytime naps. Also estimate the percentage who have been up all night without sleeping at least once during the current term.

6.9 Among seniors at your school who are looking for jobs, estimate the average annual salary they expect to earn in their first year. Do not include moving allowances or other one-time benefits.

6.10 Go to a large parking lot and use a random sample of 25 cars to estimate a 95 percent confidence interval for the fraction of the cars in this lot that were made by foreign companies. Now look at every car in this lot, and *count* the fraction that were made by foreign companies. Is this population fraction inside your 95 percent confidence interval?

6.11 Buy a 2-pound bag of plain M&M candies. With your eyes closed, remove 30 M&M's from this bag. Count the fraction in this presumably random sample that are brown, and estimate a 95 percent confidence interval for the fraction of all the M&M's in this bag that are brown. Replace these candies, use a spoon to mix them thoroughly, again close your eyes, remove 30 candies, count the number that are brown, and estimate a 95 percent confidence interval for the fraction of all the M&M's in this bag that are brown. Conduct this experiment 20 times. Now empty the bag and count the actual fraction in the entire bag that are brown. How many of your 20 confidence intervals included the actual fraction that are brown?

6.12 Buy a 1-pound bag of plain M&M candies. Pour these candies into a clean, opaque bowl, and use a spoon to mix them thoroughly. Now, with your eyes closed, remove 10 candies and count the fraction of these candies that are red. Replace these candies, use a spoon to mix them thoroughly, and again close your eyes, choose 10 candies, and count the number that are red. Conduct this experiment 25 times. Use a bar graph to summarize your results. The makers of M&M's claim that 20 percent of all plain M&M candies are red. If so, what are the theoretical mean and standard deviation for the proportion of red candies in a random sample of size 10? Compare your sample means and standard deviations with these theoretical values. Are they close? Why are they not exactly equal? Use all your sample data to estimate a 95 percent confidence interval for the fraction of all M&M plain candies that are red.

6.13 On average, how many plain M&M's are there in an 8-ounce bag?

6.14 Go to a local grocery store and ask the store manager if you can watch the checkout stands for 15 minutes to record some data for a statistics project. Once you have permission, station each team member at a checkout stand for 15 minutes and record the sales total for each customer. Do not go to special registers (such as those for fewer than 10 items). Use your data to estimate a 95 percent confidence interval for the average amount spent by a customer. Explain why you think that the time of day or day of the week may matter.

6.15 Dial 50 randomly selected telephone numbers, either in the community or on campus, and count the number of rings before the telephone is answered. If no one (or an answering machine) answers, dial a different number. When the telephone is answered, briefly thank the person for participating in a survey. Identify the population from which your data are a sample, and summarize your results. Explain why you think that the time of day or day of the week may matter.

6.16 Ask 50 randomly selected college students how many personal e-mail messages they have sent in the past week and how many personal letters they have mailed in the past week. Use each set of data to estimate 99 percent confidence intervals, and explain carefully how these intervals should be interpreted.

6.17 Ask 100 randomly selected college students whether they prefer to use a Macintosh-compatible or an IBM-compatible computer. Use these data to estimate a 95 percent confidence interval for the percentage of the student population from which this sample was drawn who prefer IBM-compatible computers.

6.18 Ask 50 randomly selected students how many biological siblings they have. Estimate 95 and 99 percent confidence intervals for the means of the populations from which your data are a sample.

6.19 Of those students at your college who regularly send personal e-mail, estimate the fraction that normally begin their e-mail messages with a salutation such as "Dear Gary" or "Gary," rather than just beginning the message.

6.20 Estimate 95 percent confidence intervals for the average answers by students at your college to these questions:

1. On average, how many hours a week do you spend watching television?
2. On average, how many hours a week do you spend surfing the internet?
3. On average, how many hours a week do you engage in vigorous physical exercise (sufficient to make you perspire)?
4. How many movies have you seen at a movie theater during the past year?
5. How many rental movies have you watched on a VCR during the past year?

HAVE YOU EVER WONDERED?

We often do things "instinctively" or "by nature." Where do instincts come from? If they are truly instinctive, how do we acquire them? Perhaps by natural selection. For instance, how do seed-bearing plants determine how and when to release their seeds to ensure future plants? They did not develop this during their lifetimes, but through countless generations of natural selection. Over time, those plants that did not release seeds at the right time or in the right manner failed to have offspring to carry on these ineffectual survival traits. Natural selection means that traits or behavior that is conducive to survival and reproduction is more likely to survive and be passed on to future generations, either in genes or in genetic programs.

Why do so many animals instinctively bark or roar when they confront strangers? This barking is intended to warn the stranger that this particular animal is not weak and is willing to fight. Similar animals that did not issue effective warnings got into too many fights and were thinned out of the gene pool. Why do so many animals have fierce teeth, horns, or other built-in weapons? Because their ancestors were able to compete effectively for food and reproductive opportunities and so perpetuate their genes. Why are gazelles, which are poor fighters, able to run fast? Those that were not fast enough did not pass on their genes. Why do some creatures that are neither strong nor swift blend so well into their natural surroundings? You know the answer.

This theory has many intriguing implications. Why, for example, are modern males, on average, larger and more muscular than females? Natural selection suggests that because young men fought with each other for innumerable generations when the weapons were hands, rocks, and clubs, the largest and strongest lived while the smallest and weakest died (and with them, their genes).

Now, consider this question carefully. If I were to hand you a baby, would you hold it against your right shoulder or your left shoulder? There is a pronounced preference. What is it and why? The surprising answer is found in this chapter.

chapter 7

hypothesis tests

A friend of mine once remarked to me that if some people asserted that the earth rotated from east to west and others that it rotated from west to east, there would always be a few well-meaning citizens to suggest that perhaps there was something to be said for both sides, and that maybe it did a little of one and a little of the other; or that the truth probably lay between the extremes and perhaps it did not rotate at all.

—Maurice G. Kendall

IN CHAP. 6, we saw how sample data can be used to estimate the value of a population parameter, such as μ or p, and determine a confidence interval that gauges the precision of this estimate. In this chapter, we will see how sample data can be used to support or refute theories about the value of μ or p, for example, whether vitamin C wards off colds, aspirin reduces the risk of heart attack, and cigarette smoking is dangerous to your health. Sample data also have been used to test whether the death penalty deters murder, the federal deficit affects interest rates, and minorities have been systematically excluded from schools, occupations, and juries.

Any theory that can be cast in terms of a population mean or success probability can be tested with appropriate sample data.

In this chapter, you will see how to make these statistical tests and interpret the results. You will learn not only how to do the tests correctly, but also how to recognize tests that are misleading. We begin by using tests of the population mean to develop a general framework. Then we apply this framework to the use of categorical data to test the success probability p. We conclude the chapter with a discussion of pitfalls to be avoided.

7.1 A General Framework for Tests of a Mean

Statistical tests are not as definitive as mathematical proofs, because sample data are subject to sampling error. There is always a chance, however small, that a new sample will discredit a previously confirmed theory. If the average net weight of a random sample of 10 boxes is 23.99 ounces, these data are consistent with the claim that these boxes were drawn from a population with an average weight of 24 ounces. But a second sample of 10 boxes might find an average net weight of 23.5 ounces. We cannot be absolutely certain of the average net weight of all the boxes in the population unless all are weighed.

Most theories are potentially vulnerable to reassessment because there never is a final tabulation of all possible data. New experiments and fresh observations continually provide new evidence—data that generally reaffirm previous studies, but occasionally create doubt or even reverse conclusions that were once thought firmly established. Theories are especially fragile in the humanities and social sciences, in which there are few data and it is difficult to control for extraneous influences. In the 1930s, John Maynard Keynes theorized that there was a very simple relationship between a nation's aggregate income and household spending. This theory was seemingly confirmed by annual data for the 1930s and early 1940s, but was found inadequate when data for the 1950s became available. In the 1960s, economists believed that there was a simple inverse relationship between a nation's unemployment rate and its rate of inflation: When unemployment goes up, inflation goes down. In the 1970s, unemployment and inflation both went up, and economists decided that they had overlooked other important influences, including inflation expectations.

Even in the physical sciences, where innumerable laboratory experiments can generate mountains of data, theories are not invulnerable to reassessment. Edward Leamer, a very talented statistician, wrote that

> All knowledge is human belief, more accurately human opinion. What often happens in the physical sciences is that there is a high degree of conformity of opinion. When this occurs, the opinion held by most is asserted to be an objective fact, and those who doubt it are labeled "nuts." But history is replete with examples of opinions losing majority status, with once objective "truths" shrinking into the dark corners of social intercourse.[1]

For centuries before Copernicus, people were certain that the sun revolved around the earth. Until Galileo's experiments at the Leaning Tower of Pisa, conventional wisdom was that heavy objects fall faster than lighter ones. Before Einstein, Newtonian physics was supreme. When scientists think of new ways to test old theories or fresh ways to interpret old data, the weakness of accepted theories may become exposed.

If we can never be absolutely sure that a theory is true or false, we might make a probability statement, such as, "Based on the available data, there is a .90 probability that this theory is true." Bayesians are willing to make such subjective assessments, but other statisticians insist that either a theory is true or it is not. A theory is not true 90 percent of the time, with its truth or falsity determined by a playful Nature's random number generator.

Statistics has followed another route. Instead of estimating the probability that a theory is true, based on observed data, statisticians calculate the reverse probability—that we would observe such data if the theory were true. An appreciation of the fact that these are two very different statements is crucial to an understanding of the meaning and limitations of hypothesis tests. Hypothesis tests are a proof by statistical contradiction. We calculate the probability that the sample data would look the way they do if the theory were true. If this probability is low, then the data are not consistent with the theory and therefore reject it, what Thomas Huxley called "the great tragedy of science—the slaying of a beautiful hypothesis by an ugly fact." This is a proof by statistical contradiction: Because these data are unlikely to occur if the theory were true, we reject the theory. Of course, we can never be 100 percent certain, because unlikely events sometimes do happen. Note, too, that for a theory not to be rejected, it need only be consistent with the data. This is a relatively weak conclusion, because many other theories may also be consistent with the observed data.

The Null and Alternative Hypotheses

We will use a specific example to make these general ideas more concrete. In 1868, a German physician, Carl Wunderlich, reported his analysis of over 1 million temperature readings from 25,000 patients. He concluded that 98.6 degrees Fahrenheit (98.6°F), which is 37.0 degrees Celsius (37.0°C), is the average temperature of healthy adults. This conclusion has become so well known that parents routinely compare their children's temperature to a "normal" temperature of 98.6. The 1990 edition of the widely used *Stedman's Medical dictionary* defines fever as "a bodily temperature above the normal of 98.6°F."

Wunderlich thought that the armpit was the best place for measuring body temperatures. The thermometers he used had to be read while they were in place, and when used in the armpit, they required 15 to 20 minutes to give a stable reading. Today, doctors believe that the mouth is a better place for measuring body temperature, and their thermometers are easier to read and stabilize more quickly. Suppose that, as doctors, we have been observing temperatures for some time, of both healthy and sick people, and we wonder if Wunderlich's 98.6 figure is really the average temperature of healthy humans.

Our *research question* is whether the average temperature is higher or lower than 98.6. To answer this research question, we can select a random sample of 50 healthy people and measure their temperatures. For a proof by statistical contradiction, we make an assumption, called the **null hypothesis** (or H_0), about the population from which the sample is drawn. Typically, the null hypothesis is a "straw assumption" that we anticipate rejecting. To demonstrate, for example, that a medication has beneficial effects, we see whether the sample data reject the hypothesis that there is no (*null*) effect.

Here, our research question suggests the natural null hypothesis that the average temperature of healthy humans is 98.6, the conclusion reached by Wunderlich more than 100 years ago. Although we may have initiated this study precisely because we suspect that Wunderlich's conclusion may be erroneous, the way to demonstrate this by statistical contradiction is to see if the evidence rejects the straw hypothesis that the average temperature is 98.6.

The **alternative hypothesis** (usually written as H_1 or H_A) describes the population if the null hypothesis is not true. Here, the natural alternative hypothesis is that the population mean is not equal to 98.6:

$$H_0\colon \mu = 98.6$$
$$H_1\colon \mu \neq 98.6$$

This alternative hypothesis is our research question, stated in terms of the value of the population parameter μ. Usually, as here, the alternative hypothesis is *two-sided,* because (even though we might have a hunch about how the study will turn out) we are reluctant to rule out beforehand the possibility that the population mean may be either lower or higher than the value specified by the null hypothesis. If, before seeing the data, we could rule out one of these possibilities, the alternative hypothesis would be *one-sided.* If, for example, we were convinced beforehand that the average temperature of healthy humans could not possibly be higher than 98.6, the one-sided alternative hypothesis would be H_1: $\mu < 98.6$.

P values

Once we have specified the null and alternative hypotheses, we collect and examine our sample data. As we are concerned with the value of μ, the population mean temperature, we take a random sample of humans and examine the sample mean—because this is what we would use to estimate the value of the population mean. The estimator that is used to test the null hypothesis is called the *test statistic.*

I am not a medical doctor, and the only thermometer I own is an inexpensive one of questionable accuracy. However, in 1992, three researchers at the University of Maryland School of Medicine reported the results of 700 readings of 148 healthy persons taken with modern thermometers.[2] Table 7.1 shows a random sample of their data. The sample mean for these 50 readings is 98.23, and the sample standard deviation is .67. We temporarily (and unrealistically) assume that the population standard deviation is known to be .67. Does this value of the test statistic provide persuasive evidence that the average temperature of healthy humans is less than 98.6, or can it be explained by the inevitable variation in the means of small samples drawn from a population with a standard deviation of .67?

If the individual observations are normally distributed, then so is the sample mean. Even if the individual observations are not normally distributed, a sample of size 50 should be sufficient to appeal to the central limit theorem. Under the null hypothesis, each individual observation in our sample is from a population with a mean of 98.6. As explained in Chap. 6, the sampling distribution for the mean of a random sample of 50 observations from a population with a mean of 98.6 and a standard deviation of .67 has a mean of 98.6 and a standard deviation of .67, divided by the square root of the sample size (here, $.67/\sqrt{50} = .095$). Thus

$$\text{Under the null hypothesis} \quad \bar{x} \text{ is } N[98.6, .095]$$

Figure 7.1 shows this sampling distribution. Remember that there is a .95 probability that the value of a normally distributed random variable will be within

Table 7.1

Oral temperatures of 50 healthy adults, degrees Fahrenheit

98.2	98.8	98.4	98.7	98.2	97.7	97.4	98.6	97.2	97.6
98.0	99.2	99.4	98.8	98.8	97.1	98.0	98.5	98.2	97.5
96.9	98.8	98.4	97.7	98.0	98.4	98.2	99.9	98.6	99.5
98.7	98.6	98.3	98.3	98.2	97.0	98.4	98.6	97.4	98.8
97.8	98.3	98.8	98.8	97.9	97.5	98.2	98.9	97.0	97.5

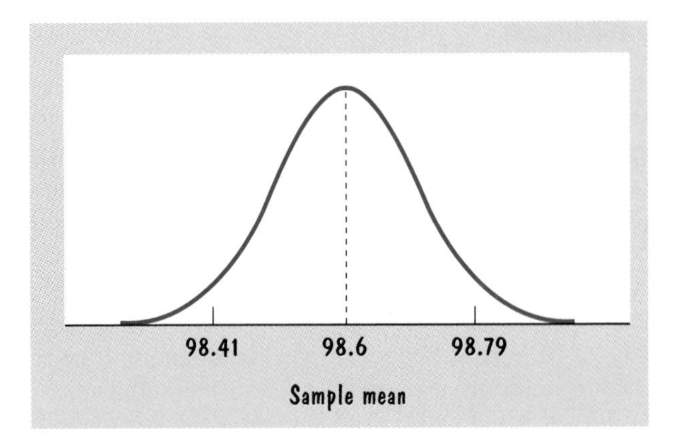

Figure 7.1 The sampling distribution for the sample mean if the null hypothesis is true.

2 (or, more precisely, 1.96) standard deviations of its mean. Here, if the null hypothesis is true, there is a .95 probability that a random sample of 25 persons will yield a sample mean in the range 98.6 ± 1.96(.095) = 98.6 ± .19, or 98.41 to 98.79. The fact that our sample mean turned out to be 98.23 certainly casts doubt on the null hypothesis.

To consider more formally the extent to which our sample mean of 98.23 provides persuasive evidence against the null hypothesis, we convert to a standardized z value in order to calculate the probability that the mean of a sample of this size would be so far from 98.6, if the null hypothesis were true:

$$
\begin{aligned}
z &= \frac{\bar{x} - \mu}{\sigma / \sqrt{n}} \\
&= \frac{98.23 - 98.6}{.67/\sqrt{50}} \\
&= -3.85
\end{aligned}
$$

We do not calculate the probability that z will be exactly equal to −3.85. The probability that a normally distributed variable will equal any single number is 0. Instead, we calculate the probability that z will be so far from 0, here the probability that z will be equal to −3.85, or even lower: $P[z \leq -3.85] = .00006$.

> The ***P* value** for a test of a null hypothesis about the population mean is the probability, if the null hypothesis is in fact true, that a random sample of this size yields a sample mean that is this far (or farther) from the value of the population mean assumed by the null hypothesis. A small *P* value casts doubt on the null hypothesis.

In calculating the *P* value, we must take into account whether the alternative hypotheses is one-sided or two-sided. For a one-sided alternative hypothesis, the *P* value is the probability that the sample mean would be so far to one side of the population mean specified by the null hypothesis. For a two-sided alternative hypothesis, the *P* value is the probability that the test statistic would be so far, *in either direction,* from the population mean. In our body-temperature example, we report the probability that the sample mean will be 3.85 standard deviations or more, in either direction, from the 98.6 value for the population mean given by the null hypothesis. Because the normal distribution is symmetric, we double the .00006 probability calculated above and report the *two-sided P value* as 2(.00006) = .00012.

Because this probability is so slight, sampling error is an unconvincing explanation for why the average temperature in our sample is less than 98.6. If it had turned out that the sample mean were 98.55 or 98.65, we could have reasonably attributed this difference to the inevitable variation in the outcomes of random samples. However, a z value of -3.85, representing a difference of 3.85 standard deviations between the sample mean and the value of the population mean under the null hypothesis, is too improbable to be explained plausibly by the luck of the draw. We have shown, by statistical contradiction, that the average temperature of healthy humans is not 98.6.

We can summarize our general procedure as follows. First specify the null and alternative hypotheses. Then use the sample data to estimate the value of the population parameter whose value is specified by the null hypothesis. If this sample statistic is approximately normally distributed, calculate the z value, which measures how many standard deviations the estimate is from the null hypothesis:

$$z = \frac{\text{observed statistic} - \text{null hypothesis parameter value}}{\text{standard deviation of statistic}} \tag{7.1}$$

The *P* value, giving the probability of observing a z value so far (or farther) from 0, is the probability (were the null hypothesis true) that the estimate would be this many standard deviations from the null hypothesis parameter value. If the alternative hypothesis is two-sided, we double this probability to obtain the two-sided P value.

For testing a null hypothesis about the value of the population mean, we use the sample mean as our test statistic and calculate this z value:

$$z = \frac{\bar{x} - \mu}{\sigma/\sqrt{n}} \tag{7.2}$$

Using an Estimated Standard Deviation

So far, we have blithely assumed that the population standard deviation is known to be .67, which is the standard deviation of our sample. In fact, this is just an estimated value. Usually, as here, we do not know the value of the population standard deviation, and we need to take into account the fact that we are using an estimate. Chapter 6 explained how the t distribution can be used in place of the normal distribution when the sample standard deviation s is used in place of the population standard deviation σ. Here, instead of the z statistic in Eq. (7.2), we use the following:

> When the standard deviation is estimated, we can test a null hypothesis about the value of the population mean by using this t statistic
>
> $$t = \frac{\bar{x} - \mu}{s/\sqrt{n}} \tag{7.3}$$
>
> and then determining the P value from the t distribution with $n - 1$ degrees of freedom.

The P value can be obtained from statistical software or (if necessary) interpolated from Table 4 at the end of this book. The P value is exactly correct if the data come from a normal distribution and, because of the power of the central limit theorem, is an excellent approximation if we have at least 15 observations from a generally symmetric distribution, or at least 30 observations from a very asymmetric distribution.

Figure 7.2 shows a histogram of our sample data. Since we have 50 observations and the histogram is not wildly symmetric, we can use the t distribution with considerable confidence. The substitution of the sample mean, the value of μ under the null hypothesis, the sample standard deviation, and the sample size into Eq. (7.3) gives a t value of -3.85:

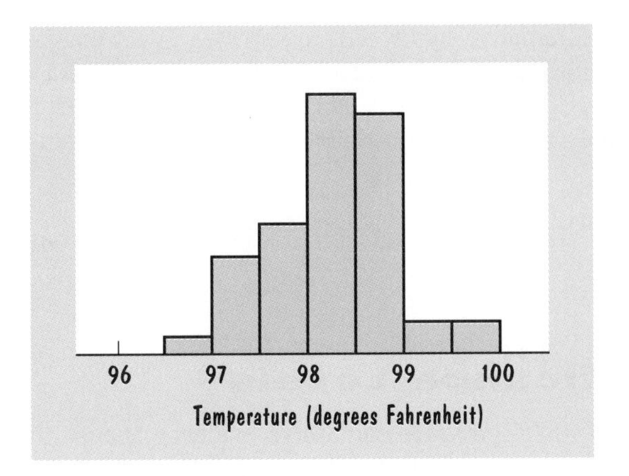

Figure 7.2 A histogram of the temperature data.

$$
\begin{aligned}
t &= \frac{\bar{x} - \mu}{\sigma/\sqrt{n}} \\
&= \frac{98.23 - 98.6}{.67/\sqrt{50}} \\
&= 3.85
\end{aligned}
$$

(This is the same as our z value, because we assumed above that the population standard deviation is equal to the sample standard deviation.)

There are $n - 1 = 50 - 1 = 49$ degrees of freedom. Table 4 does not show exactly 49 degrees of freedom. However the cutoffs for 40 and 60 degrees of freedom show that the probability of a t value below -3.85 is less than .005; statistical software gives a more precise probability, .0002, implying a two-sided P value of $2(.0002) = .0004$. These data are strong evidence that the average temperature of healthy humans is not 98.6.

A statistical software package also can be used to calculate the requisite components of a hypothesis test, as shown in the Minitab output in Table 7.2. After the 50 temperature observations are entered in column C1 of Minitab's data worksheet, the command `TTEST OF MU = 98.6 ON DATA IN C1` asks the program to determine the t value for a test of the null hypothesis $\mu = 98.6$, using the data in column C1. The next three lines are the output. The first of these lines states that this is a t test of $\mu = 98.6$ versus the two-tailed alternative that μ is not equal to 98.6. Then we have the variable name, number of observations, mean of the sample data, standard deviation of the sample data, standard error of the sample mean, t value, and two-sided P value.

Table 7.2

A Minitab hypothesis test for human temperatures

```
MTB> TTEST OF MU = 98.6 ON DATA IN C1

TTEST OF MU = 98.6 VS MU N.E. 98.6

        N      MEAN     STDEV    SE MEAN        T         P
C1     50    98.234      .672       .095    -3.85     .0004
```

Significance Levels

A *P* value of .0004 provides persuasive evidence against the null hypothesis. What about a *P* value of .01, or .10, or .25? Where do we draw a line and say that this *P* value is persuasive, but that one is not? In the 1920s, the great British statistician R. A. Fisher endorsed a 5 percent cutoff:

> It is convenient to draw the line at about the level at which we can say: "Either there is something in this treatment, or a coincidence has occurred such as does not occur more than once in twenty trials. . . ."
>
> If one in twenty does not seem high enough odds, we may, if we prefer, draw the line at one in fifty (the 2 percent point), or one in a hundred (the 1 percent point). Personally, the writer prefers to set a low standard of significance at the 5 percent point, and ignore entirely all results which fail to reach that level.[3]

Researchers today often report that their results either are or are not **statistically significant** at a specified level, such as 5 or 1 percent. What this cryptic phrase means is that the appropriate *P* value is less than this specified *significance level.* For example, because the *P* value of .0004 is less than .05, we can report that we "found a statistically significant difference at the 5 percent level between the average temperature of a sample of healthy humans and the 98.6 value that is commonly assumed." The reader is thereby told that there is less than a .05 probability of observing such a large difference between the sample mean and the value of the population mean given by the null hypothesis. An observed difference that is too large to be attributed plausibly to chance alone is statistically significant in the sense that its statistical improbability persuades us to reject the null hypothesis.

Fisher's endorsement of a 5 percent rule is so ingrained that some researchers simply say that their results are "statistically significant," with readers understanding that the *P* value is less than .05. Some say *statistically significant* if the *P* value is less than .05 and *highly significant* if the *P* value is less than .01. Simple rules are convenient and provide a common vocabulary, but we must not be blind to the fact that there is a continuum of *P* values. There is little difference between *P* values of .049 and .051. There is a big difference between P values of .049 and .0004. The most reasonable approach is to report the P value and let readers judge for themselves. They will remember Fisher's rule of thumb, but they will also recognize that a P value of .051 is not quite significant at the 5 percent level, that .049 is barely significant, and that .0004 is highly significant.

The reader's reaction may well be tempered by what is being tested, with some situations calling for more persuasive proof than others. The Food and Drug Administration may demand very strong evidence when testing a drug with modest benefits and potentially dangerous side effects. Similarly, in cases where the conventional wisdom is very strong, evidence to the contrary will have to very powerful to be convincing. An interesting example of this logic appeared in the *Journal of Experimental Psychology,* which has published a great deal of research on extrasensory perception (ESP) and other controversial topics. The editor of this journal wrote that

> In editing the *Journal* there has been a strong reluctance to accept and publish results . . . when those results were significant [only] at the .05 level, whether by one- or two-tailed test. This has not implied a slavish worship of the .01 level, as some critics have implied. Rather, it reflects a belief that it is the responsibility of the investigator in a science to reveal his effect in such a way that no reasonable man would be in a position to discredit his results by saying they were the product of the way the ball bounces.[4]

Using Confidence Intervals

Confidence intervals provide a useful alternative method of reporting the results of a two-sided statistical test. A null hypothesis will be rejected by a two-sided hypothesis test if and only if its value lies outside a corresponding confidence interval. We can illustrate this equivalence by considering our test of the null hypothesis that the average temperature of healthy humans is $\mu = 98.6$. This null hypothesis will be rejected at the 5 percent level if the two-sided P value is less than .05, which requires the sample mean to be more than (approximately) 2 standard deviations from the 98.6 value given by the null hypothesis. A 95 percent confidence interval for μ includes all values that are within 2 standard deviations of the sample mean. If the sample mean is more than 2 standard deviations from 98.6 then the two-sided P value will be less than .05 and a 95 percent confidence interval will not include 98.6. Therefore, a hypothesis test can be conducted by using the sample data to construct a confidence interval and seeing whether the parameter value specified by the null hypothesis is inside this interval.

The additional information provided by a confidence interval is a sense of the practical importance of the difference between the parameter value assumed by the null hypothesis and the value of the estimator. If we just report that the P value is .0004 or that we "found a statistically significant difference at the 5 percent level," readers will not know the actual value of the estimator. In addition to the P value, we should report a confidence interval.

For our temperature example, Eq. (6.7) gives the formula for a confidence interval using an estimated value of the standard deviation. Our sample standard deviation is $s = .67$, and with $50 - 1 = 49$ degrees of freedom, a 95 percent confidence interval uses $t^* = 2.01$. Thus we can report that a 95 percent confidence interval for the average temperature of healthy humans is

HOW TO DO IT

Hypothesis Tests for the Population Mean

We want to use a sample of size 50 to test whether the average temperature of healthy humans is 98.6°F.

1. Specify the value of μ under the null and alternative hypotheses. Here,

$$H_0: \mu = 98.6$$
$$H_1: \mu \neq 98.6$$

2. Calculate the sample mean; here it is 98.23.
3. Calculate the z value or the t value, depending on whether the standard deviation σ is known or estimated with the sample standard deviation s:

$$z = \frac{\bar{x} - \mu}{\sigma/\sqrt{n}} \qquad \text{or} \qquad t = \frac{\bar{x} - \mu}{s/\sqrt{n}}$$

In our example, the sample standard deviation is .67, and the t value is

$$t = \frac{\bar{x} - \mu}{\sigma/\sqrt{v}} = \frac{98.23 - 98.6}{.67/\sqrt{50}} = -3.85$$

4. Determine the probability of observing a z value or t value so far from 0. If the alternative hypothesis is two-sided, double this probability to

$$\begin{aligned}
95\% \text{ confidence interval for } \mu &= \bar{x} \pm t^*(\text{standard error of } \bar{x}) \\
&= \bar{x} \pm t^*\left(\frac{s}{\sqrt{n}}\right) \\
&= 98.23 \pm 2.01\left(\frac{.67}{\sqrt{50}}\right) \\
&= 98.23 \pm .19 \\
&= 98.04 \text{ to } 98.42
\end{aligned}$$

obtain the two-sided P value. In our example, with $50 - 1 = 49$ degrees of freedom, $P[t < -3.85] = .0002$, implying a two-sided P value of $2(.0002) = .0004$.

5. Alternatively, we can use computer software, as illustrated by the Minitab output in Table 7.2.

6. A P value below .05 indicates that the results are statistically significant at the 5 percent level; that is, the data provide persuasive evidence against the null hypothesis in that, were the null hypothesis true, there would be less than a 5 percent chance of observing a sample mean that is so far from the population mean assumed by the null hypothesis. If the P value is not less than .05, the data do not reject the null hypothesis at the 5 percent level.

7. To assess the practical importance of the results, also report a confidence interval (using the procedures detailed in Chap. 6). For our example, using the sample standard deviation and the t distribution with $50 - 1 = 49$ degrees of freedom, we find that

$$\begin{aligned}
95\% \text{ confidence interval for } \mu &= \bar{x} \pm t^*\left(\frac{s}{\sqrt{n}}\right) \\
&= 98.23 \pm 2.01\left(\frac{.67}{\sqrt{50}}\right) \\
&= 98.23 \pm .19
\end{aligned}$$

As anticipated, the value 98.6 is not inside this confidence interval. The general procedure for a hypothesis test regarding the value of a population mean is summarized in the how-to-do-it box.

The observation that any value inside a 95 confidence interval would not be rejected at the 5 percent level if tested as a null hypothesis reinforces our earlier remark that not rejecting a null hypothesis is a rather weak conclusion. Not rejecting a null hypothesis is not at all the same as proving the null hypothesis to be true. An unrejected null hypothesis is but one of many parameter values that are consistent with the data. Thus we must be careful to say, "The data do not reject the null hypothesis," rather than "The data prove the null hypothesis to be true."

Our 95 percent confidence interval for the mean temperature of healthy humans does not mean that an individual must have a temperature of 98.23 ± .19 to be healthy, or that 95 percent of all healthy humans have a body temperature in this range. Rather it is an expression of the confidence we have in our estimate of the average body temperature of healthy humans. For measuring the amount of variation in temperatures among healthy individuals, the more appropriate figure is the .67 standard deviation in our sample data. If the temperatures of healthy humans are normally distributed with a mean of 98.23 and a standard deviation of .67, then 95 percent of the readings will be in the range 98.23 ± 1.96(.67) = 98.23 ± 1.31, that is, from 96.92 to 99.54. The Maryland researchers found some variation in temperatures with age, race, gender, and the time of day when the reading is taken. The 98.23 ± .19 figure is our estimate of the average temperature of healthy people of different ages, races, genders, and times of day. Taking into account these variations among individuals, the Maryland researchers suggested that temperatures above 99.9 be considered feverish.

Physical Identification

Gauss was led to the normal distribution by his study of astronomical measurements. When a distance or shape is measured repeatedly, the measurements vary somewhat due to imperfections in the equipment and its use. Gauss noticed that repeated measurements conform to a normal distribution—about one-half above average and one-half below average, with most clustered in the middle and then tapering away in a bell-shaped curve. In accord with the central limit theorem, these measurement variations, called *measurement errors,* are apparently the cumulative result of a large number of small imperfections. Even today, carefully repeated measurements with the most advanced scientific equipment still vary slightly, and these variations are still normally distributed.

The normal distribution of measurement errors also underlies computerized physical identification equipment. When we see a friend, how do we recognize this person's identity? Somewhere in our minds, we have stored a record of the person's physical characteristics (height, weight, shape of the face, and so on), and when the characteristics of the person we see match our mental record, we recognize who it is. Of course, we sometimes make mistakes—when poor lighting impairs our vision or when we see only the back of the head of someone who superficially resembles a friend.

In theory, computers equipped with video cameras could be programmed to recognize people in much the same way—by storing a record of each person's physical characteristics and then matching these to the person that the camera sees. Several companies have in fact developed computerized identification systems for use in places as diverse as nuclear plants, the University of Georgia cafeteria, and day care centers (to identify parents picking up children). For instance, the Hand Scan Identimat asks a person to insert four fingers into slots and then compares the measured length, thickness, and curvature with the values stored in the computer. The measurements of any single individual vary slightly with each use, depending on the way in which the fingers are inserted, the condition of the machines, and even the temperature that day. These measurement errors conform to a normal distribution, as Gauss observed centuries ago with very different equipment and data. The figure below shows the distributions of measured index finger length that might be obtained by two different individuals. As indicated by the accept/reject cutoff line in this figure, there is some chance that a finger measurement of the authorized individual will be so large that the machine incorrectly denies access and some chance that the impostor's measurement will be so low that it is accepted. The cutoff shown in the figure is set so that the probability of accepting an impostor is much smaller than the probability of rejecting an authorized person. The optimal cutoff depends on the consequences of making these two very different kinds of errors.

The less variation there is in the measurements, the less these distributions will overlap and the smaller is the chance of identification errors. Taking into account the several measured characteristics of four fingers, the makers of the

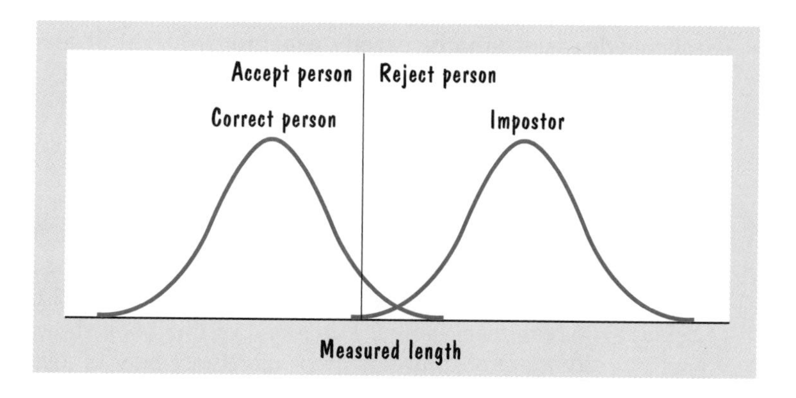

Identimat claim that there is only a 1 percent chance of rejecting an authorized person and a 1.5 percent chance of accepting an impostor.[5] Other systems are based on fingerprints, thumbprints, palmprints, and voiceprints. One of the most accurate systems is Eyedentify (and most expensive at $12,000 per machine), which uses an infrared camera to examine the pattern of retinal blood vessels and reports error rates of .1 and .0001 percent.

 Sign Test for the Median

We have seen that the mean can be a misleading statistic for describing the center of a set of data. If 99 people earn $10,000 per year and 1 person earns $4,010,000, then the $50,000 mean does not accurately describe the typical worker's pay. In such cases, the median is a popular alternative for describing the center of the data. With a bit of ingenuity, we can construct a hypothesis test for the median.

To illustrate the logic, suppose that the president of a large university claims that the median salary of full professors is $100,000, a claim that cannot be directly verified since salaries are confidential. To test this claim, we can take a random salary of, say, 20 full professors and see how many earn more than $100,000 and how many earn less. If we were to find roughly one-half above and one-half below, these data would be consistent with the president's claim. On the other hand, if all 20 earn less than $100,000, this would cast considerable doubt on the president's claim. What if it turns out that 5 of the salaries are larger than $100,000 and 15 are smaller?

Our null hypothesis is that the median is $100,000, and we will let the alternative hypothesis be two-sided:

$$H_0\text{: median} = \$100{,}000$$
$$H_1\text{: median} \neq \$100{,}000$$

Now suppose that we take a random sample of professors and discard any salaries that happen to be exactly equal to $100,000. If the null hypothesis is true and our sample is small relative to the population (not more than 10 percent of the size of the population), then we can use a binomial model with each observation independent and equally likely to be larger or smaller than $100,000.

If each observation has a .5 probability of being larger than $100,000, we can use the binomial distribution with $p = .5$ to calculate the probability that the number of salaries above or below $100,000 would be as large as (or larger than) the number actually observed. If it turns out that 15 of 20 salaries are below $100,000, then Table 2 at the end of this book shows that with $p = .5$ and $n = 20$, the probability of 15 or more successes is .0207. Since our alternative hypothesis is two-sided, we double this probability to obtain the two-sided *P* value: 2(.0207) = .0414. Although these data do not reject the president's claim decisively, they certainly cast doubt on it. (If the appropriate probability is not in Table 2, we can use computer software or the normal approximation to the binomial distribution, discussed in the next section.)

This clever procedure is called a **sign test** because it focuses on the signs of the deviations from the purported median. Those observations above the median are positive deviations, and those below are negative deviations. If the null hypothesis is true, the numbers of positive and negative deviations should be roughly equal. The binomial distribution tells us the probability that the number of positive or negative deviations would be as large as (or larger than) the number we actually observe.

HOW TO DO IT

Hypothesis Tests for the Population Median

We want to use a sample of size 20 to test whether the median salary of full professors at a large university is $100,000.

1. Specify the value of the median under the null and alternative hypotheses. Here,

$$H_0\text{: median} = \$100{,}000$$
$$H_1\text{: median} \neq \$100{,}000$$

2. Calculate how many of the observations are above the median and how many are below; here, 5 are above and 15 are below.
3. If the null hypothesis is true, then an observation that is not equal to the specified value for the median is equally likely to be larger or smaller than this specified value. Use the binomial distribution with $p = .5$ to calculate the probability that the number of observations above or below the median would be as large as (or larger than) that actually observed. If the alternative hypothesis is two-sided, double this probability to obtain the two-sided *P* value.

   Here, with $n = 20$, the probability of 15 or more successes in 20 trials is .0207, and the two-sided *P* value is 2(.0207) = .0414.
4. Alternatively, if the sample is large, we can use the normal approximation to the binomial distribution, discussed in the next section.

Most of the statistical tests used in this book assume that the data can be described by a specific probability distribution, such as the normal distribution. The sign test is an example of a **nonparametric** or *distribution-free* test which is appropriate, *no matter what* the distribution of the data. These tests work for normal distributions, but they also work for nonnormal distributions. If the data do, in fact, conform to a normal distribution, then a nonparametric test is not as powerful as a test that takes advantage of the information provided by the normal distribution. On the other hand, if the normal distribution is inappropriate, then a nonparametric test has a clear advantage over a test that makes an erroneous assumption.

7.1 A Texas A&M tradition is for members of the student body to stand during football games, indicating that each is a "12th Man," ready to play if needed. In 1983, the coach decided to let some of the regular students play:

> They all laughed when Texas A&M coach Jackie Sherrill came up with his version of the 12th Man, a kickoff return team made up entirely of students.
>
> It is no longer a joke. Going into today's game against Texas, the 12th-Man unit has allowed an average of only 13.6 yards per return.
>
> "These guys cover kickoffs better than anybody who plays against us," Sherrill said.
>
> And A&M's road kickoff-return unit, made up of regular players since the 12th-Man team doesn't travel, is allowing more than 18 yards per return.
>
> Even the White House has been impressed. After a presidential press aide attended an Aggies game, President Reagan sent each 12th Man an autographed picture along with a personal message.[6]

What other information would you like to see before concluding that the students cover kickoffs better than the regular football players?

7.2 The earlobe test was introduced in a letter to the prestigious *New England Journal of Medicine,* in which Dr. Sanders Frank reported that 20 of his male patients with creases in their earlobes had many of the risk factors (such as high cholesterol levels, high blood pressure, and heavy cigarette use) associated with heart disease. For instance, the average cholesterol level for his patients with noticeable earlobe creases was 257 (milligrams per 100 millimeters), compared to an average of 215 with a standard deviation of 10 for healthy middle-aged men. If these 20 patients were a random sample from a population with a mean of 215 and a standard deviation of 10, what is the probability their average cholesterol level would be 257 or higher? Explain why these 20 patients may, in fact, not be a random sample.

7.3 A seminal study in the 1930s investigated whether the average IQ score of blacks who had recently moved to New York City from the South differed from the average IQ score of blacks who had been born and raised in New York City.[7] Among 12-year-old black males born in New York City, IQ scores were known to be normally distributed with a mean of 87 and a standard deviation of 29. IQ tests were administered to a sample of 56 black 12-year-olds who had lived in New York City for 2 years or less. If these 56 recent migrants were a random sample from a normal distribution with a mean of 87 and a standard deviation of 29, from what distribution was the sample mean drawn? The sample mean turned out to be 64. What are the z values and P values? Provide a logical explanation, other than sampling error, of why recent migrants might have lower IQ scores, on average, than those born in New York City.

7.4 Researchers separately asked 153 husbands and their wives to state the highest school grade completed by the wives.[8] The difference, x = husband's answer − wife's answer, is a measure of the extent to which husbands exaggerate or belittle their wives' educational accomplishments. They found that x averaged .32 with a standard deviation of 1.1. Use a P value and a 95 percent confidence interval to test at the 5 percent level the null hypothesis that the population mean is 0.

7.5 A builder has contracted for the delivery of 400,000 pounds of loam. The supplier proposes making delivery in 100 truckloads that she claims will average 4000 pounds. The builder insists on weighing a random sample of 16 truckloads and obtains these data:

3820	3900	4000	3820	3780	3880	3700	3620
4380	3560	3840	3980	4040	4140	3780	3780

If you assume that these weights are from a normal distribution and use a one-tailed test, do these data reject the supplier's claim at the 5 percent level?

7.6 A researcher examined the admissions to a mental health clinic's emergency room on days when the moon was full.[9] For the 12 days with full moons from August 1971 through July 1972, the number of people admitted were as follows:

5 13 14 12 6 9 13 16 25 13 14 20

Calculate the two-sided P value for a test of the null hypothesis that these data are a random sample from a normal distribution with a population mean equal to 11.2, the average number of admissions on other days.

7.7 Table 3.2 gives 2 years of annual returns for 17 randomly selected stock mutual funds. The returns for the S&P 500 index were 1.3 percent in 1994 and 37.4 percent in 1995. For each of these 3 years, determine the t value and two-sided P value for a test of the null hypothesis that these data were drawn from a normal distribution with a population mean equal to the return on the S&P 500 index that year.

7.8 (Nonparametric) Table 7.1 shows the oral temperatures of 50 healthy adults. Use these data to calculate the two-sided P value for a test of the null hypothesis that the median temperature is 98.6.

7.2 Tests of a Success Probability

We began the chapter with this question: If I were to hand you a baby, would you hold it against your right shoulder or your left shoulder? If we were to conduct this experiment with 100 randomly selected people, our data would be categorical, as right and left do not have natural numerical values but are instead two categories of possible outcomes. To analyze such categorical data, we can use the binomial model by thinking of each trial as a random draw from a population from which there is a probability p of selecting a person who will hold the baby against, say, the left shoulder.

The principles developed in the first part of this chapter can be easily extended to tests of the binomial success probability p. First, we specify the null and alternative hypotheses about the value of p. Then we calculate the simple success proportion $\hat{p} = x/n$, which is the test statistic used to estimate the value of p. The P value is the probability that the sample success proportion x/n would be so far (or farther) from the value of p if the null hypothesis were true. The P value either can be calculated exactly by using the binomial distribution or, if the sample is large, can be approximated by the normal distribution. If the alternative hypothesis is two-sided, we double this probability to obtain the two-sided P value. That is all there is to it! We use several examples to illustrate the application of these procedures, and then we apply them to the baby on the shoulder.

Exact Binomial Probabilities

Sixteen pregnant women were asked to read a children's story aloud 3 times per day during the last 6½ weeks of pregnancy.[10] These readings were tape-recorded, and shortly after birth, each baby was allowed to choose (by varying the rhythm of its sucking on a special bottle) between a tape recording of his or her mother reading the familiar story and a tape recording of the mother reading an unfamiliar story. In 13 of 16 cases, the babies chose the familiar story. To assess the statistical significance of these results, the natural null hypothesis is that each baby has a .5 probability of choosing the familiar story. The alternative hypothesis should be two-sided, because we cannot rule out the possibility that a baby would prefer to listen to something new.

$$H_0\text{: } p = .5$$
$$H_1\text{: } p \neq .5$$

Using the binomial distribution (Table 2 at the end of this book), we find

$$\begin{aligned} P[x \geq 13] &= P[x = 13] + P[x = 14] + P[x = 15] + P[x = 16] \\ &= .0085 + .0018 + .0002 + .0000 \\ &= .0105 \end{aligned}$$

The two-sided P value is 2(.0105) = .0210, which is sufficiently low to reject the null hypothesis at the 5 percent level, but not at the 1 percent level. This procedure is summarized in a how-to-do-it box.

For a somewhat more complicated example, Dow Chemical used this procedure in auditing the invoices that it received each week for bulk mailing and duplicating.[11] The null hypothesis was that .25 percent of all the entries on these invoices contained clerical errors. Because Dow was not concerned with fewer errors, the one-sided alternative was that there were more errors:

$$H_0\text{: } p = .0025$$
$$H_1\text{: } p > .0025$$

Dow's auditors examined a random sample of 225 entries each week. If the null hypothesis is true, then the binomial model with $n = 225$ and $p = .0025$ gives the probabilities of finding x errors:

$$P[x \text{ errors}] = \binom{225}{x} .0025^x (1 - .0025)^{n-x}$$

Computer software will do the calculations for us:

$$P[x = 0] = .569 \qquad P[x = 1] = .321 \qquad P[x = 2] = .090 \qquad P[x > 2] = .020$$

If Dow rejected the week's invoices based on the discovery of more than 2 errors in the sample, the probability that a true null hypothesis would be rejected would have been .02. If, instead, Dow rejected the invoices when more than 1 error was found, this probability would have increased to .02 + .09 = .11. The 2-error cutoff is closer to .05, and this is what Dow used.

HOW TO DO IT

Tests of a Success Probability Using the Exact Binomial Probability

We want to see whether a mother reading a story aloud before a child is born affects the child's choice of a story after birth.

1. Is the binomial model appropriate? Yes. Each observation has two possible outcomes, familiar or unfamiliar story, and the observations are independent.
2. Specify the value of p under the null and alternative hypotheses. Here,

$$H_0: p = .5$$
$$H_1: p \neq .5$$

3. How many successes are there in the sample? Here, $x = 13$ in $n = 16$ observations.
4. Use the binomial distribution to calculate the probability that the number of successes would be so far from the population mean np, using the value of p given by the null hypothesis. For a two-sided alternative hypothesis, double this probability. In our example,

$$\begin{aligned} P[x \geq 13] &= P[x = 13] + P[x = 14] + P[x = 15] + P[x = 16] \\ &= .0085 + .0018 + .0002 + .0000 \\ &= .0105 \end{aligned}$$

   and the two-sided P value is $2(.0105) = .0210$.
5. A P value below .05 indicates that the results are statistically significant at the 5 percent level. Here, the .021 P value is sufficient to reject the null hypothesis at the 5 percent level.

Normal Approximation to a Binomial

In Chap. 5, we saw that if the sample is sufficiently large (np and $n(1 - p)$ both larger than 5), then the sample statistic $\hat{p} = x/n$ is approximately normally distributed:

$$\hat{p} = \frac{x}{n} \text{ is } N\left[p, \sqrt{\frac{p(1-p)}{n}}\right]$$

Therefore,

> We can test a null hypothesis about the value of the population success proportion p by using this z statistic to calculate the approximate P value:
>
> $$z = \frac{\text{observed statistic} - \text{null hypothesis parameter value}}{\text{standard deviation of statistic}}$$
>
> $$= \frac{x/n - p}{\sqrt{p(1-p)/n}} \tag{7.4}$$

Chapter 6 explained how we can also use a normal approximation to compute a confidence interval for p. Because a confidence interval does not involve a null hypothesis with a specified value for p, we use the sample statistic $\hat{p} = x/n$ when estimating the standard error:

$$\begin{aligned}\text{Confidence interval for } p &= \hat{p} \pm z^*(\text{standard error of } \hat{p}) \\ &= \hat{p} \pm z^*\sqrt{\frac{\hat{p}(1-\hat{p})}{n}} \\ &= \frac{x}{n} \pm z^*\sqrt{\frac{x/n(1-x/n)}{n}}\end{aligned}$$

For a 95 percent confidence interval, we would use $z^* = 1.96$.

As an illustration, consider a 1958 study of the racial awareness of black children.[12] The sample consisted of 253 black children, 134 from Arkansas and 119 from Massachusetts, whose ages ranged from 3 to 7 years. Each child was offered four dolls, two white and two black, and told, "Give me the doll that you would like to play with." Overall, 83 children chose a black doll, and 169 chose a white doll. (One child did not want to play with dolls and refused to choose one.) Can the fact that two-thirds of these children chose a white doll be explained plausibly by chance—the coincidental result of color-blind choices—or does it indicate the black children in 1958 were well aware of race?

For a statistical test, the null hypothesis is that the choices are color-blind, so that each child has a probability $p = .5$ of selecting a white doll. Because we have no compelling reason beforehand to rule out a preference for either white or black dolls, the alternative hypothesis is two-sided:

$$H_0\colon p = .5$$
$$H_1\colon p \neq .5$$

Evidence against the null hypothesis is provided by an observed sample proportion x/n that is far from .5. The exact probability that 169 or more of the 252 children who participated will choose a white doll is given by the binomial distribution. Using statistical software,

$$P[x \geq 169] = .000000032$$

For a two-sided P value, we double this probability: 2(.000000032) = .000000064.

Because this is a reasonably large sample, we can also calculate an approximate P value by using a normal approximation to the binomial distribution. Checking our rule of thumb, and using the value of p specified by the null hypothesis, we see that $np = 252(.5) = 126$ and $n(1 - p) = 252(.5) = 126$ are both (much) larger than 5. Using $x = 169$, $n = 252$, and the fact that $p = .5$ under the null hypothesis, we use Eq. (7.4) to calculate the z value:

$$\begin{aligned} z &= \frac{x/n - p}{\sqrt{p(1-p)/n}} \\ &= \frac{169/252 - .5}{\sqrt{.5(1 - .5)/252}} \\ &= 5.42 \end{aligned}$$

We find the approximate probability that x/n is larger than 169/252 by determining the probability that a standardized z value is larger than 5.42:

$$\begin{aligned} P\left[\frac{x}{n} \geq \frac{169}{252}\right] &= P[z \geq 5.42] \\ &= .000000031 \end{aligned}$$

The two-sided P value is 2(.000000031) = .000000062, which is very close to the exact value.

Table 7.3 shows the output from Smith's Statistical Package. After the user enters the null hypothesis value of p and the values of n and x in the three boxes, the bottom half of the computer screen shows the exact and approximate one-sided P values. (The continuity correction gives a slightly more accurate P value, but is beyond the scope of this textbook.)

Whether we calculate it by hand or by computer, the observed sample proportion 169/252 = .671 is 5.42 standard deviations from the .5 value of p under the null hypothesis, decisively rejecting the hypothesis that black children in 1958 were color-blind.

Similarly, a 95 percent confidence interval for p does not include $p = .5$:

$$\begin{aligned} 95\% \text{ confidence interval for } p &= \hat{p} \pm z^*(\text{standard error of } \hat{p}) \\ &= \frac{x}{n} \pm z^*\sqrt{\frac{x/n(1 - x/n)}{n}} \end{aligned}$$

Table 7.3

A hypothesis test for p from Smith's Statistical Package

```
A sample success proportion can be used to test a null
hypothesis about the success probability in the popula-
tion that yielded this random sample.

According to the null hypothesis,

    What is the probability of success?        p = [ .5 ]

    How many sample observations?              n = [ 252 ]

    How many successes in this sample?         x = [ 169 ]

In this sample of 252 observations, the expected number
of successes under the null hypothesis is 126. The bino-
mial probability of 169 or more successes is
.0000000323. Using a normal approximation (with a conti-
nuity correction) to the binomial distribution, the
(absolute) z value is 5.417. The probability of such a
large z value is .000000031.
```

$$= \frac{169}{252} \pm 1.96\sqrt{\frac{169/252(1 - 169/252)}{252}}$$
$$= .671 \pm .058$$

A How-To-Do-It box summarizes this procedure for using a normal approximation to the binomial distribution to test a null hypothesis about the value p of a success probability or population proportion.

How Do We Hold Babies?

We are ready at last to answer the question posed at the beginning of this chapter. While observing a rhesus monkey at a zoo, Lee Salk, a psychologist, noticed that on 40 of 42 occasions the monkey held her baby on the left side of her body, close to her heart.[13] He then observed 287 human mothers shortly after they had given birth. Of 255 right-handed mothers, 83 percent held the babies on the left sides of their bodies; of 32 left-handed mothers, 78 percent held their babies on the left side. In contrast, when 438 shoppers carrying packages that are the approximate size of a baby left a supermarket with automatic doors, exactly one-half held the package on their left sides.

In a controlled experiment, Salk compared the weight gains and amount of crying for babies in a nursery who listened to a tape recording of a human heartbeat 24 hours per day for 4 days with a control group who did not listen

HOW TO DO IT

Tests of a Success Probability Using a Normal Approximation

Is a randomly selected black child equally likely to choose a black or white doll?

1. Is the binomial model appropriate? Yes. Each observation has two possible outcomes, a child who chooses a black doll and a child who chooses a white doll, and the observations seem independent with a constant probability of selecting a child who chooses a white doll.
2. Specify the value of p under the null and alternative hypotheses. Here,

$$H_0\colon p = .5$$
$$H_1\colon p \neq .5$$

3. How many successes are in the sample? Here, $x = 169$ in $n = 252$ observations.
4. Using the value of p specified by the null hypothesis, are both np and $n(1 - p)$ larger than 5, so that a normal approximation is appropriate? Here, $np = 252(.5) = 126$ and $n(1 - p) = 252(.5) = 126$.
5. Use Eq. (7.4) to calculate the z value. In our example,

$$\begin{aligned} z &= \frac{x/n - p}{\sqrt{p(1-p)/n}} \\ &= \frac{169/252 - .5}{\sqrt{.5(1-.5)/252}} \\ &= 5.42 \end{aligned}$$

to the tape. The babies who listened to the heartbeats had a median weight gain of 409 grams and cried 38 percent of the time, compared to the control group, which had a median weight loss of 20 grams and cried 60 percent of the time. Because the babies consumed the same amount of food, he concluded that the weight gain was evidently due to decreased crying.

Salk speculated that "it is not in the nature of nature to provide living organisms with biological tendencies unless such tendencies have survival value. . . . [W]hen a baby is held on the mother's left side, not only does the baby receive soothing sensations from the mother's heartbeat but also the mother has the sensation of her heartbeat being reflected back from the baby."

6. Alternatively, we can use computer software, as illustrated by the Smith's Statistical Package output in Table 7.3.
7. Determine the probability of observing a z value so far from 0. If the alternative hypothesis is two-sided, double this probability to obtain the two-sided P value. In our example, $P[z > 5.42] = .000000031$, implying a two-sided P value of $2(.000000031) = .000000062$.
8. The results are statistically significant at the 5 percent level if the P value is less than .05. A .000000062 P value rejects the null hypothesis decisively.
9. To assess the practical importance of the results, report a confidence interval (using the procedures detailed in Chap. 6). Here, a 95 percent confidence interval for p is

$$
\begin{aligned}
95\% \text{ confidence interval for } p &= \hat{p} \pm z^* \sqrt{\frac{\hat{p}(1-\hat{p})}{n}} \\
&= \frac{169}{252} \pm 1.96 \sqrt{\frac{169/252(1 - 169/252)}{252}} \\
&= .671 \pm .058
\end{aligned}
$$

Evidently, mothers instinctively know that their infants are soothed by the sound of a heartbeat. We can use the data for $n = 287$ human mothers to test the null hypothesis that a baby is equally likely to be held on the left or right side: H_0: $p = .5$. The observed sample proportion $x/n = .82$ gives this z value:

$$
\begin{aligned}
z &= \frac{x/n - p}{\sqrt{p(1-p)/n}} \\
&= \frac{.82 - .5}{\sqrt{.5(1 - .5)/287}} \\
&= 10.84
\end{aligned}
$$

The observed .82 sample proportion is 10.84 standard deviations away from the .5 value of p (were the null hypothesis true). The P value is virtually 0, and the null hypothesis is decisively rejected.

Are Mendel's Data Too Good to Be True?

Gregor Mendel's probabilistic model of how genes pass from one generation to the next is one of our most important and elegant scientific theories. Mendel reported many experimental results supporting his theory, and other researchers have conducted numerous experiments confirming his insights. One of Mendel's original experiments involved the self-fertilization of hybrid yellow-seeded sweet peas. His theory implied that the offspring have a .75 probability of being yellow-seeded and a .25 probability of being green-seeded. Mendel reported the following results:

OFFSPRING	NUMBER
Yellow-seeded	6021
Green-seeded	2002
Total	8023

To determine if these data support or reject Mendel's theory, we can use the binomial model with p equal to the probability of a yellow-seeded offspring. The null and alternative hypotheses are

$$H_0: p = .75$$
$$H_1: p \neq .75$$

The observed sample proportion $x/n = 6021/8023 = .7505$ is almost exactly .75, the value of p under the null hypothesis. Because the sample of $n = 8023$ plants is certainly large enough to use a normal approximation, we convert to a standardized z value:

$$z = \frac{x/n - p}{\sqrt{p(1-p)/n}} = \frac{.7505 - .75}{\sqrt{.75(.25)/8023}} = .097$$

Since $P[z > .097] = .4615$, the two-sided P value is $2(.4615) = .923$, which is very far from the .05 required to reject the null hypothesis at the 5 percent level. Therefore, these data are consistent with Mendel's theory.

Under the null hypothesis, the population mean for the number of yellow-seeded plants $np = 8023(.75) = 6017.25$. Mendel's reported number of 6021 is off by fewer than 4 plants in an experiment involving 8023 plants! Mendel's data support his theory amazingly well—perhaps too well. Some 70 years later, R. A. Fisher argued that Mendel's data are literally too good to be true.

Fisher reversed the usual hypothesis-test question by asking, If the null hypothesis is correct, what is the probability of observing a value of x/n that is so close to p? This is equal to 1 minus the two-sided P value: $1 - .923 = .077$. If there is a .923 probability that a normally distributed random variable will be more than .097 standard deviation (in either direction) from its population mean, then there is a .077 probability that a normally distributed variable will be within .097 standard deviation (in either direction) of its mean. Such a close correspondence between theory and data requires a bit of luck, or something less innocent.

Mendel studied seven inherited characteristics of pea plants and, in every single case, reported this close a correspondence between his predictions and his results. The probability of such a phenomenal run of luck is about .000004. It is difficult to believe that Mendel was so lucky. Even worse, in one of his studies, he miscalculated the theoretical probability and then reported data that fit this erroneous probability almost exactly. It is hard to disagree with Fisher's conclusion that Mendel's data are too good to be true. Fisher was, however, so impressed by Mendel's scientific contributions and the fact that Mendel was a monk that he placed the blame elsewhere: "I have no doubt that Mendel was deceived by a gardening assistant, who knew only too well what his principal expected from each trial."[14]

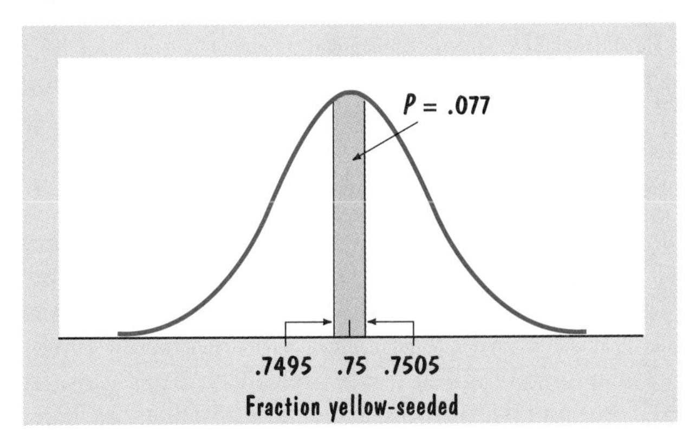

7.9 A marine biologist hypothesized that when there is a long interval between calving, female whales are more likely to have sons than daughters. Data collected by the Center for Coastal Studies found that whales with an interval of 1

to 2 years between calving had 22 sons and 20 daughters. For mothers calving at intervals of 3, 4, or 5 years, there were 16 sons and only 4 daughters. If these are independent observations with male and female calves equally likely, what is the probability that of 20 calves, 16 or more will be male? (Do not use a normal approximation.)

7.10 In February 1995, *Sports Illustrated* reported that the most serious kinds of knee injuries "are virtually epidemic in women's college basketball."[15] It found that among currently active basketball players in the six major Division I conferences, a total of 83 women and 26 men had suffered tears in their anterior cruciate ligaments (ACLs), one of the two main ligaments that support the knee. If we consider each ACL injury a binomial trial with a .5 probability of involving a female and a .5 probability of involving a male, what is the exact probability that of 109 injuries, more than 82 would be to females?

7.11 As of 1995, U.S. presidents have fathered 91 boys and 59 girls. Assuming these to be independent events, use a normal approximation to test at the 1 percent level the null hypothesis that the children of presidents are equally likely to be female or male.

7.12 A high school basketball coach who was named "Coach of the Decade" by the *Los Angeles Times* told his players that there is a 60 percent probability that a missed shot will bounce "long," to the opposite side of the basket from which the shot was taken. To test this claim, an examination of video tapes of three men's Division III college basketball games found that 84 of 169 missed shots bounced short and 85 bounced long.[16] Calculate the z value and use the two-sided P value to test the following null hypotheses:

a. A missed shot has a .6 probability of bouncing long.

b. A missed shot has a .5 probability of bouncing long.

7.13 High doses of anticancer drugs can be effective in killing malignant cells in women with breast cancer, but also destroy vital bone marrow. In May 1990, Prof. Karen Antman of Harvard Medical School reported that 150 of 259 patients with advanced breast cancer who received a combination of bone marrow transplants and high doses of anticancer drugs were free of cancer after 12 months.[17] In contrast, of those patients with advanced breast cancer who received conventional treatment, 30 percent were free of cancer after 12 months. If you let $p = .30$ be the null hypothesis, are Antman's results statistically significant, using a one-tail test at the 1 percent level and a normal approximation?

7.14 A soccer player organized a penalty-kick tournament for 10 right-footed Division III college soccer players.[18] A total of 152 penalty kicks were taken, of which 88 were shot to the right side of the goal (from the kicker's perspective) and 64 were shot to the left side.

a. Use these data to calculate the two-sided P value for a test of the null hypothesis that shots to the right and left sides of the goal are equally likely. Do not use a normal approximation.
b. Now calculate the two-sided P value by using a normal approximation.
c. Overall, 110 of 152 shots were successful. Using $p = .5$ when estimating the standard deviation, calculate a 95 percent confidence interval for the probability that a penalty kick is successful.

7.15 In 1977, the U.S. Supreme Court (*Castaneda vs. Partida*) observed that persons with Spanish surnames constituted 79 percent of the population of Hidalgo County, Texas, but only 339 of 870 recent jurors in this county. The Court calculated a z value of 29 and commented that "as a general rule for such large samples, if the difference between the expected value and the observed number is greater than two or three standard deviations, then the hypothesis that the jury drawing was random would be suspect."
a. What parameter value constitutes the null hypothesis?
b. Show how the Court calculated this z value.
c. Explain the reference to *large samples.*
d. Explain the reference to *2 or 3 standard deviations.*

7.16 Karl Pearson reported that he flipped a coin 24,000 times and obtained 12,012 heads and 11,988 tails. Do these data reject at the 5 percent level the null hypothesis that the probability of heads is .5? Do you feel that Pearson's data are too good to be true? Explain your reasoning.

7.3 The Use and Abuse of Statistical Tests

Hypothesis tests are undoubtedly one of the most widely used applications of statistics. In every academic discipline, professional journals are filled with statistical tests of interesting theories. The most useful or controversial tests find their way into the popular press. Newspapers, magazines, and television news programs report all sorts of theories that have been "proved" statistically. I have before me three newspaper clippings reporting that

- Vitamin C does help ward off colds . . . there were statistically significant differences between treatment groups.
- Statistics prove a prima facie case of discrimination against blacks and women [in Alabama jury rolls].
- There is irrefutable proof that the history of the United States is shaped and controlled by an external nonhuman mind.

To be intelligent consumers and/or producers of such tests, so that we can separate useful information from misleading rubbish, we need to be able to recognize ways in which hypothesis tests can be misleading. Chapter 4 has already alerted us to some of the pitfalls in producing sample data, and "garbage in, garbage out" is a shorthand reminder of the fact that a beautifully executed statistical test will be ruined by biased sample data. In the remainder of this chapter, we look at two additional pitfalls in interpreting the results of statistical tests.

Is It Important?

In testing hypotheses, it is easy to be confused by the distinction between statistical significance and practical importance. A statistically significant result may have little practical importance. Conversely, a researcher may find a potentially very important result that does not happen to be statistically significant. To illustrate what may appear paradoxical, let us return to our body temperature example.

We have seen that a test at the 5 percent level of the null hypothesis that $\mu = 98.6$ can be conducted by seeing whether a 95 percent confidence interval includes 98.6. To exaggerate our impending conclusion, pretend that this test garners as much national attention as the polio vaccine tests in the 1950s and that a sample of 1,000,000 persons are tested, whose average temperature turns out to be 98.598, with a standard deviation of .67. For this large a sample, the t value for a 95 percent confidence interval is 1.96. Thus, our 95 percent confidence interval is

$$
\begin{aligned}
95\% \text{ confidence interval for } \mu &= \bar{x} \pm t^*\left(\frac{s}{\sqrt{n}}\right) \\
&= 98.598 \pm 1.96\left(\frac{.67}{\sqrt{1{,}000{,}000}}\right) \\
&= 98.598 \pm .001 \\
&= 98.597 \text{ to } 98.599
\end{aligned}
$$

Because 98.6 is outside this interval, these results reject the null hypothesis at the 5 percent level. With such a large sample, 98.597 is more than 2 standard deviations from 98.6. Yet the difference between a temperature of 98.6 and 98.597 is of no practical importance.

For the reverse case, suppose that our underfunded researcher is interested in the temperature of people who are 100 years old and can only afford to locate and test four people, whose average temperature turns out to be 98.0. This is substantially below 98.6; yet, because we have so few data, a 95 percent confidence interval is too wide to rule out 98.6 (nor can it rule out 97.4):

$$
\begin{aligned}
95\% \text{ confidence interval for } \mu &= \bar{x} \pm t^*\left(\frac{s}{\sqrt{n}}\right)\\
&= 98.0 \pm 1.96\left(\frac{.67}{\sqrt{4}}\right)\\
&= 98.0 \pm .66\\
&= 97.34 \text{ to } 98.66
\end{aligned}
$$

This result is not statistically significant, but is substantial and may be important enough to warrant the collection of additional data.

These two examples are deliberate caricatures designed to demonstrate that *P* values measure only statistical significance, not practical importance. In addition to examining the *P* value, we look at a confidence interval and use our common sense to judge whether, as a practical matter, the estimate is large or small.

Consider the question of whether women or men are more likely to be right-handed. A sample of 6672 people found that 92 percent of the women and 90 percent of the men were right-handed. This difference is statistically significant, but unimportant. Or consider the conclusion of two economists studying the effect of inflation on election outcomes. They estimated that the inflation issue increased the Republican vote in the 1976 election by 7 percentage points, plus or minus 10 percentage points. Because 0 is inside this interval, they concluded that, "in fact, and contrary to widely held views, inflation has no impact on voting behavior."[19] That is not at all what their data show. The fact that they cannot rule out 0 does not prove that 0 is the correct value. Their 95 percent confidence interval does include 0, but it also includes everything from −3 to +17 percent. Their best estimate is 7 percent, plus or minus 10 percent, and 7 percent is more than enough to swing most elections one way or another.

Data Mining

A selective reporting of results is potentially one of the most serious abuses of hypothesis tests. To illustrate this pitfall, I will pretend that I am a psychic and you pretend that you are a famous statistician. I come to your office, claiming that I have an uncanny ability to predict the outcomes of coin flips. I do not pretend to be perfect, but I do get more than 50 percent right. My "secret" is that I have discovered a pattern in coin flips: after a head occurs, the next flip is likely to be a tail, and vice versa. You pull out a pencil, paper, and coin, and you dutifully test my theory.

You let the null hypothesis be that on each flip I have a $p = .5$ probability of making a correct prediction, and you let the one-sided alternative hypothesis be my claim that p is greater than .5. Pressed for time and skeptical of my claim, you make just 20 coin flips. A check of the binomial distribution in Table 2 of this book (you have kept it for reference) shows that a guesser has a .058 probability of getting 14 or more right. So you choose this cutoff for statistical significance.

Well, the truth of the matter is that I am no psychic and my theory is worthless. I make my 20 guesses and get 8 correct, and you throw me out of your office and try to go back to work. As I go out the door, another alleged psychic walks in, with her theory that coin flips have momentum: A head is likely to be followed by another head, and a tail is likely to be followed by another tail. In the interest of science, you flip the coin 20 more times and find she gets only 11 right. Out she goes and in comes another, who this time believes that a head tends to be followed by a head, but that two heads in a row tend to be followed by a tail; another test, another fraud exposed.

Somewhere along the line, one of these frauds is going to get lucky and get 14 or more right. A 5 percent chance that a true null hypothesis will be rejected implies that about 1 out of every 20 worthless theories will be found statistically significant at the 5 percent level. This reasoning is not limited to fraudulent psychics; it applies to all statistical tests of all theories. Even if you never test anything but worthless theories, you are certain to give your statistical seal of approval to some, if you test enough of them.

With mountains of data, powerful computers, and incredible pressure to produce statistically significant results, untold numbers of worthless theories get tested. The harried researcher will try 20 theories, write up the result that is the most statistically significant (lowest P value), and throw the rest in the trash. Or 20 struggling graduate students will each try a different theory; one will get a statistically significant result, a thesis, and a research grant. The problem for science and society is that we see only the tip of this statistical iceberg. We see the result that is statistically significant, but not the other 19 that did not work out. If we knew that this was 1 of 20 tries, we would be much less impressed by its statistical significance.

Three psychiatrists once set out to identify observed characteristics that distinguish schizophrenic persons from nonschizophrenic.[20] They considered 77 different characteristics, and 2 of their 77 tests turned up statistically significant differences at the 5 percent level. They emphasized the statistical significance of these two characteristics, overlooking the sobering thought that, by chance alone, about 4 out of every 77 independent tests should be statistically significant at the 5 percent level.

Not finding a statistically significant result can sometimes be as interesting as finding one. In 1887, Albert Michelson and Edward Morley conducted a famous experiment in which they measured the speed of light both parallel and perpendicular to the earth's motion, expecting to find a difference that would confirm the prevailing theory of light. Their report that they did not detect a statistically significant difference laid the groundwork for Einstein's special theory of relativity and the development of quantum mechanics by others. Their not-statistically-significant result literally revolutionized physics. Less dramatically, it is worth knowing that well-designed research did not find a statistically significant effect of a medication, campaign finance reform, or capital punishment. However, a study of psychology journals found that of 294 articles using statistical tests, only 8 reported results that were not statistically significant at the 5 percent level.[21]

This belief that only statistically significant results are worth publishing leads researchers to test many theories until the desired significance level is attained—a process known as **data mining** (or *data grubbing* or *fishing expeditions*). The eager researcher takes a body of data, tests countless theories, and reports the one that is the most statistically significant. The end result—the discovery of a low *P* value—shows little more than the researcher's energy. We know beforehand that even if only worthless theories are examined, the researcher will eventually stumble on a worthless theory that seems to be supported by the data. It is not surprising when a data miner finds a statistically significant result, and it should not convince us that the researcher has found anything more than sampling error. We cannot tell whether the end result of a data mining expedition shows the veracity of a useful theory or the perseverance of a determined researcher.

The Sherlock Holmes Inference

Many theories get rejected out of hand without a formal statistical test. Think of my 20 coin flips again. In these tests, each quack walked into your office with a duck-brained scheme for predicting coin flips, and a coin was flipped 20 times. In practice, researchers often have the data in hand before they concoct the theories to be tested. A practiced observer can look at the data and quickly see which theories are promising and which are not going to work. Maybe 5 tests are run through the computer, and 50 are run through the mind.

For example, I flipped a coin 20 times, and the results were as follows (T = tails, H = heads):

T H H T H T T T T T
H H T H T T H H T T

In less than 30 seconds, I found a pattern, a "theory," to explain these coin flips. My theory is that a flipped coin tends to give a tail followed by two heads, over and over. If we test my theory using these data, we find that it correctly predicts 15 out of 20 outcomes, a statistically significant result! These data apparently show that I am not just a hopeless guesser.

Yet, would you bet that my system will work on 15 of 20 new flips? Of course not, because I told you how I came up with my "theory," and my method is unconvincing. I looked at the data, tried many possible theories in my mind, and found one that would have been right 15 out of 20 times if only I had known the theory before the coin was flipped—but I did not.

This is pure and simple data mining. When you see how it is done and when it is applied to something like coin flips, the results are unpersuasive because your intuition resists such theories. But professional data mining is applied to situations in which we do not have much intuition and you do not get to see the data mining in action; you see only the final results.

If I came to you wearing a $2500 suit and said that I had been studying the stock market for many years and had come up with a theory that works, you might pay attention. If I told you that in 15 of the last 20 years, this theory has correctly predicted whether the stock market will go up or down, you might get very interested. Yet, I may have found my stock market system in the same way as I found my coin-flip system. If H stands for an up year and T for a down year, the 20 coin flips recorded earlier might well be a history of the stock market over some 20-year period. I could look at this history of the ups and downs and, in less than 30 seconds, find a pattern that seemed to explain the stock market: The market goes backward one step, then upward two steps. And sure enough, this theory is right in 15 out of 20 years. If we subjected it to the usual statistical tests, we would find that this 75 percent accuracy decisively rejects the null hypothesis that I have only a 50 percent chance of predicting which way the market is going. The statistical evidence sounds pretty impressive, unless you know how I did it.

The theoretical logic behind hypothesis tests is that the researcher puts forward a theory, gathers some data to test this theory, and reports the results—whether statistically significant or not. However, anxious for publishable (or salable) results, some people work in the other direction—they study the data until they stumble upon a theory that will be shown to be statistically significant. This is the *Sherlock Holmes approach:* Examine the data and use inductive reasoning to identify theories that explain the data. As Sherlock Holmes put it: "It is a capital mistake to theorize before you have all the evidence." The Sherlock Holmes method can be very useful for solving crimes and other purposes. Many important scientific theories, including Mendel's genetic theory, were discovered by identifying theories that explain data. But the Sherlock Holmes approach has also been the source of thousands of quack theories.

How do we tell the difference between a good theory and quackery? There are two good antidotes to unrestrained data mining: common sense and fresh data. Unfortunately, there are often limited supplies of both. The first thing that we should think about when confronted with statistical evidence in support of some theory is, Does it make sense? If it is a ridiculous theory, we should not be persuaded by anything less than mountains of evidence, and even then we should be skeptical. One of the reasons you laughed off my claims of predicting coin flips is that your common sense tells you that coin flips are unpredictable.

Even if I walked in with data showing that I had correctly predicted 100 out of 100 flips, you would remain unconvinced. Instead of acknowledging my coin-calling prowess, you would suspect that I cheated. As Thomas Paine once wrote, "Is it more probable that nature should go out of her course, or that a man should tell a lie?"

Most situations are not as clear-cut as coin predictions, but nonetheless we should still use common sense in evaluating statistical evidence. Unfortunately, common sense is an uncommon commodity, and many silly theories have been seriously tested by honest researchers. I will show you some examples shortly.

The second antidote is fresh data. It is not a fair test of a theory to use the very data that were mined to discover the theory. For an impartial test, we should specify the theory before we see the data that will be used to test the theory. If some data have been used to concoct the theory, then look for fresh data that have not been contaminated by data mining. If I announce that I have a system that predicted 15 out of 20 coin flips, your likely response is, "Let me flip a coin, and we will see how you do." These fresh data will most likely expose my fraudulent system. We should do the same with any theory that has been dug up by data mining.

In chemistry, physics, and the other natural sciences, this is standard procedure. Whenever someone announces an important new theory, a dozen people will rush to their laboratories to see if they can replicate the results. Unfortunately, most social scientists do not have laboratories where they can conduct

Cartoon by Rob Pudin, courtesy *The Skeptical Inquirer*

experiments and produce fresh data. Instead, they must patiently record data as events occur. If they have a theory about presidential elections, the economy, or world peace, they may have to wait years or even decades to accumulate enough fresh data to test their theories. (Thus the joke that, for some social scientists, *data* is the plural of *anecdote*.)

When further testing is done with fresh data, the results are often disappointing, in that the original theory does not explain the new data nearly as well as it explained the data that were the inspiration for the theory. The usual excuse is that there have been important institutional changes that necessitate modification of the original model. By mining the fresh data for a new theory, a provocative first paper can turn into a career. An alternative explanation of the disappointing results with the fresh data is that when a theory is uncovered by data mining, statistical tests based on these mined data typically exaggerate the success of the model. In the remainder of this chapter, we look at some theories that are statistically significant, but unpersuasive.

How Not to Win at Craps

The author of a book on how to win the dice game craps recorded the outcomes of 50,000 dice rolls at a Las Vegas casino. The numbers came up just about as often as would be expected with fair dice, but a careful scrutiny of the sequence in which the numbers appeared revealed some unusual patterns. The sequence 4–4–11 could be expected about 20 times in 50,000 roles, but it appeared 31 times. If you had bet $100 on 11 whenever 4 came up twice in a row, you would have made a $13,700 profit. The author also found that the sequence 7–12–7 occurred 38 times and that on 10 of these 38 occasions, the next number was either a 2, 3, or 12. Are you convinced by such evidence? This is exactly analogous to my finding a pattern in 20 coin flips. Your common sense should lead you to reject this system (and dissuade you from going to Las Vegas to gather fresh data).

Tall Presidents

A month before the 1972 presidential election pitting Richard Nixon against George McGovern, *Parade* magazine reported that sociologist Saul Feldman of Case Western University,

> whose research interest is height, took a reading of Presidential candidates for the past 40 years. He reports that with only one exception (Wendell Wilkie in 1940), the taller of the two candidates has always won the electoral victory. In 1968, he points out, Nixon just barely defeated Humphrey, who at 5′11″ stands a half-inch shorter than the

> president. . . . "Heightism" is a prejudice just like sexism and racism, Feldman believes.[22]

Do Feldman's 10 observations show a statistically significant relationship between height and electoral success? If height is unimportant, then there is a .5 probability that the winner will be the taller of the two candidates. The binomial distribution tells us that if $p = .5$, there is only a .01 probability of 9 or more successes in 10 observations. Thus the evidence seems to reject the null hypothesis decisively and to confirm Feldman's heightism theory.

But let us think about this evidence a little more carefully. Because candidates can run again and again, presidential elections are not the independent observations assumed by a binomial distribution. Dwight D. Eisenhower beat Adlai Stevenson in 1952 and again in 1956. The 1956 election did not provide independent evidence about randomly selected candidates of different heights; it simply confirmed that Eisenhower was more popular than Stevenson. Another 4 of Feldman's 10 elections involved Franklin Roosevelt running again and again.

In addition, we cannot help but wonder why Feldman started with the 1932 election. Why did he look back 40 years, instead of 20 years, 60 years, or some other number? Perhaps this 40-year period yielded the data that were most favorable to his theory, but it is not appropriate to discard data simply because they do not support your theory. Whenever someone uses only part of the available data, we should wonder why the rest of the values were not used.

What about common sense, the first antidote to data mining? I do not believe that a half-inch difference in height was responsible for Nixon's victory over Humphrey. Nor do I believe that height explains the four victories by Roosevelt, whose legs were paralyzed. If someone who is 6 feet 4 inches runs against someone 5 feet 4 inches, this noticeable height difference may well influence some voters. When George Bush ran against Michael Dukakis, some called it "the wimp versus the shrimp." When Ross Perot debated George Bush and Bill Clinton while sitting on bar stools, some noticed that Perot's feet did not reach the floor. But did anyone other than Feldman know that Humphrey was a half-inch shorter than Nixon?

What about the second antidote, fresh data? One reason why Feldman's theory was reported in *Parade* magazine was that he was virtually alone in predicting a McGovern victory in 1972. Because McGovern was 1½ inches taller than Nixon, Feldman predicted that McGovern "should inch out Nixon in November." McGovern lost every state but Massachusetts (some of whose residents began displaying bumper stickers saying simply, "Don't blame us"). In 1976, Jimmy Carter beat Gerald Ford despite a 2½-inch height disadvantage. In 1980, and again in 1984, the taller man, Ronald Reagan, did win. So did George Bush in 1988, although he lost in 1992 to Bill Clinton, who had a half-inch height disadvantage. In all, using fresh data, Feldman's theory has been right an unimpressive 3 out of 6 times.

The Super Bowl and the Stock Market

On Super Bowl Sunday in January 1983, both the business and sports sections of the *Los Angeles Times* carried articles on the "Super Bowl Stock Market Predictor."[23] The theory is that the stock market goes up if the National Football Conference (NFC) or a former NFL team now in the American Football Conference (AFC) wins the Super Bowl; the market goes down if the AFC wins. This theory had been correct for 15 out of 16 Super Bowls—the one exception occurred in 1970, when Kansas City beat Minnesota and the market went up .1 percent. Thus the *Times* quoted a broker for Dean Witter Reynolds: "Market observers will be glued to their TV screens . . . it will be hard to ignore an S&P indicator with an accuracy quotient that's greater than 94%." And, indeed, it was. An NFC team won, the market went up, and the Super Bowl system was back in the newspapers next year, stronger than ever. Using fresh data, from its discovery through 1996, the Super Bowl system has been right an impressive 12 out of 14 times.

The accuracy of the Super Bowl system is obviously just an amusing coincidence, since the stock market has nothing to do with the outcome of a football game. The data reflect the fact that, over these 30 years, the stock market has generally gone up and the NFC has usually won the Super Bowl. The correlation is made more impressive by the gimmick of including the Pittsburgh Steelers, an AFC team, with the NFL. The excuse is that Pittsburgh once was in the NFL; the real reason is that Pittsburgh won the Super Bowl 4 times in years when the market went up.

Inspired by the success of the Super Bowl indicator, a *Los Angeles Times* staff writer did a little data mining in 1989 and discovered several comparable coincidences.[24] The "Yo, Adrian Theory" holds that if a Rocky or Rambo movie is released, the stock market will go up that year. On average, the Dow increased by 173 points in the 4 years when a Rocky movie appeared and 245 points in the years when a Rambo movie was released. The Geraldo Rivera indicator notes that the Dow has fallen an average of 13 points on the day after seven major Geraldo Rivera specials. The George Steinbrenner indicator monitors his firing of Yankee managers. On the 5 occasions that he fired Billy Martin, the Dow rose an average of 5 points the following day; the 8 times Steinbrenner fired someone else, the Dow fell an average of 3 points the next day.

Extrasensory Perception

Professor J. B. Rhine and his associates at Duke University produced an enormous amount of evidence concerning extrasensory perception (ESP), including several million observations involving persons attempting to identify cards viewed by another person. For a stylized illustration, consider an experiment with 5 cards, each with a different picture on the face. The five cards are shuffled, and a card is selected at random. The subject attempts to identify the selected card, and the sender records the results. The experiment is repeated 25 times.

Under the null hypothesis that the subject has no special ability to identify the cards, the probability of a correct identification is $1/5 = .2$. The binomial distribution with $p = .2$ and $n = 24$ shows that there is a .047 probability of 9 or more successes in 25 observations. Thus 9 or more correct identifications would show a statistically significant ability at the 5 percent level.

A messy complication is that many people are tested. Even if only unremarkable guessers are tested, about 1 out of every 20 will get 9 or more correct. If 400 guessers are tested, we can anticipate that approximately 20 will do sufficiently well to suggest they have ESP. If millions of tests are conducted, some astounding performances should be found, even if nothing more than lucky guesses are involved. This is a form of data mining—produce a large number of samples and then report only the most remarkable results. Something with a 1-in-1000 chance of occurring is not really so remarkable if we take into account the fact that this is just 1 out of 1000 tests that were conducted.

Our first antidote to data mining is common sense, but here common sense is divided. Many people do not believe in ESP, but others do. Perhaps the safest thing to say is that if ESP does exist, the evidence is not strong enough to convince the skeptics. And even if there is statistically significant ESP, it does not seem to be of much practical importance. There is certainly no public evidence that people have such strong mental powers that they can knock airplanes out of the sky, have long-distance conversations without a telephone, or even win consistently in Las Vegas.

The second antidote to data mining is fresh data. These people who score well should be retested to see whether their success was due to ESP or data mining. Of course, with many, many subjects, a few will doubtless compile high scores through several rounds. When Rhine retested his high scorers, he reported that they almost always showed a marked decline in ability after their initial success. He wrote, "This fatigue is understandable . . . in view of the loss of original curiosity and initial enthusiasm." An alternative explanation is that the high scores on the early rounds were just lucky guesses.

The large number of subjects is not the only difficulty in assessing the statistical significance of Rhine's results. He also looked for specific test periods in which the subjects did well: at the start of the test, in the middle, or at the end. This is very similar to the statistical pitfall of testing several people and reporting only the highest scores. A remarkable performance by 1 of 20 subjects or by 1 of 5 subjects on one-fourth of a test is not really that remarkable.

Rhine also looked for either "forward displacement" or "backward displacement," in which the subjects' choices did not match the contemporaneous cards, but did match the next card, the previous card, two cards hence, or two cards previous. In this way, one sample can be multiplied into several tests, increasing the chances of finding coincidental matches. Rhine also considered it remarkable when there was "avoidance of target" or "negative ESP," in which the subject got an unusually low score. Rhine explained how a subject

> may begin his run on one side of the mean and swing to the other side just as far by the time he ends a run; or he may go below in the middle of the run and above it at both ends. The two trends of deviations may cancel each other out and the series as a whole average close to "chance."

He described the results of one test as follows:

> [T]he displacement was both forward and backward when one of the senders was looking at the card, and only in the forward direction with another individual as sender; and whether the displacement shifted to the first or second card away from the target depended on the speed of the test.[25]

With so many subjects and so many possibilities, it should be easy to find patterns, even in random guesses.

ESP research is interesting and provocative. But the understandable enthusiasm of the researchers has unfortunately led to a variety of forms of data mining that undermine their tests. They would be less controversial and more convincing—one way or the other—if there were more uses and fewer abuses of statistical tests.

The FTC's Listerine Ruling

Warner-Lambert once tested its medicinal claims for Listerine mouthwash in a 4-year study involving 3000 students at St. Barnabas Catholic School.[26] The severity of the students' cold symptoms was recorded by the school's doctor on a scale of 0 to 6 and then averaged over the 4 years of the study:

	LISTERINE	CONTROL GROUP
Overall severity	2.19	2.31
Symptom severity		
Nasal discharge	2.34	2.46
Nasal congestion	2.66	2.79
Post nasal drip	2.02	2.18
Sneezing	1.81	1.95
Sore throat	1.32	1.47
Cough	1.92	2.11

All the observed differences were statistically significant at the 5 percent level, but the Federal Trade Commission (FTC) ruled that this study did not validate Listerine's advertised medicinal benefits. The FTC argued that the experimental design was flawed because the control group used no placebo for the first 2 years of the study and, during the last 2 years, gargled with colored water that did not taste or smell like Listerine. The students knew that they were participating in a study to test Listerine and they knew whether they were gargling with Listerine, knowledge that may well have influenced what they told the school doctor about the severity of their colds.

The FTC also observed that statistical significance "does not indicate the size of the difference or how much benefit can be expected because of the difference. If a large enough sample is used, a very, very small difference can be found to be statistically significant." No doctor or patient can distinguish between a cold with a 2.19 severity from one with a 2.31 severity.

The FTC consequently ruled that Warner-Lambert could no longer advertise that its product prevented, cured, or alleviated colds and that the company had to refute the claims it made in past advertisements by spending $10 million publicizing the fact that Listerine does not prevent colds or sore throats or reduce their severity. (The $10 million figure represented Warner-Lambert's average annual spending on Listerine advertisements during the preceding 10 years.)

7.4

Beating the Lottery

Lotteries, especially the multimillion-dollar Lotto jackpots, offer people a chance to change their lives completely—and the popularity of Lotto games indicates that many are very eager to do so. The only catch is that the chances of winning a Lotto game are minuscule. This chasm between dreams and reality gives entrepreneurs an opportunity to sell gimmicks that purportedly enhance one's chances of winning. One mail-order company sells a Millionaire Maker for $19.95, using this pitch: "Lotto players face a dilemma each time they buy a ticket. What numbers to pick? Studies have shown that most Lotto *winners* don't use any sort of special system to select their numbers. Instead, they tap the power of *random selection*."[27] Their $19.95 product for tapping the power of random selection is a battery-powered sphere filled with numbered balls that "mixes the balls thoroughly" and chooses a "perfectly random set of numbers."

Gail Howard sells a very different system—a report on how to handicap lottery numbers, just like handicapping horses.[28] She has reportedly appeared on the "Good Morning America" television show, written an article in *Family Circle* magazine, and publishes a monthly report ("Lottery Buster")—"But you'd better hurry, while distribution of this controversial information is still permitted." Howard's report says, "You don't have to worry about the state lottery games being crooked or rigged. . . . The winning numbers are selected through a completely random process." Nonetheless, she offers several tips for "greatly improving your chances of winning." For instance, do not bet on six consecutive numbers or six numbers that have won before, because these are highly unlikely to win (as are any six numbers, a skeptic would point out). Her main tip is to chart the winning numbers, look for trends, and bet on hot numbers because *"About half of the winning numbers drawn have had a hit within the previous **three** games!"* Not coincidentally, she sells an encyclopedia of winning numbers. (When 6 numbers are drawn from 44, the probability that a particular number will be picked at least once in three games is about .36.)

No doubt, just as careful study can uncover worthless patterns in coin flips or dice rolls, so tedious scrutiny can discern worthless patterns in lottery numbers. The most persuasive reason for skepticism about lottery-beating systems is that anyone who had a system that really worked would become rich buying lottery tickets, rather than peddling books and battery-powered gadgets.

exercises

7.17 Willard H. Longcor tested some inexpensive and precision-made dice.[29] With the precision-made dice, he made 2 million rolls, recording whether an even or odd number appeared. A new die was used after every 20,000 rolls, to guard against imperfections from repeated tosses. The same experiment was conducted with inexpensive dice, but Longcor stopped after 1,160,000 rolls. For each type of dice, use a normal approximation to test at the 1 percent level the null hypothesis that even and odd numbers are equally likely to be rolled. Why, even if the null hypothesis is rejected, might you continue to play dice games?

	ROLLS	EVEN	ODD
Precision-made	2,000,000	1,000,900	999,100
Inexpensive	1,160,000	588,410	571,590

7.18 Federal courts have distinguished between statistical significance and practical importance. In *United States vs. Goff* (1975), two defendants appealed their mail fraud conviction, arguing that the 23-person grand jury was selected from lists of registered voters of the eastern district of Louisiana, of whom 21.06 percent were black, rather than from the voting-age population, of whom 26.33 percent were black. The district and appellate courts ruled that this difference "was not substantial." If you were a judge, how would you decide whether this difference was substantial?

7.19 On the eve of the 1987 Super Bowl, *Sports Illustrated* advised that

> there may be something to the J theory, as elucidated by San Francisco accountant Steve Carroll. Carroll expected the Redskins or 49ers to come out on top because they fit the trend of this decade: "All six 1980s Super Bowls were won by teams with starting quarterbacks whose first names began with the letter J and had three letters," says Carroll. The Raiders' Jim Plunkett was on the winning side in 1981 and '84; Joe Montana led the 49ers to titles in '82 and '83; Joe Theismann's Redskins won in '83; and Jim McMahon's Bears won last year.[30]

How would you, as a statistician, respond to this evidence in support of the J theory?

7.20 Critically evaluate this news article:

> Bridgeport, Conn.—Christina and Timothy Heald beat "incredible" odds yesterday by having their third Independence Day baby.
>
> Mrs. Heald, 31, of Milford delivered a healthy 8-pound 3-ounce boy at 11:09 a.m. in Park City Hospital where her first two children, Jennifer and Brian, were born on Independence Day in 1978 and 1980, respectively.
>
> Mrs. Heald's mother, Eva Cassidy of Trumbull, said a neighbor who is an accountant figures the odds are 1-in-484 million against one couple having three children born on the same date in three different years.[31]

7.21 Table 1 at the end of this book is supposed to contain random numbers. But look at the 22d through 24th numbers in the 25th row (20, 73, 17) and the 8th through 5th numbers in the 30th row—they are the same! The chances that 3 randomly selected 2-digit numbers will match 3 other randomly selected numbers are 1 in 1 million. Does this discovered match show that these are not random numbers after all?

7.22 A study of the relationship between socioeconomic status and juvenile delinquency tested 756 possible relationships and found 33 to be statistically significant at the 5 percent level.[32] What statistical reason is there for caution here?

7.23 In 1989, *The New York Times* reported that if the Dow Jones Industrial Average increases between the end of November and the time of the Super Bowl, the football team whose city comes second alphabetically will probably win the Super Bowl.[33] Thus the Dow increased from 2115 to 2235 between November 30, 1989, and the day of the January 1990 Super Bowl, and San Francisco beat Cincinnati. How would you explain the success of this theory? Why did the newspaper choose the end of November for its starting date?

7.24 There were three Friday-the-13ths during 1987. That year, the chief economist at a Philadelphia bank reported that in the past 40 years there had been 6 other years with three Friday-the-13ths and that a recession started in 3 of those years.[34] Explain why this evidence does not convince you that recessions are more likely in years with three Friday-the-13ths.

summary

To conduct a hypothesis test, we specify the null and alternative hypotheses and then use the sample data to estimate the population parameter whose value is specified by the null hypothesis. If this sample statistic is approximately normally distributed, we calculate the z value,

$$z = \frac{\text{observed statistic} - \text{null hypothesis parameter value}}{\text{standard deviation of statistic}}$$

and the corresponding probability of observing a z value so far (or farther) from 0. If the alternative hypothesis is two-sided, we double this probability to obtain the two-sided P value.

For testing a null hypothesis about the value of the population mean, we use the sample mean as our test statistic and calculate this z value:

$$z = \frac{\bar{x} - \mu}{\sigma/\sqrt{n}}$$

The sample mean is normally distributed if the individual observations are normally distributed or (approximately) if the sample is large enough to invoke the central limit theorem.

If we do not know the standard deviation σ, we can estimate it from the sample standard deviation s and calculate this t statistic

$$t = \frac{\bar{x} - \mu}{s/\sqrt{n}}$$

determining the P value from the t distribution with $n - 1$ degrees of freedom. This P value is exact if the data come from a normal distribution and a good approximation if we have at least 15 or 30 observations.

Researchers generally report that their results either are or are not statistically significant at a specified level, such as 5 or 1 percent, depending on whether the appropriate P value is less than this specified significance level. A null hypothesis is rejected by a two-sided hypothesis test if and only if it lies outside a corresponding confidence interval. The additional information provided by a confidence interval is a sense of the practical importance of the difference between the null hypothesis and the value of the estimator.

For tests of the binomial success probability p, we first specify the null and alternative hypotheses about the value of p and then use the sample success proportion x/n as the test statistic to estimate the value of p. The P value can be either calculated exactly by using the binomial distribution or, if the sample is large, approximated by the normal distribution using this z statistic:

$$z = \frac{x/n - p}{\sqrt{p(1-p)/n}}$$

We need to remember that statistical significance is not the same as practical importance and to be skeptical of selectively reported results uncovered by data mining—the testing of many theories, some of which may have been suggested by the data. Two antidotes to unrestrained data mining are common sense and fresh data.

review exercises

7.25 The assistant manager of the Yale Cafe at Commons suspected that the supplier of hamburger patties was delivering hamburgers that weighed less, on average, than the advertised 16 ounces. For his statistics paper, he weighed a random sample of 48 patties very carefully on a balance scale and found them to have an average weight of 15.978 ounces with a variance of .028. Using a two-tailed test, do these data reject at the 1 percent level the null hypothesis that the population mean is 16 ounces?

7.26 A 1985 editorial in *The Wall Street Journal* stated:

> We wrote recently about a proposed Environmental Protection Agency ban on daminozide, a chemical used to improve the growth and shelf life of apples. EPA's independent scientific advisory panel last week ruled against the proposed prohibition. "None of the present studies are considered suitable" to warrant banning daminozide, the panel said. It also criticized the "technical soundness of the agency's quantitative risk assessments" for the chemical. EPA can still reject the panel's advice, but we see some evidence that the agency is taking such evaluations seriously. There is no better way to do that than to keep a firm grip of the scientific method, demanding plausible evidence when useful substances are claimed to be a danger to mankind.[35]

Does the *Journal*'s interpretation of the scientific method implicitly take the null hypothesis to be that daminozide is safe or that it is dangerous?

7.27 To settle a lawsuit brought by minority and female firefighters, the San Francisco fire chief in 1988 mandated that firefighters be 40 percent minorities and 10 percent female. As of 1992, of the city's 1448 firefighters 28 percent were minorities and 3.2 percent were female.[36] If the 1448 firefighters were a random sample from a pool of eligible workers that are 40 percent minority, what is the probability that 28 percent or fewer would be minorities? What other information would you like to see before concluding, as one statistician did, "These data show that the Fire Department is not moving very aggressively to hire minorities and females."

7.28 Seymour Siwoff, Steve Hirdt, and Peter Hirdt compiled major league batting averages from 1975 to 1984 during late-inning pressure situations—seventh inning or later with the batter's team tied or trailing by three runs or fewer (four if the bases are full).[37] Here are the data for two reputed clutch hitters:

	PRESSURE SITUATIONS		
	AT-BATS	HITS	OVERALL BATTING AVERAGE
Reggie Jackson	649	174	.263
Steve Garvey	904	282	.302

For each batter, assume the binomial model to be appropriate and calculate the *P* value for a one-tailed test of the null hypothesis that his probability of getting a hit in a pressure situation is equal to his overall batting average.

7.29 A high school basketball coach said that a missed free throw by a right-handed shooter is more likely to bounce to the right, while the reverse is true of a left-hander. To investigate this theory, a student asked two college basketball players to shoot 50 free throws apiece.[38] These players missed 21 of 100 free throws, of which 15 landed on the same side as their shooting hands and 6 bounced to the opposite side. Assuming the binomial model to be applicable, calculate the two-sided *P* value for testing the null hypothesis that missed free throws are equally likely to bounce to either side. Do not use a normal approximation.

7.30 Young children who play ice hockey are separated by age. In 1991, for example, children born in 1984 were placed in the 7-year-old league, and children born in 1983 were placed in the 8-year-old league. A student with a December 11 birthday observed that children with birth dates early in the year are months older than those with later birth dates—someone born in January 1984 is 11 months older than someone born in December 1984.[39] Because coaches give more attention and playing time to better players, this student suspected that children with early birth dates have an advantage when they are young that might cumulate over the years. To test this theory, he looked at the birth dates of 1487 National Hockey League (NHL) players in 1991 and found that 934 of these players had birth dates during the first 6 months of the year. Assuming these players to be a random sample from the population of NHL players for all years, use a normal approximation to calculate the z value and two-sided P value for the null hypothesis that there is a .5 probability that an NHL player has a birth date in the first 6 months of the year. Is this null hypothesis rejected at the 5 percent level?

7.31 *The Wall Street Journal* sponsors a monthly stock-picking contest in which four stocks are picked by four investment professionals and four stocks are selected by throwing four darts at the *Journal*'s financial pages. After 6 months, the average return for the four professionally selected stocks is compared to that of the four dartboard stocks and to the Dow Jones Industrial Average, a widely followed index of stock prices. Between July 1990, when these rules were adopted, and September 1996, the pros beat the darts 38 out of 63 times and beat the Dow 34 out of 63 times.[40] Is this performance sufficient to reject at the 5 percent level the null hypothesis that, in each contest, the pros have a 50 percent chance of beating the darts? A 50 percent chance of beating the Dow? Use the exact binomial probabilities and one-tailed tests.

7.32 In *Hazelwood School District vs. the United States* (1977), the federal government argued that this district in a rural area in St. Louis County had followed a "pattern or practice" of employment discrimination, as evidenced by the fact that only 15 of 405 (3.7 percent) of the Hazelwood teachers hired during the preceding 2 years were black, while 15.4 percent of the teachers in St. Louis County, including the city of St. Louis, were black. The school district argued that the 3.7 percent figure in Hazelwood should be compared either with data showing that only 2 percent of the Hazelwood students were black, or with the fact that blacks constituted 5.7 percent of the teachers in St. Louis County, excluding the city of St. Louis (which was attempting to maintain a 50 percent black teaching staff). The Supreme Court ruled that the student body was irrelevant, but instructed the lower courts to gather more information in order to determine the appropriate labor market. Use a normal approximation to calculate the z values and two-sided P values for these two different values of p (.154 and .057) under the null hypothesis that, with regard to race, hiring is a binomial process with a success probability p.

7.33 The U.S. Postal Service claims that 94 percent of all first-class letters and postcards mailed from Los Angeles County to Contra Costa County, California, are delivered within 2 days. To test this claim, 5 postcards were mailed in 1990 from each of 10 locations on 2 separate days—100 postcards in all.[41] Of these cards, 89 were delivered within 2 days, and 11 were not. Test the U.S. Postal Service's claim, assuming that each postcard is an independent observation with a constant probability of being delivered within 2 days. Use a one-tailed test, and do not use a normal approximation. Are you persuaded to reject the U.S. Postal Service's claim?

7.34 The results of hypothesis tests are often misinterpreted in the popular press. For instance, explain how this 1985 *Business Week* summary of a Data Resources Incorporated (DRI) study of insider trading in the stock market is misleading:

> DRI also compared the action of a sample of takeover stocks in the month before the announcement with the action of those same stocks in other, less significant one-month intervals. The conclusion: There was only 1 chance in 20 that the strength of the takeover stocks was a fluke. In short, the odds are overwhelming that inside information is what made these stocks move.[42]

7.35 After a family Thanksgiving dinner, a statistics student conducted an ESP experiment.[43] He shuffled 10 cards, each labeled with a single number from 1 to 10, drew 1 card, and concentrated on it while each of the 23 family members present wrote down a guess. He reshuffled the cards and repeated the process 20 times. One family member got 7 out of 20 correct and another got 5 right. What is the exact probability that random guessing will get 5 or more correct? What is the exact probability that 2 or more people out of 23 random guessers will get 5 or more correct? When he repeated this experiment for the two persons who got more than 5 correct, one got 2 right and the other 0. How do you explain this curious reversal?

7.36 In December 1984, *Sports Illustrated* reported, "Mindy, 6, a dolphin at the Minnesota Zoo, predicts the outcomes of NFL games by choosing among pieces of Plexiglas, each bearing a different team's name, that are dropped into the pool. This season, Mindy has a 32–21 record." Is this record sufficient to reject at the 5 percent level the null hypothesis that Mindy has a .5 probability of picking a winner? Use a two-tailed test and calculate the exact P value. Explain why this report might be considered data mining.

7.37 It has been claimed that pickles cause cancer, war, communism, and traffic accidents: 99.9 percent of all cancer victims have eaten pickles during their lifetimes, as have 100 percent of all soldiers, 96.8 percent of all communists, and 99.7 percent of those involved in automobile and airplane accidents.[44] Explain why this evidence either is or is not persuasive.

7.38 A 1991 experiment at the emergency room of the University of California at San Diego Medical Center compared the accuracy of doctors and a computer program in quickly diagnosing patients who complained of chest pains and may have experienced a heart attack.[45] The computer program used the same patient information available to the doctors—electrocardiogram test results, other physical symptoms, and health history—and used a database that weighed the importance of this information. The accuracy of the diagnoses made by the doctors and the computer program was determined by the results of blood tests of enzyme levels. Because these enzyme tests take a few hours to process, an earlier diagnosis can be the basis for the application of life-saving procedures.

In this experiment, 36 of the 331 patients who complained of chest pains had actually experienced a heart attack. The computer program made the correct diagnosis for 35 of these 36 heart attack victims, while the doctors were right 28 out of 36 times. What additional data do we need before concluding that the computer program outperformed the doctors?

7.39 Camilla Benbow and Julian Stanley identified 300 U.S. children under the age of 14 who scored 700 or higher on the mathematics part of the SAT. Only about 5 percent of first-year college students score that high and about 1 in 10,000 13-year-olds. Benbow and Stanley reported the four characteristics listed below for these 300 mathematically gifted children. For each of these four characteristics, use a normal approximation to find the probability that 300 children selected at random would show the disparities reported.

a. 280 were boys, compared to 50 percent of the population as a whole.

b. 20 percent were left-handed, compared to 8 percent of the population as a whole.

c. 60 percent had allergies or asthma, compared to 10 percent of the population as a whole.

d. 70 percent were nearsighted, compared to 15 percent of the population.

7.40 An advertisement for a computer program claimed that this program had selected 14 stocks, 13 of which went up during a 4-month period:

> Each disk is valid for only four months. You pay only $199 for the program . . . and purchasing the disk entitles you to renew the disk every four months thereafter for the same $199 per disk. . . .
>
> Our guarantee of satisfaction is very compelling. If, after the four months, you are not satisfied with the program or have not seen it pay for itself many times over, please return it and get a full refund of your $199 investment. You can't lose.[46]

Is it true that you cannot lose? Why do the vendors want you to renew the disk every 4 months? How could an unscrupulous company write a very simple program for predicting the direction of the stock market, guaranteeing that one-half its customers will be satisfied?

7.41 Seventy randomly selected male college students were asked to guess their weight and then were weighed: 23 were within 1 pound of their correct weight, 35 underestimated their weight by more than 1 pound, and 12 overestimated their weight by more than 1 pound.[47] Use these data to test the null hypothesis that male college students who misestimate their weight by more than 1 pound are as likely to underestimate as to overestimate. Do not use a normal approximation. Explain why you either are or are not persuaded to reject the null hypothesis.

7.42 The following letter appeared in *Sports Illustrated*:

> In regard to the observations made by *Sports Illustrated* correspondent Ted O'Leary concerning this year's bowl games (Scorecard, January 16), I believe I can offer an even greater constant. On Jan. 2 the team that won in all five games started the game going right-to-left on the television screen. I should know, I lost a total of 10 bets to my parents because I had left-to-right in our family wagers this year.[48]

Under the null hypothesis that the winning team is equally likely to have started the game from the left or right side of the television screen, what is the probability that all five winners would start on the right side? Are you persuaded by this statistical evidence?

7.43 In the 1977 case of *Ballew vs. Georgia,* the Supreme Court opinion, written by Justice Blackmun, stated, "Statistical studies suggest that the risk of convicting an innocent person rises as the size of the jury diminishes. Because the risk of not convicting a guilty person increases with the size of the panel, an optimal jury size can be selected as a function of the interaction between the two risks." Explain his reasoning.

7.44. An article in the business section of *The New York Times* suggested that the Chinese calendar might be helpful in predicting whether stock prices will go up or down: "Trying to discern any sort of seasonality in the stock market may be a futile exercise. Perhaps the most ingenious 'analysis' was done by the Rothschild firm, which noted that February begins the Year of the Dragon in the Chinese calendar. . . . During the 20th century . . . there were twice as many up Dragon years [4] as down Dragon years [2]."[49] What other data would you need to determine whether Dragon years are especially good years to buy stocks? Why are you confident that the relationship is not statistically significant? Why, even if the relationship were statistically significant, would you be unpersuaded?

7.45 In a 1982 racial discrimination lawsuit, the court accepted the defendant's argument that racial differences in hiring and promotion should be separated into eight job categories.[50] In hiring, it turned out that blacks were underrepresented by statistically significant amounts (at the 5 percent level) in four of the eight job categories. In the other four categories, whites were underrepresented in two cases, and blacks were underrepresented in two cases, though the differences were not statistically significant at the 5 percent level. The court concluded that four of eight categories were not sufficient to establish a prima facie case of racial discrimination. Assume that the data from these eight job categories are independent random samples.

a. What is the null hypothesis?

b. If the null hypothesis is true and the court requires statistical significance at the 5 percent level in all eight job categories, what is the probability that the null hypothesis will be rejected?

c. If the null hypothesis is true and the court requires statistical significance at the 5 percent level in at least one of the eight job categories, what is the probability that the null hypothesis will be rejected?

d. Explain why data that are divided into eight job categories might not show statistical significance in any of these job categories, even though there is a statistically significant relationship when the data are aggregated.

e. Explain why data that are divided into eight job categories might show statistical significance in each of the eight categories, even though there is not a statistically significant relationship when the data are aggregated.

7.46 Under its original guidelines, the Environmental Protection Agency (EPA) would have concluded that second-hand smoke poses a cancer risk if a 95 percent confidence interval for the difference in the incidence of cancer in control and treatment groups excluded 0. In 1996 it was reported that the EPA had changed this criterion to a 90 percent confidence interval.[51] If in fact, second-hand smoke does not pose a cancer risk, would a switch from a 95 percent to a 90 percent confidence interval make it more or less likely that the EPA would conclude that second-hand smoke poses a cancer risk? Explain your reasoning.

7.47 A treatment group was given a cold vaccine, while the control group received a placebo. Doctors then recorded the fraction of each group that caught a cold and calculated the two-sided P value to be .08. Explain why you either agree or disagree with each of these interpretations of these results:

a. "There is an 8 percent probability that this cold vaccine works."

b. "If a randomly selected person takes this vaccine, the chances of getting sick fall by about 8 percent."

c. "These data do not show a statistically significant effect at the 5 percent level; therefore, we are 95 percent certain that this vaccine doesn't work."

7.48 Four medical researchers compared the actual calories in 40 health and diet foods purchased in Manhattan, New York, with the calories listed on the label.[52] The following data are the percentage differences between the actual and labeled calories (a positive value means that the measured calories exceeded the amount shown on the product label). For each of the three groups shown, determine the two-sided P value for a test of the null hypothesis that the population mean is 0.

NATIONALLY ADVERTISED		REGIONALLY DISTRIBUTED		LOCALLY PREPARED	
Noodles and alfredo	2	Meatless sandwich	41	Chinese chicken	15
Cheese curls	−28	Oatmeal cookie	46	Gyoza	60
Green beans	−6	Lemon pound cake	2	Jelly diet candy	250
Mixed fruits	8	Banana cake	25	Fruit diet candy	145
Cereal	6	Brownie	39	Florentine manicotti	6
Fig bars	−1	Butterscotch bar	16.5	Egg foo young	80
Oatmeal cookie	10	Blondie	17	Hummus with salad	95
Crumb cake	13	Oak bran snack bar	28	Baba ghanoush	3
Crackers	15	Granola bar	−3		
Blue cheese dressing	−4	Apricot bar	14		
Imperial chicken	−4	Chocolate cookie	34		
Vegetable soup	−18	Carrot muffin	42		
Cheese	10				
Chocolate pudding	5				
Sausage biscuit	3				
Lasagna	−7				
Spread cheese	3				
Lentil soup	−.5				
Pasta with shrimp	−10				
Chocolate mousse	6				

7.49 (Nonparametric) For each of the three food categories in Exercise 7.48, determine the two-sided P value for a test of the null hypothesis that the population median is 0.

7.50 (Nonparametric) Use the data in Exercise 2.2 to calculate the two-sided P value for testing the null hypothesis that the population mean for the expected number of biological children is 2, and for testing the null hypothesis that the population median is 2.

projects

For each of the following projects, type a report in ordinary English, using clear, concise, and persuasive prose. Any data that you collect for this project should be included as an appendix to your report. Data used in your report should be presented clearly and effectively.

7.1 Find someone who claims to have extrasensory perception, and test this claim.

7.2 Roughly 25 percent of the cars registered in the United States were made by foreign companies, such as Toyota, Honda, and BMW. (This tabulation includes all cars made by these companies, including those assembled in the United States.) Choose a large parking lot (or lots) at your college (at least 100 cars), and count the percentage of these cars manufactured by foreign companies. Do not include trucks or motorcycles in your tabulation. Determine a 95 percent confidence interval for the fraction of all cars at your college that were made by foreign companies. Do these data reject at the 5 percent level the null hypothesis that 25 percent of all cars at your college were made by foreign companies? Is there any reason to think that the parking lot (or lots) you examined may have yielded biased data?

7.3 Roughly 20 percent of the passenger cars registered in the United States are red (either bright or medium). Choose a large parking lot or lots at your college (at least 100 cars), and count the percentage of these cars that are red. Determine a 95 percent confidence interval for the percentage of all cars at your college that are red. Do these data reject at the 5 percent level the null hypothesis that 20 percent of all cars at your college are red?

7.4 A study by Camilla Benbow and Julian Stanley of children younger than age 14 who scored 700 or higher on the mathematics part of the SAT found that 20 percent were left-handed, compared to 8 percent of the entire population. Are 8 percent of the students at your college left-handed? What about the faculty? How could you obtain data for the faculty without questioning them directly?

7.5 Ask a random sample of either female or male students at your college to estimate their weight, and then weigh each person and calculate the difference between the actual weight and the estimate. (You can encourage participation by giving a small prize to those who are within 1 pound of their actual weight.) Use your data to test the null hypothesis that the average error in the population is 0.

7.6 Follow the instructions in Project 7.5, but ask the students in your sample to estimate their heights.

7.7 Test the null hypothesis that the average length of a person's last name is eight letters. Decide beforehand either to omit hyphenated last names, such as Aschari-Lincoln, or to treat them as two separate last names, Aschari and Lincoln.

7.8 Find an oral thermometer that can be easily read, and take the temperatures of a random sample of 50 healthy persons. Use your data to test the null hypothesis that the average body temperature of healthy humans is 98.6°. Also display your data in a histogram, and construct a 95 percent confidence interval for the population mean.

7.9 Find a morning newspaper that gives a daily forecast for your area of the high and low temperatures for that day and reports the high and low temperatures for the preceding day. By looking at back issues of this newspaper, obtain at least 100 daily observations. For both the high and low temperatures, calculate the daily prediction errors (actual temperature that day minus predicted temperature for that day). Use these data to estimate a 99 percent confidence interval for the prediction error and to test at the 1 percent level the null hypothesis that the population mean is 0.

7.10 From the most recent February issue of *Money* magazine, randomly select 30 stock mutual funds and record each fund's return that year. Now determine the two-sided *P* value for a test of the null hypothesis that these data were drawn from a population with a mean equal to the return on the S&P 500 index that year. Also use these data to determine the two-sided *P* value for a test of the null hypothesis that a randomly selected mutual fund has a .5 probability of doing better than the S&P 500 index.

7.11 Conduct a taste test of either Coke versus Pepsi or Diet Coke versus Diet Pepsi. Survey at least 50 students who identify themselves beforehand as cola drinkers with a definite preference for one of the brands you are testing. Calculate the fraction of your sample whose choice in the taste test matches the brand identified beforehand as their favorite. (Do not tell your subjects that this is a test of their ability to identify their favorite brand; tell them it is a test of which tastes better.) Determine the two-sided *P* value for a test of the null hypothesis that there is a .5 probability that a cola drinker will choose his or her favorite brand.

7.12 Ask at least 50 randomly selected college students whether they expect to receive more or less money from social security than their parents will receive. (Discard those who say they will receive the same amount as their parents.) Calculate the two-sided *P* value for a test of the null hypothesis that half of this population believe they will receive more than their parents.

7.13 Find 5 avid basketball players and ask each of them to shoot 100 free throws. Do not tell them the purpose of this exercise, which is to determine if a missed free throw is equally likely to bounce to the same or opposite side as their shooting hand, as described in Exercise 7.29. Use your data to calculate the two-sided *P* value for testing the null hypothesis that missed free throws are equally likely to bounce to either side.

7.14 Apply the test described in Exercise 7.30 to another group of professional athletes.

7.15 College students are said to experience the Frosh 15—an average weight gain of 15 pounds during their first year at college. Test this folklore by asking at least 100 randomly selected students how much weight they gained or lost during their first year at college. Determine the two-sided P value for testing the null hypothesis that the population mean is a 15-pound gain, and also determine a 95 percent confidence interval for the population mean.

HAVE YOU EVER WONDERED?

In the 1890s, two newspaper reporters, Lincoln Steffens and Jacob Riis, created the illusion of a major crime wave in New York by writing increasingly sensational stories that were intended to boost circulation. Even though there had been no increase in criminal activity, frightened readers demanded that the city leaders do something about the crime problem. Theodore Roosevelt, then president of the city's Board of Police Commissioners, ended the fictitious crime wave simply by asking Steffens and Riis to write about something else.

Reporters everywhere tend to focus on bad news—robberies, bankruptcies, and wars. If someone has a pleasant, carefree day, this is not considered a newsworthy event that will sell newspapers or attract television viewers. Evidently, people do not want to open the newspaper in the morning or turn on the television in the evening and see lots of stories about people who had nice days.

How might we move beyond this anecdotal evidence and investigate whether when there are two stories of equal importance, one good and one bad, the news media really do devote more coverage to the bad? This chapter explains how we can compare two populations by looking at samples from each.

chapter 8

comparing two samples

Nothing is good or bad but by comparison.
—Thomas Fuller

CHAPTER 7 EXPLAINED how a random sample can be used to test a null hypothesis about a specified value of the population mean μ or the success probability p. Often, instead of a specific value for μ or p, we are interested in comparing two populations: whether nonsmokers live longer than cigarette smokers; whether children who watch television frequently are more violent than children who seldom watch television; whether people who take large doses of vitamin C have fewer colds than those who take less vitamin C. Educators compare the performance of students taught in different ways; engineers compare the reliability of different designs; pollsters compare the opinions of different demographic groups.

In this chapter, we extend the tests that you have just learned to encompass comparisons of two population means and comparisons of two success probabilities.

8.1 The Difference Between Two Means

When we looked at 50 body temperature readings, we compared the sample mean to a benchmark of 98.6, the widely known figure set forth by Carl Wunderlich in the 1860s. Often, however, we have no benchmark, since we want to compare two groups and do not know the population mean for either. We may want to compare the durability of two kinds of thermometers. Or we may want to compare the effectiveness of two medications for dissolving blood clots in the arteries of heart attack victims. When we compare two sample means to each other, rather than comparing one sample mean to an assumed population mean, we must modify the details of our test, but not the logic.

 General Framework

To illustrate the general methodology, consider the question posed at the beginning of this chapter. How might we investigate whether the news media give equal coverage to good news and bad news that is of equal importance? The challenge is to find good and bad news of comparable importance. An economist had the ingenious idea of looking at television reporting of changes in the unemployment rate.[1] Every month, the federal government releases its latest estimate of the unemployment rate, and because it is an important measure of the health of the economy, this figure is invariably reported on network news programs. A decrease in the unemployment rate is just as important as an increase. But if the media are preoccupied with bad news, newscasts may spend more time on this story when the unemployment rate increases than when it decreases.

Writing in 1985, he examined data for 1973 (the earliest available) through 1984 on the amount of time that network newscasts devoted to reporting increases or decreases in the unemployment rate. As shown in Table 8.1, this happened to be a period in which the number of increases in the unemployment

Table 8.1

Television network reporting of changes in the unemployment rate

	INCREASE IN UNEMPLOYMENT	DECREASE IN UNEMPLOYMENT
Number of observations	171	170
Average news time, seconds	161.8	123.6
Standard deviation	110.8	103.9

rate was almost exactly equal to the number of decreases. The average amount of time devoted to bad news (increasing unemployment) was indeed larger than the average time devoted to good news (declining unemployment), 161.8 versus 123.6 seconds. The statistical question is whether this observed difference in the sample means can be explained by the large standard deviations and limited number of observations. If not, then these data provide persuasive evidence of the tendency of network news programs to spend different amounts of time on good and bad economic news.

To answer this question statistically, we need to assume that the amount of time that a network news program spends reporting an increase in the unemployment rate can be described by a probability distribution with a population mean μ_1 and standard deviation σ_1, from which these 171 observations are a random sample, and that the amount of time spent reporting a decrease in the unemployment rate can be described by a probability distribution with a population mean μ_2 and standard deviation σ_2, from which these 170 observations are a random sample. (The sampling assumption might be justified by arguing that these particular newscasts from this particular period are a sample from a much larger population of newscasts.)

Our null hypothesis is that the two population means are equal, so that any observed difference in the sample means is due to sampling error:

$$H_0 : \mu_1 = \mu_2$$
$$H_1 : \mu_1 \neq \mu_2$$

The alternative hypothesis is two-sided because, before looking at the data, we cannot rule out the possibility that network news programs spend more time on good news than on bad, or vice versa.

After specifying the null and alternative hypotheses, we need to choose an appropriate estimator to use as a test statistic. It is natural to use the difference between the two sample means to estimate the size of the difference between the population means. If the underlying populations are normal (or if both samples are sufficiently large), the difference in the sample means is normally distributed with a population mean equal to the difference in the population means and a standard deviation equal to the square root of the *sum* of the population variances:

$$\text{Mean of } \bar{x}_1 - \bar{x}_2 = \mu_1 - \mu_2$$
$$\text{Standard deviation of } \bar{x}_1 - \bar{x}_2 = \sqrt{\frac{\sigma_1^2}{n_1} + \frac{\sigma_2^2}{n_2}}$$

We can summarize these relations in this shorthand notation:

$$\bar{x}_1 - \bar{x}_2 \text{ is } N\left[\mu_1 - \mu_2, \sqrt{\frac{\sigma_1^2}{n_1} + \frac{\sigma_2^2}{n_2}}\right] \tag{8.1}$$

Chapter 7 showed how to convert a normally distributed test statistic to a standardized z statistic:

$$z = \frac{\text{observed statistic} - \text{null hypothesis parameter value}}{\text{standard deviation of statistic}}$$

Here, the observed statistic is the difference in the two sample means; the population value under the null hypothesis is the specified value of $\mu_1 - \mu_2$ (which is 0); and the standard deviation of the statistic is shown in Eq. (8.1). Thus, our z statistic is

$$\begin{aligned} z &= \frac{(\bar{x}_1 - \bar{x}_2) - 0}{\text{standard deviation of } (\bar{x}_1 - \bar{x}_2)} \\ &= \frac{\bar{x}_1 - \bar{x}_2}{\sqrt{\sigma_1^2/n_1 + \sigma_2^2/n_2}} \end{aligned} \tag{8.2}$$

We can also determine a confidence interval for the difference in the population means by using the probability distribution of the difference in the sample means given in Eq. (8.1). First, we choose a confidence level, such as 95 percent, and we consult the normal distribution in Table 3 to determine the z value z^* that corresponds to this probability; for example, $z^* = 1.96$ for a 95 percent confidence interval. The corresponding confidence interval is equal to our estimate, plus or minus the requisite number of standard deviations of this estimator:

$$\text{Confidence interval} = \text{estimate} \pm z^*(\text{standard deviation of estimator})$$

Thus a confidence interval for the difference in the population means is

$$\begin{aligned} \text{Confidence interval for } (\mu_1 - \mu_2) &= \bar{x}_1 - \bar{x}_2 \pm z^*[\text{standard deviation of } (\bar{x}_1 - \bar{x}_2)] \\ &= \bar{x}_1 - \bar{x}_2 \pm z^* \sqrt{\frac{\sigma_1^2}{n_1} + \frac{\sigma_2^2}{n_2}} \end{aligned} \tag{8.3}$$

The z statistic and confidence interval in Eqs. (8.2 and (8.3) require values for the population standard deviations σ_1 and σ_2. We seldom know these values, but the next two sections explain how the sample standard deviations can be used in their place.

Unequal Standard Deviations

Our first option is simply to use our two sample standard deviations as estimates of the two population standard deviations and then use a t distribution in place of the normal distribution to determine statistical significance and to estimate a confidence interval. The second option, discussed below, is to assume that both samples are from normal distributions with the *same* standard deviation σ.

If we use the two separate estimated standard deviations from each sample s_1 and s_2, the z statistic in Eq. (8.2) is replaced by this t statistic:

If we do not assume that the population standard deviations are equal, then the null hypothesis $\mu_1 = \mu_2$ can be tested with this t statistic:

$$t = \frac{(\bar{x}_1 - \bar{x}_2) - 0}{\text{standard error of } (\bar{x}_1 - \bar{x}_2)}$$

$$= \frac{\bar{x}_1 - \bar{x}_2}{\sqrt{s_1^2/n_1 + s_2^2/n_2}} \tag{8.4}$$

Although this statistic does not have an exact t distribution, a t distribution generally gives an accurate approximation, using either of two methods for calculating the degrees of freedom. The recommended approach is to use a value (usually not a whole number) estimated from the data. The estimated value will be at least as large as the smaller of $n_1 - 1$ and $n_2 - 1$ and will never be larger than $n_1 + n_2 - 2$. The calculation of the estimated degrees of freedom is quite complicated and should be done by statistical software. If you do not have access to such software, a simpler, and more conservative, approach is to use the smaller of $n_1 - 1$ and $n_2 - 1$.

Because of the power of the central limit theorem, the t distribution is an excellent approximation if each sample has as least 5 observations from distributions that are generally similar and not extremely asymmetric, or at least 20 observations from dissimilar or very asymmetric distributions.[2]

For a confidence interval, we use the estimated standard deviations and replace the z value z^* in Eq. (8.3) with the appropriate t value t^*, using statistical software that will estimate the degrees of freedom, or, in the absence of such software, letting the degrees of freedom equal the smaller of $n_1 - 1$ and $n_2 - 1$.

If we do not assume that the population standard deviations are equal, then a confidence interval for $\mu_1 - \mu_2$ is

$$\text{Confidence interval for } (\mu_1 - \mu_2) = \bar{x}_1 - \bar{x}_2 \pm t^*[\text{standard error of } (\bar{x}_1 - \bar{x}_2)]$$

$$= \bar{x}_1 - \bar{x}_2 \pm t^* \sqrt{\frac{s_1^2}{n_1} + \frac{s_2^2}{n_2}} \tag{8.5}$$

We first use the study of the television reporting of changes in the unemployment rate to illustrate these calculations. By substituting the appropriate sample data from Table 8.1 into Eq. (8.4), the value of the t statistic is

$$t = \frac{161.8 - 123.6}{\sqrt{110.8^2/171 + 103.9^2/170}}$$
$$= 3.284$$

Statistical software shows the estimated degrees of freedom to be 337.9 and the corresponding two-sided P value to be .001. (If we did not have statistical software and used $170 - 1 = 169$ as a conservative estimate of the degrees of freedom, Table 4 at the end of this book shows that the cutoff for a two-sided t test at the 1 percent level with 169 degrees of freedom is approximately 2.6; the observed t value of 3.284 is well beyond this cutoff and consequently rejects the null hypothesis at the 1 percent level.)

If these data are random samples from populations with the same mean, there is only about a 1-in-1000 chance of observing such a large disparity in the average time devoted to good and bad news about the unemployment rate. This researcher concluded that there is a statistically significant difference in the television reporting of increases and decreases in the unemployment rate, a difference in the direction that he had suspected beforehand—that bad news is emphasized and good news is slighted. And the difference is substantial. Thirty-eight seconds is a lot of time on a network news program, and it represents 30 percent more time spent on bad news than on good.

For a 95 percent confidence interval with 337.9 (or the more conservative 169) degrees of freedom, the appropriate t value is approximately 1.97, and our confidence interval is

$$\text{95\% confidence interval for } (\mu_1 - \mu_2) = \bar{x}_1 - \bar{x}_2 \pm t^*[\text{standard error of } (\bar{x}_1 - \bar{x}_2)]$$
$$= \bar{x}_1 - \bar{x}_2 \pm t^* \sqrt{\frac{s_1^2}{n_1} + \frac{s_2^2}{n_2}}$$
$$= 161.8 - 123.6 \pm 1.97 \sqrt{\frac{110.8^2}{171} + \frac{103.9^2}{170}}$$
$$= 38.2 \pm 22.9$$

Notice that, consistent with rejection of the null hypothesis $\mu_1 = \mu_2$ at the 5 percent level, 0 is not inside this 95 percent confidence interval.

For an example with much smaller samples, consider an experimental test of a possible cure for cancer, a disease that kills half a million people in the United States each year. In this laboratory test, cancerous tumors were implanted in nine mice, three of which were given a promising anticancer drug.[3] Each tumor was later removed and weighed to see if the drug had slowed its growth. Here are the tumor weights (in grams):

	CONTROL	TREATMENT
	1.29	.96
	1.60	1.59
	2.27	1.14
	1.31	
	1.88	
	2.21	
Mean	1.76	1.23
Standard deviation	.4303	.3245

Using Eq. (8.4), we find the t value for testing the null hypothesis that these sample data are from populations with the same means:

$$t = \frac{1.76 - 1.23}{\sqrt{.4303^2/6 + .3245^2/3}}$$
$$= 2.064$$

Statistical software estimates the degrees of freedom as 5.4 and the two-sided P value as .090. Thus these samples are too small and the standard deviations are too large to reject decisively the possibility that the observed difference in mean tumor weights may be due to sampling error.

On the other hand, if we think of P values as a continuum, .090 is not much larger than .05, and while the drug did not stop the growth of these cancerous tumors, it did slow their growth substantially. The samples were very small, but the experiment had promising results in the neighborhood of statistical significance at the 5 percent level. Additional tests can provide supplementary data that will either replicate and reinforce these results or show them to be a fluke.

Equal Standard Deviations

In contrast to these approximate probability calculations, an exact t distribution can be used if we are willing to assume that both random samples are from normal distributions with the *same* standard deviation σ. In this case we use the **pooled variance** s_P^2 as an estimator of the two populations' common variance σ^2:

$$s_P^2 = \frac{(n_1 - 1)s_1^2 + (n_2 - 1)s_2^2}{n_1 + n_2 - 2} \tag{8.6}$$

The pooled variance lies between the two sample variances, depending on the relative size of the samples. If the samples are the same size, the pooled variance is exactly halfway between the two sample variances. Otherwise, it is closer to the variance of the larger sample.

In our t statistic, we replace the separate estimates of the two sample standard deviations with a common, pooled estimate:

If we assume the population standard deviations are equal, the null hypothesis $\mu_1 = \mu_2$ can be tested with this t statistic, with $n_1 + n_2 - 2$ degrees of freedom:

$$t = \frac{\bar{x}_1 - \bar{x}_2}{\sqrt{s_P^2/n_1 + s_P^2/n_2}} \tag{8.7}$$

Equivalently, we can compute the t value this way:

$$t = \frac{\bar{x}_1 - \bar{x}_2}{s_P\sqrt{1/n_1 + 1/n_2}} \tag{8.8}$$

For a confidence interval, we choose a confidence level, such as 95 or 99 percent, and consult the appropriate t distribution in Table 4 to determine the t value t^* that corresponds to this probability. The confidence interval for the difference in the population means is equal to the difference in the sample means, plus or minus the requisite number of standard deviations:

If we assume that the population standard deviations are equal, then a confidence interval for $\mu_1 - \mu_2$ is

$$\begin{aligned}\text{Confidence interval for } \mu_1 - \mu_2 &= \bar{x}_1 - \bar{x}_2 \pm t^*\sqrt{\frac{s_P^2}{n_1} + \frac{s_P^2}{n_2}} \\ &= \bar{x}_1 - \bar{x}_2 \pm t^* s_P\sqrt{\frac{1}{n_1} + \frac{1}{n_2}}\end{aligned} \tag{8.9}$$

Returning to our network news example, we see that the pooled variance is

$$\begin{aligned}s_P^2 &= \frac{(171-1)(110.8^2) + (170-1)(103.9^2)}{171 + 170 - 2} \\ &= 11{,}538.11\end{aligned}$$

Because of the nearly equal sample sizes, the t value is equal to the value obtained earlier:

$$\begin{aligned}t &= \frac{161.8 - 123.6}{\sqrt{11{,}538.11/171 + 11{,}538.11/170}} \\ &= \frac{38.2}{11.63} \\ &= 3.284\end{aligned}$$

With 171 + 170 − 2 = 339 degrees of freedom, statistical software shows the two-sided P value to be .001, as before.

For the laboratory test of the potential anticancer drug, the pooled sample variance is

$$s_P^2 = \frac{(6-1)(.43^2) + (3-1)(.32^2)}{6+3-2}$$
$$= .162$$

and the t value is

$$t = \frac{1.76 - 1.23}{\sqrt{.162/6 + .162/3}}$$
$$= 1.860$$

This t value is less than the 2.365 required for statistical significance at the 5 percent level using a two-tailed test with 6 + 3 − 2 = 7 degrees of freedom. (The two-sided P value is .105.)

Table 8.2 summarizes the two possible test statistics and the values for our two, very different examples. These calculations illustrate the general rule that the t statistics in Eqs. (8.4) and (8.7) are equal when the samples are the same size and are very close if the samples are approximately the same size. If one sample is much larger than the other, we may need to decide whether we are willing to assume that the population standard deviations are equal. Fortunately, in practice, the P values often turn out to be not very sensitive to this decision.

Table 8.3 shows the Minitab computer output for the unemployment news example. The data were entered in columns C1 and C2 of Minitab's data

Table 8.2

Testing the difference between two means

		UNEMPLOYMENT NEWS ($n_1 = 171$, $n_2 = 170$)	ANTICANCER DRUG ($n_1 = 6$, $n_2 = 3$)
$\sigma_1 \neq \sigma_2$:	$t = \dfrac{\bar{x}_1 - \bar{x}_2}{\sqrt{s_1^2/n_1 + s_2^2/n_2}}$	$t = 3.284$ df = 337.9 $P = .001$	$t = 2.064$ df = 5.4 $P = .090$
$\sigma_1 = \sigma_2$:	$t = \dfrac{\bar{x}_1 - \bar{x}_2}{\sqrt{s_P^2/n_1 + s_P^2/n_2}}$	$t = 3.284$ df = 339 $P = .001$	$t = 1.860$ df = 7 $P = .105$

df: degrees of freedom
P: two-sided P value

Table 8.3

Minitab hypothesis test of the unemployment news

```
MTB> TWOSAMPLE T 95% CONFIDENCE FOR DATA IN C1 AND C2

TWOSAMPLE T FOR C1 VS C2

          N      MEAN     STDEV    SE MEAN

C1      171     161.8     110.8      8.473

C2      170     123.6     103.9      7.969

95 PCT CI FOR MU C1 - MU C2: (15.286, 61.114)

TTest MU C1 = MU C2: T = 3.28 P = 0.0011 DF = 338


MTB> TWOSAMPLE T 95% CONFIDENCE FOR DATA IN C1 AND C2;
SUBC> POOLED.

TWOSAMPLE T FOR C1 VS C2

          N      MEAN     STDEV    SE MEAN

C1      171     161.8     110.8      8.473

C2      170     123.6     103.9      7.969

95 PCT CI FOR MU C1 - MU C2: (15.281, 61.119)

TTest MU C1 = MU C2: T = 3.28 P = 0.0011 DF = 339

POOLED STDEV = 107.416
```

worksheet. The first command shown, `TWOSAMPLE T 95% CONFIDENCE FOR DATA IN C1 AND C2`, asks the program to determine the *t* value and a 95 percent confidence interval, using the data in columns C1 and C2. Unless *pooled* is specified, the program allows the population standard deviations to be different. The Minitab output in the second half of Table 8.3 does just this.

Table 8.4 shows what the computer output from the SAS software package looks like, this time using the anticancer drug data. I saved the 9 observations as the variable TUMORS, with one group labeled TREATMENT and the other CONTROL. After the variable name, the first three lines of output give the sample sizes, sample means, sample standard deviations, and standard errors of the sample means. The next three lines give the results of two *t* tests of the null hypothesis that the population means are equal—one test that allows the population variances to be unequal and one that assumes they are equal. Shown are the *t* values, degrees of freedom, and two-sided *P* values.

Table 8.4

SAS two-sample hypothesis test of experimental anti-cancer drug

TTEST PROCEDURE

Variable: TUMORS

GROUP	N	Mean	Std Dev	Std Error
TREATMENT	6	1.76000000	0.43034870	0.17569812
CONTROL	3	1.23000000	0.32449961	0.18734994

Variances	T	DF	Prob> |T|
Unequal	2.0635	5.4	0.0898
Equal	1.8601	7.0	0.1052

HOW TO DO IT

Hypothesis Tests for the Difference between Two Population Means Using Estimated Standard Deviations

We want to use two samples of $n_1 = 6$ and $n_2 = 3$ observations to test whether a drug affects the growth rate of a cancerous tumor.

1. Specify the relationship between the population means under the null and alternative hypotheses (usually whether they are equal). In our example,

$$\text{H}_0 : \mu_1 = \mu_2$$
$$\text{H}_1 : \mu_1 \neq \mu_2$$

2. Calculate the sample means, here 1.76 and 1.23.
3. Either allow the population standard deviations to be unequal, or assume that they are equal; then calculate the t value

$$t = \frac{\overline{x}_1 - \overline{x}_2}{\sqrt{s_1^2/n_1 + s_2^2/n_2}} \qquad \text{or} \qquad t = \frac{\overline{x}_1 - \overline{x}_2}{\sqrt{s_P^2/n_1 + s_P^2/n_2}}$$

where

$$s_P^2 = \frac{(n_1 - 1)s_1^2 + (n_2 - 1)s_2^2}{n_1 + n_2 - 2}$$

In the former case, the degrees of freedom can be estimated by software from the data (or by using the smaller of $n_1 - 1$ and $n_2 - 1$); in the latter case, there are $n_1 + n_2 - 2$ degrees of freedom. In our example, with $s_1 = .4303$ and $s_2 = .3245$, the t value is either

$$t = \frac{1.76 - 1.23}{\sqrt{.4303^2/6 + .3245^2/3}} = 2.064 \quad \text{or} \quad t = \frac{1.76 - 1.23}{\sqrt{.162/6 + .162/3}} = 1.860$$

4. Determine the probability of a t value so far from 0. If the alternative hypothesis is two-sided, double this probability to get the two-sided P value. In our example, with unequal population variances, there are 5.4 degrees of freedom, $P[t > 2.064] = .0449$, and the two-sided P value is $2(.0449) = .090$; with equal variances, there are $6 + 3 - 2 = 7$ degrees of freedom, $P[t > 1.860] = .0526$, and the two-sided P value is $2(.0526) = .105$.
5. Alternatively, we can use computer software, as illustrated by the Minitab output in Table 8.3 and the SAS output in Table 8.4.
6. A P value below .05 indicates that the results are statistically significant at the 5 percent level; this is not the case in our example.
7. To assess the practical importance of the results, also report a confidence interval for $(\mu_1 - \mu_2)$,

$$\bar{x}_1 - \bar{x}_2 \pm t^* \sqrt{\frac{s_1^2}{n_1} + \frac{s_2^2}{n_2}} \quad \text{or} \quad \bar{x}_1 - \bar{x}_2 \pm t^* \sqrt{\frac{s_P^2}{n_1} + \frac{s_P^2}{n_2}}$$

Matched-Pair Samples

A possible sample design for a company that wants to test consumer reaction to two colas is to select two random samples of 10 persons and have the people in the first sample taste and rate one cola, while the people in the second sample taste and rate the other cola. Perhaps the ratings turn out as shown in Table 8.5, with the first cola rated higher among those surveyed. A statistical test needs to take into account the possible sampling error when we choose two small random samples from a population with diverse cola preferences. The second cola may have received lower ratings because, by the luck of the draw, it was tasted by a disproportionate number of people who prefer the other cola or who do not like cola.

To gauge this potential sampling error, our statistical test takes into account the size of the samples and the standard deviations of ratings. Allowing for the

Table 8.5

Cola ratings for two independent samples

	FIRST SAMPLE'S RATING OF FIRST COLA	SECOND SAMPLE'S RATING OF SECOND COLA
	8	6
	2	3
	6	7
	10	7
	9	6
	2	1
	8	6
	10	8
	1	2
	6	4
Mean	6.20	5.00
Standard deviation	3.4254	2.3570

population standard deviations to be unequal and using Eq. (8.4), we see that the t value is

$$t = \frac{6.2 - 5.0}{\sqrt{3.4254^2/10 + 2.3570^2/10}} = .913$$

Statistical software shows that this statistic has 16.0 degrees of freedom and a two-sided P value of .375. (If we assume that the population variances are equal, the pooled variance is 8.64, the t value is .913, there are 18 degrees of freedom, and the two-sided P value is .373.) The standard deviations are too large and the samples too small to rule out sampling error as an explanation for the observed difference in the mean ratings of these colas.

One way to reduce the sampling error inherent in the selection of two independent random samples is to use a single random sample, having everyone taste and rate both colas. What if the data in Table 8.5 were, in fact, such a comparison sample, with each pair of observations representing two ratings by a single person? Table 8.6 shows the revised statistical calculations. Each difference in an individual's ratings of these two colas can be treated as a single observation x. If the underlying ratings are normally distributed or if the sample is large, then the sample mean of x is approximately normally distributed and we can test the null hypothesis that the population mean of x is 0 by using the single-sample t statistic given by Eq. (7.3):

$$t = \frac{\bar{x} - \mu}{s/\sqrt{n}}$$

$$= \frac{1.2 - 0}{1.6193/\sqrt{10}}$$

$$= 2.34$$

With 10 − 1 = 9 degrees of freedom, Table 4 shows that this is (barely) larger than the 2.262 required for a two-tailed test for statistical significance at the 5 percent level. (Statistical software shows the two-sided P value to be .044.)

The observed 1.2 average difference in the sample ratings is statistically significant at the 5 percent level if the data are from a single comparison sample, but not if they are from two independent samples. The use of a single sample controls for one source of sampling error—that one sample may draw a disproportionate number of cola lovers who give colas very high ratings or cola haters who give very low ratings. Here, persons 4, 5, and 8 gave relatively high ratings to both colas, while persons 2, 6, and 9 gave low ratings to both colas. As a result, the ratings in each sample had a larger standard deviation (3.43 and 2.36) than did the difference in the ratings (1.62).

In two independent samples, one sample may contain more cola lovers (or cola haters) than does the other sample, so that the relative merits of the two colas are confounded with the effects of a general like or dislike for cola. Because we cannot tell whether the higher ratings in the first sample are due to the relative merits of the cola or to a disproportionate number of cola lovers, our results are not statistically significant. By using a single sample, we control

Table 8.6

Cola ratings by a single sample

PERSON	RATING OF FIRST COLA	RATING OF SECOND COLA	DIFFERENCE
1	8	6	2
2	2	3	−1
3	6	7	−1
4	10	7	3
5	9	6	3
6	2	1	1
7	8	6	2
8	10	8	2
9	1	2	−1
10	6	4	2
		Mean	1.20
		Standard deviation	1.6193

for this confounding effect. If we select a person who gives relatively high ratings, this person will be in both samples, and we can focus our attention on which cola this person prefers, without being concerned with the level of the person's ratings.

A single, comparison sample can be used not only in taste tests, but also in any situation in which a direct comparison is appropriate, for instance, the performance of 10 machines doing two different tasks, the speed of 10 runners on two different track surfaces, or the accuracy of 10 forecasters predicting two different events. Unfortunately, a comparison test using a single sample is often unfair or impractical. In a taste test, the first item may leave a lingering taste that distorts the rating of the second item. In a test measuring the useful life of a television under different conditions, the same television cannot be tested twice. In medical experiments, we cannot compare two drugs by giving both to the same patient. We cannot compare two textbooks or two teaching styles by making students take the same course twice. In comparing the opinions of females and males, whites and blacks, Republicans and Democrats, two samples are required by the very nature of the poll.

If a comparison sample is inappropriate, the product tester, medical researcher, or pollster can attempt to assemble two samples that contain **matched pairs** that are virtually identical with respect to influences that might confound the results of two independent samples. If truly identical twins can be found, it is as if a single person were tested twice, and the difference in each matched-pair result can consequently be treated as a single observation.

To illustrate the procedure, consider a test to see if a certain cholesterol-reducing drug affects the risk of heart attack. Two independent samples might be inconclusive because the effects of the cholesterol-reducing drug are confounded with other factors, including age, gender, smoking habits, and blood pressure. In small samples, there is a nonnegligible chance, for example, that the drug will be given mostly to nonsmokers, while the control group consists mostly of smokers. A lower incidence of heart problems might be attributed to the drug when it is at least partly due to the differences in smoking habits. Statistically persuasive results will consequently require large samples to reduce the probability of the results being tainted by this kind of sampling error—it is unlikely that a large, randomly selected sample will consist mostly of smokers, while another consists mostly of nonsmokers. A heart attack study needs large samples to begin with, because relatively few people suffer heart attacks. If we add to this burden the need to guard against confounding effects, our samples will have to be enormous to be statistically persuasive.

A single-sample comparison would protect against the confounding effects, but we cannot put the same person in both the treatment group and the control group. The next best thing is matched pairs—giving the drug to one person while using a very similar person as a control. This is, in fact, just what researchers at the National Heart, Lung, and Blood Institute did.[4] One sample of 3806 middle-aged males who had high cholesterol levels but no history of heart disease was divided into two samples, with each person in the treatment group matched with someone in the control group according to age, cholesterol level, smoking habits, blood pressure, and other factors that are believed to affect the chances of having a heart attack. Each matched pair served as a virtual twin, as if a single person were simultaneously in the treatment and control groups.

The patients were selected by doctors at 12 medical centers throughout the United States. Those in the treatment group were given a cholesterol-reducing drug over a 10-year period, and those in the control group were given a placebo. Neither the patients nor their doctors knew who was in the control group and who was in the treatment group. At the end of this 10-year period, the researchers found that those in the treatment group had 19 percent fewer heart attacks, 20 percent fewer chest pain attacks, and 21 percent fewer coronary bypass operations.

The most important advantage of matched-pair sampling is that it allows for a smaller sample, because controlling for the influence of confounding factors means that the researcher does not need to rely solely on sample size to reduce sampling error. In addition, as a practical matter, it defuses criticism that the samples may have biased the results. Critics cannot raise doubts by speculating that the medication might have been given mostly to people who do not smoke or who have low blood pressure if, in fact, the people taking the medication were similar to those in the control group.

As in this cholesterol study, the accurate identification of matching pairs is a time-consuming process. It requires considerable knowledge of the important factors that need to be controlled and careful attention to the matching process. Because the matching is subjective, it is subject to error. Some factors may be held constant at unrepresentative levels, for example, testing a cholesterol-reducing drug on young women who are not at risk. Other factors, such as smoking habits, might be inadvertently neglected, so that the matched pairs are not really virtual twins. However, when feasible and done carefully, matched-pair samples offer an attractive sample design.

A Victim of Incorrect Statistical Advice

In 1987, a small data processing firm tried to persuade a large grocery chain that it could reduce the chain's cost of handling workers' compensation invoices. This small firm and the firm that handled the grocery chain's claims both based their charges on complex formulas, depending on the number of lines in the doctors' bills and other factors. The only way to tell which firm was less expensive was to compare their charges for actual invoices. As a test, the established firm processed a random sample of 3800 bills and charged an average of \$9.26 per bill, with a standard deviation of \$9.70; the new firm processed a random sample of 3500 bills and charged an average of \$7.35 per bill, with a standard deviation of \$5.26. A consultant advised the grocery chain's management to stick with its current data processing firm, because the large standard deviations meant that the observed difference could easily be explained by sampling error. Do you agree?

The consultant evidently compared the observed \$1.91 difference in the sample means to the \$9.70 and \$5.26 sample standard deviations, forgetting that

HOW TO DO IT

Hypothesis Tests for a Matched-Pair Sample

Suppose that we want to use $n = 25$ matched pairs to test whether a certain drug affects cholesterol levels. The random variable x is equal to the difference between the cholesterol levels of one person in the treatment group and the matched person in the control group. A negative value for x indicates that the treatment person had a lower cholesterol level than did the control person.

1. Specify the null and alternative hypotheses; usually, as here,

$$H_0 : \mu = 0$$
$$H_1 : \mu \neq 0$$

2. Calculate the sample mean and standard deviation for x; here, the 25 observations yield these results:

$$\bar{x} = -32.7 \qquad s = 24.9$$

3. Calculate the t value, as in our example:

$$t = \frac{\bar{x} - \mu}{s / \sqrt{n}} = \frac{-32.7 - 0}{24.9 / \sqrt{25}} = -6.57$$

4. Determine the probability of observing a t value so far from 0; and if the alternative hypothesis is two-sided, double this probability to obtain the two-sided P value. Here, with $25 - 1 = 24$ degrees of freedom, $P[t < -6.57] = .0000004$, implying a two-sided P value of $2(.0000004) = .0000008$.
5. A P value below .05 indicates that the results are statistically significant at the 5 percent level. The cholesterol data decisively reject the null hypothesis.
6. To assess the practical importance of the results, also report a confidence interval, using the procedures detailed in Chap. 7. For our example, a

$$\begin{aligned} 95\% \text{ confidence interval for } \mu &= \bar{x} \pm t^* \left(\frac{s}{\sqrt{n}}\right) \\ &= 32.7 \pm 2.064 \left(\frac{24.9}{\sqrt{25}}\right) \\ &= 32.7 \pm 10.3 \end{aligned}$$

the standard deviation of a sample mean is equal to the sample standard deviation divided by the square root of the sample size. Equation (8.4) shows that, because these are very large samples, the standard error of the difference in the sample means is a mere \$0.18:

$$\begin{aligned} \text{Standard error of } (\bar{x}_1 - \bar{x}_2) &= \sqrt{\frac{s_1^2}{n_1} + \frac{s_2^2}{n_2}} \\ &= \sqrt{9.70^2/3800 + 5.26^2/3500} \\ &= .18 \end{aligned}$$

The \$1.91 difference in the sample means is more than 10 standard deviations from 0 and provides overwhelming evidence that the new firm is, on average, less expensive.

More formally, the null hypothesis is that the handling costs come from populations with the same mean: $H_0 : \mu_1 = \mu_2$. Using Eq. (8.4) we see that the t value for testing the null hypothesis is consequently

$$t = \frac{\bar{x}_1 - \bar{x}_2}{\sqrt{s_1^2/n_1 + s_2^2/n_2}}$$

$$= \frac{9.26 - 7.35}{\sqrt{9.70^2/3800 + 5.26^2/3500}}$$

$$= \frac{1.91}{.18}$$

$$= 10.6$$

The P value is minuscule, providing decisive evidence against the null hypothesis that the population means are equal. This $1.91 difference is also substantial, representing more than a 20 percent saving for the grocery chain, a potential saving that was lost because the chain's management was given incorrect statistical advice.

Testing Automobile Emissions Using New Fuels

The Environmental Protection Agency (EPA) tests new automobile fuels and fuel additives to see whether they have a substantial adverse impact on automotive emissions. In a 1981 test of a new fuel called Petrocal, the emission levels of several noxious gases were recorded as 16 automobiles were driven with a standard approved fuel and then driven with Petrocal. The nitrogen oxide data were as follows:

PETROCAL	STANDARD	DIFFERENCE	PETROCAL	STANDARD	DIFFERENCE
1.385	1.195	.190	.875	.687	.188
1.230	1.185	.045	.541	.498	.043
.755	.755	.000	2.186	1.843	.343
.775	.715	.060	.809	.838	−.029
2.024	1.805	.219	.900	.720	.180
1.792	1.807	−.015	.600	.580	.020
2.387	2.207	.180	.720	.630	.090
.532	.301	.231	1.040	1.440	−.400

These are matched-pair samples since each car used both fuels. The 16 observed differences have a mean of .0841 with a standard deviation of .1672. For testing the null hypothesis that there is no difference in the mean emissions in the population, the t value is

$$t = \frac{.0841 - 0}{.1672/\sqrt{16}}$$

$$= 2.01$$

With 16 − 1 = 15 degrees of freedom, the probability of such a large t value is .030, which is statistically significant at the 5 percent level using a one-tailed test, but not with a two-tailed test. Relative to the 1.0754 average level of emissions with the standard fuel, the observed .0841 difference represents about an 8 percent increase in nitrogen oxide emissions. Nonetheless, the EPA concluded that the environmental impact was small, and because the fuel had "narrowly missed" passing other standards, it approved Petrocal. A lawsuit by the Motor Vehicles Manufactures Association challenged this conclusion, arguing that the EPA was ignoring its own standards. An appellate court agreed and revised the EPA's decision.[8]

The Wilcoxon Rank-Sum Test*

Chapter 7 describes a nonparametric sign test for the median that does not require any assumptions about the shape of the sampling distribution. The Wilcoxon rank-sum test is a nonparametric test for comparing two independent samples. (Another nonparametric test, the *Mann-Whitney U test,* is calculated somewhat differently but is equivalent to the Wilcoxon rank-sum test.)

To illustrate this test, consider the fact that a grocery store has a limited amount of shelf space and thousands of products it might stock. The store manager must decide which products to carry and which to ignore. This is an important decision, because a store that carries the wrong products will lose sales and customers. It is also an important decision for product manufacturers, because a company cannot sell products that customers cannot find.

For those products that the store does carry, an additional decision involves where they are displayed. It is no accident that supermarkets put batteries, candy, and magazines by the checkout counters, where these products may remind customers of intended purchases or persuade them to buy impulsively. This visibility issue arises on every aisle of the store. Which items are put at the end of the aisle, where every passing shopper sees them, and which are put in the middle of the aisle, where they will be seen only by people who walk down that aisle? Which products are put at eye level, where they can be easily spotted, and which are put lower or higher, where it takes some effort to find them?

Two researchers made a formal study of the effect of shelf location on sales.[5] In one of their experiments, they arranged for a certain breakfast cereal to be displayed at different shelf heights in 24 stores. In 12 of the stores, the selected brand was displayed at eye level; in the other 12 stores, the brand was displayed either above or below eye level. Table 8.7 shows the rankings of these stores according to the sales of this brand of cereal. (The two stores that

Table 8.7

Ranking of cereal sales in 24 stores

RANK	SALES	SHELF
1	154	Eye level
2	150	Eye level
3	133	
4	130	Eye level
5	126	
6	123	Eye level
7	121	
8	112	Eye level
9	111	Eye level
10	109	
11	96	
12	93	
13	84	Eye level
14.5	71	Eye level
14.5	71	
16	67	Eye level
17	62	Eye level
18	58	
19	51	Eye level
20	49	
21	38	Eye level
22	37	
23	36	
24	27	

tied for ranks 14 and 15 are both given the average of these two ranks, 14.5.) To avoid cluttering up this table, only the eye-level shelves are labeled; the unlabeled shelves are those not at eye level.

The two stores with the highest sales both had this cereal displayed at eye level; however, two other eye-level stores ranked 19th and 21st in sales. Overall, did the eye-level stores have a higher or lower ranking than the other stores? The **Wilcoxon rank-sum test statistic** is equal to the sum of the ranks of whichever sample we arbitrarily choose to denote as sample 1. Here, the sum of the ranks for the eye-level stores turns out to be 130.5, and that for the other stores is 169.5:

$$\text{Eye level:} \quad \text{sum of ranks} = 1 + 2 + \ldots + 19 + 21 = 130.5$$

$$\text{Other shelves:} \quad \text{sum of ranks} = 3 + 5 + \ldots + 23 + 24 = 169.5$$

A comparison of the average ranks will tell us which sample had, on average, better rankings. Here,

$$\text{Eye level:} \quad \text{average rank} = \frac{\text{sum of ranks}}{\text{sample size}} = \frac{130.5}{12} = 10.875$$

$$\text{Other shelves:} \quad \text{average rank} = \frac{\text{sum of ranks}}{\text{sample size}} = \frac{169.5}{12} = 14.125$$

The eye-level stores tended to have lower rankings, signifying greater sales.

The statistical question is whether the observed difference in the sum of the ranks is sufficiently improbable to discredit the null hypothesis that shelf level is unimportant. The answer requires knowledge of the distribution of the sum of the ranks under the null hypothesis. For small samples, these probabilities have been tabulated and can be found from statistical software or in statistical reference books.[6] If each sample has at least 10 observations and the null hypothesis is true, then the sum of the ranks for sample 1 is approximately normally distributed with the following population mean and standard deviation:

$$\text{Sum of ranks of sample 1 is } N\left[\frac{n_1(n_1 + n_2 + 1)}{2}, \sqrt{\frac{n_1 n_2(n_1 + n_2 + 1)}{12}}\right] \tag{8.10}$$

Chapter 7 showed how to convert a normally distributed test statistic to a standardized z statistic:

$$z = \frac{\text{observed statistic} - \text{null hypothesis parameter value}}{\text{standard deviation of statistic}}$$

From Eq. (8.10), we have the following:

> The z statistic for the Wilcoxon rank-sum test of the null hypothesis that two independent samples are drawn from identical populations is
>
> $$z = \frac{(\text{sum of ranks of sample 1}) - n_1(n_1 + n_2 + 1)/2}{\sqrt{n_1 n_2(n_1 + n_2 + 1)/12}} \tag{8.11}$$

For the cereal shelf study, we can let the eye-level data be sample 1:

$$z = \frac{(\text{sum of ranks of sample 1}) - n_1(n_1 + n_2 + 1)/2}{\sqrt{n_1 n_2(n_1 + n_2 + 1)/12}}$$

$$= \frac{130.5 - 12(12 + 12 + 1)/2}{\sqrt{12(12)(12 + 12 + 1)/12}}$$
$$= 1.126$$

Because the z value is only -1.126, these data do not provide persuasive evidence against the null hypothesis that cereal sales do not depend on shelf heights: $P[z < -1.126] = .13$, implying a two-sided P value of $2(.13) = .26$.

HOW TO DO IT

Wilcoxon Rank-Sum Test

Suppose that we want to use the data in Table 8.7 to test the null hypothesis that cereal sales are not affected by whether the cereal is displayed at eye level.

1. Rank the data (as in Table 8.7), and calculate the sum of the rankings for each sample. By comparing the average ranks, we can see which sample was, on average, more highly ranked. Here, the eye-level stores have a sum of ranks of 130.5 and an average rank of $130.5/12 = 10.875$, while the other stores have a sum of ranks of 169.5 and an average rank of $169.5/12 = 14.125$.
2. If each sample has at least 10 observations, then arbitrarily denote one of the samples as sample 1 and calculate the Wilcoxon rank-sum z statistic

$$z = \frac{(\text{sum of ranks of sample 1}) - n_1(n_1 + n_2 + 1)/2}{\sqrt{n_1 n_2(n_1 + n_2 + 1)/12}}$$

(For smaller samples, use statistical software or a table of probabilities.)
Here, we can let the eye-level data be sample 1:

$$z = \frac{130.5 - 12(12 + 12 + 1)/2}{\sqrt{12(12)(12 + 12 + 1)/12}}$$
$$= 1.126$$

3. Determine the probability of observing a z value so far from 0, and if the alternative hypothesis is two-sided, double this probability to obtain the two-sided P value. Here, $P[z < -1.126] = .13$, implying a two-sided P value of $2(.13) = .26$.
4. A P value below .05 indicates that the null hypothesis is rejected at the 5 percent level. That is not the case here.

The Wilcoxon Signed-Rank Test*

The Wilcoxon rank-sum test explained in the preceding section is appropriate for comparing two independent samples. We have seen that sampling error in comparison tests can be reduced if some of the reasons for random variation can be controlled by choosing matched pairs—in essence, twins who receive identical treatments. If our data are matched pairs, we can use a nonparametric test based on a ranking of the absolute values of the differences between the matched pairs.

The 24 grocery stores analyzed in the previous section were, in fact, not two independent random samples, but rather 12 matched pairs of stores that were chosen for the similarity in their weekly sales figures. If the stores really are twins, then we can use the Wilcoxon signed-rank test in place of the Wilcoxon rank-sum test. Table 8.8 shows the necessary calculations. The first two columns show the sales in the 12 matched pairs of stores. The third column shows the difference in sales (eye-level sales minus other sales), and the fourth column gives the absolute values of these differences. The fifth column ranks the size of these absolute values, from 1 (smallest) to 12 (largest), and the sixth column affixes a positive or negative sign to these ranks, depending on whether the difference in sales was positive or negative. (Differences of 0 are ignored completely, reducing the sample size accordingly.)

Table 8.8

Matched-pair analysis of cereal sales

SALES IN MATCHED STORES					
EYE-LEVEL SHELF	OTHER SHELF	DIFFERENCE IN SALES	ABSOLUTE VALUE OF DIFFERENCE	RANK OF ABSOLUTE VALUE	SIGNED RANK
111	71	+40	40	10	+10
150	121	+29	29	9	+9
130	133	−3	3	2	−2
154	126	+28	28	8	+8
67	93	−26	26	6	−6
112	49	+63	63	12	+12
84	109	−25	25	5	−5
123	96	+27	27	7	+7
71	27	+44	44	11	+11
62	58	+4	4	3	+3
38	36	+2	2	1	+1
51	37	+14	14	4	+4
					+52

The **Wilcoxon signed-rank test** is a nonparametric test based on the sum of the signed ranks of the absolute values of the matched-pair differences; for our example, this sum is 52, the sum of the values in the last column of Table 8.8. For the null hypothesis that the two population are equal, we can anticipate roughly one-half of the signed ranks to be positive and one-half negative, with the magnitudes roughly equal, too; thus Wilcoxon's sum of the signed ranks should be close to 0. A large value (either positive or negative) casts doubt on the null hypothesis. For small samples, the exact P value can be found from statistical software or in statistical reference books.[7] If the number of matched pairs n is at least 10 and the null hypothesis is true, then Wilcoxon's sum of the signed ranks is approximately normally distributed with the following population mean and standard deviation:

$$\text{Sum of signed ranks is } N\left[0, \sqrt{\frac{n(n+1)(2n+1)}{6}}\right] \tag{8.12}$$

Again, we convert a normally distributed test statistic to a standardized z statistic:

$$z = \frac{\text{observed statistic} - \text{null hypothesis parameter value}}{\text{standard deviation of statistic}}$$

Using Eq. (8.12), we have the following:

> The z statistic for the Wilcoxon signed-rank test of the null hypothesis that two matched-pair samples are drawn from identical populations is
>
> $$z = \frac{(\text{sum of signed ranks}) - 0}{\sqrt{n(n+1)(2n+1)/6}} \tag{8.13}$$

For the cereal-shelf data in Table 8.8,

$$\begin{aligned} z &= \frac{(\text{sum of signed ranks}) - 0}{\sqrt{n(n+1)(2n+1)/6}} \\ &= \frac{52 - 0}{\sqrt{12(13)(25)/6}} \\ &= 2.04 \end{aligned}$$

Now, $P[z > 2.04] = .0207$, and the two-sided P value is $2(.0207) = 0.0414$, which is (barely) statistically significant at the 5 percent level.

This intriguing study is one of those borderline calls. If the stores were matched twins, as intended, then the data reject at the 5 percent level the null hypothesis that shelf height is unimportant. If, at the other extreme, the stores are independent samples, then the previous section showed the two-sided P

HOW TO DO IT

Wilcoxon Signed-Rank Test

Suppose that we want to use the matched-pair data in Table 8.8 to test the null hypothesis that cereal sales are not affected by whether the cereal is displayed at eye level.

1. Rank the absolute values of the differences and then sum the signed ranks (as in Table 8.8, where this sum is +52).
2. If there are at least 10 matched pairs, calculate the Wilcoxon signed-rank z statistic

$$z = \frac{(\text{sum of signed ranks}) - 0}{\sqrt{n(n+1)(2n+1)/6}}$$

   (For smaller samples, use statistical software or a table of probabilities.) Here,

$$z = \frac{52 - 0}{\sqrt{12(13)(25)/6}} = 2.04$$

3. Determine the probability of observing a z value so far from 0, and if the alternative hypothesis is two-sided, double this probability to obtain the two-sided P value. Here, $P[z > 2.04] = .0207$, implying a two-sided P value of $2(.0207) = .0414$.
4. A P value below .05 indicates that the null hypothesis is rejected at the 5 percent level. That is true here.

value to be .26, which is not sufficient to reject the null hypothesis at the 5 percent level. The conclusion then hinges on whether we believe the stores to be matched pairs and how much we are persuaded by P values of .04 and .26. The safest course, no doubt, is to gather additional data.

8.1 U.S. soldiers in World War II were taller, heavier, and stronger than those in World War I. To see if the same was true of U.S. women during these periods, a statistics class at Mount Holyoke College analyzed the following data.[9]

Although these first-year students were not randomly selected, we can consider them representative of students in similar years at similar colleges. Do these data show substantial and statistically significant changes at the 1 percent level in the mean heights, weights, and grips of female students in 1918 and 1943? Assuming these to be random samples from populations with possibly unequal standard deviations, calculate the t value, two-sided P value, and 99 percent confidence interval for each of these characteristics.

	1918 (250 STUDENTS)		1943 (308 STUDENTS)	
	MEAN	STANDARD DEVIATION	MEAN	STANDARD DEVIATION
Height, centimeters	161.7	6.3	164.8	5.9
Weight, pounds	118.4	16.6	128.0	17.2
Grip, kilograms	30.9	4.6	32.3	5.4

8.2 In the 1800s, sailors near the coast of Peru labeled the unusual currents and warm sea temperatures that sometimes appeared around Christmas El Niño, or the (Christ) boy. El Niño conditions have been linked to flooding in Central and South America; mild winters in Alaska, Canada, and the northern United States; and heavy rainfall in the southwestern United States. A researcher examined the annual Los Angeles rainfall (in inches) during 61 rain years (July 1 to June 30) from 1929 to 1989.[10] Here are the data in the 10 years identified by meteorologists and climatologists as moderate or extreme El Niño years:

RAIN YEAR	PRECIPITATION	RAIN YEAR	PRECIPITATION
1931–1932	16.95	1969–1970	7.77
1940–1941	32.76	1972–1973	17.45
1953–1954	11.99	1976–1977	14.97
1957–1958	21.13	1982–1983	32.19
1965–1966	20.44	1986–1987	9.11

In the other 51 years, precipitation in Los Angeles averaged 13.78 inches with a standard deviation of 6.23. Calculate the two-sided P value for a difference-in-means test comparing precipitation in El Niño years with other years, assuming these data to be random samples from populations with unequal standard deviations. Repeat with equal standard deviations. Are these really random samples?

8.3 A cold-vaccine test using 548 college students who believed themselves especially susceptible to colds was discussed in Chap. 4.[11] One-half were given the experimental vaccine, and one-half, the control group, were given plain water. The test results on the number of colds per person are shown below. Assuming the standard deviations to be equal, calculate the t value and two-sided P value for a difference-in-means test. Explain why the researchers concluded that "the effect is too small to be of practical importance."

	VACCINE ($n = 272$)	CONTROL GROUP ($n = 276$)
Mean	1.6	2.1
Standard deviation	.8	1.0

8.4 Telephone calls were made to randomly selected college students, who were told that they were going to be surveyed about environmental issues.[12] Students who agreed to participate were asked, "Hold, please." If the student agreed to hold, the researcher timed the number of seconds that elapsed before the student eventually hung up. In all, 10 males and 10 females were timed in this manner:

MALES	32	66	58	136	9	83	247	11	47	93
FEMALES	258	91	132	51	183	86	49	106	319	213

Use two box plots to compare these data. Assuming that the population standard deviations are equal, is the observed difference between male and female holding times statistically significant at the 5 percent level?

8.5 A Stanford University team examined the changes in triglyceride level in 30 blood samples that had been in frozen storage for 8 months:[13] Use a matched-pair test to see whether the observed differences are statistically significant at the 5 percent level.

BEFORE	AFTER	BEFORE	AFTER	BEFORE	AFTER	BEFORE	AFTER	BEFORE	AFTER
74	66	177	185	126	133	83	81	131	127
80	85	88	96	72	69	79	74	228	227
75	71	85	76	301	302	194	192	115	129
136	132	267	273	99	106	124	129	83	81
104	103	71	73	97	94	42	48	211	212
102	103	174	172	71	67	145	148	169	182

8.6 (Nonparametric) In a cloud-seeding experiment, 26 of 52 isolated clouds in south Florida were randomly selected for seeding with massive injections of silver iodide smoke, and the rainfall (in acre-feet) was recorded for each cloud.[14] Use the Wilcoxon rank-sum statistic to test the null hypothesis that these data are from identical populations.

SEEDED					UNSEEDED				
2745.6	430.0	242.5	115.3	7.7	1202.6	163.0	47.3	26.1	4.9
1697.8	334.1	200.7	92.4	4.1	830.1	147.8	41.1	24.4	0.0
1656.0	302.8	198.6	40.6		372.4	95.0	36.6	21.7	
978.0	274.7	129.6	32.7		345.5	87.0	29.0	17.3	
703.4	274.7	119.0	31.4		321.2	81.2	28.6	11.5	
489.1	255.0	118.3	17.5		244.3	68.5	26.3	4.9	

8.7 (Nonparametric) The National Transportation Safety Administration crashed 20 medium-size cars made from 1988 to 1991 into a wall at 35 miles per hour and measured the head injuries to dummies in the driver's and front-passenger seats. Use the Wilcoxon signed-rank statistic to test the null hypothesis that these data are from identical populations.

	DRIVER	PASSENGER		DRIVER	PASSENGER
Acura Legend LS	435	618	Lexus ES250	992	630
Audi 100	185	710	Mazda 929	273	858
Buick Park Ave.	1467	794	Mercedes 190E	800	833
Chevrolet Camaro	585	583	Mercury Sable	712	410
Chevrolet Lumina	1200	0	Mitsubishi Starion	952	377
Chrysler Le Baron	298	2043	Nissan 300zx	765	0
Eagle Premier	877	868	Oldsmobile Delta 88	710	539
Ford Taurus	480	258	Peugeot 505S	1983	2192
Honda Accord Lx	562	539	Toyota Cressida	790	554
Infiniti M-30	466	443	Volvo 740 GLE	519	445

8.2 The Difference between Two Success Probabilities

So far in this chapter, we have developed test procedures for the case where we have two samples of quantitative data that can be averaged. The difference between the sample means can be used to estimate the difference between the means of the populations from which these samples were drawn and to test the null hypothesis that these two population means are equal. An analogous procedure can be used to handle categorical data, which do not have natural numerical values that can be averaged.

To illustrate these procedures, we first analyze data from the 1971–1972 Toronto tests of Linus Pauling's claim that large doses of vitamin C help prevent colds.[15] Eight hundred volunteers were randomly divided into a group that received placebos and a second group that received 1 gram of vitamin C each day and 4 grams per day if cold symptoms appeared. At the end of the experiment, 72 (18 percent) of those taking the placebo had not had a cold, while 104 (26 percent) of those taking megadoses of vitamin C had been cold-free. Data on the number of colds or the number of sick days would be quantitative data that can be averaged and analyzed by the statistical procedures already developed in this chapter. But information on whether a person made it through the winter cold-free is categorical data. The person either had a cold or did not. We need slightly different statistical procedures to analyze these data.

We saw in Chap. 7 that categorical data from a single sample can be used to calculate the sample success proportion x/n, which can then be used to test a hypothesized value of a success probability (or population proportion) p. In one of our examples, 16 pregnant women read a children's story aloud during the last 6½ weeks of pregnancy, and after birth, their babies were allowed to choose between tape recordings of their mother reading the familiar story and reading an unfamiliar story. In 13 of 16 cases, the babies chose the familiar story. The sample success proportion $x/n = 13/16$ was used to test the null hypothesis that each baby has a .5 probability of choosing the familiar story.

In our two-sample vitamin C example, we can let p_1 be the probability that a randomly selected person taking the placebo would make it through this Toronto winter cold-free, and we let p_2 be the probability that a randomly selected person taking megadoses of vitamin C would be cold-free. The natural null hypothesis is that these two probabilities are equal:

$$\text{H}_0 : p_1 = p_2$$
$$\text{H}_1 : p_1 \neq p_2$$

If the data reject this null hypothesis, then we will have shown that taking megadoses of vitamin C does affect the probability of being cold-free (positively or negatively, depending on the results). The alternative hypothesis is two-sided unless we can rule out beforehand the possibility that p_1 is either larger or smaller than p_2.

A natural test statistic for the difference in two success probabilities is the difference between the success proportions in random samples drawn from each population: $x_1/n_1 - x_2/n_2$. This test statistic is unbiased, in that the population mean of the difference in the sample success proportions is equal to the difference in the population probabilities:

$$\text{Mean of } \frac{x_1}{n_1} - \frac{x_2}{n_2} = p_1 - p_2$$

This standard deviation is given by this formula:

$$\text{Standard deviation of } \frac{x_1}{n_1} - \frac{x_2}{n_2} = \sqrt{\frac{p_1(1-p_1)}{n_1} + \frac{p_2(1-p_2)}{n_2}}$$

Finally, if the samples are reasonably sized, the central limit theorem implies that the difference between the two sample success proportions is approximately normally distributed. Therefore, a shorthand summary is

$$\frac{x_1}{n_1} - \frac{x_2}{n_2} \text{ is } N\left[p_1 - p_2, \sqrt{\frac{p_1(1-p_1)}{n_1} + \frac{p_2(1-p_2)}{n_2}}\right] \tag{8.14}$$

Again, we use the following general formula to convert our normally distributed test statistic to a standardized z statistic:

$$z = \frac{\text{observed statistic} - \text{null hypothesis parameter value}}{\text{standard deviation of statistic}}$$

Here, the observed statistic is the difference between the two sample success proportions, and the population value under the null hypothesis is the assumed value of $p_1 - p_2$ (which is 0). Thus, our z statistic is

$$z = \frac{(x_1/n_1 - x_2/n_2) - 0}{\text{standard deviation of } (x_1/n_1 - x_2/n_2)} \tag{8.15}$$

Equation (8.14) shows that the standard deviation depends on the unknown values of the success probabilities p_1 and p_2. For a test of the null hypothesis that $p_1 = p_2$, the best estimate of these two equal success probabilities is provided by the pooled estimator:

$$\hat{p} = \frac{x_1 + x_2}{n_1 + n_2} \tag{8.16}$$

When we estimate the standard deviation by using this common, pooled estimator for p_1 and p_2, we have two equivalent ways of computing the z statistic:

For categorical data, the null hypothesis $p_1 = p_2$ can be tested with this z statistic:

$$z = \frac{x_1/n_1 - x_2/n_2}{\sqrt{\hat{p}(1-\hat{p})/n_1 + \hat{p}(1-\hat{p})/n_2}} = \frac{x_1/n_1 - x_2/n_2}{\sqrt{\hat{p}(1-\hat{p})(1/n_1 + 1/n_2)}} \tag{8.17}$$

However, for a confidence interval for $p_1 - p_2$, we do not assume that p_1 is equal to p_2, and it is better to use the two separate sample proportions to estimate the standard deviation:

For categorical data, a confidence interval for $p_1 - p_2$ is

$$\frac{x_1}{n_1} - \frac{x_2}{n_2} \pm z^* \sqrt{\frac{x_1/n_1(1 - x_1/n_1)}{n_1} + \frac{x_2/n_2(1 - x_2/n_2)}{n_2}} \tag{8.18}$$

where z^* is the z value corresponding to our desired confidence level, for example, $z^* = 1.96$ for a 95 percent confidence interval.

Let us apply these admittedly intimidating formulas to our vitamin C example.

The pooled sample success proportion is the total number of people who were cold-free divided by the total number of people participating in the study:

$$\hat{p} = \frac{x_1 + x_2}{n_1 + n_2} = \frac{72 + 104}{400 + 400} = \frac{176}{800} = .22$$

Using Eq. (8.17), we see that the z statistic is

$$z = \frac{72/400 - 104/400}{\sqrt{.22(1 - .22)/400 + .22(1 - .22)/400}} = 2.73$$

which Table 3 shows to have a two-sided P value of $2(.0032) = .0064$. Thus, the Toronto study showed a statistically significant effect at the 1 percent level.

Table 8.9 shows how these calculations can be done with Smith's Statistical

Table 8.9

The Smith's Statistical Package hypothesis test for the Toronto vitamin C study

```
The success proportions in two random samples can be
used to test the null hypothesis that these samples
were drawn from populations that have the same success
probability.

                                          First      Second
                                          sample     sample

How many sample observations?             [400]      [400]

How many successes in this sample?        [72]       [104]

The difference between the observed success proportions

      72/400 = .18000 and 104/400 = .26000

is .0800. Using a normal approximation and the pooled
proportion

         (72 + 104)/(400 + 400) = .2200

when calculating the standard deviation, the z value is
2.731. The probability of a z value this large (or
larger) is .0032.
```

Package. After the user enters the sample sizes and successes in the four boxes, the software shows the difference in the sample success proportions, pooled success proportions, z value, and P value.

Finally, we can calculate a 95 percent confidence interval for the difference in the population probabilities of not getting a cold by using Eq. (8.18):

$$\left(\frac{72}{400} - \frac{104}{400}\right) \pm 1.96\sqrt{\frac{72/400(1 - 72/400)}{400} + \frac{104/400(1 - 104/400)}{400}} = .080 \pm .057$$

Whether the .08 difference between .26 and .18 is substantial (something to sneeze at?) is a matter of opinion. For some, the cost of vitamin C (and the possibility of side effects) is not worth the modest improvement in the chances of being cold-free. For others, it is worth the cost.

HOW TO DO IT

Hypothesis Tests for the Difference between Two Success Probabilities

We want to use two samples of $n_1 = 400$ and $n_2 = 400$ to see if taking megadoses of vitamin C affects the probability of being cold-free.

1. Is the binomial model appropriate? Yes. Each observation has two possible outcomes, cold-free and not, and these are independent random samples.
2. Specify the null and alternative hypotheses, usually, as here,

$$H_0 : p_1 = p_2$$
$$H_1 : p_1 \neq p_2$$

3. Determine the number of success x_1 and x_2 in each sample; here, 72 and 104.
4. Calculate the pooled estimate of the two equal success probabilities, as here:

$$\hat{p} = \frac{x_1 + x_2}{n_1 + n_2} = \frac{72 + 104}{400 + 400} = \frac{176}{800} = .22$$

5. Calculate the z value, as here:

$$z = \frac{x_1/n_1 - x_2/n_2}{\sqrt{\hat{p}(1-\hat{p})/n_1 + \hat{p}(1-\hat{p})/n_2}}$$

$$= \frac{72/400 - 104/400}{\sqrt{.22(1 - .22)/400 + .22(1 - .22)/400}}$$

$$= 2.73$$

6. Find the probability of a z value so far from 0. For a two-sided alternative hypothesis, double this to obtain the two-sided P value. In our example, $P[z > 2.73] = .0032$, and the two-sided P value is $2(.0032) = .0064$.
7. Alternatively, we can use computer software, as illustrated by the Smith's Statistical Package output in Table 8.9.
8. A P value below .05 indicates that the results are statistically significant at the 5 percent level; this is the case in our example.
9. To assess the practical importance of the results, also report a confidence interval:

$$\frac{x_1}{n_1} - \frac{x_2}{n_2} \pm z^* \sqrt{\frac{x_1/n_1(1 - x_1/n_1)}{n_1} + \frac{x_2/n_2(1 - x_2/n_2)}{n_2}}$$

For our example, a 95 percent confidence interval is

$$\frac{72}{400} - \frac{104}{400} \pm 1.96 \sqrt{\frac{72/400(1 - 72/400)}{400} + \frac{104/400(1 - 104/400)}{400}} = .080 \pm .057$$

Do Headstart Programs Help?

In the early 1960s, 123 black children aged 3 to 4 years from low-income families in Ypsilanti, Michigan, participated in a study of the effectiveness of preschool programs for disadvantaged children. Coin flips were used to assign the children randomly to either the experimental group (which received intensive high-quality preschooling) or the control group (which received no preschooling). Because siblings were kept together and because of the randomness of coin flips, the two groups were slightly different sizes (58 in the experimental group and 65 in the control group). Here are some data collected when these youngsters were 19 years old (two members of the control group could not be located).[16]

	PRESCHOOL (n = 58)	NO PRESCHOOL (n = 63)
Completed high school	39	31
Currently on welfare	10	20
Arrested at some time	18	32
Currently employed	29	20

To evaluate whether these observed differences are substantial and statistically significant, we can calculate the success proportions and then test the null hypothesis that an intensive preschool program does not affect the probability of each outcome described here.

For example, a fraction 39/58 = .67 of the preschoolers completed high school, compared to 31/63 = .49 of the control group. The difference between a 67 percent and 49 percent graduation rate is substantial. The pooled sample success proportion is

$$\hat{p} = \frac{39 + 31}{58 + 63} = \frac{70}{121} = .579$$

For testing the null hypothesis that the probability of completing high school is not affected by participation in an intensive preschool program, the z statistic is

$$z = \frac{39/58 - 31/63}{\sqrt{.579(1 - .579)/58 + .579(1 - .579)/63}} = 2.01$$

which is barely larger than the 1.96 cutoff for statistical significance at the 5 percent level. Using Table 3, we see that the two-sided P value is 2(.0222) = .0444.

Table 8.10 shows the sample success proportions, z values, and two-sided P values for all four outcomes. In each case, the outcome proportions are distressing for both groups, but substantially less so for those who participated in an intensive preschool program. All the results are in the neighborhood of statistical significance at the 5 percent level, with three of the four two-sided P values a little less than .05 and one a little above. The cumulative force of the four tests strengthens the statistical case that this program has long-term beneficial effects.

One of the reasons that intensive preschooling programs are supported by politicians from all parts of the spectrum is that these expenditures are an

Table 8.10

Ypsilanti data

	SAMPLE SUCCESS PROPORTION				
	PRESCHOOL	NO PRESCHOOL	OVERALL	z VALUE	TWO-SIDED P VALUE
Completed high school	.67	.49	.59	2.01	.044
Currently on welfare	.18	.32	.25	−1.85	.065
Arrested at some time	.31	.51	.41	−2.21	.027
Currently employed	.50	.32	.41	2.04	.041

investment with long-term financial benefits. This study estimated that a $4000 investment in preschooling would return the government $29,000 in reduced welfare payments, diminished jail expenses, and increased tax revenue.

Job Discrimination: Statistical Significance and the 80 Percent Rule

Statistical tests have frequently been accepted by U.S. courts as prima facie evidence of employment discrimination that violates Title VII of the Civil Rights Act of 1964. In *Griggs vs. Duke Power Company* (1971), the Supreme Court ruled that it is illegal to use a job qualification test that has a "disparate impact" on minority applicants unless the employer can show a business necessity for the test (for example, an eye examination for pilots). In *Hazelwood School District vs. United States* (1977), the Supreme Court stated that "gross statistical disparities" can be sufficient to prove discrimination. However, courts have also recognized that a statistically significant difference may not be substantial. For instance, in *Moore vs. Southwestern Bell* (1979), a promotion test for clerks was passed by 453 of 469 whites (96.6 percent) and by 248 of 277 blacks (89.5 percent). Using the pooled success proportion (94.0 percent), the z value is 3.91, which has a two-sided P value of .000092:

$$z = \frac{.966 - .895}{\sqrt{.9401(1 - .940)/469 + .940(1 - .940)/277}} = 3.91$$

The district court ruled that although the difference between the 96.6 and 89.5 percent pass rates was statistically significant, it was not substantial enough to prove discrimination.

The U.S. Equal Employment Opportunity Commission (EEOC) has argued, and the Supreme Court has endorsed (in *Connecticut vs. Teal,* 1982), the use of an *80 percent rule* to supplement tests of statistical significance in evaluating evidence of employment discrimination against members of a specified ethnic, racial, or gender group. The 80 percent rule considers whether the selection rate for this group is less than 80 percent of the selection rate for the most favored group.

For instance, in the case of *Moore vs. Southwestern Bell* cited above, the ratio of the selection rate for blacks to the selection rate for whites is .895/.966 = .927. Although the difference is statistically significant, it is not, by the 80 percent rule, substantial. The opposite situation occurred in *Chicano Police Officers Association vs. Stover,* 1975, where 3 of 26 Spanish-surnamed officers passed a promotion test, a selection rate of 3/26 = .115, while 14 of 64 other officers passed, a selection rate of 14/64 = .219. The ratio .115/.219 = .53 is less than .80, indicating a substantial difference in selection rates. However, because the samples are so small, this difference is not statistically significant at the 5 percent level. Using the pooled-sample success proportion (3 + 14)/(26 + 64) = .189, the z value is 1.14 and the two-sided P value is .256:

$$z = \frac{.219 - .115}{\sqrt{.189(1 - .189)/64 + .189(1 - .189)/26}}$$
$$= 1.14$$

Courts have found statistical evidence of discrimination to be the most persuasive when the observed differences are both statistically significant and substantial.

8.8 In 1973, the U.S. Public Health Service studied the effects of vitamin C on the incidence of colds at a Navajo boarding school in Arizona during a 14-week period, from February to May.[17] Daily doses of from 1 to 2 grams of vitamin C were given to 321 children, while 320 children received placebos that were identical in taste and appearance to the vitamin C tablets. They found that 143 of those who took vitamin C and 92 of those taking placebos had no sick days during this 14-week period. Is this difference substantial and statistically significant at the 1 percent level?

8.9 In 1983, 22,000 male doctors volunteered to participate in an experiment testing the effects of aspirin on the incidence of heart attacks.[18] One-half of these doctors were given a single aspirin tablet every other day; the other half took a placebo. Neither the volunteers nor the researchers knew which volunteers were taking aspirin. After 5 years of what was intended to be a 7-year test, the experiments were stopped by an outside committee monitoring the results. During this period, fatal heart attacks were suffered by 18 of those taking placebos and by 5 of those taking aspirin. Nonfatal heart attacks were experienced by 171 of those taking placebos and by 99 of those taking aspirin. Use the z value and the two-sided P value to determine if the difference in fatal heart attacks is statistically significant at the 1 percent level. What about the difference in nonfatal heart attacks? Why do you suppose that the outside committee stopped the experiment 2 years early?

8.10 A 1983 federal court case (*Elison vs. City of Knoxville*) concerned a candidate at the Knoxville Police Academy who claimed that the physical qualification tests used by the academy were discriminatory because they had a disparate impact on the pass rates of males and females.[19] In her class, 6 of 9 females passed and 34 of 37 males passed. The judge accepted Knoxville's argument that the court should consider data for all classes, not just the plaintiff's class. For all classes, 16 of 19 females passed and 64 of 67 males passed. Using each set of data, is the 80 percent rule satisfied, and is the difference in pass rates statistically significant at the 5 percent level?

8.11 In a penalty-kick tournament for 10 right-footed Division III college soccer players, a total of 152 penalty kicks were taken, of which 88 were shot to the right side of the goal (from the kicker's perspective) and 64 were shot to the left side.[20] Of 88 shots to the right, 59 were successful; of 64 shots to the left, 51 were successful. Use these data to calculate the z value and two-sided P value for testing the null hypothesis that a shot to the left or the right is equally likely to be successful.

8.12 A 1980 poll asked women, "If you had enough money to live as comfortably as you would like, would you prefer to work full time, work part time, do volunteer work, or work at home, caring for the family?"[21] Overall, 39 percent wanted to work at home. Among women working outside the home, 28 percent wanted to work at home. Assume 1000 women were surveyed, 500 of whom work outside the home. Use a two-sided P value to test the null hypothesis that a preference for working outside the home is equally prevalent among women who do and women who do not work outside the home.

summary

For quantitative data, which have natural numerical values that can be averaged, the null hypothesis that two population means are equal, $\mu_1 = \mu_2$, can be tested by comparing the means of random samples from each population, a difference that is normally distributed if the underlying populations are normal (or if both samples are sufficiently large):

$$\bar{x}_1 - \bar{x}_2 \text{ is } N\left[\mu_1 - \mu_2, \sqrt{\frac{\sigma_1^2}{n_1} + \frac{\sigma_2^2}{n_2}}\right]$$

Our test statistic and confidence interval depend on whether we estimate the separate (and possibly unequal) population standard deviations σ_1 and σ_2 with the sample standard deviations s_1 and s_2 or we assume the population distributions to be equal and use a pooled estimate s_P:

	TEST STATISTIC	CONFIDENCE INTERVAL
Estimate s_1 and s_2:	$t = \dfrac{\bar{x}_1 - \bar{x}_2 - 0}{\sqrt{s_1^2/n_1 + s_2^2/n_2}}$	$x_1 - \bar{x}_2 \pm t^* \sqrt{\dfrac{s_1^2}{n_1} + \dfrac{s_2^2}{n_2}}$
Estimate s_P:	$t = \dfrac{\bar{x}_1 - \bar{x}_2 - 0}{\sqrt{s_P^2/n_1 + s_P^2/n_2}}$	$\bar{x}_1 - \bar{x}_2 \pm t^* \sqrt{\dfrac{s_P^2}{n_1} + \dfrac{s_P^2}{n_2}}$

When the sample standard deviations s_1 and s_2 are used, the degrees of freedom are estimated with statistical software. (The smaller of $n_1 - 1$ and $n_2 - 1$ is a conservative value that does not require software). When the standard deviations are assumed equal and estimated with the pooled value s_P, there are $n_1 + n_2 - 2$ degrees of freedom.

A matched-pair sample attempts to reduce the sampling error inherent in the selection of two independent random samples by comparing matched pairs that are virtually identical with respect to influences that might confound the results, for example, giving a cholesterol-reducing drug to one person and using a matched person of similar age, gender, smoking habits, and blood pressure as a control. Instead of comparing the means of these two samples, we see whether the mean of the differences between the matched pairs is statistically significantly different from zero.

To test whether two success probabilities p_1 and p_2 are equal, we examine the difference between the success proportions in random samples from each population: $x_1/n_1 - x_2/n_2$. If the samples are reasonably sized, the central limit theorem implies that the difference between the two sample success proportions is approximately normally distributed:

$$\frac{x_1}{n_1} - \frac{x_2}{n_2} \text{ is } N\left[p_1 - p_2, \sqrt{\frac{p_1(1-p_1)}{n_1} + \frac{p_2(1-p_2)}{n_2}}\right]$$

Our test statistic is

$$z = \frac{x_1/n_1 - x_2/n_2}{\sqrt{\hat{p}(1-\hat{p})/n_1 + \hat{p}(1-\hat{p})/n_2}}$$

where the two equal success probabilities are estimated with the pooled estimator

$$\hat{p} = \frac{x_1 + x_2}{n_1 + n_2}$$

For a confidence interval for $p_1 - p_2$, we do not assume that p_1 is equal to p_2, and therefore we use the two separate sample proportions when estimating the standard deviation:

$$\frac{x_1}{n_1} - \frac{x_2}{n_2} \pm z^* \sqrt{\frac{x_1/n_1(1 - x_1/n_1)}{n_1} + \frac{x_2/n_2(1 - x_2/n_2)}{n_2}}$$

review exercises

8.13 A 1986 newspaper headline read "Barnstable Auto Fatalities Drop 600 Percent," referring to the fact that there had been 12 fatalities in 1980 and 2 in 1985.[22] How large a percentage drop would you report? The accompanying article also noted, "Of particular significance is that last year neither of the two fatalities involved alcohol, [the selectman] said. That compares with alcohol being a factor in three of the 12 deaths in 1980." The selectman's daughter died in an alcohol-related automobile accident in 1982, and there was subsequently a communitywide campaign to reduce drunken driving. The article used these data to show the effects of this campaign:

	BEFORE CAMPAIGN			AFTER CAMPAIGN		
	1980	1981	1982	1983	1984	1985
Fatalities	12	12	6	4	5	2
Alcohol-Related	3	4	2	0	1	0

Assuming the population standard deviations to be equal, is the observed difference between the average number of fatalities in 1980–1982 and 1983–1985 statistically significant at the 5 percent level? What about the difference in the alcohol-related fatalities?

8.14 One hundred randomly selected students were asked to rate the qualifications of two job candidates.[23] Fifty were given a resume for a mythical John Benson, and 50 were given another resume, identical in all respects except that the name was Mary Benson. Assuming the population standard deviations to be equal, use a two-sided P value to determine if the difference in the average ratings is statistically significant at the 5 percent level.

	MARY BENSON	JOHN BENSON
Average rating	8.02	8.23
Standard deviation	2.18	1.15

8.15 To see whether persons with glaucoma have abnormally thick corneas, the corneas were measured in eight persons with glaucoma in one eye but not in the other.[24] Here are the thicknesses, in micrometers:

GLAUCOMATOUS EYE	488	478	480	426	440	410	458	460
OTHER EYE	484	478	492	444	436	398	464	476

In what sense are these a matched-pair sample? Are the observed differences in corneal thickness statistically significant at the 1 percent level?

8.16 A random sample of students at Pomona College obtained these data on grade-point average (GPA) by race (other races are not shown):[25]

	NUMBER	GPA (A = 4, B = 3, . . .)	
	SURVEYED	MEAN	STANDARD DEVIATION
Caucasian/White	127	3.391	.339
Pacific/Asian American	46	3.303	.331

a. Calculate the overall average GPA when the two categories are combined, giving a single sample of 173 students.

b. Explain why a difference-in-means test comparing the average GPA of the 127 Caucasian/white students to the average GPA of all 173 students is inappropriate.

c. Assuming the population standard deviations are equal, use an appropriate difference-in-means test for statistical significance at the 5 percent level. Is the difference substantial?

8.17 A study of changes over a 100-year period in the lead content of people's hair, measured in micrograms of lead per gram of hair, analyzed the sample data shown below.[26] If you allow the sample standard deviations to be unequal, are there statistically significant differences at the 5 percent level in the average lead content between children from 1871 to 1923 and adults from 1871 to 1923? Children from 1924 to 1971 and adults from 1924 to 1971? Children from 1871 to 1923 and children from 1924 to 1971? Adults from 1871 to 1923 and adults from 1924 to 1971?

	SAMPLE SIZE	MEAN	STANDARD DEVIATION
Children (1871–1923)	36	164.2	124.0
Adults (1871–1923)	20	93.4	72.9
Children (1924–1971)	119	16.2	10.6
Adults (1924–1971)	28	6.6	6.2

8.18 Before coming to college, a student watched a videotape claiming to show proven strategies for guaranteeing A grades. One strategy is to sit in the front of the class. To test this theory, 50 randomly selected college students were asked to write down their grade-point average and to indicate where they typically sit in large classrooms. Using the data below, he calculated the average GPA of the 26 students who sit either in the front or toward the front and used a difference-in-means test to compare this average to the average GPA of all 50 students. What crucial assumption is violated by this procedure?

The 26 students who sit in the very front or toward the front had an average GPA of 9.68 with a standard deviation of 1.1468; the 13 students who sit in the very back or toward the back had an average GPA of 8.79 with a standard deviation of 1.1445. Assuming these data are from normal distributions with the same standard deviations, determine if the difference in the sample means is substantial and statistically significant at the 5 percent level. Why, even if the difference is substantial and statistically significant, do these data not ensure that students can raise their GPAs by sitting in the front of the class?

	NUMBER OF STUDENTS	AVERAGE GPA (12-POINT SCALE)
In the very front	5	10.94
Toward the front	21	9.38
In the middle	11	9.38
Toward the back	10	8.37
In the very back	3	10.20

8.19 Data were obtained for the finals of the women's 800-meter run at the Southern California Intercollegiate Athletic Conference Championships.[27] In 3 of the years studied (1983, 1986, and 1990), the meet was held on a dirt track; in 3 other years (1987, 1988, and 1989), an artificial surface was used. Although these data are not a random sample, we can plausibly consider them to be representative of similar female athletes in similar years. If you allow the population standard deviations to be unequal, is the difference in the average times statistically significant at the 1 percent level?

	DIRT TRACK	ARTIFICIAL TRACK
Number of runners	18	18
Average time, seconds	144.90	143.11
Standard deviation	4.83	4.33

8.20 A survey of single and double majors obtained these grade-point-average data:[28]

	NUMBER SURVEYED	MEAN	STANDARD DEVIATION
One major	20	3.257	.354
Two majors	16	3.525	.230

a. Not assuming a common standard deviation, calculate the t value and two-sided P value for testing at the 1 percent level the null hypothesis that the average GPAs of all single and double majors are equal.

b. Assuming normality and a common standard deviation, calculate the pooled sample variance and determine the t value and two-sided P value for testing the null hypothesis that the average GPAs of all single and double majors are equal.

c. Does the difference in GPAs seem substantial to you?

d. Provide an interpretation of the observed difference in GPAs other than "The way to raise your GPA is to major in two subjects, instead of one."

8.21 In a September 18, 1983, article, "The Honeymoon Might Be over for Steve Garvey," the *Los Angeles Times* quoted the manager of the San Diego Padres baseball team as saying, "I'm not taking anything away from Steve [Garvey], but we've played better won-lost percentage ball without him." At the time, the Padres had won 50 and lost 52 with Garvey and won 24 and lost 22 without him. Using a normal approximation to the binomial distribution, is this difference substantial or statistically significant at the 5 percent level?

8.22 Approximately 300,000 people in the United States—two-thirds male and one-third female—undergo coronary angioplasty each year, a surgical procedure in which small balloons are inserted into arteries to unclog them. A study of 546 women and 1590 men who underwent coronary angioplasty found that 14 of the women and 4 of the men died during the operation.[29] Is this observed difference in gender mortality statistically significant at the 1 percent level? Is it substantial? Is a one-tailed or two-tailed test appropriate?

Women tend to develop heart disease later than men, and older patients are more at risk during surgical procedures. In this study, the average age of the 546 women was 4½ years older than the average age of the 1590 men, and 12 of the 14 women who died were over age 65. How might this age difference bias the results? How could you correct for it?

8.23 In the Pacific Crest Outward Bound program, students travel through the wilderness for 7 to 9 days, carrying food and equipment in backpacks. A study during the 1992 summer season compared the ratio of backpack weight to student weight for a random sample of 50 students who did not suffer injuries and 40 students who suffered knee injuries.[30] Is the difference substantial and statistically significant at the 5 percent level? (Use a pooled variance.)

	INJURIES	NO INJURIES
Mean	.383	.370
Standard deviation	.029	.030

8.24 In 1991 the National Education Association (NEA) surveyed 2000 of the 1,750,000 kindergarten-through-6th-grade teachers in the United States and reported that only 12 percent were male, down from 13.8 percent in 1986 and 17.7 percent in 1981.[31]

a. If one-half of all 1,750,000 teachers are male, what is the probability that a random sample of 2000 teachers would include so few males? Use a normal approximation.

b. If the 1981 and 1991 surveys each included 2000 teachers, is the reported decline from 17.7 to 12 percent statistically significant at the 5 percent level?

8.25 *Connecticut vs. Teal,* 1982, was concerned with a promotion test that was passed by 206 of 259 white employees and by 26 of 48 black employees. Is the observed difference statistically significant at the 1 percent level? Is it substantial by the 80 percent rule?

8.26 A study was made of 10,590 men who had vasectomies, each being matched to a man of approximately the same age, race, and martial status who had not been vasectomized.[32] Over the 5 to 14 years covered by the study, 212 of the vasectomized men and 326 of the nonvasectomized men died. If these are assumed to be two independent random samples, is the observed difference statistically significant at the 5 percent level? If we took into account that these were matched pairs, do you think that the calculated z value would be higher or lower? The researchers caution, "Exposure to vasectomy cannot be considered random; the men chose to be vasectomized." Explain why this is a reason for caution.

8.27 A 3-year study in the 1960s tested the effectiveness of sodium monofluorophosphate (MFP) as a decay-fighting agent in toothpaste.[33] The 208 children in the MFP group had an average of 19.98 cavities and a standard deviation of 10.61; the 201 children in the control group had an average of 22.39 cavities and a standard deviation of 11.96. If you assume the population standard deviations are equal, is this difference substantial and statistically significant at the 5 percent level?

8.28 A compilation of over 12,000 times at bat by major league baseball players in 1951 and 1952 separated the data into four categories, depending on whether the pitcher and batter were right- or left-handed.[34] Calculate the batting averages for each of these four groups. Then combine the data into two groups, depending on whether the pitcher and batter are same-handed or opposite-handed. Assuming all these data to be independent observations, test at the 1 percent level the null hypothesis that the probability of getting a hit does not depend on whether the pitcher and batter are same-handed or opposite-handed.

BATTER	PITCHER	TIMES AT BAT	HITS
Right	Right	5197	1201
Left	Left	1164	270
Left	Right	4002	1055
Right	Left	2245	590

8.29 The first five columns in Table 7.1 are female temperatures; the last five columns are male temperatures. Use these data to find the two-sided P value for a test of the null hypothesis that the average body temperatures of females and males are the same. Do not assume that the population standard deviations are equal, but do use two histograms to check for any evidence that these data came from dissimilar or very asymmetric distributions.

8.30 To compare the age at death of left-handed and right-handed persons, researchers examined historical data for major league baseball players who both threw and batted with the same hand.[35] Of 1472 such strongly right-handed players, the average age at death was 64.64 years, with a standard deviation of 15.5 years. Of 236 strongly left-handed players, the average age at death was 63.97 years, with a standard deviation of 15.4 years. Assuming these data to be a random sample for the purposes of estimating death ages, is there a statistically significant difference at the 1 percent level in the average age at death of right-handers and left-handers? Do not assume that the population standard deviations are equal.

8.31 A 1996 survey of 33 female and 31 male randomly selected college students obtained these data on the number of biological children they expected to have during their lifetimes:[36]

SAMPLE	FEMALES	MALES
Mean	2.1818	2.0968
Standard deviation	1.0141	1.2742

Allowing the population standard deviations to differ, calculate the two-sided P value for testing the null hypothesis that the population means are equal.

8.32 A study of 4 years of accident reports by the Dallas fire department found that fire trucks painted red made 153,348 runs and had 20 accidents, while identical fire trucks that had been painted yellow made 135,035 runs and had 4 accidents.[37] Assuming these to be independent random samples, is this difference substantial and statistically significant at the 1 percent level?

8.33 To investigate the relationship among plaque, gingivitis, and periodontis, dental assessments were made of 747 undergraduate students (average age of 21.4 years) at the University of Indonesia, a group with high socioeconomic status and substantial access to dental services, and of 592 young-adult Indonesian tea pickers (average age of 24.2 years) who have no nonemergency access to dental care.[38] The dentists compiled plaque and gingivitis indexes for each person and also noted whether the person had periodontis:

	PLAQUE (SCALE 0 TO 6)		GINGIVITIS (SCALE 0 TO 2)		
	MEAN	STANDARD DEVIATION	MEAN	STANDARD DEVIATION	CASES OF PERIODONTIS
Students	1.49	.80	.30	.28	29
Tea pickers	2.48	1.01	.55	.38	25

For each of these three sets of dental data, see if there is a statistically significant difference at the 5 percent level between the students and the tea pickers. For the plaque and gingivitis data, do not assume that the student and tea-picker data come from populations with the same standard deviation.

8.34 A study of the Brown University corpus of 1 million words of written U.S. English and the Lancaster-Oslo-Bergen (LOB) corpus of 1 million words of British English found that 37 percent of the Brown corpus were nouns and 36 percent of the LOB corpus were nouns.[39] Assuming the binomial model to be appropriate, are these differences substantial and statistically significant at the 1 percent level?

8.35 A random sample of 70 male college students were asked to guess their weight and then were weighed.[40] Of 51 who self-identified as exercising regularly, 21 guessed within 1 pound of their correct weight; of 19 who did not exercise regularly, only 2 were so accurate. Use these data to calculate the two-sided P value for testing the null hypothesis that male college students who exercise regularly and those who do not are equally likely to guess within 1 pound of their correct weight.

8.36 To see whether major league baseball players bat better during day or night games, a researcher randomly selected 75 players (not including pitchers) who batted at least 100 times during the 1994 baseball season.[41] The mean batting average for the 75 players was .2841 during day games (with a standard deviation of .0503) and .2761 during night games (with a standard deviation of .0383). Assuming these to be independent samples from populations with possibly unequal variances, test at the 5 percent level the null hypothesis that the population mean batting averages are the same for day and night games. Why might a matched-pair test be appropriate for these data? Make a matched-pair test, using the sample mean of .008 and standard deviation of .0572.

8.37 A price comparison was made of 40 everyday items at L grocery stores and R grocery stores.[42] (The items were chosen beforehand at a different grocery chain.) Here are the results, in dollars:

	L	R
Mean	2.63	3.01
Standard deviation	.61	.66

Are the observed differences statistically significant at the 5 percent level? Assume that the population standard deviations are equal.

8.38 On average, men are taller than women and therefore take longer strides. One way to adjust for this height advantage in comparing male and female track records is to divide the distance run by the runners' heights, giving a proxy for the number of strides it would take this runner to cover the distance. The division of this figure by the runner's time gives an estimate of the runner's strides per second. For instance, in 1996 Leroy Burrell was the world record holder for men for 100 meters. Dividing 100 meters by his height (1.828 meters) gives 54.70, showing that 100 meters represents approximately 54.70 strides for him. For his world record time of 9.85 seconds, this converts to 54.70/9.85 = 5.55 strides per second. In comparison, Florence Griffith Joyner, the women's record holder, is 1.7018 meters tall, so that 100 meters is 100/1.7018 = 58.76 strides for her, and her world record time of 10.42 seconds converts to 58.76/10.42 = 5.64 strides per second. A study of the top three runners in several events during 7 recent years yielded the data below on strides per second.[43] In every event, there were 21 observations for each gender.

	MEN		WOMEN	
	MEAN	STANDARD DEVIATION	MEAN	STANDARD DEVIATION
100 meters	5.52	.20	5.38	.18
200 meters	5.49	.12	5.30	.13
400 meters	4.90	.13	4.68	.17
800 meters	4.24	.12	4.11	.10
1,500 meters	4.04	.15	3.77	.11
5,000 meters	3.71	.15	3.33	.15
10,000 meters	3.57	.14	3.26	.13

Treating these data as random samples from a population of top runners in comparable years, see if any of these seven differences in male and female strides per second is statistically significant at the 5 percent level. Do not assume that the population standard deviations are the same for males and females.

8.39 A 1976 study of residents of a Florida retirement community (all of whom were over the age of 65) obtained the data below on cholesterol level.[44] Use these data and a pooled variance to test at the 5 percent level the null hypothesis that elderly females and males have the same average cholesterol level.

	FEMALES	MALES
Sample size	538	351
Age, years		
Mean	73.7	74.1
Standard deviation	5.3	5.3
Cholesterol, milligrams per deciliter		
Mean	228.9	202.4
Standard Deviation	37.1	36.1

8.40 A random sample of 248 HIV-negative gay men from the mailing list of a gay and lesbian center in southern California were asked to indicate "what you believe are *your* chances of getting HIV at some point in your life" by putting a slash through a line labeled *0 percent chance* at one end, *50 percent chance* in the middle, and *100 percent chance* at the other end.[45] These men were also asked to put a slash through another, identical line to indicate "what you believe are the chances that the average person of your gender, sexual orientation, and age has of getting HIV at some point in his/her life."

By measuring the position on the line, these slash marks were converted to probabilities with the following means and standard deviations:

	MEAN	STANDARD DEVIATION
Your chances	25.29	22.07
Other's chances	47.79	19.21

Assuming that these are independent samples from populations with the same standard deviations, is the observed difference between the sample means substantial and statistically significant at the 1 percent level? What is the two-sided P value? Would a matched-pair test be appropriate here?

8.41 In the 1980s Kahneman and Tversky reported that when 200 people were asked the following question, 92 answered yes and 108 answered no:[46] "Imagine that you have decided to see a play and paid the admission price of $10 per ticket. As you enter the theater, you discover that you have lost the ticket. The seat was not marked and the ticket cannot be recovered. Would you pay $10 for another ticket?"

When 183 people were asked the following question, 161 answered yes and 22 answered no: "Imagine that you have decided to see a play where admission is $10 per ticket. As you enter the theater, you discover that you have lost a $10 bill. Would you still pay $10 for a ticket for the play?"

Assuming these to be independent random samples, are the observed differences statistically significant at the 1 percent level?

8.42 A study of survey response rates mailed 357 surveys containing a $2 bill and mailed a control group 368 surveys with no financial incentive. Responses were received from 225 of the first group and 162 of the latter.[47] Are these observed differences substantial and statistically significant at the 1 percent level?

8.43 The first 4 columns in Exercise 6.47 are female data; the last 4 columns are male data. Determine the two-sided P value for a test of the null hypothesis that the male and female population mean hours of sleep are equal. Do not assume that the population standard deviations are equal. To see if there is any evidence that these data came from dissimilar or very asymmetric distributions, draw male and female histograms using the intervals 4.0–4.9, 5.0–5.9, 6.0–6.9, 7.0–7.9, and 8.0–8.9.

8.44 A study found that male derelicts are less often balding than are male college professors.[48]

	NUMBER STUDIED	MEDIAN AGE	NUMBER BALDING
Derelicts	141	47.5	51
Professors	49	47.5	35

Assuming these to be random samples, test at the 5 percent level the null hypothesis that male derelicts and professors are equally likely to be balding. Are the differences substantial?

8.45 Writer's Workbench is a computerized writing analysis program developed at Bell Labs to assist their technical writers. It has been used extensively at Colorado State University (CSU) to help students in first-year composition classes and in English as a Second Language (ESL) classes. The program has also been used by CSU instructors to analyze student writing styles. For example, an analysis was made of 2 different kinds of 30-minute essays written by ESL students.[49] One essay asked the students to describe a bar chart or pie chart. A second essay asked them to discuss a compare/contrast topic, such as the two ways to spend leisure time.

	DESCRIBE CHART	COMPARE/CONTRAST
Number of essays	64	69
Number of words in essay		
Mean	151.05	198.23
Standard Deviation	51.25	67.11
Average sentence length		
Mean	22.18	19.78
Standard Deviation	6.43	5.48

Are the observed differences between the chart description and compare/contrast essay statistically significant at the 5 percent level? Do not assume that the population standard deviations are equal. Summarize your findings.

8.46 In a 1990 study, half of 20,800 patients in thirteen countries who had been hospitalized for heart attacks were given streptokinase to dissolve clots in the arteries and half were given TPA.[50] By the end of the average two-week hospitalization, 8.5 percent of the streptokinase group had died, compared to 8.9 percent of the TPA group. Estimate the standard deviations with these sample frequencies and use a normal approximation to calculate the z value to determine if this difference is statistically significant at the 1 percent level.

8.47 Seven psychologists tried 2 different approaches to reducing the cigarette consumption of heavy smokers who professed a desire to quit smoking.[51] One group was given verbal encouragement and supportive counseling while those in a matched-pair group were given electrical shocks while they smoked. Here are the post-treatment cigarette consumption data, expressed as a percentage of pre-treatment consumption:

PSYCHOLOGIST	1	2	3	4	5	6	7
SUPPORTIVE	125.0	59.0	43.0	0.0	50.0	18.5	66.0
PUNITIVE	59.5	14.0	72.0	27.5	80.5	17.5	83.0

Which approach reduced cigarette consumption the most, on average? Use a matched-pair test to see whether the observed difference is statistically significant at the 1 percent level.

8.48 To see whether runners tend to improve during their high school years, a researcher looked at the season-best times (in seconds) for the 3000 meters by the top ten sophomore females, senior females, sophomore males, and senior males in Oregon in 1990:[52]

FEMALE SOPHOMORES	FEMALE SENIORS	MALE SOPHOMORES	MALE SENIORS
635	617	542	516
636	631	548	516
637	640	552	520
645	645	555	525
653	646	559	528
657	654	562	529
663	657	563	529
665	658	565	530
666	663	567	530
667	674	568	534

Assume that these data are random samples from populations consisting of the season-best times of the nation's best high school 3000-meter runners and that the population standard deviations are not necessarily equal. Is there a substantial and statistically significant difference (at the 5 percent level) in the sophomore and senior female times? In the sophomore and senior male times?

8.49 (Nonparametric) Thirteen patients with advanced stomach cancer and 11 patients with advanced breast cancer were treated with ascorbate.[53] Use the Wilcoxon rank-sum statistic to test the null hypothesis that the survival times (in days) are from identical populations.

STOMACH			BREAST		
124	412	103	1235	727	719
42	51	876	24	3808	
25	1112	146	1581	791	
45	46		1166	1804	
340	396		40	3460	

8.50 (Nonparametric) Reanalyze the data in Exercise 8.5, this time using the Wilcoxon signed-rank statistic to test the null hypothesis that these data are from identical populations.

projects

For each of the following projects, type a report in ordinary English, using clear, concise, and persuasive prose. Any data that you collect for this project should be included as an appendix to your report. Data used in your report should be presented clearly and effectively.

8.1 Ask 50 randomly selected students, 25 female and 25 male, to name their favorite singer. (Be sure that they name an individual, not an entire group; it is okay to say "the lead singer of. . . .") What fraction of the males surveyed named a male singer? What fraction of the females surveyed named a male singer? Is there a statistically significant difference in the gender preferences of female and male students?

8.2 Do students and professors at your college have similar political beliefs? (Restrict yourself to one or two simple questions, such as, Are your views generally more consistent with those of the Democratic or Republican Party?)

8.3 Ask 100 randomly selected students at your school if they exercise regularly, phrasing the question so that it is specific and unambiguous. Also ask their gender. Is there a statistically significant difference in the male and female responses?

8.4 What percentage of the seniors at your college expect to be married within 5 years of graduation? What percentage expect to have children within 5 years of graduation? How many biological children do the seniors at your college expect to have during their lives? Do males and females differ in their answers to these questions?

8.5 Estimate the average height and weight of first-year and fourth-year females and males at your school. Estimate a 95 percent confidence interval for the difference between the average first-year and fourth-year female heights, male heights, female weights, and male weights.

8.6 Find someone with an extensive collection of recent music CDs and use random sampling to find at least 25 CDs that show the playing time on the back of the CD case (where it would be visible to a shopper examining a shrink-wrapped CD) and 25 CDs that do not show the playing time on the back of the CD. Try to restrict yourself to a single genre; don't, for example, mix classical and rock music. Test at the 5 percent level the null hypothesis that there is no difference in the average playing times of CDs that do and do not display the playing time on the back of the CD case.

8.7 Do English majors at your school get better grades in English courses or in their other courses? Do mathematics majors at your school get better grades in mathematics courses or in their other courses?

8.8 Using your local Yellow Pages, take a random sample of at least 20 barber shops or beauty salons that cut both men's and women's hair. Do not choose more than one store from a chain. Have a female telephone each store in your sample and ask the price for the least expensive female haircut. A few hours later, have a male telephone this store and ask the price for the least expensive male haircut. Are the observed differences statistically significant at the 1 percent level?

8.9 Go to a local grocery and select 25 nationally advertised cereals. You will have to be very specific about the cereal and the size of the box. Now go to two different stores and record the prices for each cereal you selected. Are the observed differences at these two stores statistically significant at the 5 percent level?

8.10 Every day (except Sunday) for 2 weeks, mail two letters—one with the Zip code and one without—to a residential address at least 500 miles away. Make sure that the letters are otherwise identical and are mailed at the same time of day; inside each letter, note the day it was mailed. Have the recipient open the letter and record the day it was received. Do your data provide persuasive evidence that the Zip code reduces the delivery time?

8.11 Ask 50 males and 50 females this question: "You were engaged to be married to A, but broke off the engagement a month before the wedding when you found A in bed with your best friend. Two years have now passed, and you are engaged to be married to B. One month before the wedding, A calls you and wants to spend the night together for old-times' sake. Do you say yes or no?" Devise a procedure that will allow each person to respond anonymously, but will still allow you to distinguish male and female responses (for example, different-color paper). Is there a statistically significant difference in the male and female responses?

8.12 Ask 50 females and 50 males to write down a brief answer to this question: "You are engaged to be married to A and while you are both at a party, a former lover who is unknown to A flirts with you outrageously. You rebuff these advances, but afterward A asks you who this person is who was flirting with you. What do you say?" Devise a procedure that will allow each person to respond anonymously, but will still allow you to distinguish female and male responses. Now ask someone who is unaware of your system for identifying male and female responses to classify each response as either truthful or deceitful. Is there a statistically significant difference in the male and female responses?

8.13 Conduct a taste test of either Coke versus Pepsi or Diet Coke versus Diet Pepsi. Ask at least 50 persons who identify themselves beforehand as cola drinkers to drink each unlabeled cola and to rate each cola on a scale of 1 to 10; use a matched-pair test to assess the statistical significance of your results.

8.14 Do more expensive cookies taste better than less expensive ones? Choose two brands of cookies that appear to be similar but cost quite different amounts. Ask at least 50 persons to taste an unlabeled cookie from each brand and to rate each cookie on a scale of 1 to 10; use a matched-pair test to assess the statistical significance of your results.

8.15 Ask 50 randomly selected people the first question in Exercise 8.41, and ask 50 different randomly selected people the second question in this exercise. Is there a statistically significant difference in the responses to these 2 questions?

8.16 Ask a random sample of at least 50 students the following question: "During the school year, how many hours a week do you spend, on average, on school-related work; for example, reading books, attending class, doing homework, and writing papers?" Ask a random sample of at least 25 professors this question: "During the school year, how many hours a week do you spend, on average, on school-related work; for example, preparing lectures, teaching, grading, advising, serving on committees, and doing research?" Determine the P value for a test at the 5 percent level of the null hypothesis that the 2 population means are equal.

8.17 A stock's dividend-price ratio is its annual dividend divided by its current market price; its price-earnings ratio is the current price divided by its annual earnings. Find a copy of *The Wall Street Journal* from approximately one year ago and, in the financial section, find the listing of the 30 stocks in the Dow Jones Industrial Average. From among these 30 stocks, determine: (a) the 10 stocks with the highest dividend-price ratios; and (b) the 10 with the lowest dividend-price ratios. Now look in the most recent issue of the *Journal* and calculate the percentage price changes for each stock. Did the stocks in group (a) or (b) have the higher average percentage price increase? Use a difference-in-means test to determine if the observed differences between these two means is statistically significant at the 5 percent level. Redo this assignment, this time having group (a) be the 10 Dow stocks with the lowest price-earnings ratios and group (b) be the 10 Dow stocks with the highest price-earnings ratios.

8.18 Follow the procedure described in Exercise 8.4, but tell the person who answers the telephone that they are going to be surveyed about a political issue that you have chosen. Obtain data for at least 25 males and 25 females, and in your statistical analysis do not assume that the population standard deviations are equal.

8.19 Make up a fictitious resume and follow the procedure described in Exercise 8.14. You only need to survey 50 people, and in your statistical analysis do not assume that the population standard deviations are equal.

8.20 Ask at least 50 randomly selected college students to write down their grade point average (GPA) and to indicate where they typically sit in large classrooms: towards the front, in the middle, or towards the back. If feasible, restrict your sample to students taking the same class or similar classes. Use a difference-in-means test to compare the average GPAs of those who typically sit towards the front with those who usually sit towards the back.

HAVE YOU EVER WONDERED?

The U.S. poet Robert Frost once wrote, “Some say the world will end in fire, some say in ice.” Among scientists today, there is ominous debate about whether the planet will soon endure another ice age or whether the accumulation in the atmosphere of carbon dioxide from the burning of fossil fuels will raise the earth’s average temperature, melting the polar icecaps and flooding the world. The debate continues because the data do not clearly reveal whether there has been a systematic change in the earth’s temperature in recent years, let alone whether such a trend, if it exists, will continue or even accelerate.

It is hard to discern a trend because, in most parts of the United States, there is considerable variation in temperatures throughout the year and from one year to the next. In the northeastern United States, the winter of 1994 was much milder than usual (global warming?), while the winter of 1995 was unusually harsh (global cooling?).

Have there been changes in recent years in seasonal temperature patterns in the United States? In addition to asking whether temperatures, overall, are rising or falling, we can inquire about seasonal variations. Have spring and summer become more alike or more different? What about summer and fall? Or fall and winter? We answer these questions in this chapter by using a statistical procedure that can handle more than two samples.

chapter 9

using ANOVA to compare several means

And that's the news from Lake Wobegon, where all the women are strong, all the men are good-looking, and all the children are above average.
—Garrison Keillor

OUR STATISTICAL TESTS to this point have been limited to a single mean or success probability, or to comparisons of two means or two success probabilities. We tested whether the average body temperature of humans is 98.6°F and whether there is a 50 percent chance that a baby will choose a familiar or an unfamiliar story. We compared the average amount of time devoted to good and bad economic news, and we compared the percentage of vitamin-C takers who caught colds with the percentage of those who took placebos. Often, however, we are interested in the fine detail that an average or percentage neglects—we do not want to suffer the fate of the statistician who drowned while crossing a river that had an average depth of 3 feet.

If we suspect that a certain casino is using loaded dice, we are less interested in the average value of the numbers rolled than in the frequencies with which the individual numbers occur. If we look at the months in which people are born, marry, or die, we want to look at all 12 months, not just 1 month or a pair of months. If we look at seasonal temperatures in the United States, we want to look at all four seasons. In this chapter, we discuss a very useful statistical procedure developed by the great British statistician Sir Ronald A. Fisher (1890–1962) for handling multiple

samples of quantitative data that can be averaged. In Chap. 10, we look at a test that can be used for several samples of categorical data.

9.1 The Logic Behind an ANOVA *F* Test

The logic behind Fisher's approach can be explained as follows. In many situations we suspect that a certain medication, diet, equipment, or procedure may make a difference. The ill may get well sooner with certain medicines, people may be more productive with certain computer software, students may learn more from certain teachers. However, variations in the outcomes can be caused by other factors, too. Some patients get well faster than others, even if all receive identical treatment. Some workers are more productive than others, even if they use identical software. Some students learn more than others, even if taught by the same teacher. How do we distinguish differences in the effects of the medication, software, or teacher from those caused by variations among the people participating in the experiment? That is what you will learn to do in this chapter.

For concreteness, consider the question of whether there have been changes in seasonal temperature patterns in the United States. Table 9.1 shows annual temperature data for 1970 to 1989 for the following four seasons: winter (December to February); spring (March to May); summer (June to August); and fall (September to November).[1] These data are annual averages of temperatures from more than 1200 national weather stations for the contiguous United States (Alaska and Hawaii are omitted). Because temperatures vary considerably during the course of the year, a simple comparison of seasonal temperatures simply shows that temperatures rise in the spring and summer and drop in the fall and winter.

Instead, we want to investigate the more interesting question of seasonal irregularities relative to the past. Have winters recently been unusually cold and summers exceptionally hot? Or have winters and summers both been cooler than usual while spring and fall have become warmer? Or are there no statistically persuasive patterns? To answer these questions, the data in Table 9.1 have been seasonally adjusted relative to the base period 1900 to 1969. Each winter temperature in Table 9.1 shows how much warmer or colder winter was that year than the average winter temperature from 1900 through 1969; each summer temperature shows how much warmer or colder summer was that year in comparison to the base period of 1900 through 1969.

Table 9.1

Seasonal temperature anomalies in the contiguous United States (°F relative to 1900–1969)

	WINTER	SPRING	SUMMER	FALL
1970	−.24	−.60	.70	−.07
1971	−.08	−1.59	.14	1.06
1972	.43	.66	−.22	.15
1973	−.35	1.04	.52	1.67
1974	1.00	1.52	−.46	−.65
1975	1.20	−1.23	.17	−.30
1976	1.65	.03	−.85	−3.04
1977	−3.44	2.12	1.20	1.28
1978	−2.29	.61	.28	.43
1979	−5.19	.35	−.26	−.07
1980	1.36	−.26	.89	−.02
1981	1.69	1.63	1.15	.70
1982	−1.48	.25	−.08	.16
1983	4.26	−.46	1.16	1.75
1984	−1.34	−.35	.68	.18
1985	−1.34	2.42	−.29	.15
1986	.46	2.39	.93	.94
1987	2.80	1.94	.71	−.25
1988	−.40	.08	1.15	.13
1989	−.44	.84	.34	−.43

Table 9.2 shows the sample means and standard deviations for the four seasons. During the period from 1970 through 1989, winter on average was .087 degree cooler than during the preceding 70 years, whereas spring, summer, and fall were somewhat warmer than in the past. These differences in means are suggestive, but the substantial standard deviations show that within each seasonal group, there is considerable variation in the year-to-year data. In 1979, the winter temperature was 5.19 degrees below normal; in 1983 it was 4.26 degrees above.

For statistical purposes, we can view each year's seasonal temperature as a random draw from a population. For each of the four seasons, the data for

Table 9.2

Seasonal means and standard deviations, 1970–1989

	WINTER	SPRING	SUMMER	FALL
Mean	−.087	.570	.393	.189
Standard deviation	2.133	1.168	.612	1.016

1970 to 1989 constitute 20 observations, from which we can calculate the sample mean and standard deviation. Can the observed differences among the four seasonal sample means plausibly be attributed to sampling error—the coincidental differences that occur when we average temperatures that vary greatly from one year to the next? Or do these differences among the sample means reflect underlying shifts in the means of populations from which these data came? Are these data persuasive evidence that winters have become cooler in comparison to the other seasons?

Figure 9.1 illustrates the fact that there are two very different explanations for the observed differences among the seasonal means. (To unclutter the figure, we temporarily look at just the winter and spring temperatures.) The seasonally adjusted temperature data may come from probability distributions with different means, perhaps $-.1$ and $+.6$. Or they may come from a single population with a large standard deviation.

How can we choose between these competing explanations for the observed differences among the means? The key is to use the observed variation within each seasonal category to estimate the extent to which the temperatures vary from year to year. If, on one hand, annual temperatures within each seasonal group show little variation, then a .7 degree difference in the mean temperatures is persuasive evidence that the seasonally adjusted data do not come from

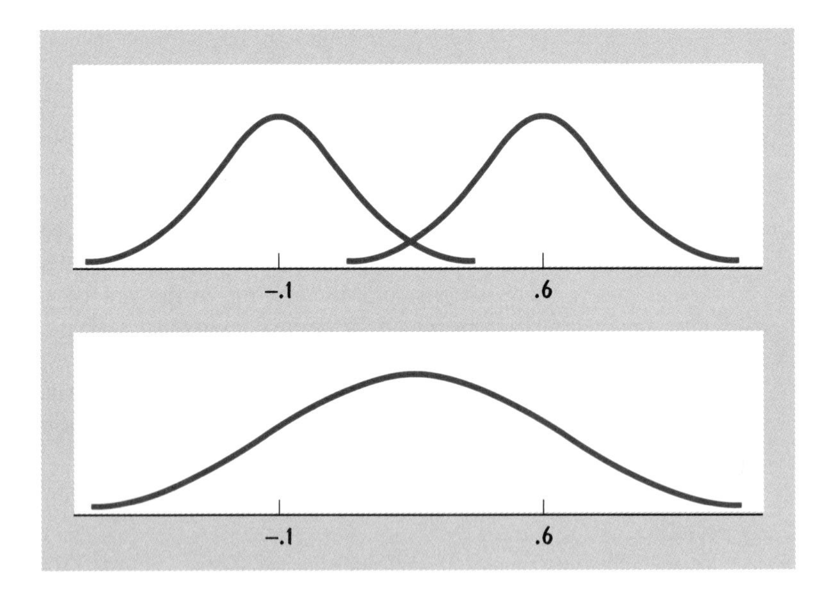

Figure 9.1 Two unequal sample means may have come from two populations with different means or from a single population with a large standard deviation.

the same population—that seasonal weather patterns have changed. If, on the other hand, there is considerable variation from one year to the next, then a .7 degree difference in the mean temperatures can be explained plausibly by sampling variation.

If we only had two means to compare, we could use the difference-in-means *t* test described in Chap. 8. It might therefore be tempting to use a sequence of *t* tests here on the six possible pairings: winter with spring, winter with summer, winter with fall, spring with summer, spring with fall, and summer with fall. However, as mentioned in the Chap. 7 discussion of data mining, as we increase the number of tests, so do we increase the probability of incorrectly rejecting at least one of the many null hypotheses we are testing. The *F* statistic described in the next section provides an attractive alternative: a single test of the null hypothesis that all the population means are equal.

9.1 To investigate the growth of women between the ages of 18 and 21, we might take four random samples of 100 women aged 18, 19, 20, and 21 years. Why might the sample means turn out to be different even if the populations from which these heights came all have the same mean and standard deviation? Use a graph of a sampling distribution to illustrate your reasoning.

9.2 Suppose that we were to look at the annual average temperatures in the contiguous United States from 1970 through 1989, but that instead of recording the temperatures relative to a base period (as is done in Table 9.1), we simply record the temperatures, for example, 55.47° F. Draw rough sketches of what seem to you to be reasonable histograms for the winter temperatures and for the summer temperatures. Be sure to label the horizontal axis in each graph.

9.3 For each year in Table 9.1, add the four seasonal temperatures and divide by 4. The result is the average temperature that year, relative to the base period of 1900 to 1969. Assuming that these 20 annual temperatures are a random sample from a normal distribution, determine the two-sided *P* value for a test of the null hypothesis that the population mean is zero. Interpret your results.

9.4 Suppose that in place of the four seasons in Table 9.1 we have monthly temperature data for these 20 years. If we were to apply a sequence of difference-in-means *t* tests to all possible pairs of months, how many pairings would we have?

9.5 Draw four box plots for the seasonal temperatures in Table 9.1. Is the visual appearance of these box plots consistent with the seasonal standard deviations given in Table 9.2?

9.2 The *F* Statistic

To describe our statistical procedure in general terms, suppose we have random samples from each of k populations. The null hypothesis is that all k of the population means are equal; the alternative hypothesis is that at least one of these population means is not equal to the others:

$$H_0 : \mu_1 = \mu_2 = \cdots = \mu_k$$
$$H_1 : \text{At least one population mean differs from the others}$$

Although we compare the sample means to test the equality of the population means, our procedure is called **analysis of variance (ANOVA)** because our analysis is based on a comparison of two kinds of variation: the variation among the sample means and the variation within each sample. In honor of Sir Ronald A. Fisher's pioneering work, the test statistic is called the F statistic:

The ***F*** **statistic** for analysis of variance is

$$F = \frac{\text{variation among sample means}}{\text{variation within samples}} \tag{9.1}$$

The numerator of the ANOVA F statistic measures the variation in the sample means about the overall mean $\bar{x}$ of all the data:

$$\text{Variation among sample means} = \frac{n_1(\bar{x}_1 - \bar{x})^2 + n_2(\bar{x}_2 - \bar{x})^2 + \cdots + n_k(\bar{x}_k - \bar{x})^2}{k - 1} \tag{9.2}$$

If all the sample means turn out to be the same, so that the variation among the sample means is equal to zero, then the data provide no evidence whatsoever against the null hypothesis. In contrast, a large value for the numerator (and the F statistic) is evidence against the null hypothesis.

The denominator of the ANOVA F statistic is a weighted average of the k sample variances:

$$\text{Variation within samples} = s_P^2 = \frac{(n_1 - 1)s_1^2 + (n_2 - 1)s_2^2 + \cdots + (n_k - 1)s_k^2}{n_1 + n_2 + \cdots + n_k - k} \tag{9.3}$$

This measure of the variation within the samples is a generalization of the pooled variance introduced in Chap. 8 for a difference-in-means test when two samples are taken from populations with the same standard deviation:

$$s_P^2 = \frac{(n_1 - 1)s_1^2 + (n_2 - 1)s_2^2}{n_1 + n_2 - 2}$$

Equation (9.3) is the pooled variance for k samples from populations with the same standard deviation. In fact, the square of the t value for a difference-in-means test using the pooled variance is exactly equal to the F value for an ANOVA test with two samples, and the P values are identical! The advantage of an F test is that it can be used when we have more than two samples.

The observed differences among the sample means are more telling evidence against the null hypothesis when there is little variation within the samples (when the pooled variance is small). If, on one hand, the data hardly vary at all within each sample, then large differences among the sample means cannot be explained persuasively by sampling error. If, on the other hand, there is a lot of variation within the samples, then sampling error can well explain why the sample means happen to differ somewhat from one another. Thus a small value for the denominator of Eq. (9.1) (and a large value for the F statistic) is evidence against the null hypothesis.

Figure 9.2 uses box plots to illustrate this logic. Because these are meant to be roughly symmetric data from the normal distributions, we assume that the sample means are very close to the medians, indicated by the vertical line in the middle of each box. Look first at the box plots for the three groups of data in Fig. 9.2*a*. Because there is a little variation within each group, the observed differences among the sample means suggest that these data did not come from populations with the same means. Figure 9.2*b* tells a very different story. The means are exactly the same as in Fig. 9.2*a*, but there is much more variation within each sample. Now it is conceivable that the data came from populations with identical means and that the proverbial luck of the draw caused the observed differences among the sample means.

Matters are actually more complicated than this, because we need to take the sample sizes into account in order to judge the amount of variation to be expected in the sample means. The F statistic does this. It considers the differences among the group means, the variation within each group, and the sample sizes.

The calculation of the F statistic from Eqs. (9.1), (9.2), and (9.3) is obviously complicated (and tedious) and best done with statistical software. Nonetheless, we work through one calculation by hand, to give you a feel for the messy details. The overall mean for the seasonally adjusted temperatures in Table 9.1 is obtained by adding all 80 observations and dividing by 80:

$$\bar{x} = \frac{(-.24) + (-.08) + \cdots + .13 + (-.43)}{80}$$

$$= .266$$

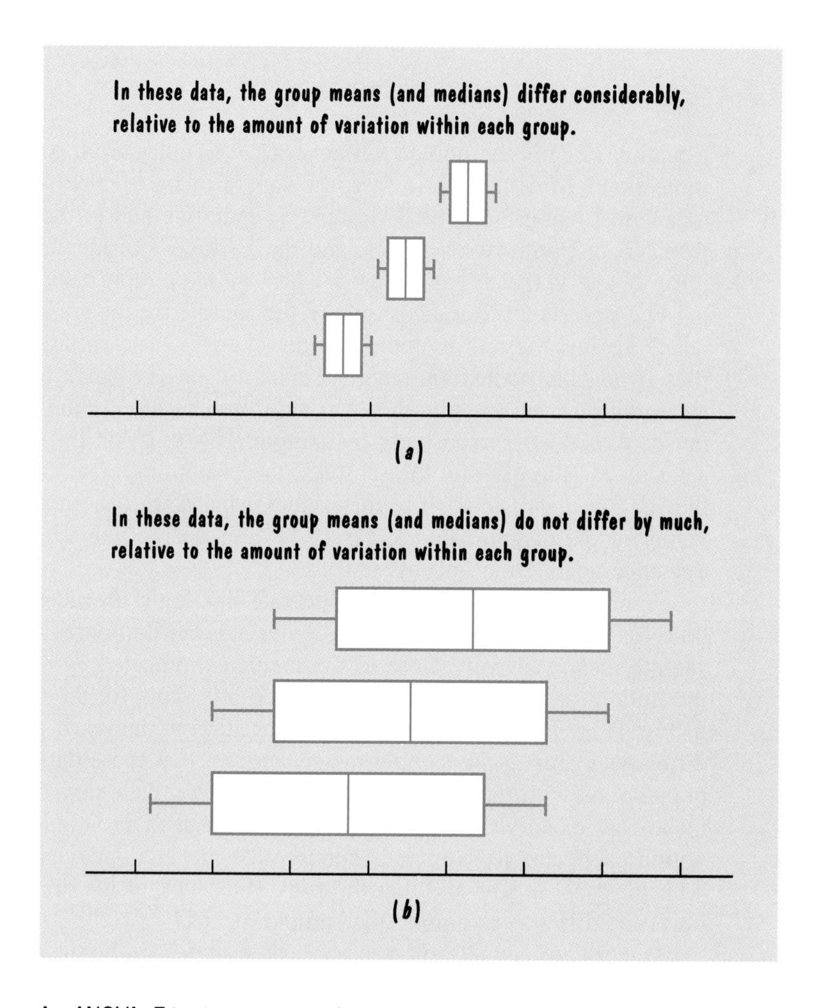

Figure 9.2 An ANOVA F test compares the variation among the group means to the variation within each group.

(With equal sample sizes, the overall mean is just a simple average of the four group means $-.087$, $.570$, $.393$, and $.189$; this is not the case with unequal sample sizes.) The numerator of the F statistic is

$$\text{Variation among sample means} = \frac{n_1(\bar{x}_1 - \bar{x})^2 + n_2(\bar{x}_2 - \bar{x})^2 + \cdots + n_k(\bar{x}_k - \bar{x})^2}{k-1}$$

$$= \frac{20(-.087 - .266)^2 + \cdots + 20(.189 - .266)^2}{4-1}$$

$$= 1.5939$$

To calculate the denominator of the F statistic, we need to use the sample standard deviations given in Table 9.2:

$$\text{Variation within samples} = \frac{(n_1 - 1)s_1^2 + (n_2 - 1)s_2^2 + \cdots + (n_k - 1)s_k^2}{n_1 + n_2 + \cdots + n_k - k}$$
$$= \frac{(20 - 1)(2.133)^2 + \cdots + (20 - 1)(1.016)^2}{20 + 20 + 20 + 20 - 4}$$
$$= 1.8302$$

Finally, the value of the F statistic is

$$F = \frac{\text{variation among sample means}}{\text{variation within samples}}$$
$$= \frac{1.5939}{1.8302}$$
$$= .87$$

Even in this relatively simple example, with equal sample sizes, the calculations are complex and taxing; statistical software provides a welcome alternative.

exercises

9.6 This section went through a step-by-step calculation of the variation among the sample means and the variation within the samples, from which the value of the F statistic was computed. Repeat this calculation, this time using the data in Table 9.1 for the 10 years from 1980 to 1989.

9.7 Without doing any calculations, explain why you believe the F value to be either larger than 1 or less than 1 for these data:

GROUP A	5.1	5.6	4.9	5.5	5.2	5.4	4.4	4.9	4.8
GROUP B	4.0	4.8	3.4	3.5	4.1	4.6	4.0	4.4	3.9
GROUP C	7.2	6.7	5.8	6.5	6.7	5.9	5.8	5.8	6.0

9.8 Without doing any calculations, explain why you believe the F value to be either larger than 1 or less than 1 for these data:

GROUP A	2.0	5.6	2.3	3.7	6.6	5.2	7.6	8.9	2.1
GROUP B	3.6	1.8	5.5	4.9	5.0	7.2	5.6	8.1	5.5
GROUP C	5.6	3.9	2.1	4.9	5.1	4.9	6.8	9.2	6.4

9.9 Exercise 8.4 gives the number of seconds for which 10 males and 10 females held the phone while waiting to be surveyed about environmental issues.

Calculate the t value for a difference-in-means t test using a pooled variance. Now determine the value of the F statistic by calculating the variation among the sample means and the variation within the samples, as described in Sec. 9.2. Is there any evident relationship between the t value and the F value?

9.10 Table 2.1 shows stock returns for the years from 1971 through 1995. Separate these stock return data into two groups: the odd-numbered years (1971, 1973, . . .) and the even-numbered years (1972, 1974, . . .). Now calculate the t value for a difference-in-means t test using a pooled variance. What is your null hypothesis? Also calculate the variation among the sample means and the variation within the samples, as described in Sec. 9.2, and from these determine the value of the F statistic. How large is the F value in relation to the t value?

9.3 The *F* Distribution

To assess the value of the F statistic, we need to know its probability distribution so that, as with any hypothesis test, we can determine whether the observed F value is sufficiently unlikely to be persuasive evidence against the null hypothesis.

If all k random samples come from normal distributions with the same standard deviation and if the null hypothesis is true, so that all the population means are equal, then the ANOVA F statistic has the ***F* distribution** with the following degrees of freedom:

Numerator: $k - 1$

Denominator: $n_1 + n_2 + \cdots + n_k - k$

The F distribution assumes that the populations have normal distributions. However, because of the power of the central limit theorem, the ANOVA F test is still reliable even if the underlying populations are not exactly normal. What is important is that the sample means be approximately normally distributed. If the distributions are generally symmetric without extreme outliers, the ANOVA F test works well with as few as five observations in each sample. Even with asymmetric distributions and outliers, the test is acceptable with reasonably large samples.

The assumption that all the populations have the same standard deviation does not have to be exactly true either. The ANOVA F test still works well if the samples are roughly the same size and not too small, and if the largest sample standard deviation is not more than twice the size of the smallest sample standard deviation. Our seasonally adjusted temperature data do not pass this test, in that the largest standard deviation is more than 3 times the smallest: $2.133/.612 = 3.5$. As a consequence, our P value may be only roughly accurate. If the P value turns out to be very small or very large, we can safely reject or not reject the null hypothesis; however, for borderline cases, we must interpret the results cautiously.

The third thing to notice about the F distribution is that it has two separate measures of degrees of freedom—one for the numerator and one for the denominator. Figure 9.3 shows three very different shapes for the F distribution. (Notice that the F distribution becomes bell-shaped as both degrees of freedom increase.) Because the F statistic is the ratio of two sums of squares, it cannot be negative. As noted in our earlier discussion, a large value of the ANOVA F statistic is evidence against the null hypothesis since it indicates that there are large differences among the sample means relative to the variation within each sample.

In our seasonal temperature example, there are $k = 4$ samples, and sample sizes n_1, n_2, n_3, and n_4 are all equal to 20, giving these degrees of freedom:

Numerator: $k - 1 = 4 - 1 = 3$

Denominator: $n_1 + n_2 + n_3 + n_4 - k = 20 + 20 + 20 + 20 - 4 = 76$

Figure 9.4 shows this F distribution. Table 5 at the end of this book shows the cutoff values of F for statistical significance at the 5 percent level. With 3

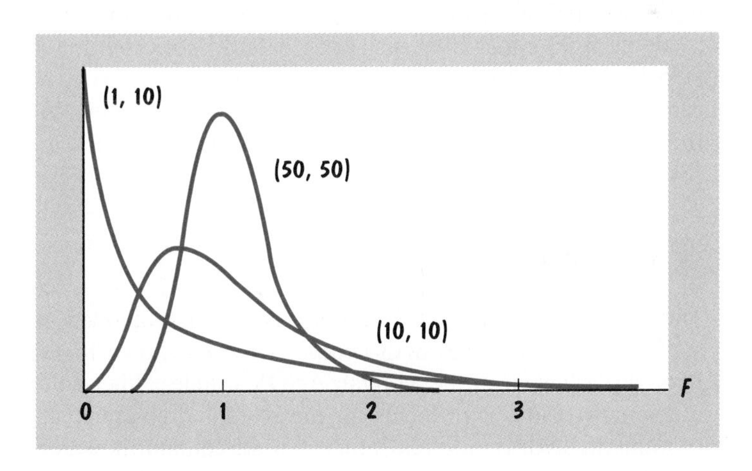

Figure 9.3 Three varied F distributions, with differing degrees of freedom (numerator, denominator).

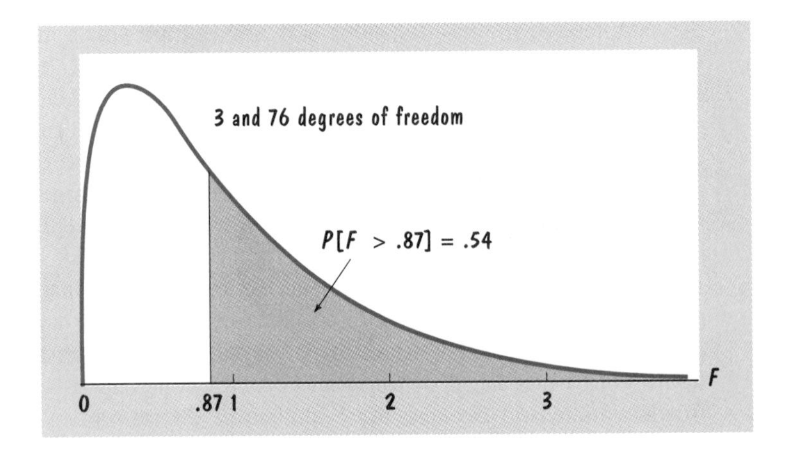

Figure 9.4 The F distribution for the seasonal temperature data.

degrees of freedom for the numerator and 76 for the denominator, the cutoff is $F > 2.7$. Our observed .87 F value is not nearly large enough for statistical significance at the 5 percent level. Statistical software shows the P value to be .54. If the null hypothesis is correct, there is a .54 probability that four samples of size 20 will yield an F value larger than .87. In these seasonal temperature data, the variation within each sample is sufficiently large to explain the observed differences among the group means.

This P value is so big that the fact that one sample standard deviation is 3.5 times the size of another (violating our 2-to-1 rule of thumb) should not dissuade us from concluding that these data do not provide remotely persuasive evidence against the null hypothesis that these seasonally adjusted temperatures came from populations with the same population mean. We cannot conclude from the data for 1970 through 1989 that there has been a systematic change in seasonal temperature patterns.

The ANOVA Table

The results of an ANOVA F test are usually summarized in an **ANOVA table,** as with the Minitab and SAS computer output shown in Table 9.3. We will go through each item in the Minitab ANOVA table. Other software packages might use somewhat different labels for the rows and columns and might arrange the columns in slightly different order, but all contain the same information shown in Table 9.3.

Minitab's first column (*source*) uses labels *factor, error,* and *total* to show that we are separating the total variation in the data into two parts: the variation among the samples means and the variation within each sample. The factor

Table 9.3

Minitab and SAS ANOVA table for the seasonal temperature data

MINITAB

ANALYSIS OF VARIANCE

SOURCE	DF	SS	MS	F	P
FACTOR	3	4.78	1.59	.87	.537
ERROR	76	139.06	1.83		
TOTAL	79	143.84			

SAS

SOURCE	DF	SUM OF SQUARES	MEAN SQUARE	F VALUE
MODEL	3	4.78	1.59	.87
ERROR	76	139.06	1.83	PR > F
CORRECTED TOTAL	79	143.84		.537

row (or *treatments* or *groups* in other software) refers to the variation among the groups, as measured by the variation among the sample means. The variation among the samples means is said to be explained by our model in that it is attributable to the different groups. The error row in the table refers to the remaining variation, the variation within each sample. It is not an error in the sense that we have made a mistake, rather it refers to the fact that some of the variation in the data (the variation within each group) is not explained by the existence of different groups.

The second column in Minitab's ANOVA table shows the degrees of freedom (DF). The third column (SS, for sum of squares) shows the separation of the total sum of squared deviations of the data about the overall mean into two sums of squares: the variation among the sample means and the variation within each sample. The fourth column (MS, for mean sum of squares) divides each sum of squares in the third column by the degrees of freedom in the second column: $4.78/3 = 1.59$ and $139.06/76 = 1.83$. The MS values in the fourth column are the numerator and denominator of the ANOVA F statistic, $F = 1.59/1.83 = .87$. The fifth column shows the value of this statistic; and the sixth column gives the P value, the probability of obtaining an F value so far from zero if the null hypothesis were true.

Taste Tests of a New Food Product

In developing a new "easy-to-prepare, nutritious, on-the-run" food product, General Foods compared the taste of two versions of this product—one liquid and the other solid—to that of a competitive product that was being sold successfully.[2] For this taste test, General Foods paid 75 females and 75 males to

HOW TO DO IT

ANOVA F Tests of the Equality of Several Means

We want to use data from k independent random samples to test the null hypothesis that the samples were drawn from populations with the same mean:

H_0: $\mu_1 = \mu_2 = \cdots = \mu_k$
H_1: At least one population mean differs from the others

In the seasonal temperature example, there are $k = 4$ samples, and the $n_1 = n_2 = n_3 = n_4 = 20$ observations give these sample means and standard deviations:

	WINTER	SPRING	SUMMER	FALL
MEAN	−.087	.570	.393	.189
STANDARD DEVIATION	2.133	1.168	.612	1.016

1. Except in the unlikely case in which we are confident that the data come from normal distributions with the same standard deviation, we should try to obtain samples that are reasonably large and roughly the same size. If it turns out that one sample has a standard deviation that is more than twice the size of another, then the ANOVA F test is only a rough approximation. In our example, the samples are the same size, but $s_1/s_3 = 2.133/.612 = 3.5$.
2. Use computer software to calculate the value of the ANOVA F statistic:

$$F = \frac{\text{variation among sample means}}{\text{variation within samples}}$$

The farther the F value is from zero, the stronger is the evidence against the null hypothesis. In our example, $F = .87$.

taste the products and be interviewed for their reactions. These 150 people were divided randomly into three groups of 50, each with 25 females and 25 males. The first group tasted the new liquid product, the second group tasted the new solid product, and the third group tasted the established product. Each person

3. The cutoffs or *P* values can be determined from the fact that the ANOVA *F* statistic has the *F* distribution with the following degrees of freedom:

   Numerator: $k - 1$

   Denominator: $n_1 + n_2 + \cdots + n_k - k$

   Our example has $4 - 1 = 3$ and $20 + 20 + 20 + 20 - 4 = 76$ degrees of freedom.
4. If the *P* value is less than .05, the data reject at the 5 percent level the null hypothesis that all the population means are equal. Here the .54 *P* value is far from statistical significance. Alternatively, the .87 *F* value in Table 5 is far from the 2.7 cutoff for statistical significance at the 5 percent level.
5. An ANOVA table can be used to summarize the results:

SOURCE	DF	SS	MS	F	P
FACTOR	3	4.78	1.59	.87	.537
ERROR	76	139.06	1.83		
TOTAL	79	143.84			

was asked to show her or his reaction to the product by marking a secret ballot, as in Fig. 9.5, showing faces ranging from delight to disgust.

For an ANOVA *F* test, we must assume that the seven boxes on this ballot correspond to a numerical scale, with the difference between any two boxes

Figure 9.5 Pictorial taste-test ballot (the scores +3 to −3 were assigned from left to right).

reflecting a constant difference in taste. General Foods assigned numerical scores to each face, ranging from +3 for delight to −3 for disgust, with the results summarized in Table 9.4. General Foods was naturally pleased to find its product more highly rated, on average, than the established competitor. The statistical question is, Can the observed differences among the average ratings be explained by taste variations from one person to the next? Perhaps, by the luck of the draw, the established product happened to be tasted mostly by people who tend to give low ratings to this type of food product. This sort of luck is always possible; what we can do statistically is to determine the probability of such bad luck.

The null hypothesis is that the three samples come from populations with the same mean:

$$H_0 : \mu_1 = \mu_2 = \mu_3$$

Table 9.4

Ratings of three food products

SCORE	NEW LIQUID	NEW SOLID	ESTABLISHED
+3	7	5	1
+2	13	11	5
+1	16	19	15
0	11	13	17
−1	2	1	8
−2	1	1	3
−3	0	0	1
Mean	1.18	1.06	.22
Variance	1.38	1.16	1.44

The overall mean of the data is .82. Fisher's F statistic can be calculated from

$$\text{Variation among sample means} = \frac{n_1(\bar{x}_1 - \bar{x})^2 + n_2(\bar{x}_2 - \bar{x})^2 + n_3(\bar{x}_3 - \bar{x})^2}{k - 1}$$
$$= \frac{50(1.18 - .82)^2 + 50(1.06 - .82)^2 + 50(.22 - .82)^2}{3 - 1}$$
$$= 13.68$$

and

$$\text{Variation within samples} = \frac{(n_1 - 1)s_1^2 + (n_2 - 1)s_2^2 + (n_3 - 1)s_3^2}{n_1 + n_2 + n_3 - k}$$
$$= \frac{49(1.38) + 49(1.16) + 49(1.44)}{50 + 50 + 50 - 3}$$
$$= 1.33$$

so that

$$F = \frac{\text{variation among sample means}}{\text{variation within samples}}$$
$$= \frac{13.68}{1.33}$$
$$= 10.32$$

Here are the results in an ANOVA format:

ANALYSIS OF VARIANCE

SOURCE	DF	SS	MS	F	P
EXPLAINED	2	27.36	13.68	10.32	.000
UNEXPLAINED	147	194.78	1.33		
TOTAL	149	222.14			

With $k = 3$ samples and each sample of size 50, the degrees of freedom are $k - 1 = 2$ and $n_1 + n_2 + n_3 - k = 147$, and the ANOVA table above shows the probability of an F value larger than 10.3 to be less than .001 (it is actually .0002). Bad luck is always possible, but here that explanation for the observed variation among the ratings is not very plausible. The differences among the sample means are too large to be credibly explained by sampling error. Evidently, the population does, on average, give higher ratings to General Foods' new product.

They Call It Stormy Monday

Mondays have a reputation on Wall Street for often being bad days for the stock market. To investigate this folklore, an empirical study analyzed these data on daily percentage returns over the period from July 3, 1963, to December 18, 1978, with 826 observations for each day of the week.[3]

	MONDAY	TUESDAY	WEDNESDAY	THURSDAY	FRIDAY
MEAN	−.134	.002	.096	.028	.084
VARIANCE	.670	.551	.644	.483	.479

(The −.134 return is 134 thousandths of 1 percent, not 13.4 percent.) During this period, the average daily return was .0152. Only Monday had a negative average return, and it works out to be a staggering −33.5 percent on an annual basis. To see if the differences among the daily returns are statistically significant, Fisher's F statistic can be calculated from

$$\begin{aligned}\text{Variation among sample means} &= \frac{n_1(\bar{x}_1 - \bar{x})^2 + n_2(\bar{x}_2 - \bar{x})^2 + \cdots + n_k(\bar{x}_k - \bar{x})^2}{k-1} \\ &= \frac{826(-.134 - .0152)^2 + \cdots + 826(-.844 - .0152)^2}{5-1} \\ &= 6.992\end{aligned}$$

and

$$\begin{aligned}\text{Variation within samples} &= \frac{(n_1 - 1)s_1^2 + (n_2 - 1)s_2^2 + \cdots + (n_k - 1)s_k^2}{n_1 + n_2 + \cdots + n_k - k} \\ &= \frac{825(.670) + 825(.55) + 825(.644) + 825(.483) + 825(.479)}{826 + 826 + 826 + 826 + 826 - 5} \\ &= .5654\end{aligned}$$

so that

$$F = \frac{\text{variation among sample means}}{\text{variation within samples}} = \frac{6.992}{.5654} = 12.37$$

The numerator has $k - 1 = 5 - 1 = 4$ degrees of freedom, and the denominator has $n_1 + n_2 + n_3 + n_4 + n_5 - k = 5(826) - 5 = 4125$ degrees of freedom. Table 5 shows the cutoff for a test at the 5 percent level to be approximately 2.37. The observed 12.37 F value shatters this cutoff, decisively rejecting the null hypothesis that the days of the week do not matter. Statistical software shows the P value to be .00000081.

Why is Monday so stormy? The researchers who compiled these data tried to find a reasonable explanation and came up empty, concluding that it is "an obvious and challenging empirical anomaly." Until it is proved otherwise, be cautious on Monday.

```
ANALYSIS  OF  VARIANCE
SOURCE       DF         SS       MS         F        P
FACTOR        4      27.97     6.99     12.38     .000
ERROR      4125    2332.27      .57
TOTAL      4129    2360.24
```

exercises

9.11 The Educational Testing Service claims that its SAT tests are a useful predictor of college performance. To test this theory, data were collected on 30 randomly selected first-year students enrolled at the same college, with 10 students from each of the following categories of predicted first-year grade-point average (GPA): 2.33 to 2.66, 2.67 to 2.99, and 3.00 to 3.13. Their actual first-year GPAs are as shown:

2.33–2.66:	2.53	2.90	2.43	2.37	2.60	2.16	2.03	2.80	3.17	2.70
2.67–2.99:	2.43	3.20	2.57	2.80	2.10	3.50	3.00	2.70	3.30	2.90
3.00–3.13:	2.40	3.07	3.40	2.76	3.27	3.83	2.86	3.60	3.17	3.50

What is the average first-year GPA for each of these groups? Calculate the F value and P value for a statistical test using these data. Identify the null hypothesis that you are testing, and interpret your results.

9.12 Forty-five randomly selected college students were each given 5 minutes to study a two-page article from a professional journal.[4] Fifteen of these students studied while listening to hard rock (Nirvana's "Smells Like Teen Spirit"); 15 studied while listening to classical music (Beethoven's Symphony No. 9, Opera 125); and the last 15 studied in quiet. After the 5-minute study period, the music was turned off and the researcher attempted to remove the article from the subjects' short-term memory by having them fill out forms identifying their gender, age, class, and major. Each student was then asked five questions about the article, and the number of right answers was recorded:

ROCK	3	3	4	4	4	3	4	3	2	2	3	2	4	3	2
CLASSICAL	3	3	2	3	4	3	4	4	4	2	3	4	3	2	4
QUIET	3	3	3	5	4	3	3	3	4	5	3	4	4	5	3

Assuming that these data are from populations with normal distributions and identical standard deviations, test at the 5 percent level the null hypothesis that the population means for the rock, classical, and quiet environments are equal.

9.13 Use the data below to investigate whether cigarette brands with high nicotine content also tend to have high carbon monoxide content (both data are in milligrams).[5] To do this, separate the cigarette brands into three categories based on the nicotine content: less than .7, .7 to 1.0, and more than 1.0. Now record the carbon monoxide content of the cigarettes in each group, and use three box plots to display these carbon monoxide data. Finally, use an ANOVA *F* test at the 5 percent level to compare the carbon monoxide means for these three groups.

	NICOTINE	CARBON		NICOTINE	CARBON
Alpine	.86	13.6	MultiFilter	.78	10.2
Benson & Hedges	1.06	16.6	Newport Lights	.74	9.5
Camel Lights	.67	10.2	Now	.13	1.5
Carlton	.40	5.4	Old Gold	1.26	18.5
Chesterfield	1.04	15.0	Pall Mall Light	1.08	12.6
Golden Lights	.76	9.0	Raleigh	.96	17.5
Kent	.95	12.3	Salem Ultra	.42	4.9
Kool	1.12	16.3	Tareyton	1.01	15.9
L&M	1.02	15.4	True	.61	8.5
Lark Lights	1.01	13.0	Viceroy Rich Light	.69	10.6
Marlboro	.90	14.4	Virginia Slims	1.02	13.9
Merit	.57	10.0	Winston Lights	.82	14.9

9.14 A statistics student made a study of the weekly office hours posted by professors in each of three divisions (natural sciences, social sciences, and humanities) at his college.[6]

	NATURAL SCIENCES	SOCIAL SCIENCES	HUMANITIES
MEAN	3.66	3.12	2.91
STANDARD DEVIATION	1.42	1.21	1.68

Here is some computer output:

```
ANALYSIS  OF  VARIANCE
SOURCE     DF        SS      MS       F       P
FACTOR      2      13.0    6.50    2.92    .056
ERROR     148    329.75    2.23
TOTAL     150    342.75
```

Write a brief paragraph summarizing this student's findings.

9.15 At a certain large university, students in an introductory statistics course are randomly assigned to sections taught by different persons. Each section uses the same textbook, does the same homework exercises, and takes the same tests.

Interpret these results of a study of the final examination scores for four sections in the fall of 1983:

ANALYSIS OF VARIANCE

SOURCE	DF	SS	MS	F	P
FACTOR	3	1,543.69	514.56	5.08	.003
ERROR	116	11,748.90	101.28		
TOTAL	119	342.75			

9.4 Simultaneous Confidence Intervals and *t* Values

If we are persuaded to reject the null hypothesis that the sample data come from populations with identical means, then it is natural to look more closely at the differences among the means. Are all the sample means quite dissimilar? Or is one sample mean very different from the rest, which are themselves very similar? And is there a logical, plausible explanation for why the means differ in the way that they do?

One way to proceed is to use our sample means and the standard errors for these means to calculate confidence intervals for each population mean. Recall the general formula, Eq. (6.9):

$$\bar{x} \pm t^*(\text{standard error of } \bar{x}) = \bar{x} \pm t^*\left(\frac{s}{\sqrt{n}}\right)$$

where t^* is the appropriate t value for a given confidence level, such as 95 or 99 percent, and depends on the degrees of freedom.

When we have a single sample, as in Chap. 6, we use that sample's standard deviation in our calculation of the standard error of the sample mean. Here, using ANOVA, we have several samples, and our F test is based on the assumption that the population standard deviations are all equal. The pooled standard deviation s_P, defined in Eq. (9.3), uses data from all k of the samples and is consequently a better estimator of the common population standard deviation σ than is the standard deviation from a single sample. Thus,

A confidence interval for a population mean μ_i is based on the mean of the sample from that population and the pooled standard deviation from all the samples.

$$\bar{x}_i \pm t^*(\text{standard error of } \bar{x}_i) = \bar{x}_i \pm t^*\left(\frac{s_P}{\sqrt{n_i}}\right) \tag{9.4}$$

The appropriate t value can be determined from Table 4 at the end of this book, using the t distribution with $n_1 + n_2 + \cdots + n_k - k$ degrees of freedom.

Instead of calculating the pooled standard deviation directly from Eq. (9.3), we can let our software do the calculations for us. The pooled standard deviation is the square root of the denominator of the ANOVA F statistic, which is shown as the mean sum of squares in an ANOVA table. In the Minitab output in Table 9.3, the pooled standard deviation is the square root of the 1.83 value given in the MS column and ERROR row:

$$s_P = \sqrt{1.83} = 1.353$$

Some software packages (including Minitab) also show the pooled standard deviation separately, in another part of the ANOVA output.

In practice, we would not look closely at the patterns among the individual sample means for our seasonal temperature data, because we were not able to reject the null hypothesis that the population means are equal. Nonetheless, we can use these data as a convenient way of illustrating the computation of confidence intervals. Plus, these intervals are usually shown in the computer output, and it is good to know what they mean, even in those cases in which we decide not to use them. With four samples of size 20, there are $n_1 + n_2 + n_3 + n_4 - k = 20 + 20 + 20 + 20 - 4 = 76$ degrees of freedom, and Table 4 shows the appropriate t value for a 95 percent confidence interval to be approximately $t^* = 2.0$. Substituting $t^* = 2.0$ $s_P = 1.353$, the sample means in Table 9.2, and the common sample sizes $n_i = 20$ into Eq. (9.4), gives our 95 percent confidence intervals:

$$\text{Winter:} \quad -.087 \pm 2.0\left(\frac{1.353}{\sqrt{20}}\right) = .087 \pm .605$$

$$\text{Spring:} \quad .570 \pm 2.0\left(\frac{1.353}{\sqrt{20}}\right) = .570 \pm .605$$

$$\text{Summer:} \quad .393 \pm 2.0\left(\frac{1.353}{\sqrt{20}}\right) = .393 \pm .605$$

$$\text{Fall:} \quad .189 \pm 2.0\left(\frac{1.353}{\sqrt{20}}\right) = .189 \pm .605$$

In this example, the confidence intervals are equally wide because all the samples are the same size; in general, the larger samples will yield narrower confidence intervals. Consistent with our inability to reject the null hypothesis, these confidence intervals all overlap one another. The format used by Minitab to display these confidence intervals is shown in Fig. 9.6.

In those cases in which we reject the null hypothesis, we can also use a sequence of t tests to compare all possible pairings of the sample means.

```
                                       INDIVIDUAL 95 PCT CI FOR MEAN
                                       BASED ON POOLED ST DEV
LEVEL         N    MEAN   ST DEV   ---+---------+---------+---------+---
Winter       20   -.087    2.133   (-----------*-----------)
Spring       20    .570    1.168             (-----------*-----------)
Summer       20    .393     .612           (-----------*-----------)
Fall         20    .189    1.016       (-----------*-----------)
                                   ---+---------+---------+---------+---
POOLED ST DEV =    1.353             -.5        .0        .5       1.0
```

Figure 9.6 Display of ANOVA confidence intervals.

> Equation (8.8) shows a way of calculating the t statistic for a difference-in-means test when the population standard deviations are assumed equal:
>
> $$t = \frac{\bar{x}_1 - \bar{x}_2 - 0}{s_P\sqrt{1/n_1 + 1/n_2}} \tag{9.5}$$

In analysis of variance, we have more than two samples, all of which are assumed to come from populations with identical standard deviations. As with simultaneous confidence intervals, we should use the pooled sample standard deviation s_P based on all k of the samples in our ANOVA F test, and there are $n_1 + n_2 + \cdots + n_k - k$ degrees of freedom.

Although our ANOVA F test did not reject the null hypothesis, we can again use the four seasonal temperature groups for illustrative purposes. In all, there are six possible pairs of comparisons:

Winter/spring: $$t = \frac{-.087 - .570}{1.353\sqrt{1/20 + 1/20}} = -1.54$$

Winter/summer: $$t = \frac{-.087 - .393}{1.353\sqrt{1/20 + 1/20}} = -1.12$$

Winter/fall: $$t = \frac{-.087 - .189}{1.353\sqrt{1/20 + 1/20}} = -.65$$

Spring/summer: $$t = \frac{.570 - .393}{1.353\sqrt{1/20 + 1/20}} = .42$$

Spring/fall: $$t = \frac{.570 - .189}{1.353\sqrt{1/20 + 1/20}} = .89$$

Summer/fall: $$t = \frac{.393 - .189}{1.353\sqrt{1/20 + 1/20}} = .48$$

With four samples of size 20, there are $n_1 + n_2 + n_3 + n_4 - k = 20 + 20 + 20 + 20 - 4 = 76$ degrees of freedom, and Table 4 shows the cutoff for statistical significance at the 5 percent level to be a t value whose absolute value exceeds 2.0. None of these six t values exceeds this threshold.

This procedure, called the *least-significant differences* method, has the weakness described at the beginning of this chapter. By doing a large number of paired comparisons, we increase the chances that at least one t test will falsely reject the null hypothesis. Even more unfortunately, there is no general agreement among statisticians about exactly how to take into account the multiplicity of confidence intervals and tests.

It is clear that to reduce the probability of at least one false rejection of the null hypothesis, we need to raise the t value cutoff for rejecting the null hypothesis. But by how much? A number of competing alternatives (including the Bonferroni and Tukey methods) have been proposed, and none have yet won general acceptance. One widely used statistical program offers more than a dozen choices. More advanced textbooks describe these alternatives in detail and offer suggestions for choosing among them.

My recommendation is to begin with the ANOVA F test to test the null hypothesis that all the population means are equal. If the data do not persuade you to reject this hypothesis, then do not scrutinize the simultaneous confidence intervals and t tests. If you do reject the null hypothesis, then use the simultaneous confidence intervals described here to get a general idea of which specific differences among the sample means are responsible for the rejection of the null hypothesis.

The Stroop Effect

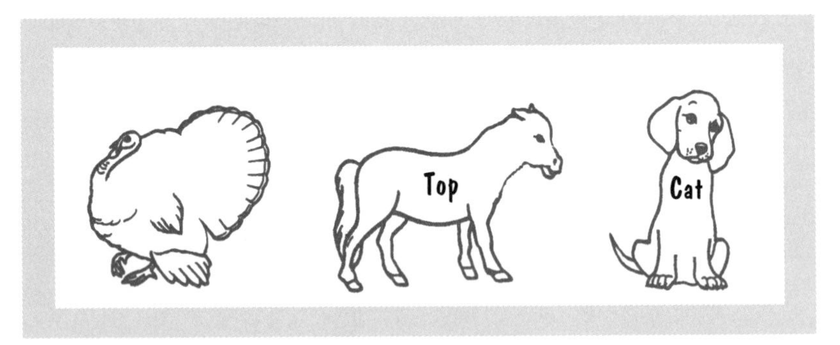

Name the 3 pictures above as quickly as you can. It is easiest to name the turkey and most difficult to name the dog, because we have trained our minds to read words when we see them and this ingrained habit interferes with our efforts to name pictures. When a picture and word send competing signals to the brain, it is difficult to respond quickly and accurately, particularly if the competing word is closely associated with the picture. In the previous figure, it is much easier to identify the horse than to identify the dog. Furthermore, because we can read words faster than we can name pictures, it is easier to read the word and ignore the picture than it is to name the picture and ignore the word.

This mental interference created by our trained inclination to read words was demonstrated in a series of classic experiments in the 1930s by a psychologist named J. R. Stroop.[7] He found that it is hard for people to name the color that letters are written in if the letters spell the name of another color; for example, it is difficult to state that letters are printed in yellow if the letters spell the word GREEN. Although we try to ignore the word and just identify the ink color, we cannot help reading the word—it happens automatically and interferes with our response.

This well known phenomenon—the Stroop effect—demonstrates how highly practiced psychological processes can occur without conscious control. A familiar example is when we are trying to ignore background conversations in a noisy room, and find that it is much harder to ignore someone who says our name than it is to ignore less familiar words.

In one test of the Stroop effect, 30 college students were timed for how fast they could name the color of a string of letters.[8] The different conditions can be illustrated by the case of letters written in the color yellow, with the subject timed in the number of seconds it takes to respond "yellow":

NEUTRAL: A string of O's: OOOOO
UNRELATED: A word unrelated to any color: SHOOT
RELATED: A word related to a non-yellow color: GRASS
COLOR NONRESPONSE: A word that is a color that is not used anywhere in the experiment: BROWN when no letters were ever colored brown.
COLOR RESPONSE: A word that is a color used elsewhere in the experiment: BLACK

Each student was tested several times, with the student's mean response time for each condition considered to be a single observation. Here are the sample means and standard deviations, in milliseconds:

CONDITION	EXAMPLE	MEAN	STANDARD DEVIATION
NEUTRAL	OOOOO	654.09	61.05
UNRELATED	SHOOT	673.50	59.44
RELATED	GRASS	696.38	67.88
COLOR NONRESPONSE	BROWN	755.06	89.54
COLOR RESPONSE	BLACK	793.65	100.09

As expected, the response time is slowed substantially by the extent to which the string of letters is related to the color yellow. To determine if these observed differences are statistically persuasive, we can use an ANOVA test:

ANALYSIS OF VARIANCE

SOURCE	DF	SS	MS	F	P
FACTOR	4	407,045	101,764	17.02	0.000
ERROR	145	867,197	5,981		
TOTAL	149	1,274,242			

The null hypothesis that the population mean response times are equal is rejected decisively. If this null hypothesis were true, there would be less than a .001 probability that the F value would be 17.02 or larger. (The P value is, in fact, less than .0000000001.)

Using the pooled variance, here are 95 percent confidence intervals for the 5 population means:

$$\text{neutral (OOOOO): } 654.09 \pm 1.98\left(\frac{77.33}{\sqrt{30}}\right) = 654.09 \pm 27.96$$

$$\text{unrelated (SHOOT): } 673.50 \pm 1.98\left(\frac{77.33}{\sqrt{30}}\right) = 673.50 \pm 27.96$$

$$\text{related (GRASS): } 696.38 \pm 1.98\left(\frac{77.33}{\sqrt{30}}\right) = 696.38 \pm 27.96$$

$$\text{color nonresponse (BROWN): } 755.06 \pm 1.98\left(\frac{77.33}{\sqrt{30}}\right) = 755.06 \pm 27.96$$

$$\text{color response (BLACK): } 793.65 \pm 1.98\left(\frac{77.33}{\sqrt{30}}\right) = 793.65 \pm 27.96$$

Here is a graphical representation of these confidence intervals:

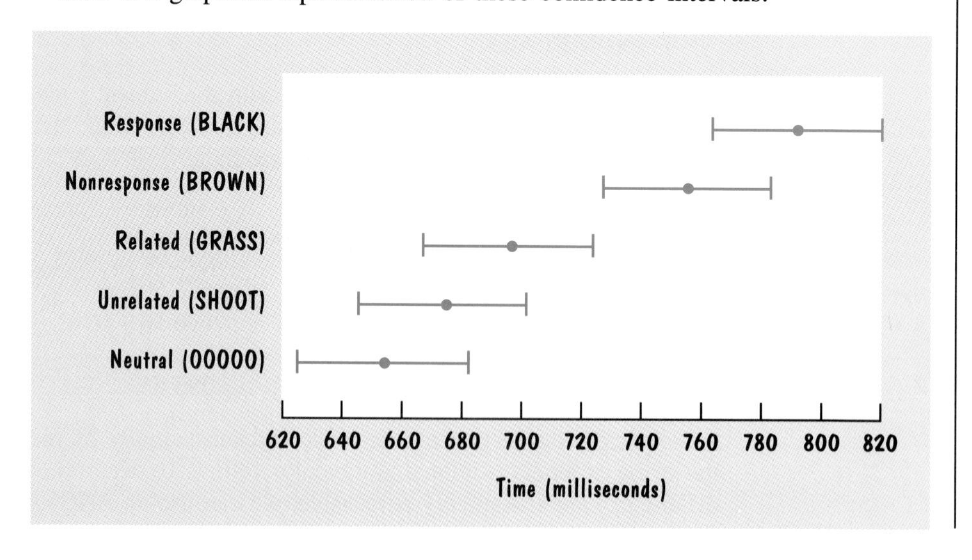

While many of the adjacent confidence intervals overlap, there are clear differences between the extremes. The lower limit for the confidence interval for a color-response word (such as BLACK) is 84 milliseconds above the upper limit for neutral letters (OOOOO) and 64 milliseconds above the upper limit for unrelated letters (SHOOT).

Casual Sleep Is Not Good for You

In the fall of 1995, the final examination for my introductory statistics course was at 9:00 in the morning. Just before handing out the examination, I asked 30 students in this course to write on slips of paper their names and the number of hours they had slept in the past 24 hours. I told them that these slips of paper would be placed in a sealed envelope and not opened until the beginning of the following semester, when the data would be analyzed by that semester's statistics class.

After the envelope was opened the following semester, the class analyzed these data in a variety of ways, including drawing a histogram and a box plot and calculating the mean, median, and standard deviation. I also looked up each student's grade on the final examination and matched that to the number of hours of sleep. These examination scores are shown below, with the data grouped into three categories based on the number of hours of sleep.

The null hypothesis for an ANOVA F test is that if all students are separated into these three sleep categories, the population mean of the final examination score for a student picked at random does not depend on the sleep group from which the student is selected. Here are the sample means and standard deviations for each group:

	LESS THAN 4 HOURS	4 TO 8 HOURS	MORE THAN 8 HOURS
MEAN	74.286	83.412	90.500
STANDARD DEVIATION	9.895	7.133	6.656

Those students who slept more got better grades on average. For a statistical analysis, the ANOVA F test should be appropriate here, since the samples are reasonably sized and no standard deviation is twice the size of another. Here are the results:

```
ANALYSIS  OF  VARIANCE
SOURCE    DF         SS        MS       F       P
FACTOR     2     869.25    434.63    7.23    .003
ERROR     27    1623.05     60.11
TOTAL     29    2492.30
```

If the null hypothesis were true, there is only a .003 probability that the observed differences among the sample means would be so large relative to the variation within each sample.

For determining confidence intervals for the population means, the pooled standard deviation is the square root of the error mean sum of squares (MS):

$$s_P = \sqrt{60.11} = 7.753$$

There are $n_1 + n_2 + n_3 - k = 7 + 17 + 6 - 3 = 26$ degrees of freedom, and Table 4 shows the appropriate t value for a 95 percent confidence interval to be $t^* = 2.056$. Substituting these values and the sample means and standard deviations above into Eq. (9.4) gives our 95 percent confidence intervals:

$$\text{Less than 4:} \qquad 74.286 \pm 2.056\left(\frac{7.753}{\sqrt{7}}\right) = 74.3 \pm 6.0$$

$$\text{4 to 8:} \qquad 83.412 \pm 2.056\left(\frac{7.753}{\sqrt{17}}\right) = 83.4 \pm 3.9$$

$$\text{More than 8:} \qquad 90.500 \pm 2.056\left(\frac{7.753}{\sqrt{6}}\right) = 90.5 \pm 6.5$$

Here is a graphical representation of these confidence intervals:

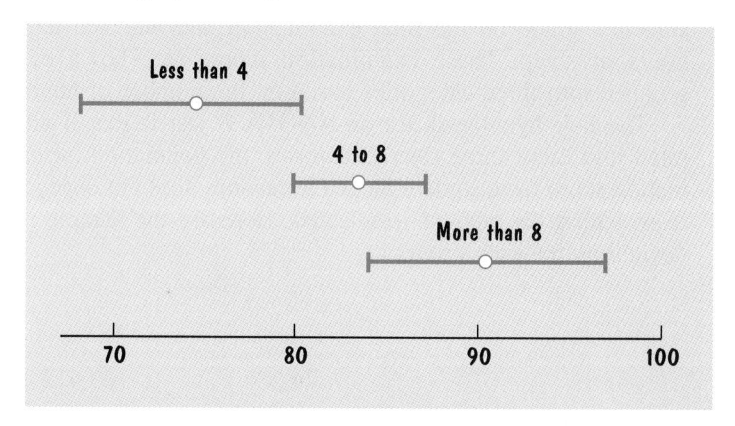

The confidence interval for the middle category (4 to 8 hours) overlaps the intervals for the two extreme categories, but the 95 percent confidence intervals for the two extreme categories are distinct. The evidence is not decisive that those who get 4 to 8 hours of sleep do better, on average, than those who get less than 4 hours, or that those who get more than 8 hours of sleep tend to do better than those who get 4 to 8 hours. However, it is clear that those who get more than 8 hours of sleep do a lot better, on average, than those who get less than 4 hours.

These results do not prove that additional sleep causes students to get better grades; it may be that good students who have worked conscientiously all semester do not need to cram the night before the final and consequently can get a full night's sleep. My advice is to study hard during the semester and sleep well before the final examination, not during it.

LESS THAN 4 HOURS (n = 7)	4 to 8 HOURS (n = 17)		MORE THAN 8 HOURS (n = 6)
60	83	72	97
73	91	90	93
84	69	80	92
75	79	88	78
56	85	92	94
78	84	94	89
85	79	79	
	92	79	
	82		

exercises

9.16 Example 2.4 shows a time-series graph of the number of persons treated for hornet stings at the emergency room of a Cape Cod hospital before, during, and after Hurricane Bob. Here are the data used in that graph, grouped into three categories.

0 to 14 days before:	1	2	3	3	3	2	3	3	1	5	2	2	6	2
0 to 14 days after:	24	23	31	23	25	26	10	10	20	11	14	13	3	5
15 to 28 days after:	4	5	6	2	8	20	7	4	2	2	3	1	0	3

Use three box plots to compare these data, and then apply an ANOVA F test. What is your null hypothesis, and what is your conclusion? If you reject the null hypothesis at the 5 percent level, calculate 95 percent confidence intervals for the 3 population means.

9.17 A study of the language proficiency of international students attending college in the United States asked professors at Pitzer College to rate on a scale of 1 to 5 (with 1 being most proficient and 5 least proficient) the grammatical abilities of international students in their classes whose native language was not English.[9] The following data were obtained for 34 students in these 3 language groups: 16 Asian (native speakers of Chinese, Japanese, or Korean),

7 Austronesian (native speakers of Malay or Tagalog), and 11 Indo-European (native speakers of French, German, Italian, or Spanish).

ASIAN:	2.8	3.9	3.9	2.1	2.2	3.0	1.0	3.0	3.1	2.7	2.4
	2.1	2.7	2.6	2.5	2.7						
AUSTRONESIAN:	3.0	3.5	1.0	3.9	4.4	2.3	3.4				
INDO-EUROPEAN:	2.1	2.1	1.6	1.3	3.0	1.6	1.1	3.1	3.2	2.4	1.0

Use box plots to display the data, and then use an ANOVA F test to see if the differences among these sample means are statistically significant at the 5 percent level. If the P value is less than .05, calculate 95 percent confidence intervals for the three population means. Which language group was, on average, judged the most proficient in grammatical abilities?

9.18 A highway department conducted an experiment with 3 different brands of paint for traffic lanes. Each paint was used on parts of 5 different streets and then measured for brightness 12 months later. Use Fisher's F test to see if there are statistically significant differences at the 1 percent level among these 3 brands. If there are, calculate 99 percent confidence intervals for the population means.

BRAND 1	BRAND 2	BRAND 3
72	80	65
67	71	59
58	65	58
72	71	63
65	66	58

9.19 Twenty-one industrialized countries were divided into 3 groups, based on the percentage of their citizens' 1988 diets that consisted of vegetables.[10] Shown below for each of these countries is the 1988 age-adjusted mortality rate (per 100,000 people) from coronary heart disease for persons aged 35 to 74:

PERCENTAGE OF DIET CONSISTING OF VEGETABLES					
1.9 TO 2.8		1.5 TO 1.8		0.8 TO 1.4	
Australia	210.9	Canada	190.6	Austria	166.6
Belgium	131.2	Denmark	220.4	Finland	296.8
France	71.3	Ireland	299.8	Iceland	211.4
Italy	107.3	Japan	35.6	Israel	182.9
Spain	86.4	Netherlands	166.5	Norway	226.5
United Kingdom	284.7	Switzerland	115.1	Sweden	207.1
New Zealand	265.6	West Germany	171.9	United States	198.6

Summarize these data with three box plots, and then use an ANOVA F test to see if the differences among the 3 group means are statistically significant at the 5 percent level. If so, calculate 95 percent confidence intervals for the three population means.

9.20 I divided the 50 states in the United States plus the District of Columbia into three groups, depending on the percentage of children living in single-parent families in 1994. For the 17 areas in each group, I then recorded the percentage of all births to single teens:

PERCENTAGE OF CHILDREN IN SINGLE-PARENT FAMILIES					
14.4 TO 21.5		21.6 TO 25.7		25.8 TO 57.3	
5.4	7.8	6.3	9.3	7.5	11.3
5.4	8.2	6.6	9.8	8.0	11.5
6.6	8.2	6.8	10.0	8.2	12.0
6.7	8.3	7.0	10.3	8.5	12.0
6.8	8.3	7.0	10.6	9.1	12.4
6.9	8.3	8.0	11.1	9.9	13.3
7.3	8.8	8.1	11.9	10.2	16.2
7.6	9.0	8.2	12.2	10.3	16.5
7.8		8.7		10.5	

Use three box plots to summarize these data. Now use an ANOVA *F* test to see if the differences among the group means are statistically significant at the 5 percent level. If so, calculate 95 percent confidence intervals for the three population means.

summary

To test the null hypothesis that several independent random samples were drawn from populations with the same mean, we can use the ANOVA *F* statistic:

$$F = \frac{\text{variation among sample means}}{\text{variation within samples}}$$

The farther the *F* value is from 0, the stronger is the evidence against the null hypothesis, since the observed differences among the sample means are then large relative to the variation within each group—making it unlikely that sampling variation is responsible for these differences.

If the null hypothesis is true and the underlying populations have normal distributions with the same standard deviation, this *F* statistic has the *F* distribution with the following degrees of freedom:

Numerator:	$k - 1$
Denominator:	$n_1 + n_2 + \cdots + n_k - k$

An ANOVA table can be used to summarize the results.

This F test is fairly robust with regard to the assumptions that the populations have normal distributions with the same standard deviation, as long as (1) the samples are roughly the same size and fairly symmetric without extreme outliers and (2) the largest sample standard deviation is not more than twice the size of the smallest.

If the data reject the null hypothesis that the population means are equal, then we can use simultaneous confidence intervals and t values to analyze further the sources of this rejection.

review exercises

9.21 A chemical firm tested the durability of four kinds of rubber-covered fabric. The following data show the weight loss from rubbing tests on 4 samples of each fabric:[11]

TEST	FABRIC 1	FABRIC 2	FABRIC 3	FABRIC 4
1	268	218	235	236
2	251	241	229	227
3	274	226	273	234
4	270	195	230	225

Use Fisher's ANOVA F test to see if the observed differences are statistically significant at the 1 percent level.

9.22 Table 3.2 shows the 1994 and 1995 annual returns for 17 randomly selected stock mutual funds. Adjust these data to show how well the fund did relative to the market as a whole, by subtracting the market's 1.3 percent return in 1994 from each fund's 1994 return and subtracting the market's 37.4 percent return in 1995 from each fund's 1995 return. For example, American Leaders A's adjusted returns are $.0 - 1.3 = -1.3$ in 1994 and $37.0 - 37.4 = -.4$ in 1995. Now calculate the P value for testing the null hypothesis that the population means for the adjusted returns are the same in 1994 and 1995. First, use a difference-in-means t test, assuming the population standard deviations are equal. Then use an ANOVA F test. Compare your results.

9.23 Three Girl Scout troops operating in very similar areas recorded the following annual sales of cookie boxes. Use box plots to display these data, and then test at the 5 percent level the null hypothesis that these data were drawn from the same normal distribution.

	TROOP 287	TROOP 319	TROOP 403
1986	1238	815	1436
1987	1159	1478	1152

1988	906	1253	1282
1989	1246	1208	1414
1990	1648	1314	1234
1991	1280	1337	1543
1992	1113	948	1438
1993	1114	1215	1282
1994	852	1207	1548
1995	1197	992	1502

9.24 Three long jumpers at a track meet were each given 5 jumps. Do these data (in feet and inches) indicate a statistically significant difference at the 5 percent level in their long-jumping ability? What is the null hypothesis being tested here?

A	23′6″	22′10″	24′2″	24′4″	23′11″
B	24′9″	24′7″	23′5″	25′6″	25′0″
C	25′3″	26′8″	25′8″	24′9″	26′8″

9.25 Apply an ANOVA test to the annual returns on Treasury bonds, corporate bonds, and corporate stock shown in Table 2.1. What is your null hypothesis, and do these data reject it?

9.26 A weather researcher suspected that the heat produced by workers in Melbourne during the workweek might affect the city's temperature.[12] For each Melbourne winter (May through October) from 1964 to 1990, he calculated the average daily maximum temperature for each of the 7 days of the week. Interpret this Minitab output (the temperatures are in degrees Celsius):

```
ANALYSIS  OF  VARIANCE
SOURCE     DF        SS        MS       F       P
FACTOR      6     5.405      .901    1.90    .082
ERROR     182    86.100      .473
TOTAL     188    91.505

                                   INDIVIDUAL  95  PCT  CI'S  FOR  MEAN
                                   BASED  ON  POOLED  STDEV

LEVEL       N      MEAN     STDEV    -------+------+------+------
Mon        27    16.350      .655               (------*------)
Tue        27    16.280      .686             (------*------)
Wed        27    16.234      .553            (------*------)
Thu        27    16.130      .684           (------*------)
Fri        27    15.992      .786      (------*------)
Sat        27    15.871      .712    (------*------)
Sun        27    15.945      .717     (------*------)
                                     -------+------+------+------
POOLED  STDEV  =   .688                   15.90   16.20   16.50
```

9.27 Exercise 3.33 gives the heights of soprano, alto, tenor, and bass singers in the New York Choral Society in 1979. Use an ANOVA *F* test to see if the differences among the average heights for these 4 groups are statistically significant at the 5 percent level. If they are, calculate 95 percent confidence intervals for the population means.

9.28 Below are data from a grocery store for 75 breakfast cereals: cereal name; cereal manufacturer (A = American Home Food Products, G = General Mills, K = Kelloggs, N = Nabisco, P = Post, Q = Quaker Oats, R = Ralston Purina); grams of sugar; and the shelf location (1 = bottom, 2 = middle, 3 = top). Group the data by shelf location, and use 3 box plots to compare the sugar content by shelf location. Now apply an ANOVA *F* test. What is your null hypothesis, and what is your conclusion?

100% Bran	N	6	3	Frosted Flakes	K	11	1	Nutrigrain W.	K	2	3
100% Nat. Bran	Q	8	3	Frosted Mini Wht.	K	7	2	Oatmeal Raisin	G	10	3
All Bran	K	5	3	Fruit & Fibre	P	10	3	Post Nat. R. Bran	P	14	3
All-Bran Extra	K	0	3	Fruitful Bran	K	12	3	Product 19	K	3	3
Almond Delight	R	8	3	Fruity Pebbles	P	12	2	Puffed Rice	Q	0	3
App. Cin. Cheerio	G	10	1	Golden Crisp	P	15	1	Puffed Wheat	Q	0	3
Apple Jacks	K	14	2	Golden Grms.	G	9	2	Quaker Squares	Q	6	3
Basic 4	G	8	3	GrapeNut Flakes	P	5	3	Raisin Bran	K	12	2
Bran Chex	R	6	1	Grape-Nuts	P	3	3	Raisin Nut G.	K	8	3
Bran Flakes	P	5	3	Great Grain Pecan	P	4	3	Raisin Squares	K	6	3
Cap'n Crunch	Q	12	2	Honey Gram Os	Q	11	2	Rice Chex	R	2	1
Cheerios	G	1	1	HoneyNut Chs.	G	10	1	Rice Krispies	K	3	1
Cinn. Toast Crun.	G	9	2	Honey-comb	P	11	1	Shredded Wheat	N	0	1
Clusters	G	7	3	Just Rt. Nuggets	K	6	3	Shred Wht. Bran	N	0	1
Cocoa Puffs	G	13	2	Just Rt. Fruit	K	9	3	Smacks	K	15	2
Corn Chex	R	3	1	Kix	G	3	2	Special K	K	3	1
Corn Flakes	K	2	1	Life	Q	6	2	Strawberry Fruit	N	5	2
Corn Pops	K	12	2	Lucky Charms	G	12	2	Total Corn	G	3	3
Count Chocula	G	13	2	Maypo	A	3	2	Total Raisin	G	14	3
Cracklin Oat Bran	K	7	3	Muesli RD&A	R	11	3	Total Whole	G	3	3
Cream of Wheat (Qk)	N	0	2	Muesli RP&P	R	11	3	Triples	G	3	3
Crispix K	C	3	3	Mueslix Blend	K	13	3	Trix	G	12	2
Crisp Wht. & Rsns.	G	10	3	Multi Cheerios	G	6	1	Wheat Chex	R	3	1
Double Chex	R	5	3	Nut & Honey Cr.	K	9	2	Wheaties	G	3	1
Froot Loops	K	13	2	NutriGrain A-R	K	7	3	Wheaties Honey	G	8	1

9.29 An atomic absorption spectrophotometry analysis of samples of Romano-British pottery at 3 sites in the United Kingdom yielded the data below on the percentage of aluminum oxide in each sample.[13] Apply an ANOVA *F* test and, if the differences among the sample means are statistically significant at the 5 percent level, calculate 95 percent confidence intervals for the 3 population means.

LLANDERYN		ISLAND THORNS	ASHLEY RAILS
14.4	11.6	18.3	17.7
13.8	11.1	15.8	18.3
14.6	13.4	18.0	16.7
11.5	12.4	18.0	14.8
13.8	13.1	20.8	19.1
10.9	12.7		
10.1	12.5		

9.30 Below are measurements of the maximum skull breadth of samples of male Egyptians in 3 different time periods ranging from 4000 B.C. to 150.[14] (The researchers theorized that change in skull size over time would be evidence of the interbreeding of the Egyptians with immigrant populations.) Compare these 3 samples with box plots and an ANOVA F test. If the P value is less than .05, calculate 95 percent confidence intervals for the three population means.

4000 B.C.			1850 B.C.			150		
131	129	126	137	137	132	137	143	138
125	134	135	129	137	133	136	141	131
131	126	134	132	136	138	128	135	143
119	132	128	130	137	130	130	137	134
136	141	130	134	129	136	138	142	132
138	131	138	140	135	134	126	139	137
139	135	128	138	129	136	136	138	129
125	132	127	136	134	133	126	137	140
131	139	131	136	138	138	132	133	147
134	132	124	126	136	138	139	145	136

9.31 The first and fifth columns in Exercise 6.47 are data for first-year students; the second and sixth columns are second-year students; the third and seventh columns are third-year students; and the fourth and eighth columns are fourth-year students. Use an ANOVA F test to see if the differences among the hours of sleep for these 4 groups are statistically significant at the 5 percent level. If they are, calculate 95 percent confidence intervals for the population means.

9.32 A standardized test of logical reasoning was given on the first day of class to students enrolled in introductory mathematics courses at a California State University campus.[15] The test results were not made available to the professors teaching these courses until after the courses were completed and grades had been assigned. An ANOVA test was then used to see if the logical reasoning scores were related to the course grades. The F value was 12.85, with a P value of .000000001. Here are additional results:

GRADE	SAMPLE SIZE	MEAN LOGICAL REASONING SCORE	STANDARD DEVIATION	95% CONFIDENCE INTERVAL
A	85	16.90	2.53	16.90 ± 0.68
B	95	15.65	3.26	15.65 ± 0.65
C	85	14.01	3.27	14.01 ± 0.68
D	18	12.83	4.33	12.83 ± 1.48
F	14	13.57	4.38	13.57 ± 1.68

Summarize these results.

9.33 Thirty randomly selected college students were asked to taste 3 brands of chocolate chip cookies (Pepperidge Farms, SnackWell, and Chips Ahoy) and rate each on a scale of 1 to 10, with 1 the worst possible rating and 10 the best.[16] An ANOVA test of the data yielded the results below. Write a brief summary of the most important conclusions that you draw from these results.

ANALYSIS OF VARIANCE

SOURCE	DF	SS	MS	F	P
FACTOR	2	44.60	22.30	8.65	.0004
ERROR	87	224.30	2.58		
TOTAL	89	268.90			

VARIABLE	N	MEAN	STDEV	95 PCT CI'S FOR MEAN
Pepperidge	30	5.4667	1.7953	5.4667 ± .5830
SnackWell	30	6.0667	1.4606	6.0667 ± .5830
Chips Ahoy	30	4.3667	1.5421	4.3667 ± .5830

9.34 In 1991, 48 college sophomores (24 females and 24 males) were surveyed regarding their grade point average and the typical number of hours spent studying each week.[17] Here are the means and standard deviations for the grade data:

	FEMALES	MALES
MEAN	3.39	3.22
STANDARD DEVIATION	0.311	0.386

Here is an ANOVA table:

ANALYSIS OF VARIANCE

SOURCE	DF	SS	MS	F	P
FACTOR	1	.348	.348	2.83	.100
ERROR	46	5.653	.123		
TOTAL	47	6.001			

Summarize these results.

9.35 Here are the mean and standard deviations for the data described in the preceding exercise on the typical number of hours spent studying each week:

	FEMALES	MALES
MEAN	17.2	14.8
STANDARD DEVIATION	7.30	8.90

Here is an ANOVA table:

```
ANALYSIS  OF  VARIANCE
SOURCE    DF         SS       MS       F       P
FACTOR     1      69.12    69.12    1.04   0.312
ERROR     46    3047.50    66.25
TOTAL     47    3116.62
```

Summarize these results.

9.36 Reanalyze the data in Exercises 9.34 and 9.35, this time using 2 difference-in-means t tests with pooled variances.

9.37 Shown below are the annual growth rates of real per capita income in 41 developing countries classified by the degree to which they restricted international trade.[18] Summarize these data with three box plots, and then use an ANOVA F test to see if the differences among the 3 group means are statistically significant at the 1 percent level. Summarize your findings.

FREE-TRADE		MODERATE RESTRICTIONS			STRONG RESTRICTIONS		
6.5	1.5	5.6	1.3	−1.0	2.0	−1.1	−3.2
6.3	1.4	4.0	1.1	−1.2	2.0	−1.6	−3.4
5.4	0.4	3.3	0.3	−3.5	1.2	−2.0	
4.1	0.4	3.1	−0.1	−3.9	0.5	−2.3	
3.8	0.1	2.7	−0.8		0.5	−2.5	
2.9		1.8	−1.0		−0.4	−3.1	

9.38 Use a computer random number generator to select 25 random 2-digit numbers. (Numbers can be selected more than once.) Do this 4 times and then apply an ANOVA test to these 4 sets of data. What is your null hypothesis, and do these data reject it at the 5 percent level? If every student in your class does this test, how often do you anticipate the null hypothesis being rejected at the 5 percent level? (If you do not have access to a computer random number generator, use Table 1 in this textbook, being careful that your 4 sets of numbers do not overlap.)

9.39 Exercise 8.48 gives the season-best times for the 3000 meters for 4 categories of runners. Use an ANOVA F test to see if the differences among the average times for these 4 groups are statistically significant at the 5 percent level. If they are, calculate 95 percent confidence intervals for the population means.

9.40 Samples from a can of well-mixed dried eggs were sent to 6 commercial laboratories to be analyzed for the percentage fat content.[19] The laboratories did not know that all of the samples came from the same can and consequently had the same fat content. Use an ANOVA *F* test to see if the differences among the laboratory measurements are statistically significant at the 1 percent level. If they are, calculate 99 percent confidence intervals for the population means, and display these intervals in a graph.

LAB1	LAB2	LAB3	LAB4	LAB5	LAB6
.62	.30	.46	.18	.35	.37
.55	.40	.38	.47	.39	.43
.34	.33	.27	.53	.37	.28
.24	.43	.37	.32	.33	.36
.80	.39	.37	.40	.42	.18
.68	.40	.42	.37	.36	.20
.76	.29	.45	.31	.20	.26
.65	.18	.54	.43	.41	.06

For each of the following projects, type a report in ordinary English, using clear, concise, and persuasive prose. Any data that you collect for this project should be included as an appendix to your report. Data used in your report should be presented clearly and effectively.

9.1 Among students at your college, is there a statistical relationship between birth order (first-born or only child, middle-born, or last-born) and grade-point average?

9.2 Among students at your college, is there a statistical relationship between year in college (first, second, third, or fourth) and grade-point average?

9.3 Among seniors at your college, is there a statistical relationship between major (humanities, natural sciences, or social sciences) and the number of hours per week spent on school work (in classes, in laboratories, or studying)?

9.4 Identify at least 4 categories of athletic shoes; for example, aerobic, basketball, cross-training, racket sports, running. From a catalog or large store, determine the prices of at least 5 female shoes in each category and 5 males shoes in each category. For each gender, see if the differences in shoe prices among your categories are statistically significant at the 5 percent level.

9.5 Do more-expensive chocolate chip cookies taste better than less-expensive ones? Restrict your comparison to three brands, either all soft or all crispy, and ask your subjects to rate each brand on a scale of 1 to 10.

9.6 Exercise 9.22 describes an ANOVA test using the stock return data for 1994 and 1995 in Table 3.2. Follow the instructions in this exercise, but also use the data for 1996, 1997, and any other years that have since been reported in the annual February issue of *Money* magazine.

9.7 Follow the instructions for Exercise 9.25, but this time use the reference given in Table 2.1 to extend your data back to 1926.

9.8 Follow the instructions in Exercise 9.28, but this time go to a local grocery store and collect your own data.

9.9 Have each team member select a teacher, and record the number of times that this teacher says “Uh” or “Um” during the class period. Do this for at least 10 class periods. Use ANOVA to test the null hypothesis that the population means are equal.

9.10 Ask 50 randomly selected female students whose mother and father are both alive to estimate the number of minutes during the past week they spoke (either by telephone or in person) to their mother and the number of minutes during the past week they spoke to their father. Repeat this survey for 50 randomly selected male students. Now divide your data into 4 categories: female/father, female/mother, male/father, and male/mother. What patterns do you see in these data? Use ANOVA to test the null hypothesis that the 4 population means are equal.

9.11 Use a computer program to generate random stock price changes by simulating a sequence of flips of a fair coin. Assume that (1) stock prices go up $1 if a head is flipped and down $1 if a tail is comes up and (2) 100 flips represent a day’s trading. Set up a calendar of 10 five-day weeks—Monday, Tuesday, Wednesday, Thursday, and Friday—and record the daily price changes for these 50 days. Do not record each of the 100 flips each day, just the net price change that day. Now calculate the average price change on each of the 5 days of the week. Which day of the week was the most profitable? Using an F test, calculate the P value of the null hypothesis that the population mean of the stock return is the same for each day of the week.

9.12 In this project, you will compare the prices for music compact disks (CDs) in 3 local stores. Begin by making a list of at least 50 CDs that you consider worth owning. In compiling this list, you can look through the CDs that you and your friends already own, or you can browse at a store that won’t be included in your shopping comparison. Do not include any multiple-CD collections. After you have your list, go to each of the 3 selected stores and write down the price of each CD (or indicate that it was not in the store). Are there important differences in the availability of these CDs? For those CDs that are sold by all three stores, use ANOVA to see if there are statistically significant differences among the stores' prices.

9.13 Find where franks (hot dogs) are sold at a local grocery store. Divide the franks into at least 2 categories (for example, beef and turkey) and record the cholesterol, fat and sodium in at least 10 brands in each category. If one brand has a regular variety and a low-cholesterol, low-fat, or low-sodium variety, use the regular variety. Use an ANOVA test to see if the differences in cholesterol among the frank categories are statistically significant at the 5 percent level. Do the same for fat and sodium.

9.14 Look at the 3 most recently completed major league baseball seasons. In each league (American and National), identify the 10 players with the highest batting averages in the middle season. Taking these 10 players as your sample, record their batting averages in all 3 seasons. If, for example, you choose the 1995, 1996, and 1997 seasons, identify the ten best batters in 1996 and then record their average batting average in 1995, 1996, and 1997. For each league, use an ANOVA test to see if their batting averages vary by season.

9.15 Follow the directions for the preceding project, using earned run averages instead of batting averages.

9.16 Ask at least 100 randomly selected college students to write down their grade point average (GPA) and to indicate where they typically sit in large classrooms: in the very front, towards the front, in the middle, towards the back, or in the very back. If feasible, restrict your sample to students who are taking the same class or similar classes. Use an ANOVA F test to see if the differences in the GPAs among these 5 categories are statistically significant at the 5 percent level.

9.17 Divide the departments at your school into 3 divisions; for example, humanities, natural sciences, and social sciences. Now ask a random sample of fourth-year students who plan on working immediately after graduation to tell you their major and their predicted first-year salary. Use an ANOVA F test to see if there are statistically significant differences among the salaries by academic division.

9.18 Divide the departments at your school into 3 divisions; for example, humanities, natural sciences, and social sciences. Now ask a random sample of students to tell you their major and their predicted annual salary 10 years after graduation. Use an ANOVA F test to see if there are statistically significant differences among the salaries by academic division.

9.19 From a daily newspaper, collect data for 1 year on the volume of trading on the New York Stock Exchange each day of the week: Monday, Tuesday,

Wednesday, Thursday, and Friday. Use an ANOVA F test to see if the differences in the daily volume of trading are statistically significant at the 1 percent level.

9.20 From a daily newspaper, collect data for 1 year on the percentage change in the Dow Jones Industrial Average of stock prices each day of the week: Monday, Tuesday, Wednesday, Thursday, and Friday. Use an ANOVA F test to see if the differences in the daily percentage changes are statistically significant at the 1 percent level.

9.21 Identify 3 categories of radio stations (for example, easy listening, rock, and talk), making sure that there are at least 5 radio stations in your area in each category. Listen to each station during a specified time interval (for example, from 10:15 to 10:45 on a weekday morning) and record the amount of time spent on commercials during this interval. Are the differences in commercial time among these categories statistically significant at the 5 percent level?

9.22 Divide the departments at your school into 3 divisions; for example, humanities, natural sciences, and social sciences. Now use data for a random sample of courses to see if class sizes vary by division.

9.23 Ask a random sample of 100 college students to tell you their year in college and the average number of students in the classes they are taking this term. Use an ANOVA test to see if the differences in class sizes grouped by year in college are statistically significant at the 1 percent level.

9.24 For each month for the past 10 years, randomly select a day and use an old newspaper or other source to determine the amount of precipitation in a nearby city on that day. The 12 months are the categories and the sample for each category consists of the precipitation data on your 10 randomly selected days. Use an ANOVA test to see if the differences in monthly precipitation are statistically significant at the 5 percent level.

9.25 Follow the instructions for the preceding project, but, instead of the daily precipitation, use the difference between the high and low temperatures on each randomly selected day.

HAVE YOU EVER WONDERED ?

Since psychiatrist Alfred Adler's seminal work in the 1920s, there have been more than 2000 studies of behavioral differences among first-born and later-born children. Many are contradictory and unpersuasive; some are rigorous and compelling.

The Harvard Preschool Project, a major study lasting more than a decade, found that parents spent twice as much time with a first-born child than with a later-born, and that the first-born scored higher at 3 years of age on standardized tests of language and intelligence. Other studies emphasize how personality might be influenced by the dynamics among children of different ages interacting with one another and with their parents. First-born and only children often seem to have more self-control and self-esteem and to be driven to succeed. Middle children seek attention and do not respond well to punishment. The last-born are said to be carefree, charming, and manipulative.

A 25-year study by Frank Sulloway, supported in part by a $192,000 "genius grant" from the MacArthur Foundation, argues that genetic differences evolved during eons of favoritism toward oldest children, who were the biggest, strongest, and most likely to survive. An extreme example is that the practice of infanticide when resources are scarce never involves the first-born; less extreme is the labeling of young children in ancient Japan as "cold rice," because they were fed after the oldest son. Sulloway reasons that the first-born carry genes primordially programmed to respect parents and the established order, while the later-born are biologically inclined to take risks and challenge authority. John Adams, Calvin Coolidge, Ayn Rand, and Rush Limbaugh were first-borns; Charles Darwin, Karl Marx, Lenin, Fidel Castro, and Bill Gates were later-borns. By his calculations, in the 1850s, later-borns were 5 times as likely as first-borns to accept Darwin's revolutionary theories of evolution.

All these theories—time with parents, family dynamics, and genetics—suggest that first-borns are more likely than later-borns to be law-abiding. Is it true? In this chapter, we use a new statistical tool to answer this question.

chapter 10

chi-square tests for categorical data

When I was a young man about to go out into the world, my father says to me a very valuable thing. He says to me like this: "One of these days in your travels, a guy is going to come up to you and show you a nice brand-new deck of cards on which the seal is not yet broken, and this guy is going to offer to bet you that he can make the jack of spades jump out of the deck and squirt cider in your ear. But, son, do not bet this man, for as sure as you stand there, you are going to wind up with an earful of cider."

—*Sky Masterson, in Damon Runyon's* Guys and Dolls

THE ANOVA *F* test described in Chap. 9 can be used with numerical data that can be averaged; for example, we used ANOVA to compare changes in seasonal temperatures in the United States. This chapter introduces a test that can be used when the data are categorical: the plant lived or died, the person is a convicted felon or not.

This test can be used for two very different situations. We begin with the case of qualitative data that have been separated into several categories, for instance, putting politicians' last names into these alphabetical groups: A to G, H to S, or T to Z. Our statistical procedure is able to test any null hypothesis about how the data fall into these categories. For our political example, the null hypothesis might be that the alphabetical order of last names is politically unimportant, so that if 30 percent of the population has a last name beginning with a letter in the category A to G, then there is a 30 percent chance that a politician selected at random will have a last name in this category.

The second situation that we consider involves data that have been categorized in two different ways, for example, by political views and gender, or by juvenile delinquency and birth order. Here, our statistical procedure enables us to test the null hypothesis that the two categorizations are independent, that political views are independent of gender, that juvenile delinquency is unrelated to birth order.

10.1 Tests for Several Categories

We begin with an example from a student's statistics paper.[1] While mulling over the fallacious law of averages, this student realized that the common belief that the average outcome is the most likely outcome might explain a tendency he had noticed for people who are asked to predict a randomly selected number to avoid extremes. If asked to choose a number from 1 to 4, few people will choose 1 or 4; if asked to choose a number from 1 to 10, very few will choose 1 or 10. Perhaps people mistakenly believe that because the average is in the middle, a middle number is more likely to occur.

To test this theory, he set up a mock extrasensory perception (ESP) experiment in which he thoroughly shuffled four cards numbered 1 to 4, closed his eyes and selected a card, and concentrated on the number written on the card while the participant tried to identify the card selected. His real interest was whether they would avoid the extreme numbers 1 and 4. He repeated this charade for 40 randomly selected people and obtained these results:

CARD NUMBER	TIMES SELECTED
1	2
2	13
3	21
4	4

The experiment turned out as he had anticipated, with very few people choosing the extreme values. But are the tendencies shown in this sample of 40 people persuasive, or might they be explained reasonably well by chance?

Our null hypothesis is that people are equally likely to choose any of the four numbers, just as if their selections were made by rolling a four-sided die, with the numbers 1, 2, 3, and 4 on the sides. If so, what is the probability that the number 1 will appear just twice, or that the number 3 will appear 21 times? We might be tempted to do a series of binomial tests for some of (or all) the individual numbers, but there are several reasons why we should not. First, the outcomes are related, in that if there are few 1s, there must be a large number of 2s, 3s, or 4s. Second, if we make several tests, it is more likely that we will incorrectly reject at least one of these null hypotheses. Third, if we look at the data and pick out anomalies, such as the large number of 3s, this is data mining. We should specify the null hypothesis before we see the data, not after. All these reasons argue for a statistical test that looks at the number of 1s, 2s, 3s, and 4s simultaneously.

The Chi-Square Statistic

In 1900, Karl Pearson developed a test statistic that simultaneously compares all the observed and expected numbers when the possible outcomes are divided into mutually exclusive categories.

The **chi-square statistic** is

$$\chi^2 = \sum \frac{(\text{observed} - \text{expected})^2}{\text{expected}} \tag{10.1}$$

where χ is the lowercase Greek letter chi and the summation, designated by capital Greek sigma Σ, is over the categories of possible outcomes. A large value for χ^2 provides statistical evidence against the null hypothesis.

The calculation of this test statistic is laid out in Table 12.1, and a step-by-step explanation will help explain Pearson's logic. First, we specify the expected value for the number of observations in each category if the null hypothesis is true. In our example, the null hypothesis is that each of the four possible numbers has a .25 probability of being selected by a participant, so that the expected value of the number of times it will appear in 40 observations is $np = 40(.25) = 10$. Second, we record the observed number of observations in each category. Third, we measure the extent to which the null hypothesis is contradicted by the data, by squaring the deviation between the observed number and the expected number in each category. This squaring keeps positive and negative deviations from canceling and, in addition, emphasizes large deviations because the improbability of large deviations makes their presence more convincing evidence that the null hypothesis is not true. Fourth, we divide each squared deviation by the expected number for that category. This scaling takes into account the fact that a deviation of 10 when 10 are expected provides more evidence against the null hypothesis than does a deviation of 10 when 1000 are expected. Finally, we add the scaled squared deviations. Because the chi-square statistic is the sum of squared deviations, it cannot be negative. A chi-square value close to 0 shows that the observed values are very close to what we expect if the null hypothesis is true. A large chi-square value is evidence against the null hypothesis.

Table 10.1 shows that the chi-square value is 23.0 for our card-picking example. Is this value large enough to discredit the null hypothesis? To answer this question, we need to know the probability that a chi-square value would be so far from 0, were the null hypothesis true. We need to know the probability distribution for the chi-square statistic.

The Chi-Square Distribution

It can take a substantial amount of time, even with a computer, to calculate the exact probabilities for various values of the chi-square statistic. Not having any computer at all, Pearson figured out an approximation, based on the ubiquitous central limit theorem. Because each of the deviations in Eq. (10.1) is approximately normally distributed, the probability distribution for the chi-square statistic can be approximated by the **chi-square distribution,** which is the distribution for the sum of the squared values of normally distributed variables.

The chi-square distribution can be calculated readily and is shown in Table 6 at the end of this book. Like the t distribution and the F distribution, the chi-square distribution depends on the number of degrees of freedom. Figure 10.1 shows 3 chi-square distributions, for 4, 6, and 10 degrees of freedom. Notice

Table 10.1

Calculation of a chi-square value

CARD	OBSERVED	EXPECTED	(OBSERVED − EXPECTED)^2	$\frac{(\text{OBSERVED} - \text{EXPECTED})^2}{\text{EXPECTED}}$
1	2	10	$(-8)^2 = 64$	6.4
2	13	10	$3^2 = 9$	.9
3	21	10	$11^2 = 121$	12.1
4	4	10	$(-6)^2 = 36$	3.6
Total	40	40		23.0

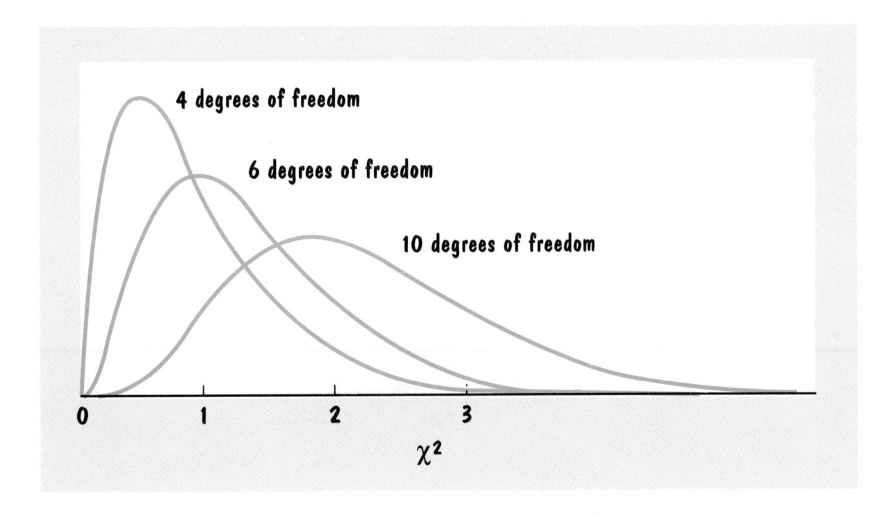

Figure 10.1 Three chi-square distributions.

that negative values are not possible and that as the number of degrees of freedom increases, the chi-square distribution approaches a normal distribution. A general rule of thumb is that the chi-square distribution is a good approximation as long as none of the expected numbers are smaller than 5. If some categories have very small expected numbers, these can be combined with other categories. For example, if 50 participants had been asked to choose a number from 1 to 100, we might group all the responses into 10 categories: 1 to 11, 11 to 20, and so on.

When there is a single list of possible categories, the degrees of freedom is equal to the number of categories minus 1. This can be explained by the fact that if we know the values in all but one category, then we know the value in the last category, too. For our card example, there are 4 categories and consequently $4 - 1 = 3$ degrees of freedom. Table 6 shows that, with 3 degrees of freedom, there is only a .05 probability that the value of the chi-square statistic will exceed 7.81. Figure 10.2 shows this chi-square distribution and the 7.81 cutoff for a test at the 5 percent level. The actual chi-square value of 23.0 is far beyond this cutoff and consequently rejects the null hypothesis decisively. Statistical software shows the P value to be a minuscule .00014. The tendency to choose the middle numbers (2 and 3) and avoid the extremes (1 and 4) yielded deviations between the observed and expected numbers under the null hypothesis that are too large to be explained plausibly by sampling error. (Incidentally, I now repeat this experiment every semester on the first day of class, and there is invariably a statistically significant aversion to the numbers 1 and 4.)

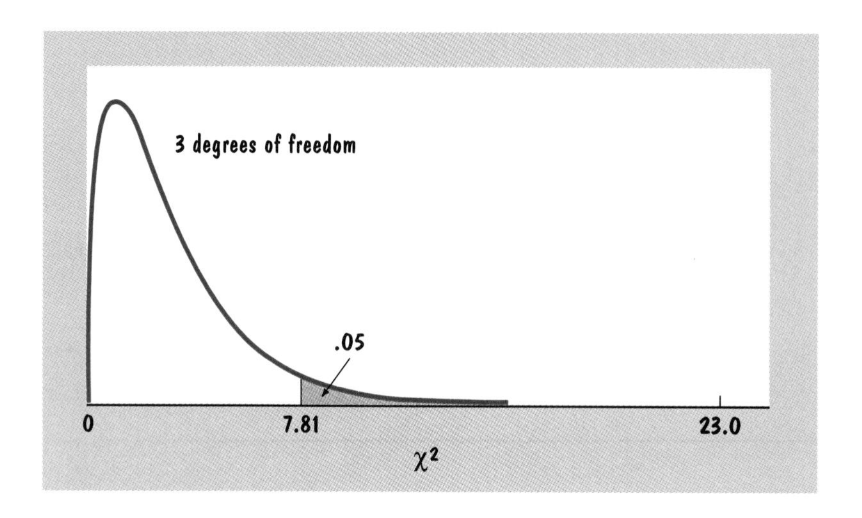

Figure 10.2 A chi-square test.

EXAMPLE 10.1

Alphabetical Listing of Political Candidates

Many political observers believe that election results are influenced by the order in which candidates are listed on the ballot. Some indecisive citizens reputedly vote for the first name that sounds reasonable, or at least familiar, thus giving candidates whose names appear early on a list an advantage over those listed later. In cases where names are listed alphabetically, candidates whose surname (last name) begins with a letter early in the alphabet presumably have an advantage. A political party wanting to exploit this advantage could deliberately nominate candidates with early-letter names. (This is the same logic that causes some firms that advertise in telephone directories to name their company AA Plumbing or AAAAA Appliance Repair.) Concerned that order may matter, many governments now draw names out of a hat or use some other random process to determine the order in which candidates' names are listed on ballots.

Researchers investigating the effects of alphabetical listing of candidates in the 1973 general elections in the Republic of Ireland analyzed the data in Table 10.2.[2] The surnames have been alphabetically divided into five categories that encompass an approximately equal number of registered voters. (Last names that begin with M are especially common in Ireland.) First, we use the chi-square statistic to test the null hypothesis that the nominated candidates are selected without regard for surname. If so, we expect roughly 20.3 percent of

HOW TO DO IT

Chi-Square Tests for a Single List of Categories

We want to test whether people asked to identify a selected card numbered 1, 2, 3, or 4 are equally likely to pick each of the numbers.

1. Divide the data into mutually exclusive categories, here the four possible card numbers.
2. Specify the null hypothesis and the expected number of observations in each category if the null hypothesis is true. Here, each number has an equal probability of being selected. With 4 categories and 40 observations, the expected number in each category is $np = 40(.25) = 10$.
3. Check that no expected number is smaller than 5. This is okay here.
4. Determine the observed number in each category; here, 2, 13, 21, and 4.
5. Calculate the value of the chi-square statistic, as here:

$$\chi^2 = \sum \frac{(\text{observed} - \text{expected})^2}{\text{expected}}$$

$$= \frac{(2-10)^2}{10} + \frac{(13-10)^2}{10} + \frac{(21-10)^2}{10} + \frac{(4-10)^2}{10}$$

$$= 23.0$$

6. Use statistical software or Table 6 to find the probability that the value of a chi-square statistic with (number of categories − 1) degrees of freedom would be so far from 0. Here, with $4 - 1 = 3$ degrees of freedom, the P value is .00014.
7. A P value below .05 indicates that the results are statistically significant at the 5 percent level. These data decisively reject the null hypothesis.
8. We interpret our results by considering whether the observed differences from the expected numbers make sense. Here, we anticipated that people would avoid the extreme values 1 and 4, and this turned out to be the case.

Table 10.2

Irish surnames and elections

FIRST LETTER OF SURNAME	FRACTION OF IRISH VOTERS	NUMBER OF CANDIDATES	NUMBER ELECTED
A–C	.203	91	47
D–G	.179	63	32
H–L	.172	59	23
M–O	.253	75	22
P–Z	.194	47	20
Total	1.001	335	144

the 335 candidates to have last names that begin with the letter A, B, or C; 17.9 percent to begin with D, E, F, or G; and so on. Thus, we have the following data on the observed and expected numbers under the null hypothesis:

SURNAME	OBSERVED	EXPECTED	OBSERVED − EXPECTED
A–C	91	.203(335) = 68.0	23.0
D–G	63	.179(335) = 59.9	3.1
H–L	59	.172(335) = 57.6	1.4
M–O	75	.253(335) = 84.7	−9.3
P–Z	47	.194(335) = 65.0	−18.0
Total	335	335.2	.2

There are many more candidates with A to C surnames and many fewer with P to Z surnames than would be expected if candidates were chosen without regard for the alphabetical ordering of names. From Eq. (10.1), the value of the chi-square statistic turns out to be 14.1:

$$\begin{aligned}\chi^2 &= \frac{23.0^2}{68.0} + \frac{3.1^2}{59.9} + \frac{1.4^2}{57.6} + \frac{(-9.3)^2}{84.7} + \frac{(-18.0)^2}{65} \\ &= 14.1\end{aligned}$$

With 5 − 1 = 4 degrees of freedom, Table 6 shows that the cutoffs are 9.49 for a 5 percent test and 13.28 for a 1 percent test. These data reject at the 1 percent level the null hypothesis that candidates were selected without regard for surname. (Statistical software shows the P value to be .007.)

We can also ask whether, given the candidates presented to them, Irish voters are influenced by the alphabetical listing of names on the ballot. The null hypothesis is that they are not. Because 91 of 335 candidates had A to C surnames, the null hypothesis implies that we expect a fraction 91/335 of the 144 people elected to have A to C surnames. Using similar logic for the other four categories, we have the following:

SURNAME	OBSERVED	EXPECTED	OBSERVED − EXPECTED
A–C	47	(91/335)144 = 39.1	7.9
D–G	32	(63/335)144 = 27.1	4.9
H–L	23	(59/335)144 = 25.4	−2.4
M–O	22	(75/335)144 = 32.2	−10.2
P–Z	20	(47/335)144 = 20.2	−.2
Total	144	144.0	.0

These results are much less clear-cut. The differences are noticeably smaller, and the P to Z surnames won almost exactly as many elections as expected if the null hypothesis were true. The chi-square value works out to be 5.94, which is substantially less than the 9.49 cutoff for statistical significance at the 5 percent level (the *P* value is .20). While these data reject at the 1 percent level the null hypothesis that the candidates are selected without regard for surname, they do not reject the null hypothesis that those elected are selected from the candidates without regard for surname. Perhaps the voters care less about alphabetical listing than the politicians do.

exercises

10.1 Two hundred randomly selected college students[3] were asked to pick a number from 1 to 10. Use the data shown below to test at the 1 percent level the null hypothesis that each of these 10 numbers is equally likely to be chosen. What pattern do you see in the responses?

NUMBER	1	2	3	4	5	6	7	8	9	10
RESPONSES	4	8	31	37	13	16	57	24	6	4

10.2 A study of the relationship between automobile weight and accident frequency used the data below.[4] Test at the 5 percent level the hypothesis that the probability that an accident will fall into a weight class is equal to its registration frequency. (Show all calculations.)

AUTOMOBILE WEIGHT (POUNDS)	REGISTRATION FREQUENCY (PERCENT)	NUMBER OF ACCIDENTS
Less than 3000	21.04	162
3000–3999	46.13	318
4000–4999	31.13	689
5000 or more	1.70	35

10.3 In 1992, a student counted the number of candies of each color in a 1-pound bag of plain M&M's.[5] Assume that these data are a random sample from the population of all M&M's.

COLOR	Brown	Green	Orange	Red	Tan	Yellow
NUMBER	154	39	48	86	61	134

Test at the 1 percent level the following null hypotheses about the population:

a. All colors are equally likely.

b. The manufacturer claims that 30 percent of all plain M&M's are brown, 20 percent are red, 20 percent are yellow, 10 percent are green, 10 percent are orange, and 10 percent are tan. (In September 1995, tan M&M's were replaced by blue.)

10.4 A study compared the birth and death dates of 348 deceased people listed in the book *Four Hundred Notable Americans,* to see whether famous people might be able to postpone death until after the celebration of a birthday.[6] The data were divided into 12 categories:

CATEGORY	−6	−5	−4	−3	−2	−1	0	1	2	3	4	5
NUMBER	24	31	20	23	34	16	26	36	37	41	26	34

Here −6 means that the person's death was 6 months before the birth month, and 0 means that the person's death was during the birth month. Test at the 1 percent level the null hypothesis that a death is equally likely to occur in any of the 12 categories.

10.5 The random walk hypothesis states that changes in stock prices cannot be predicted from previous changes. To test this theory, a student randomly selected 128 five-day periods and, for each 5-day period, calculated the number of days that the Dow Jones Industrial Average of stock prices went up.[7] Use her data to test the null hypothesis that daily changes in the Dow Jones Industrial Average are independent binomial observations with $p = .5$.

DAYS MARKET WENT UP	5	4	3	2	1	0
NUMBER OF OCCURRENCES	3	27	48	31	17	2

10.2 Contingency Tables

Pearson's chi-square test is a remarkably versatile way of gauging how closely the data agree with the detailed implications of a specified null hypothesis—a procedure so general that it is often called a *goodness-of-fit test.* We divide the possible outcomes into mutually exclusive categories and then compare the observed number of outcomes in each category with the expected number if the null hypothesis were true. One application, as we have seen, is to a single list of possible outcomes, such as four numbered cards or five categories of surnames. In another common application, the data are labeled in more than one way.

Consider a study that investigated whether juvenile delinquency among males is related to birth order.[8] Based on answers to questions regarding school truancy, theft, and gang fights, 1137 boys were separated into two categories—most delinquent and least delinquent—and then separated into four categories based on birth order. In Chap. 5, we first encountered a **contingency table** (or *two-way table*) in which data are classified in one way by the rows and in another way by the columns:

	OLDEST	IN-BETWEEN	YOUNGEST	ONLY CHILD	TOTAL
MOST DELINQUENT	127	123	93	17	360
LEAST DELINQUENT	345	209	158	65	777
TOTAL	472	332	251	82	1137

What patterns are revealed in these data? One way of helping to describe the data is to show the column percentages:

	OLDEST	IN-BETWEEN	YOUNGEST	ONLY CHILD	TOTAL
MOST DEL.	127 (26.9%)	123 (37.0%)	93 (37.1%)	17 (20.7%)	360 (31.7%)
LEAST DEL.	345 (73.1%)	209 (63.0%)	158 (62.9%)	65 (79.3%)	777 (68.3%)
TOTAL	472	332	251	82	1137

Overall, 31.7 percent of these boys were classified as most delinquent; the percentages are somewhat smaller for the oldest and only children, and somewhat larger for the in-between and youngest children. Are these observed differences statistically significant, or might they be explained by sampling error—the inevitable variation that occurs in random samples?

Our null hypothesis is that there is no relationship between birth order and delinquency. If so, we expect the percentage of delinquent children in each birth-order category to be equal to the overall percentage: 31.7 percent. Thus, among the 472 oldest children, the expected number of "most delinquent" children is $.317(472) = 149.4$. Similar calculations give these expected numbers, were the null hypothesis true:

	OLDEST	IN-BETWEEN	YOUNGEST	ONLY CHILD	TOTAL
MOST DELINQUENT	149.4	105.1	79.5	26.0	360
LEAST DELINQUENT	322.6	226.9	171.5	56.0	777
TOTAL	472	332	251	82	1137

Instead of determining the expected numbers by first calculating the column percentages, we can simply multiply the appropriate row and column totals and then divide by the total number of observations; for example, the expected number in the oldest-most delinquent cell is 360(472)/1137 = 149.4. The easiest way, of course, is to use statistical software.

For the chi-square distribution to give a good approximation to the exact probabilities, none of the expected numbers should be less than 5. This is certainly the case here.

A comparison of the observed and expected numbers shows the same pattern revealed by a comparison of the column percentages. There are fewer oldest-child and only-child delinquents than expected if the null hypothesis were true, and there are more in-between and youngest-child delinquents than expected. To test the null hypothesis statistically, we use the chi-square statistic to measure how far the observed values are from the expected numbers:

$$\begin{aligned}\chi^2 &= \frac{(127-149.4)^2}{149.4}+\frac{(345-322.6)^2}{322.6}+\frac{(123-105.1)^2}{105.1}+\frac{(209-226.9)^2}{226.9}\\ &\quad+\frac{(93-79.5)^2}{79.5}+\frac{(158-171.5)^2}{171.5}+\frac{(17-26.0)^2}{26.0}+\frac{(65-56.0)^2}{56.0}\\ &= 17.3\end{aligned}$$

We saw earlier that when the data are labeled in one direction, the number of degrees of freedom is equal to the number of categories minus 1. When, as here, the data are labeled in two directions, the number of degrees of freedom is

$$(\text{Number of rows} - 1)(\text{Number of columns} - 1)$$

In our birth-order example, there are $(4 - 1)(2 - 1) = 3$ degrees of freedom. (Given the row and column totals, three numbers in the first (or second) row will determine all the other entries.) Table 6 shows that a chi-square distribution with 3 degrees of freedom has a 7.81 cutoff for a test at the 5 percent level and an 11.34 cutoff for a test at the 1 percent level. Thus these deviations between the observed and expected numbers (were birth order unrelated to delinquency) are sufficiently large for us to reject the null hypothesis at the 1 percent level. (Statistical software shows that the P value is .00093.) These data indicate that oldest and only children are less likely to be delinquent.

For the simplest possible contingency table, with two columns and two rows, the chi-square test is exactly equivalent to a z test of the difference in the sample proportions, using Eq. (8.17). The advantage of the chi-square test is that it can handle situations with more than two categories of outcomes.

Table 10.3 shows how the SAS software program displays the results of a chi-square test. Each cell shows five pieces of information: the observed frequency, the expected numbers if the rows and columns are independent, the percentage of the total number of observations, the percentage of that row's

Table 10.3

SAS chi-square test for birth order and delinquency

STATISTICS FOR TABLE OF DELINQ BY BRTHORD

DELINQ	BRTHORD				
FREQUENCY EXPECTED PERCENT ROW PCT COL PCT	OLDEST	INBETWEEN	YOUNGEST	ONLY	TOTAL
MOST-DEL	127	123	93	17	360
	149.4	105.1	79.5	25.9	
	11.17	10.82	8.18	1.50	31.66
	35.28	34.17	26.67	4.72	
	26.91	37.05	37.05	20.73	
LEAST-DEL	345	209	158	65	777
	322.6	226.9	171.5	56.0	
	30.34	18.38	13.90	5.72	68.34
	44.40	26.90	20.33	8.37	
	73.09	62.95	62.95	79.27	
TOTAL	472	332	251	82	1137
	41.51	29.20	22.08	7.21	100.00

STATISTICS FOR TABLE OF DELINQ BY BRTHORD

STATISTIC	DF	VALUE	PROB
CHI-SQUARE	3	17.282	.001

total, and the percentage of that column's total. Below this information-packed table are the results of the chi-square test: the degrees of freedom, chi-square value, and P value.

The Unshuffled Draft Lottery

A national lottery was held on December 1, 1969, to determine the order in which U.S. males born in 1952 would be drafted in 1970 to fight in the Vietnam war. The federal government prepared 366 capsules, each containing a birthdate—January 1, January 2, and so on. Each month's capsules were carried to the drawing in a separate box and emptied into a large drum, one after another—first January, then February, and on through the months, with December's capsules poured in last. After the drum was rotated a few times, the capsules

HOW TO DO IT

Chi-Square Test for Two-Way Contingency Tables

We want to test whether birth order is related to juvenile delinquency.

1. Specify the row and column categories, and determine the number of observations in each category.

	OLDEST	IN-BETWEEN	YOUNGEST	ONLY CHILD	TOTAL
MOST DEL.	127	123	93	17	360
LEAST DEL.	345	209	158	65	777
TOTAL	472	332	251	82	1137

2. Specify the null hypothesis (usually that the row and column classifications are independent), and use statistical software to determine the expected number of observations in each cell if the null hypothesis is true. Here:

	OLDEST	IN-BETWEEN	YOUNGEST	ONLY CHILD	TOTAL
MOST DEL.	149.4	105.1	79.5	26.0	360
LEAST DEL.	322.6	226.9	171.5	56.0	777
TOTAL	472	332	251	82	1137

   The expected numbers can also be determined by multiplying the appropriate row and column totals and then dividing by the total number of observations; for example, $360(472)/1137 = 149.4$.

3. Alternatively, enter the observations in a statistical software program, as illustrated by the SAS output in Table 10.3.

were selected, with those drawn first drafted first and those drawn last most likely not drafted at all. The results are summarized in Table 10.4.

A casual inspection suggests that the late months—August through December—got more than their share of the "first-drafted" numbers, while the early months—January through May—got a lot of the "last-drafted" numbers. A plausible explanation is that a few rotations of the drum did not do a very good job of mixing up the capsules. The early months mostly stayed at the bottom,

4. Check that no expected number is smaller than 5.
5. Calculate the value of the chi-square statistic, as here:

$$\chi^2 = \sum \frac{(\text{observed} - \text{expected})^2}{\text{expected}}$$

$$= \frac{(127 - 149.4)^2}{149.4} + \frac{(345 - 322.6)^2}{322.6} + \cdots + \frac{(65.0 - 56.0)^2}{56.0}$$

$$= 17.3$$

6. Use statistical software or Table 6 to find the probability that the value of a chi-square statistic with (number of rows − 1)(number of columns − 1) degrees of freedom would be so far from zero. Here, with (4 − 1)(2 − 1) = 3 degrees of freedom, the *P* value is .00093.
7. A *P* value below .05 indicates that the results are statistically significant at the 5 percent level. These data allow us to reject the null hypothesis decisively.
8. Do the observed differences between the observed and expected numbers make sense? Here, it is plausible that delinquency might be less likely among first-born and only children.

while the later months mostly stayed on top, ready to be picked first. Is this suspicion just sour grapes from a statistician with a December birthday? After all, even with a perfectly random selection, some days have to be picked first, and these may turn out to be December days. Are the apparent patterns in Table 10.4 so strong that they cannot be explained by the luck of the draw?

This is the sort of question that the chi-square test is designed to answer. The null hypothesis is that the capsules were well shuffled, so that each day

Table 10.4

The 1969 draft lottery

	1 TO 122 (FIRST DRAFTED)	123 TO 244 (MIDDLE)	245 TO 366 (LAST DRAFTED)	TOTAL
January	9	12	10	31
February	7	12	10	29
March	5	10	16	31
April	8	8	14	30
May	9	7	15	31
June	11	7	12	30
July	12	7	12	31
August	13	7	11	31
September	10	15	5	30
October	9	15	7	31
November	12	12	6	30
December	17	10	4	31
Total	122	122	122	366

was equally likely to be selected. If true, we expect each month to have its days equally divided among the three categories 1 to 122, 123 to 244, and 245 to 366. Months with 30 days would average $30/3 = 10$ days per category in the long run, over many such drawings. For months with 31 days, the expected number in each category is $31/3 = 10.33$. February, with 29 days in 1952, would average $29/3 = 9.67$ days. To measure how severely the observed numbers differ from these expected numbers, we calculate the chi-square value:

$$\chi^2 = \frac{(9 - 10.33)^2}{10.33} + \frac{(12 - 10.33)^2}{10.33} + \cdots + \frac{(4 - 10.33)^2}{10.33}$$

$$= 37.2$$

Because Table 10.4 has 12 rows and 3 columns, there are $(12 - 1)(3 - 1) = 22$ degrees of freedom. A chi-square statistic with 22 degrees of freedom has a 5 percent chance of exceeding 33.9 and a 1 percent chance of exceeding 40.3. There is only about a .02 probability of observing a chi-square value as large as 37.2. Thus these data reject at the 5 percent level, but not the 1 percent level, the null hypothesis that the capsules were well shuffled. In response to this statistical evidence, the government was more careful the following year, when it conducted a draft lottery for males born in 1953. In this lottery, two drums were used—one with dates and one with numbers—and each drum was loaded in random order.[9]

Nicotine Replacement Therapies

Several types of nicotine replacement therapy (NRT) have been developed to help people stop smoking cigarettes. Table 10.5 shows data from an aggregation of the results of 50 independent tests of the efficacy of four different therapies (gum, inhalation, intranasal spray, and transdermal patches) and the control groups for these studies.[10]

Overall, 14.2 percent of the people studied were able to quit smoking. For those in the control groups, 10.6 percent quit; for those using one of the four therapies, 15.2 to 25.9 percent quit. For each therapy, the differences between the NRT group and control group seem substantial. Among the four therapies, the nasal spray and inhaler seem to have considerably different success rates. Whether the other differences are substantial is perhaps debatable.

We can use a chi-square test to determine if the differences among the five categories are statistically persuasive. Here are the expected numbers if there is no relationship between the five categories and the results (quitting or not quitting):

	GUM	INHALER	NASAL SPRAY	PATCHES	CONTROL	TOTAL
QUIT	897.3	20.6	16.4	176.5	1,361.2	2,472
DID NOT	5,430.7	124.4	99.6	1,068.5	8,238.8	14,962
TOTAL	6,328	145	116	1,245	9,600	17,434

A comparison of the observed with the expected numbers confirms our interpretation of the column percentages. The number of people in the control groups who quit smoking was considerably smaller than would be expected if the therapies were ineffective. Those using a nasal spray or patches quit much more often than would be expected; those using gum or an inhaler quit somewhat more often than would be expected.

For a formal statistical test, the chi-square value works out to be 238.1; with $(2-1)(5-1) = 4$ degrees of freedom, the P value is virtually 0. Thus, these studies provide overwhelming evidence that nicotine replacement therapies do

Table 10.5

Four nicotine replacement therapies and the control groups

	GUM	INHALER	NASAL SPRAY	PATCHES	CONTROL	TOTAL
Quit	1,149	22	30	255	1,016	2,472
	(18.2%)	(15.2%)	(25.9%)	(20.5%)	(10.6%)	(14.2%)
Did not	5,179	123	86	990	8,584	14,962
	(81.8%)	(84.8%)	(74.1%)	(79.5%)	(89.4%)	(85.8%)
Total	6,328	145	116	1,245	9,600	17,434

affect one's chances of quitting cigarette smoking. It is far from a sure thing that someone who uses an NRT will quit smoking, but the empirical evidence that these therapies raise the chances of quitting from 11 percent to 15 to 26 percent is substantial and statistically persuasive.

Statistical Evidence on the Death Penalty

Example 8.3 discussed the use of statistical evidence in enforcing the Civil Rights Act of 1964. Courts have been more reluctant to accept statistical evidence in cases involving important constitutional issues. In the 1987 case of *McCleskey vs. Kemp,* the Supreme Court upheld the constitutionality of Georgia's death penalty, despite statistical evidence that blacks who murder whites are disproportionately sentenced to death. Here are some of the data considered by the court regarding 2475 Georgia murder convictions between 1973 and 1980:

	BLACK DEFENDANT		WHITE DEFENDANT		
	BLACK VICTIM	WHITE VICTIM	BLACK VICTIM	WHITE VICTIM	TOTAL
DEATH SENTENCE	18 (1.3%)	50 (21.9%)	2 (3.1%)	58 (7.8%)	128 (5.2%)
NO DEATH SENTENCE	1420 (98.7%)	178 (78.1%)	62 (96.9%)	687 (92.2%)	2347 (94.8%)
TOTAL	1438	228	64	745	2475

For a chi-square test, the expected numbers are as follows:

	BLACK DEFENDANT		WHITE DEFENDANT		
	BLACK VICTIM	WHITE VICTIM	BLACK VICTIM	WHITE VICTIM	TOTAL
DEATH SENTENCE	74.37	11.79	3.31	38.53	128
NO DEATH SENTENCE	1363.63	216.21	60.69	706.47	2347
TOTAL	1438	228	64	745	2475

Compared to these expected numbers, there were 56.37 fewer death sentences for black defendants with black victims, 38.21 more death sentences for black defendants with white victims, 1.31 fewer death sentences for white defendants with black victims, and 19.47 more death sentences for white defendants with white victims. The value of the chi-square statistic is 186.6 which, with $(2 - 1)(3 - 1) = 2$ degrees of freedom, is highly statistically significant.

If we group the data by the race of the defendant, the differences between the observed values

	BLACK DEFENDANT	WHITE DEFENDANT	TOTAL
DEATH SENTENCE	68 (4.1%)	60 (7.4%)	128 (5.2%)
NO DEATH SENTENCE	1598 (95.9%)	749 (92.6%)	2347 (94.8%)
TOTAL	1666	809	2475

and the expected numbers

	BLACK DEFENDANT	WHITE DEFENDANT	TOTAL
DEATH SENTENCE	86.16	41.84	128
NO DEATH SENTENCE	1579.84	767.16	2347
TOTAL	1666	809	2475

seem less pronounced, and now it is white defendants who are more likely to be sentenced to death. The value of the chi-square statistic is 12.35, which is statistically significant at the 1 percent level.

A confounding influence here is that a death sentence is much more likely when the victim is a police officer, and most murders of police officers involved black defendants and white victims. The Supreme Court ruled, by a narrow 5-to-4 margin, that there was no direct evidence that the defendant in this particular case, a black man convicted of killing a white Atlanta policeman, had been treated unfairly. Although the court acknowledged that the data suggested imperfections in Georgia's system of capital punishment, the majority ruled that this evidence did "not demonstrate a constitutionally significant risk of racial bias."

10.6 A study investigated whether a mouse raised by a foster mother would be more likely to fight with other mice than a mouse raised by its natural mother.[11] Show the column percentages and see whether these data show a statistically significant relationship at the 1 percent level.

	FIGHTERS	NONFIGHTERS
NATURAL	27	140
FOSTER	47	93

10.7 When the Titanic sank on April 14, 1912, it was carrying 1315 passengers: 402 adult females, 805 adult males, and 108 children.[12] Of the 498 passengers who were saved, 296 were adult females, 146 were adult males, and 56 were children.

In which of these three groups was the number of people saved disproportionately large, and in which was it disproportionately small? Are these patterns statistically significant at the 1 percent level?

10.8 Flu vaccines are usually effective and safe in children and adults, but less effective with infants under the age of 6 months, the most vulnerable group. After a nose-drip vaccine was shown to be effective in a sample of 9000 children and adults, doctors tested this vaccine in 30 infants aged 2 to 5 months.[13] Six babies got a low dose, 15 got a stronger dose, and a control group of 9 infants got drops with no virus. Two-thirds of the first group and nearly 90 percent of the second group developed antibodies that protect against the flu. The control group, as predicted, developed no antibodies. Use a chi-square test to determine if these results are significant at the 5 percent level. What is your null hypothesis?

	DEVELOPED ANTIBODIES	NO ANTIBODIES
LOW DOSE	4	2
STRONGER DOSE	13	2
CONTROL	0	9

10.9 A University of Southern California study examined the relationship between a woman's shoe size and whether the woman frequently experienced foot pain.[14] Do the data show a statistically significant relationship at the 5 percent level?

SHOE SIZE	NUMBER SURVEYED	NUMBER WITH FOOT PAIN
SMALLER THAN 8	150	87
BETWEEN 8 AND 10	130	104
10 OR LARGER	120	101

10.10 A 50-percent sample of live births in the United States between 1969 and 1971 obtained the data below on the gender of babies in relation to their birth order.[15] (Thus, a birth order of 1 signifies a mother's first child, while a birth order of 2 indicates her second child.) Is there a statistically significant relationship at the 1 percent level between birth order and a baby's gender?

BIRTH ORDER	MALES	FEMALES	TOTAL
1	849,710	798,600	1,648,310
2	620,978	586,935	1,207,913
3	365,000	345,317	710,317
4	201,484	191,524	393,008
5	110,339	105,085	215,424
6	60,579	58,137	118,716
7	34,322	32,876	67,198
8 or higher	50,459	48,611	99,070

summary

When the possible outcomes are divided into mutually exclusive categories, a variety of null hypotheses can be tested with the chi-square statistic

$$\chi^2 = \sum \frac{(\text{observed} - \text{expected})^2}{\text{expected}}$$

which compares the observed and expected number of occurrences in each category. The value of the chi-square statistic cannot be negative. A value close to zero is most consistent with the null hypothesis; a large chi-square value provides statistical evidence against the null hypothesis.

If none of the expected numbers are smaller than 5, then the P value for a given chi-square value can be determined from tabulated values for the chi-square distribution. When there is a single list of possible categories, the degrees of freedom is equal to the number of categories minus 1. For a contingency table, with the data classified in one way by row and in another way by column, we can use the chi-square statistic to test whether these classifications are independent; in this case, the number of degrees of freedom is equal to (number of rows − 1)(number of columns − 1).

review exercises

10.11 A random sample of 239 college students obtained these data on birth order and grade-point average (GPA):[16]

	GPA > B+	GPA ≤ B+
ONLY CHILD	12	10
OLDEST CHILD	45	78
NOT OLDEST	31	63

(The data on oldest and not oldest are for families with more than one child.) Make a chi-square test at the 1 percent level of the null hypothesis that birth order and GPA are independent.

10.12 As part of an investigation of the inheritance of various traits, the athletic skills of brothers were evaluated.[17] Test at the 1 percent level the null hypothesis that the athletic skills of brothers are independent.

SECOND BROTHER	FIRST BROTHER ATHLETIC	BETWIXT	NONATHLETIC
ATHLETIC	906	20	140
BETWIXT	20	76	9
NONATHLETIC	140	9	370

10.13 In March 1992, 880 children were awaiting adoption in Texas. At the time, Texas adoption procedures did not allow parents to adopt children of races different from the parents' race.[18] Of the 880 children awaiting adoption, 266 were black and 255 were Hispanic; of 654 families on waiting lists to adopt children, 132 were black and 86 were Hispanic. Use a chi-square test at the 1 percent level to determine if there is a statistically significant difference in the racial composition of the children awaiting adoption and the families waiting to adopt.

10.14 In a 1984 lawsuit (*Bazemore vs. Friday*), racial discrimination was alleged in the merit pay raises awarded by the North Carolina Agricultural Extension Service.[19] At the trial, the data were divided into 6 geographic districts:

	PAY RAISE		NO PAY RAISE	
	WHITE	BLACK	WHITE	BLACK
NORTHCENTRAL	47	24	12	9
NORTHEAST	35	10	8	3
NORTHWEST	57	5	9	4
SOUTHEAST	54	16	10	7
SOUTHWEST	59	7	12	4
WEST	47	0	13	0

The court analyzed five 2-by-2 contingency tables (one for each geographic district in which blacks were employed) and found that only in the northwestern region was there close to statistical significance at the 5 percent level. In the exercises below, assume that the data for these five geographic districts are independent random samples.

a. If the null hypothesis is true and the court requires statistical significance at the 5 percent level in all five districts in order to reject the defendant's claim that pay raises were independent of race, what is the probability that the null hypothesis will be rejected?

b. If the null hypothesis is true and the court requires statistical significance at the 5 percent level in at least one of the five districts, what is the probability that the null hypothesis will be rejected?

c. Combine the five geographic areas to give a single 2-by-2 contingency table relating pay raises to race. Is there a statistically significant relationship at the 5 percent level?

10.15 Yale University admitted its first women undergraduates in 1969. For the first few years, male and female students were admitted separately, to preserve Yale's tradition of "graduating 1000 fine young men each year." However, a separate admission policy can cause disparities between the caliber of the male and female students. A University Committee on Coeducation compared the grades of male and female Yale undergraduates:

	MALES	FEMALES
HONORS	6591	1593
HIGH PASS	7573	1655
PASS	3108	539
OTHER	590	117

What patterns do you see in these data? Is there a statistically significant relationship at the 5 percent level between grades and gender?

10.16 After attempting to imitate a popular television character, a young man concluded that whether one is right-handed or left-handed affects how far apart the two middle fingers on each hand can be spread. If a wider V is made with the right hand, the person is probably left-handed; if a wider V is made with the left hand or if there is no difference, the person is probably right-handed. After an article explaining this theory appeared in the February 20, 1974, issue of *Current Science,* 3225 readers reported the following results:

	TEST PREDICTIONS	
READERS	RIGHT-HANDED	LEFT-HANDED
RIGHT-HANDED	2057	783
LEFT-HANDED	56	289

Use a chi-square test to determine if there is a statistically significant relationship at the 1 percent level between handedness and test predictions.

10.17 Partway through the 1994 National Basketball Association season, the publicity department for the San Antonio Spurs released these data on the relationship between the hair color of one of the players (Dennis Rodman) and the team's performance:[20]

	WIN	LOSE
BLONDE	22	3
PURPLE	9	2
RED OR BLUE	6	7

Is there a statistically significant relationship at the 5 percent level? Why, even if there is a statistically significant relationship, should we treat these results cautiously?

10.18 A study of videotapes of four of Monica Seles' Grand-Slam tennis matches categorized Seles' first serves as being to the opponent's forehand, body, or backhand and as to whether the serve was effective or ineffective.[21] A serve was considered effective if the opponent was unable to return it or returned it so weakly that Seles immediately hit a winning shot. Test at the 1 percent level and null hypothesis that the effectiveness of Seles' first serve is independent of whether she serves to the forehand, body, or backhand.

	FOREHAND	BODY	BACKHAND
EFFECTIVE	16	9	22
INEFFECTIVE	26	12	136

10.19 Each of 12 college intramural basketball players was asked to take 30 shots 15 feet from the basket—10 shots from the free-throw line, 10 shots from a 45° angle to the basket, and 10 shots from the baseline.[22] Use the data below to test at the 5 percent level the null hypothesis that hits and misses are independent of location:

	FREE-THROW LINE	45° ANGLE	BASELINE
HIT	84	62	53
MISS	36	58	67

10.20 A study of 2000 major league baseball games during the 1959 season tabulated the number of runs scored by inning.[23] The data below are for the first inning of all 2000 games and for the ninth inning of 1576 games in which both teams batted. (The game ends if the home team is ahead after the visiting team has batted in the ninth inning.) Use these data to make a chi-square test (at the 1 percent level) of the null hypothesis that run scoring does not depend on whether it is the first or ninth inning of a baseball game.

	RUNS SCORED					
	0	1	2	3	4	>4
FIRST INNING	1400	318	162	62	36	22
NINTH INNING	1213	203	95	43	16	6

10.21 Two surveys, conducted 20 years apart, asked five questions to gauge people's trust in government (for example, "Are quite a few of the people running the government a little crooked?"). Based on their responses, those surveyed were labeled as follows:[24]

	CYNICAL	MIDDLE	TRUSTING
1958	200	456	1057
1978	1198	599	438

What patterns do you see in these data? Is there a statistically significant difference at the 5 percent level between the 1958 and 1978 responses?

10.22 To study social mobility in Great Britain, a 1954 sample compared the occupations of 3497 male workers and their fathers.[25]

	SON'S STATUS		
FATHER'S STATUS	UPPER	MIDDLE	LOWER
UPPER	588	395	159
MIDDLE	349	714	447
LOWER	114	320	411

If there was no social mobility whatsoever, what would the data look like? If there were unrestricted social mobility, what would you expect the data to look like? Compute the column percentages, and then test at the 1 percent level the null hypothesis that son's and father's occupational statuses are independent.

10.23 A student asked a random sample of 54 of her classmates whether they were nearsighted and whether their grade-point average (GPA) was above or below B+:

	NEARSIGHTED	NOT NEARSIGHTED
GPA $<$ B+	14	15
GPA $\geq$ B+	17	8

Do these data reject (at the 1 percent level) the null hypothesis that GPA and nearsightedness are independent? This student then multiplied all her data by 3:

	NEARSIGHTED	NOT NEARSIGHTED
GPA $<$ B+	42	45
GPA $\geq$ B+	51	24

"In doing this, I assumed that I had originally picked a perfectly random sample of students, and that if I were to have polled 3 times as many people, my data would have been greater in magnitude, but still distributed on the same normal distribution." Does this procedure make sense to you? How do you think it affected her chi-square value?

10.24 Shown below are 1995 data on the total U.S. population and the number of reported AIDS cases by ethnicity:[26]

	MALES		FEMALES	
	POPULATION	AIDS	POPULATION	AIDS
WHITE	77,492,000	211,856	82,903,000	15,570
BLACK	11,124,000	121,017	12,963,000	35,372
HISPANIC	9,940,000	68,051	9,680,000	12,293
ASIAN/PACIFIC	3,456,000	2,902	3,756,000	325
AMERICAN INDIAN	697,000	1,010	736,000	173

Among males, how does the actual ethnic breakdown of AIDS cases differ from what would be expected if AIDS were independent of ethnicity? How about among females? (Do not do a statistical test, as these are population values, not a sample.)

10.25 Ninety-nine break shots on a full rack of billiard balls resulted in 54 sunk balls, allocated among the 15 balls as follows:[27]

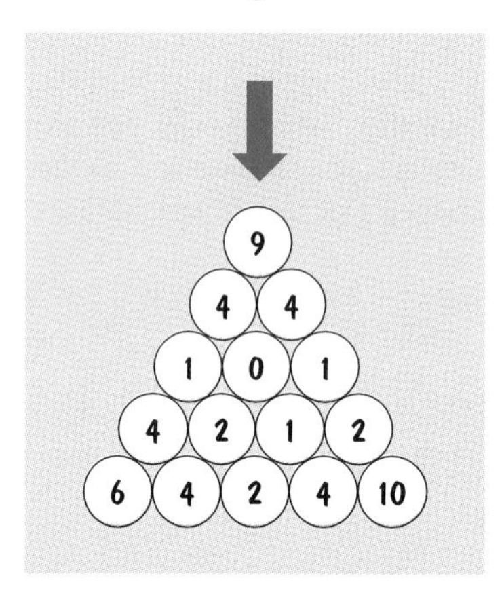

Thus the balls in the three corners of the rack were sunk 9, 6, and 10 times. If we divide the rack into three categories—corners (C), adjacent (A), and other (O),

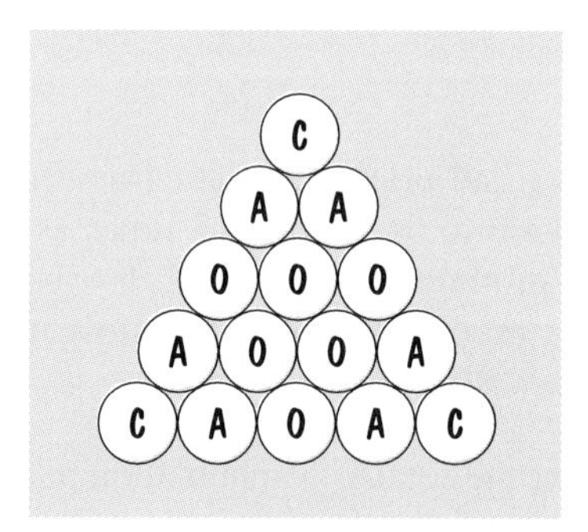

is there a statistically significant relationship between a ball's position in the rack and its likelihood of being sunk on the break?

10.26 Using the data in Exercise 10.25, divide the rack according to the placement of the 8 ball, the seven striped balls (white in the diagram), and the seven solid balls (teal blue in the diagram):

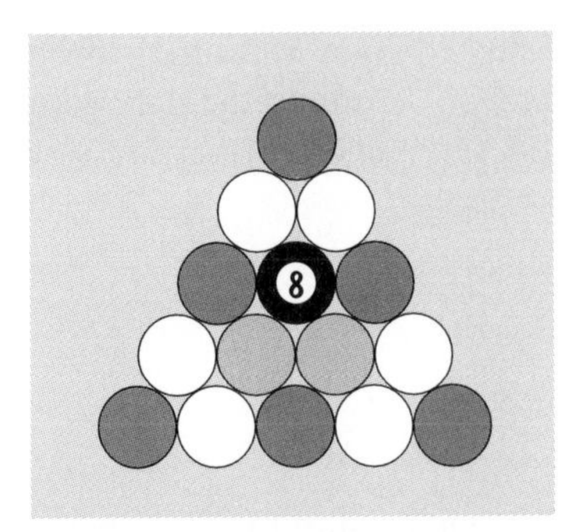

The two light teal blue balls indicate that the seventh striped ball and seventh solid ball can be placed in either of these positions. For your analysis, assume that of the 3 balls sunk from these positions, 1.5 was solid and 1.5 was striped. Omit the 8 ball entirely from your analysis. Is there a statistically significant relationship between a ball's being solid or striped and its likelihood of being sunk on the break?

10.27 A study of severe heart attacks suffered by German workers (either on or off the job) grouped the data according to the day of the week when the heart attack occurred:

MONDAY	TUESDAY	WEDNESDAY	THURSDAY	FRIDAY	SATURDAY	SUNDAY
145	105	111	120	97	115	98

Is there a statistically significant relationship between the day of the week and the incidence of severe heart attacks?

10.28 Table 10.5 shows some data on the effectiveness of 4 different kinds of nicotine replacement therapy. Omit the control group data so that we can concentrate on differences among the 4 therapies. Are the differences in effectiveness statistically significant at the 1 percent level?

10.29 A 1-year study was made of injuries at a large British engineering company that used a weekly rotation of 8-hour shifts beginning at 6:00 a.m., 2:00 p.m., and 10:00 p.m. so that every worker worked the same number of morning, afternoon, and night shifts.[28] Do the following data on the number of injuries that occurred on each shift reject at the 5 percent level the null hypothesis that the incidence of injuries does not depend on the shift time?

SHIFT	NUMBER OF INJURIES
Morning	1372
Afternoon	1578
Night	1686

10.30 An animal-behavior specialist studied John McEnroe's behavior as he won the 1983 Wimbledon tennis tournament.[29] Among the data gathered were whether McEnroe grunted audibly on his serve and whether the serve was an ace (winner), error, or other:

	GRUNT	SILENT
ACE	61	35
ERROR	32	8
OTHER	144	53

Is there a statistically significant relationship (at the 5 percent level) between his grunting and the outcomes of his serves?

10.31 Suspecting that a disproportionate number of people might be born in November, which is 9 months after Valentine's Day, a survey of 148 college students obtained the following birth months:[30]

January	7	May	7	September	11
February	15	June	12	October	12
March	16	July	8	November	21
April	6	August	16	December	17

Determine the P value for a test of the null hypothesis that all birth months are equally likely. Summarize your findings.

10.32 Basketball players are often told that a missed shot usually bounces long, to the opposite side of the basket from which the shot is taken; for example, a missed shot from the right side of the basket usually bounces to the left of the basket. A college player suspected that this tendency might be reversed at the end of a game, when the missed shots of tired players bounce off the front of the rim. Videotapes of four Division III college women's basketball games were studied to see whether missed shots bounced short (towards the shooter) or bounced long (away from the shooter).[31] The rebounds were further characterized by whether they occurred during the first or last 10 minutes of the game:

	SHORT	LONG
FIRST TEN MINUTES	16	25
LAST TEN MINUTES	20	10

A similar study of videotapes of three different Division III men's basketball games yielded the data below from the first and last 10 minutes of each half:

	SHORT	LONG
FIRST TEN MINUTES	49	39
LAST TEN MINUTES	35	46

For each set of data, determine the *P* value for a chi-square test of the null hypothesis that short and long rebounds are independent of whether the shot is taken during the first or last 10 minutes.

10.33 A 1986 study by the National Transportation Safety Board of 30 frontal automobile crashes involving 139 people used the following data:[32]

	KILLED	INJURED	UNINJURED
NO RESTRAINT	4	49	4
LAP BELT	13	36	1
LAP AND SHOULDER BELT	1	29	2

Calculate the *P* value for a test of the null hypothesis that the injury outcomes are unrelated to the restraint categories. Summarize your conclusions. Why is this calculated *P* value only a rough approximation?

10.34 One hundred and twenty college students (62 male and 58 female) participated in a study in which they were given a sheet of paper with the following instructions:[33]

> Please make up a story (a few sentences) creating a fictional character who fits the following theme (Please do not write about yourself): *In a large coeducational institution the average student will feel isolated in his introductory courses.*

The students were then asked to write down the name of their fictional character. In half of the cases, the instructions replaced "his introductory courses" with "his or her introductory courses." The researcher was interested in whether this change would affect the subject's choice of a male or female character. Here are the results:

	MALE SUBJECTS (62)		FEMALE SUBJECTS (58)	
	MALE CHARACTER	FEMALE CHARACTER	MALE CHARACTER	FEMALE CHARACTER
HIS	27	4	23	6
HIS OR HER	17	14	13	16

a. Why did the study ask the subjects not to write about themselves?
b. Combine the male and female subjects and make a chi-square test at the 1 percent level of the null hypothesis that the choice of a male or female character is independent of whether "his" or "his or her" are used in the instructions.
c. What patterns do you see in the results?

10.35 A study of all National Basketball Association players who had attempted at least 100 free throws during the 1993-1994 regular season determined how many players in each of 6 height intervals made more than 74.3% of their free throws (the NBA average):[34]

HEIGHT (INCHES)	MORE THAN 74.3%	TOTAL
MORE THAN 82	8	27
81 OR 82	18	45
79 OR 80	19	38
76, 77, OR 78	24	36
LESS THAN 76	28	33

Thus there were 27 players taller than 82 inches, of whom 8 made more than 74.3 percent of their free throws. Test at the 1 percent level the null hypothesis that free-throw percentage and height are unrelated.

10.36 Seventy randomly selected college students were asked to taste 3 unlabeled chocolate chip cookies from a local grocery, and identify the one they liked best.[35] Twelve students chose Chips Ahoy!; 11 chose Lady Lee, and 47 chose supermarket bakery cookies. Are these results statistically significant at the 1 percent level? What is your null hypothesis?

10.37 Infants typically begin to crawl at 6-to-8 months of age. For those born in the summer and fall in places with cold winters, crawling may be restricted and delayed by layers of clothing. To investigate this possibility, the University of Denver Infant Study Center compiled data for 425 infants on the birth month and the age at which the infant was first able to crawl four feet in one minute.[36] The median age at which these infants first began to crawl was 30.89 weeks. For each season of birth, they then calculated the number of infants who began crawling before 30.89 weeks and the number who began crawling later:

SEASON OF BIRTH	BEFORE 30.89 WEEKS	AFTER 30.89 WEEKS
WINTER (DECEMBER-FEBRUARY)	64	51
SPRING (MARCH-MAY)	43	36
SUMMER (JUNE-AUGUST)	44	53
FALL (SEPTEMBER-NOVEMBER)	50	84

Use these data to determine the P value for a statistical test of the null hypothesis that there is no relationship between birth month and crawling age. Summarize your results.

10.38 Fifty college students were asked whether the amount of time that elapsed between making their beds was more than or less than 15 days.[37] What patterns do you see in these data? Use the chi-square distribution to see if there is a statistically significant relationship at the 5 percent level between gender and bed making:

	MORE THAN 15 DAYS	LESS THAN 15 DAYS
FEMALES	18	7
MALES	12	13

10.39 Use the data in the preceding exercise to make a difference-in-means test at the 5 percent level of the null hypothesis that male and female students are equally likely to allow more than 15 days to elapse before making their beds. Are your results consistent with those obtained in the preceding exercise? Are you surprised?

10.40 Twenty-three-thousand pregnant New England women were asked whether they had used a hot tub during the first two months of pregnancy.[38] When these women gave birth, the researchers recorded whether their babies had brain and spinal cord defects caused by the failure of the neural tube to close:

	HOT TUB	NO HOT TUB
NEURAL TUBE DEFECTS	7	41
NO NEURAL TUBE DEFECTS	1,247	21,404

Use the chi-square distribution to calculate the *P* value for a test of the null hypothesis that the incidence of neural tube defects is unrelated to hot tub use. Summarize your conclusions. Why is the calculated *P* value only a rough approximation?

projects

For each of the following projects, type a report in ordinary English, using clear, concise, and persuasive prose. Any data that you collect for this project should be included as an appendix to your report. Data used in your report should be presented clearly and effectively.

10.1 Set up a mock ESP experiment by writing the numbers 1 through 4 on 4 identical pieces of paper and placing these in a hat or other opaque container. Now tell a randomly selected person, the subject, that you are going to select one of these pieces of paper and concentrate on the number, while the subject tries to read your mind. Be very careful to ensure that the subject cannot see the number on the paper. Record the answer, and then tell the subject the selected number. Repeat this experiment with 100 different subjects. When you have all your data, test at the 1 percent level the null hypothesis that each number is equally likely to be chosen by the subjects.

10.2 Follow the instructions for Project 10.1, but use 10 identical slips of paper numbered 1 through 10.

10.3 The manufacturer of plain M&M candies claims that 30 percent are brown, 20 percent red, 20 percent yellow, 10 percent blue, 10 percent green, and 10 percent orange. Use a 1-pound bag of M&M's and a chi-square test to confirm or reject the manufacturer's claim.

10.4 To investigate whether birth order is related to a preference for an individualist sport or one emphasizing teamwork, make a list of sports that fall into these two categories. Sports like running, swimming, and tennis should be counted as individualist sports; baseball, basketball, and soccer are team sports. Now ask 100 randomly selected students: "What is your favorite sports to play?" After they answer this question, ask them whether they were a first-born or only child, middle child, or last-born child. Is there a statistically significant relationship?

10.5 Among students at your college, is there a statistical relationship between birth order (first-born or later-born) and political affiliation (Democrat, Republican, or independent)?

10.6 Among students at your college, is there a statistical relationship between birth order (first-born or only child, middle-born, or last-born) and support for the death penalty?

10.7 Do you think that first-year students or seniors are more likely to be dissatisfied with their choice of college? To find out, ask a random sample of students at your college: "If you had it do over again, would you enroll at this college or at another college?" Also ask whether they are first-, second-, third-, or fourth-year students.

10.8 Ask 100 college students: "If you could have only 1 pet for the rest of your life, would you choose a cat, a dog, or neither?" Record the person's gender, too; and when you have finished collecting your data, see if there is a statistically significant relationship at the 1 percent level between gender and pet choice.

10.9 Select three well-known automobile models with roughly the same price. (Among the places you can find this information is the May issue of *Consumer Reports* or AutoVantage on the World Wide Web.) Now ask a random sample of at least 25 students and 25 professors which of these 3 cars they would choose, were they to win a contest that entitled them to have any one of these cars for free. Identify the patterns you see in these responses, and determine if they are statistically significant at the 5 percent level.

10.10 Exercise 10.16 describes a finger-spreading prediction of whether a person is left-handed or right-handed. Ask at least 50 randomly selected students to spread the middle two fingers on each hand to make a V. Record whether the wider V is made with: (a) the right hand or (b) either the left hand or no difference. Then ask the person whether he or she is left-handed or right-handed. How well does the wider V predict handedness? Use a chi-square test to determine if there is a statistically significant relationship at the 5 percent level between handedness and the test predictions.

10.11 Ask 100 randomly selected people this question: "If you could only watch 1 television show a week, what would it be?" Separate their answers into several categories (such as comedy, drama, or sports) and see if there are statistically significant differences in the male and female responses.

10.12 Ask at least 100 college students, "Do you usually make sure you look good before leaving your room?" Also record the person's gender and year in college. Determine the P values for a test of these null hypotheses: (a) the responses are unrelated to gender, and (b) the responses are unrelated to year in college.

10.13 Ask randomly selected college students if they have had a serious romantic relationship in the past 2 years and, if so, to identify the month in which the most recent relationship began. When you have found 120 students who answer yes and can identify the month, make a chi-square test of the null hypothesis that each month is equally likely for the beginning of a romantic relationship.

10.14 Roll a 6-sided die 120 times, recording the outcome of each roll. Use these data to test the null hypothesis that each number is equally likely to be rolled.

10.15 Use a computer random number generator to select 1000 random 1-digit numbers. Use these data to test the null hypothesis that each of the 10 digits (0 through 9) is equally likely to be selected.

10.16 Exercise 6.32 describes the game Roshambo (rock-scissors-paper). Play this game against at least 30 different people, recording the initial move of each opponent. Use these data to test the null hypothesis that rock, scissors, and paper are used equally often on the initial move.

10.17 Ask 50 randomly selected students this question and then compare the male and female responses: "You have a coach ticket for a nonstop flight from Los Angeles to New York. Because the flight was overbooked, randomly selected passengers will be allowed to sit in open first-class seats. You are the first person selected. Would you rather sit next to: (a) the U.S. president; (b) the president's wife; or (c) Michael Jordan?

10.18 Use the obituaries in a book of famous people, *The New York Times,* or a local newspaper to find the birth and death dates of at least 100 persons. Divide these data into 3 categories: deaths that occurred during the 30 days preceding the birthday, deaths that occurred on the birthday or during the 29 days following the birthday, and on other days. Test at the 5 percent level the null hypothesis that a person's death date is not related to the birth date.

10.19 Ask a random sample of at least 50 students to identify their year in college and to answer a simple political question such as, "Are your views generally more consistent with those of the Democratic or Republican party?" Is there a statistically persuasive relationship?

10.20 Do students and professors at your college have similar religious beliefs? Collect the data for your statistical test by asking a random sample of students and professors to choose among at least 3 categories for describing their religious beliefs; for example, "Please choose the category that best describes your religious beliefs: deeply religious, moderately religious, not very religious."

10.21 Make a list of 30 countries and then ask a political science professor to separate these into 3 groups of 10 countries based on their degree of political freedom: most free, middle, and least free. Now ask an economics professor to separate your list into 3 groups of 10 countries based on their economic standard of living: highest, middle, and lowest. Is there a statistical relationship?

10.22 Is there a statistical relationship between the states that the Democratic and Republican candidates won in the most recent presidential election and the states that the parties' candidates won in the preceding presidential election?

10.23 Ask 50 college students and 50 professors, "If you had to give up music, television, or sports for the rest of your life, which would you choose?" Are there statistically persuasive differences between the students and professors?

10.24 Make a list of 10 well-known books (including one that you feel will be controversial) and ask at least 30 professors and 30 students to separate these books into 2 groups of 5, based on how important it is that college students read these books: most important and least important. Are there statistically persuasive differences between how the students and professors rate the book you felt would be controversial?

10.25 Follow the instructions for project 10.24, but this time ask at least 100 students (and no professors) and determine each student's year in college. Determine the P value for a test of the null hypothesis that student ratings of the book you felt would be controversial are independent of the year in college.

HAVE YOU EVER WONDERED?

The Columbia River begins in British Columbia, Canada, flows southward through the state of Washington, and then turns westward, forming a boundary between Washington and Oregon, to the Pacific Ocean. Plutonium has been produced in Hanford, Washington, since the 1940s, and some radioactive waste has leaked into the Columbia River, possibly contaminating the water supplies of Oregon residents living near the river.

Is there a statistical relationship between proximity to the Columbia River and the incidence of cancer? Do Oregonians living near the river die of cancer more often than people living farther away? To answer these questions, we can use the statistical tools described in this chapter.

chapter 11

simple regression

"The cause of lightning," Alice said very decidedly, for she felt quite sure about this, "is the thunder—no, no!" she hastily corrected herself. "I meant it the other way."

"It's too late to correct it," said the Red Queen. "When you've once said a thing, that fixes it, and you must take the consequences."

—Lewis Carroll, Alice in Wonderland

THROUGHOUT THE HUMANITIES, social sciences, and natural sciences, simple models successfully explain complex phenomena. Linguistic models use aptitude and motivation to explain language proficiency. Economic models use income and price to explain the demand for a product. Political models use incumbency and the state of the economy to explain the outcomes of presidential elections. Agricultural models use weather and soil nutrients to explain crop yields. Physics models use mass and acceleration to explain the force applied by a body in motion.

In this chapter, we analyze the simplest type of model, which has only one identified explanatory variable. We look at the assumptions that underlie this model and see how statistical methods can be used (1) to estimate the relationship between an explanatory variable and the variable being explained or predicted and (2) to test

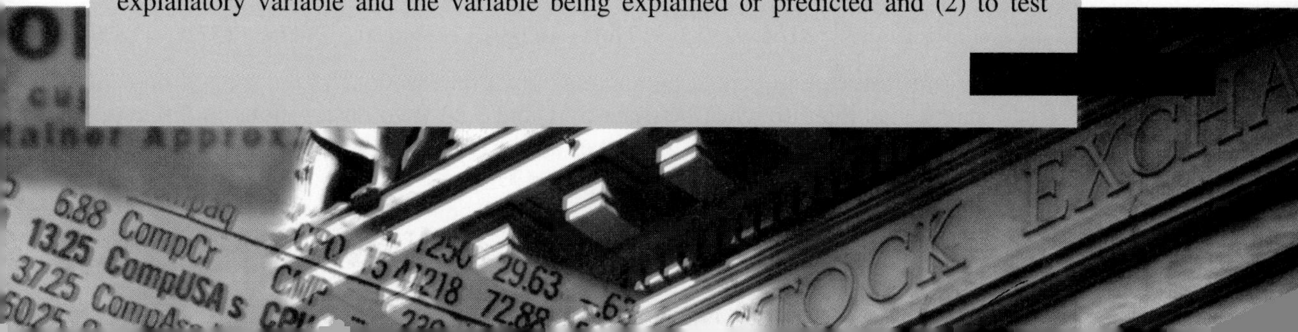

null hypotheses about the size of this relationship. More-advanced courses are devoted to more-complex models with multiple explanatory variables.

11.1 The Regression Model

In Chap. 4, we saw how John Snow established a link between sewage-contaminated water and cholera by comparing cholera mortality rates for households served by two different water companies: one that pumped water from a contaminated section of the Thames River and one that obtained its water far upstream. We might use comparable observational data to investigate the link between Columbia River water contaminated by radioactive waste and cancer mortality. However, in this case, there is no clear distinction between contaminated and uncontaminated water that allows us to group our data into two separate and distinct categories. Instead, it is a matter of degree. There is a continuum along which the water supplies for households living near the Columbia River had varying degrees of exposure to radioactive wastes. Thus, a 1965 study of the statistical relationship between proximity to the Columbia River and cancer mortality compared an exposure index and the annual cancer mortality rate per 100,000 residents for nine Oregon counties near the Columbia River:[1]

EXPOSURE INDEX	11.64	8.34	6.41	3.83	3.41	2.57	2.49	1.62	1.25
CANCER MORTALITY	207.5	210.3	177.9	162.3	129.9	130.1	147.1	137.5	113.5

We could group our data into two categories, high exposure and low exposure, and use a difference-in-means test to compare the average cancer mortality for these two categories. But where do we draw the line between high and low exposure? Suppose that we draw it between 6.41 and 3.83:

	HIGH EXPOSURE			LOW EXPOSURE					
EXPOSURE INDEX	11.64	8.34	6.41	3.83	3.41	2.57	2.49	1.62	1.25
CANCER MORTALITY	207.5	210.3	177.9	162.3	129.9	130.1	147.1	137.5	113.5

The difference between 11.64 and 6.41 (two observations in the same group) is larger than the difference between 6.41 and 3.83 (two observations in different groups)—forcing us to ask ourselves why 11.64 and 6.41 are considered equivalent while 6.41 and 3.83 are considered different. This predicament did not arise from a poor choice of where to draw the line. No matter how we try to separate these data into two artificially distinct groups, we will find differences within the categories that are bigger than differences between the categories. More generally, we would like to employ a statistical procedure that reflects the fact that the data lie along a continuum—that 11.64 is larger than 6.41, which is larger than 3.83, which is larger than 1.25.

Chapter 2 showed how we can explore the empirical relationship between two quantitative variables with a **scatter diagram,** in which the explanatory variable is plotted on the horizontal axis and the dependent variable on the vertical axis. A scatter diagram is also called a *relational graph,* since it is meant to show how the dependent variable is related to the explanatory variable. Figure 11.1 shows a scatter diagram of these data on radioactive exposure and cancer mortality.

To analyze statistically the empirical relationship shown in Fig. 11.1, we need to have a model of how the cancer mortality data are related to the data on exposure to radioactive waste. The *linear* (or straight-line) model is

$$y = \alpha + \beta x$$

where y is the dependent variable, x is the explanatory variable, and α and β are two parameters that describe the relationship between x and y. This model is shown in Fig. 11.2.

The parameter α is the y *intercept* for the line—the value of y when $x = 0$. The parameter β is the *slope* of the line—the change in y when x increases by

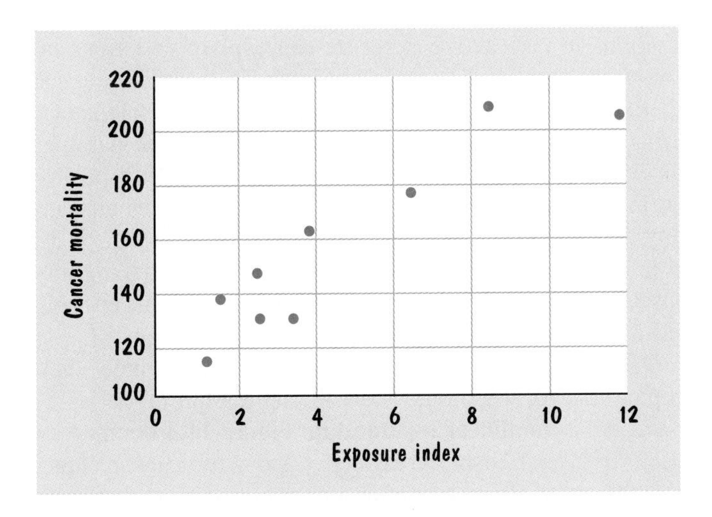

Figure 11.1 Cancer mortality and exposure to radioactive waste.

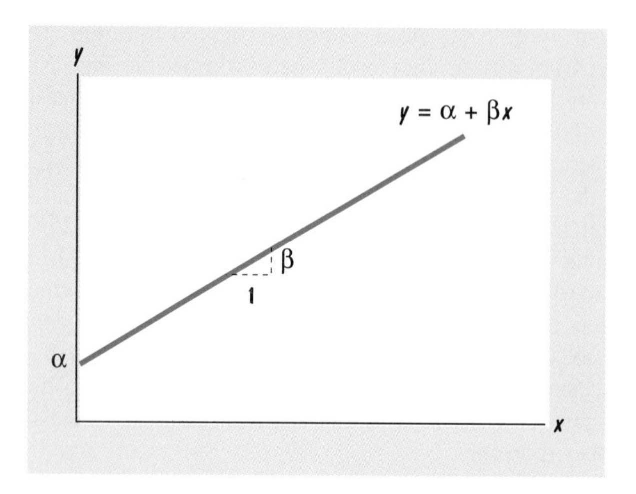

Figure 11.2 A linear relationship between x and y.

1. If, for instance, x is household income and y is household spending (both in dollars), a .9 value for β means that a \$1.00 increase in household income is accompanied by a \$.90 increase in spending. If x is the number of hours per week spent studying for a test and y is the test score (from 0 to 100), a 3.5 value for β means that 1 additional hour per week of studying is accompanied by a 3.5 increase in the test score.

In our example, cancer mortality would be the dependent variable y and radioactive exposure the explanatory variable x. Our model then states that variations in radioactive exposure can explain variations in cancer mortality. As here, regression models are often intended to depict a causal relationship, in that the researcher believes that changes in the explanatory variable cause changes in the dependent variable. The sign of the parameter β tells us whether this effect is positive or negative. In our example, a positive value for β means that increased exposure to radioactive waste causes an increase in cancer mortality. In other cases, we may use a regression model to investigate the empirical relationship between two variables, without making any assumptions about causality. The term *explanatory variable* is neutral, encompassing causation as well as mere statistical association.

Our model is conveniently linear, but it is unlikely that the real world is this simple. Instead, we should view the regression model as a linear approximation to a possibly nonlinear relationship. Figure 11.3 compares a hypothetical nonlinear "true relationship" between x and y to a linear approximation that is satisfactory as long as we confine our attention to realistic and relatively modest variations in the explanatory variable. It would be nonsensical to extend our

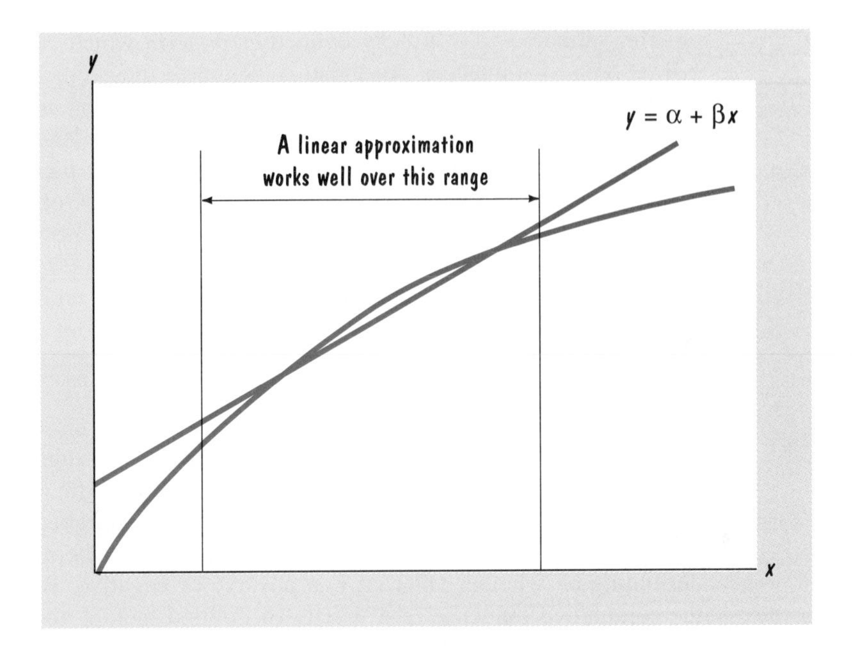

Figure 11.3 A linear approximation of a nonlinear relationship.

linear model to negative values of radioactive exposure or to more than 100 percent mortality. We should also be cautious about values that are not impossible but nonetheless are far beyond the boundaries of our data. Since our largest value of x is 11.64, we should be cautious about extrapolating our model to $x = 20$, 50, or 1000.

Similarly, we should be wary of using a model for a $7 trillion economy to predict behavior if this economy's national income were $0. Or using a model of studying and test scores to predict the test score of someone who studies 168 hours per week. Or using a model relating the income and education of high school and college graduates to predict the income of someone with 40 years of education. Or using a model of farm output over the past 20 years to predict agricultural output 2000 years ago or 2000 years from now. Linear approximations are very useful as long as we do not try to extrapolate them to situations that are drastically different from the intended application, situations about which we have little knowledge or statistical evidence.

The data in Fig. 11.1 do not lie perfectly on a straight line, because cancer mortality depends on more than exposure to contaminated water. In recognition of these other factors, the **simple linear regression model** is

$$y = \alpha + \beta x + \epsilon \tag{11.1}$$

The variable ϵ is a broadly defined *error term* which encompasses the effects on y of influences other than x. Some of these other factors are intentionally neglected because we do not think they are very important; others may be omitted because we simply do not have any useful data for measuring them; and still others may be left out because we overlooked their importance. In multiple-regression models, which are beyond the scope of this book, other important influences can be incorporated into the model by expanding the number of explanatory variables.

To reinforce the fact that variables y, x, and ϵ take on different values, while α and β are fixed parameters, the regression model can also be written with i subscripts:

$$y_i = \alpha + \beta x_i + \epsilon_i$$

For notational simplicity, we will generally omit these subscripts.

The inevitable omitted influences encompassed by the error term ϵ imply that our data will seldom, if ever, lie exactly on a straight line. For any given value of x, the value of y may be somewhat above or below the line $\alpha + \beta x$, depending on whether value of ϵ is positive or negative. Because ϵ represents the cumulative influence of a variety of omitted factors, the central limit theorem suggests that the values of ϵ can be described by a normal distribution. In addition, we make four assumptions:

1. The population mean of ϵ is 0.
2. The standard deviation of ϵ is a constant σ.
3. The values of ϵ are independent of each other.
4. The values of ϵ are independent of x.

The first assumption is innocuous, since the value of the intercept is set so that the population mean of ϵ is 0. The second assumption is appealing, but sometimes is violated. With times-series data, for instance, the omitted influences may depend on the population, which grows over time, causing the standard deviation of ϵ to grow, too. One way to guard against such effects is to scale the data, for example, by using per capita values, as explained in Chap. 2. The third assumption states that the values of the error term for different observations are independent of each other. However, the error term involves omitted variables whose values may not be independent. Our error term here may include the effects of airborne emissions from factories, which affect a number of adjacent counties. Advanced courses show how correlations among errors can be detected and how our statistical procedures can be modified to take this additional information into account. Fortunately, violations of the second and third assumptions do not bias our estimates of α and β; however, their standard deviations will be unnecessarily large and underestimated, leading us to overestimate the statistical significance of our results.

The fourth assumption is probably the most crucial. It states that the values of the error term ϵ are independent of the values of x. Perhaps the factories that emit airborne pollutants happen to be near the Columbia River, so that people living close to the river not only are drinking contaminated water, but also are breathing polluted air. If so, the fourth assumption is violated, and the effects of radioactive waste in the river will be confounded with the effects of carcinogens in the air. If we look at only the contaminated water and neglect the polluted air, we will overestimate the effect of the contaminated water on cancer mortality. The best way to deal with this problem is to use the multiple-regression models analyzed in more advanced courses so that we can include both explanatory variables and thereby take both factors into account. Here, we are using the simple model to introduce the principles of regression analysis, and we assume that all four assumptions are correct.

If so, the effects of the error term ϵ on the observed values of y can be depicted as in Fig. 11.4. For any given value of x, such as x_1, the population mean of ϵ is 0 and the mean of y is given by the line $y = \alpha + \beta x$. If the value of ϵ turns out to be positive, the value of y will be above the line; if the value of ϵ happens to be negative, the value of y will be below the line. Of

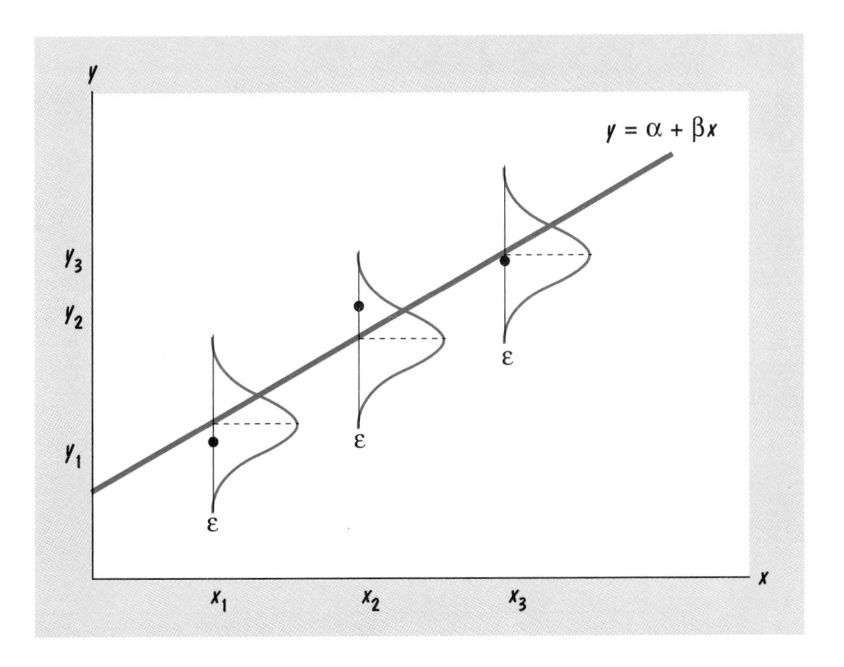

Figure 11.4 The inevitable error term.

course, we have no way in practice of knowing which values of ϵ are positive and which are negative. All we see are the observed values of x and y, as in Fig. 11.5, which we want to use to estimate the slope and intercept of the unseen line. The next section shows how to do so.

Using Time as an Explanatory Variable

Although a regression model is often based on the researcher's belief that there is a causal relationship between the explanatory variable and the dependent variable, it can also be used for merely descriptive purposes—to explore the statistical relationship between two variables. A common example is the fitting of a line to a time-series graph that shows how something varied over time. For instance, Fig. 11.6 shows the annual precipitation (in millimeters) in the contiguous United States from 1900 through 1989.[2] This graph shows considerable year-to-year variation, from a low of 602.2 in 1910 to a high of 865.6 in 1983, but no apparent long-term trend.

We can test statistically whether there is a trend by using the regression model

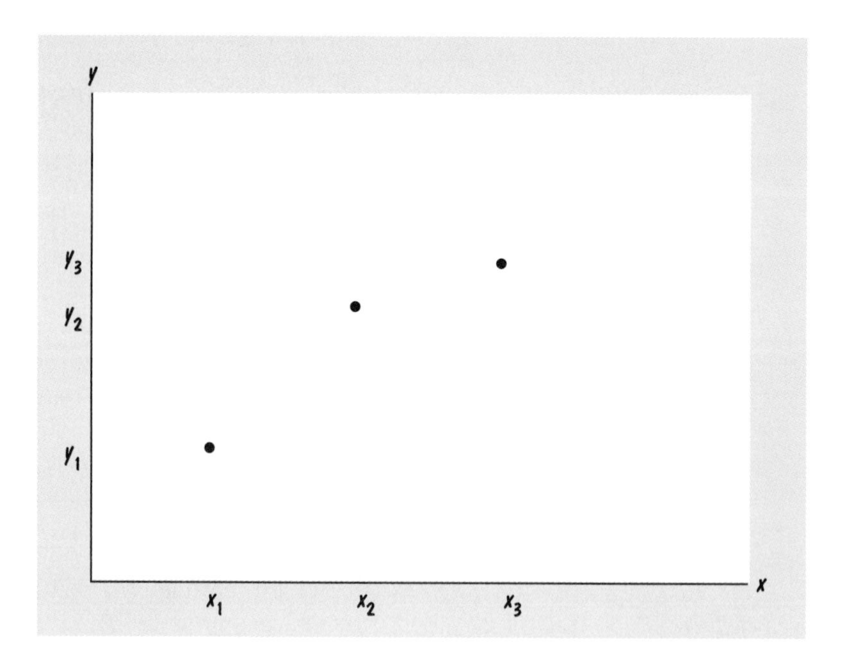

Figure 11.5 The observed data.

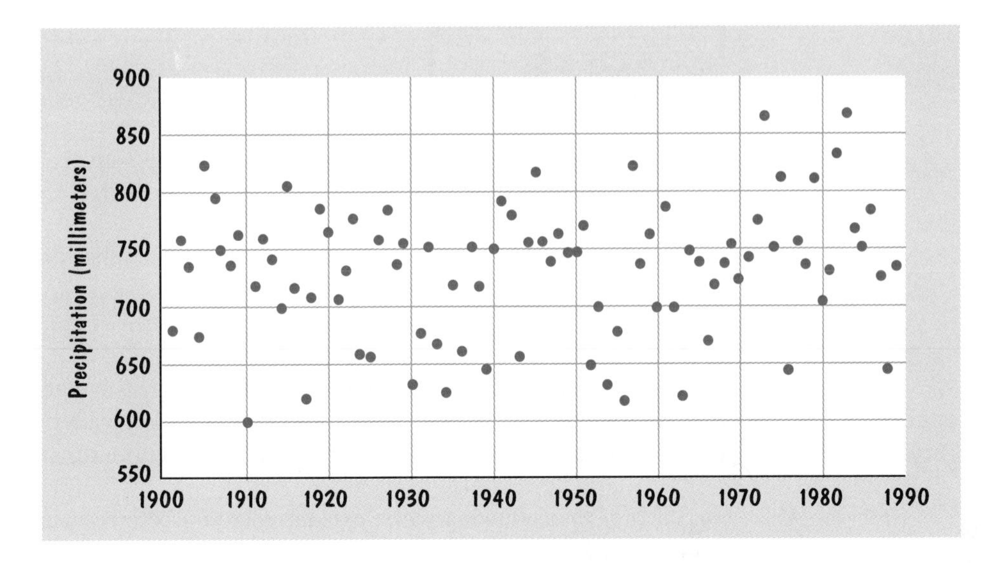

Figure 11.6 Annual precipitation in the contiguous United States from 1900 to 1989.

$$y = \alpha + \beta x + \epsilon$$

where y is the annual precipitation and x is the year (1900, 1901, and so on). Using the procedures described in the next section, a line fit to these data has a slope of .287 (but the difference from zero is not statistically significant at the 5 percent level):

$$y = 173.12 + .287x$$

For $x = 1900$, the predicted value of annual precipitation is $y = 173.12 + .287(1900) = 718.42$; for $x = 2000$, the predicted value is $y = 173.12 + .287(2000) = 747.12$. This difference represents a 4 percent increase over 100 years: $(747.12 - 718.42)/718.42 = .040$. A closer examination of the data reveals a general decrease from 1900 to the mid-1930s (the dust bowl period) and an upward trend since the mid-1930s. (A line fit to data from 1934 through 1989 has a slope of .87 and is nearly statistically significant.)

Trends in annual precipitation are *caused* not by the passage of time, but rather by a complex web of meteorological forces that scientists are trying to understand, so that they can predict whether the observed trend will continue and how it might be affected by changes in human behavior. The purpose of fitting regression lines to time-series graphs is to identify patterns and quantify their magnitudes. It remains for experts to supply causal explanations.

exercises

11.1 Explain why each of the following pairs of data is positively related, negatively related, or essentially unrelated. If there is a relationship, which is the dependent variable?

a. A child's Little League batting average and the number of hours spent practicing

b. The temperature in Houston, Texas, and attendance at the city's Fourth of July parade

c. The number of automobiles on a freeway and their average speed

d. The weight of a car and its fuel usage in miles per gallon

e. The mileage on a 5-year-old Chevrolet and its current market price

11.2 For each of the following pairs, explain why you believe that they are positively related, negatively related, or essentially unrelated; if they are related, identify the dependent variable.

a. Among students in a chemistry class, hours spent studying and course grade

b. Among children under 18, height and time for the 100-meter dash

c. Among men over age 30, age and number of hairs on head

d. Among women over age 30, age and number of grandchildren

e. The year and the population of the United States, for 1900 through 1990

11.3 Here is a scatter diagram of the homework and final examination scores of Alice, Bill, Carrie, David, and Elaine:

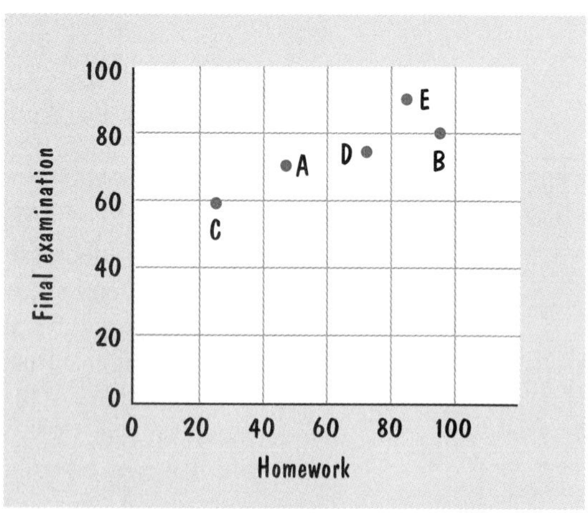

a. Which student had the highest grade on the homework?

b. Which student had the highest grade on the final examination?

c. Was the average score higher on the homework or on the final examination?

d. Was there a greater dispersion of scores on the homework or on the final examination?

e. Are homework and final scores positively related, negatively related, or unrelated?

11.4 Professor Smith claims that a student's score on the final examination can be predicted from homework scores. Here are some randomly selected homework and final examination scores:

	HOMEWORK x	FINAL EXAMINATION y
FRED	92	87
GAIL	65	71
HANNAH	75	75
IRENE	83	84
JOSHUA	95	93

a. What are the average homework and final examination scores?
b. Plot these data with homework scores on the horizontal axis and final examination scores on the vertical axis.
c. Does there seem to be a positive or negative relationship?
d. Draw a straight line that appears to fit these data, and then calculate rough estimates of the slope and intercept of the line you drew.
e. In plain English, explain what these values for the slope and intercept mean.

11.5 There are two competing theories regarding the relationship between student course evaluations and the number of years that a professor has been teaching. One theory holds that experienced teachers tend to get the highest ratings, while the other theory says that young professors, who are closest in age to their students, generally score highest. Make a rough sketch of what a scatter diagram of evaluations (y) and experience (x) would look like if the first theory were correct. Make a rough sketch of what a scatter diagram would look like if the second theory were correct. In either case, why might we anticipate that the data will not lie perfectly on a line $y = \alpha + \beta x$? If there is no relationship at all between course evaluations and experience, what would a scatter diagram look like?

11.2 Least-Squares Estimation

The use of relational graphs to understand natural and social phenomena was pioneered by J. H. Lambert, a Swiss-German mathematician and scientist, in a 1765 book that used scatter diagrams to show, for example, how the rate of evaporation of water depends on the water's temperature. In a remarkably prescient explanation of what was, at that time, a novel procedure, he wrote

> We have in general two variable quantities, x, y, which will be collected with one another by observation, so that we can determine for each

> value of x, which may be considered an abscissa, the corresponding ordinate y. Were the experiments or observations completely accurate, these ordinates would give a number of points through which a straight or curved line should be drawn. But as this is not so, the line deviates to a greater or lesser extent from the observational points. It must therefore be drawn in such a way that it comes as near as possible to its true position and goes, as it were, through the middle of the given points.[3]

Drawing a line by hand "through the middle of the given points" was the only procedure available to Lambert in 1765. Two hundred years later, in the 1970s, the chief economic forecaster at one of the largest U.S. banks still fit lines to scatter diagrams with a pencil and transparent ruler. Fitting lines by hand is generally a hopelessly amateurish and crude method for estimating the slope and intercept of a regression model. It is often hard to know exactly where to draw the line and even harder to identify the intercept and calculate the slope. In addition, two-dimensional scatter diagrams cannot be used with multiple-regression models. It is much better to use explicit mathematical formulas for estimating the slope and intercept, analogous to using the sample mean to estimate the population mean.

If our fitted line will be used to predict the values of y for specified values of x, it is logical to draw a line that will do this best for the observed values of x and y. If we let the intercept of our fitted line be a and the slope be b, then for any specified value of x, the predicted value of y is

$$\hat{y} = a + bx \tag{11.2}$$

The variable $\hat{y}$ is written with a caret ("hat" sign) over it to distinguish this, the predicted value, from the actual observed value y. Similarly, a and b are the intercept and slope, respectively, of the fitted line and are not necessarily equal to the values of the unknown parameters α and β. Figure 11.7 shows that the prediction errors $y - \hat{y}$ are equal to the vertical distances from the observations to the fitted line.

We cannot measure the overall predictive accuracy by adding the prediction errors, because the positive and negative deviations from the line will cancel and the sum of the prediction errors will be close (or even equal) to zero. Instead, the German scientist Karl Gauss (1777–1855) argued that we should measure the fitted line's overall prediction accuracy by adding the squared values of the prediction errors.

> A **least-squares estimator** of a model's parameters minimizes the model's sum of squared prediction errors.

By squaring the prediction errors, we do not allow positive and negative errors to offset each other. We do not care whether our prediction error is positive or

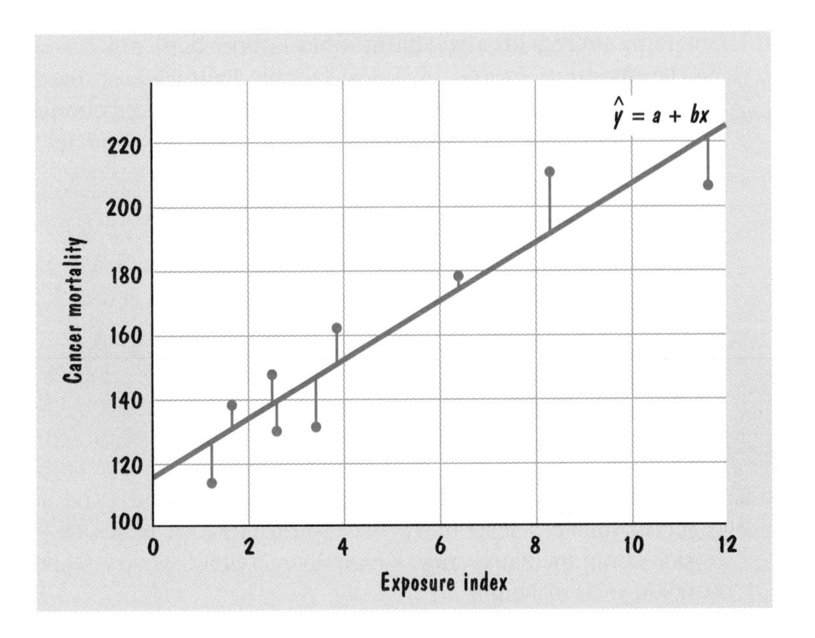

Figure 11.7 The prediction errors are the vertical distances from the fitted line.

negative, only how large it is. The squaring of errors implies that large errors are more worrisome than small errors. We prefer a fitted line that has ninety 1-inch errors to a line that has one 10-inch error. This logic is appealing. In addition, Gauss showed that least-squares mathematics is very tractable and that least-squares estimators are quite sensible. For example, he showed that when several measurements are used to estimate a distance, the least-squares estimator is the mean of these measurements.

For the linear regression model, Gauss showed that the least-squares estimator of the slope β is

$$b = \frac{\Sigma(x_i - \overline{x})(y_i - \overline{y})}{\Sigma(x_i - \overline{x})^2}$$

where the summations (designated by Σ) are over the n observations. After the slope estimate b is calculated, the estimate of the intercept α is easily computed from the mathematical fact that the least-squares line goes through the average values of x and y: $\overline{y} = a + b\overline{x}$. Therefore,

$$a = \overline{y} - b\overline{x}$$

Statistical software obviates the need to memorize these formulas or do the tedious calculations by hand. After we enter the observed values of y and x, the computer program almost immediately shows us the least-squares estimates a

and b, along with a wealth of other information, which we shall soon explore. The fitted line in Fig. 11.7 is in fact the least-squares line for our plutonium example. No line has a smaller sum of squared prediction errors. The least-squares estimate of the slope is 9.23, and the least-squares estimate of the intercept is 114.7:

$$\hat{y} = 114.7 + 9.23x$$

In the next section, we see how confidence intervals and hypothesis tests can be used to gauge the precision of these estimates and make inferences about the values of α and β.

The Relationship between Height and Weight

The data below give the average weights (in pounds) and heights (measured in inches above 5 feet) of 18- to 24-year-old U.S. males. A least-squares regression confirms that average weight increases by about 4½ pounds for every additional inch of height:

$$\hat{y} = 120.7 + 4.55x$$

The American Heart Association says that the ideal male weight is $110 + 5.0x$, putting the average 18- to 24-year-old U.S. male 5 to 10 pounds overweight. (The ideal female weight is $100 + 5.0x$.)

HEIGHT x	2	3	4	5	6	7	8	9	10	11	12	13	14
WEIGHT y	130	135	139	143	148	152	157	162	166	171	175	180	185

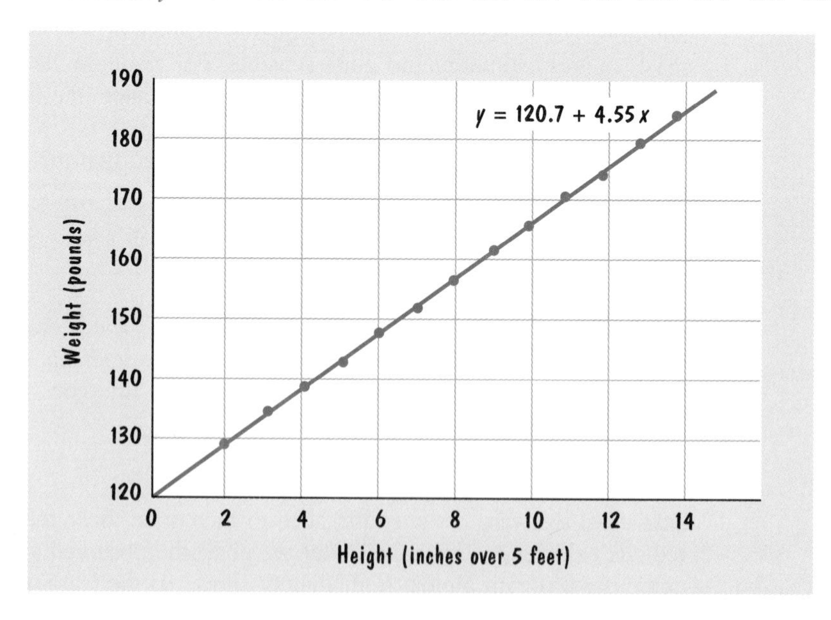

GUINDON By Guindon

Harry has a height problem. According to the weight chart he should be 7 feet tall.

exercises

11.6 The U.S. National Center for Health Statistics reported the following average weights (in pounds) and heights (measured in inches above 5 feet) of 18- to 24-year-old U.S. females:

HEIGHT x	−3	−2	−1	0	1	2	3	4	5	6	7	8
WEIGHT y	111	114	118	121	124	128	131	134	137	141	144	147

Use these data to draw a scatter diagram and to obtain the least-squares estimates of the equation $y = \alpha + \beta x + \epsilon$. Briefly compare this empirical relationship for females with that for males in Example 11.2. Be sure to compare the slopes and interpret any difference between them. Also explain why you either agree or disagree with this reasoning: "A person with zero height has zero weight. Therefore, a regression of weight on height must predict zero weight for zero height."

11.7 Below are some data on cricket chirps per second and temperature in degrees Fahrenheit for the striped ground cricket.[4] Which is the dependent and which is the explanatory variable? Draw a scatter diagram with the cricket chirps axis running from 0 to 20 and the temperature axis running from 60 to 100. Does the relationship seem linear? Estimate the equation $y = \alpha + \beta x + \epsilon$ by least squares, and draw this fitted line in your scatter diagram. In ordinary English, interpret your estimate of β.

CHIRPS	TEMPERATURE	CHIRPS	TEMPERATURE	CHIRPS	TEMPERATURE
20.0	88.6	15.5	75.2	17.2	82.6
16.0	71.6	14.7	69.7	16.0	80.6
19.8	93.3	15.4	69.4	17.0	83.5
18.4	84.3	16.2	83.3	17.1	82.0
17.1	80.6	15.0	79.6	14.4	76.3

11.8 In a 1993 election in the second senatorial district in Philadelphia, which would determine which party controlled the Pennsylvania State Senate, the Republican candidate won according to the votes cast on election day, 19,691 to 19,127. However, the Democrat won the absentee ballots, 1391 to 366, thereby winning the election by 461 votes, and Republicans charged that many absentee ballots had been illegally solicited or cast. A federal judge ruled that the Democrats had engaged in a "civil conspiracy" to win the election and declared the Republican the winner. Among the evidence presented was the graph below of

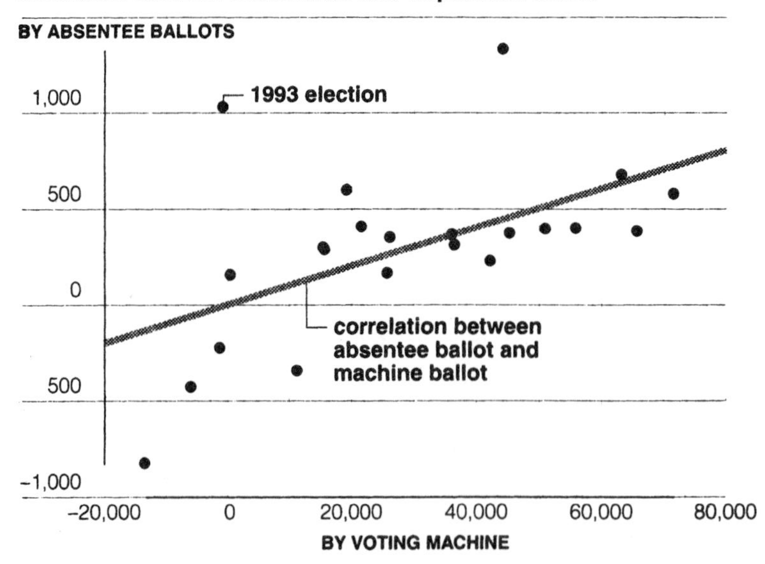

22 state senatorial elections in Philadelphia during the years 1982 to 1993.[5] On the horizontal axis is the difference between the numbers of votes for the Democratic and Republican candidates cast on election day; on the vertical axis is the difference between the numbers of votes for the Democratic and Republican candidates cast by absentee ballots.

a. What criterion was used to draw a line through the points?

b. Explain why this line's positive or negative slope is either plausible or counterintuitive.

c. Was this evidence cited by the Republican or Democrat? Explain your reasoning.

11.9 Letting the horizontal axis go from 60 to 200 and letting the vertical axis go from 0 to 40, draw a scatter diagram, using the data below[6] on peak wind velocity x in miles per hour for nine hurricanes and estimates of the percentage y of the trees in the hurricane's path that were killed or severely damaged. Use least squares to estimate the model $y = \alpha + \beta x + \epsilon$, and draw the least-squares line in your scatter diagram. Interpret the estimated value of β.

	x	y
ANDREW, FLORIDA	150	30
BOB, CAPE COD	100	2
GILBERT, JAMAICA	130	7
GILBERT, YUCATAN	190	13
HUGO, PUERTO RICO	100	5
HUGO, SOUTH CAROLINA	80	10
HUGO, VIRGIN ISLANDS	140	10
OFA, SAMOA	120	28
VAL, SAMOA	150	30

11.10 Here are the prepregnancy weights of 25 mothers and the birth weights of their children, both in kilograms.[7] Which would you use as the dependent variable and which as the explanatory variable in the regression model $y = \alpha + \beta x + \epsilon$? Letting the horizontal axis go from 40 to 80 and letting the vertical axis go from 2 to 5, draw a scatter diagram and then use least squares to see if there is a positive or negative relationship; interpret your estimates of α and β.

MOTHER	CHILD	MOTHER	CHILD	MOTHER	CHILD	MOTHER	CHILD	MOTHER	CHILD
49.4	3.52	70.3	4.07	73.5	3.23	65.8	3.35	63.5	4.15
63.5	3.74	50.8	3.37	59.0	3.57	61.2	3.71	59.0	2.98
68.0	3.63	73.9	4.12	61.2	3.06	55.8	2.99	49.9	2.76
52.2	2.68	65.8	3.57	52.2	3.37	61.2	4.03	65.8	2.92
54.4	3.01	54.4	3.36	63.1	2.72	56.7	2.92	43.1	2.69

11.3 Confidence Intervals and Hypothesis Tests

It is always tempting to think that a set of data is definitive. This is another of those temptations that should be resisted. When we use a sample mean to estimate a population mean, we must recognize that our sample is but one of many that might have been selected, and consequently we must use a confidence interval to gauge the potential sampling error. The same is true of least-squares estimates of the regression model. Our estimates of the intercept α and slope β depend on the particular values of the error term ϵ that happen to occur.

Look back at Fig. 11.4, which shows how the observed values of y depend on the values of ϵ. As drawn, a line fit to these three points will be fairly close to the true relationship that generated these data. However, if the error term for x_3 were substantially above the line, rather than slightly below, a fitted line would be much steeper than the true relationship, giving an estimate of the slope β that was too high and an estimate of the intercept α that was too low. On the other hand, if the error term for x_1 were substantially above the line, a fitted line would be flatter than the true relationship, giving an estimate of β that was too low and an estimate of α that was too high.

Because we see only the observed values of x and y, and not the true relationship, we cannot say whether the particular estimates we obtain are too high or too low, just as we cannot say whether a particular sample mean is higher or lower than the unknown population mean. However, we can gauge the reliability of our estimates by calculating standard errors and confidence intervals for our estimators. If the error term ϵ is normally distributed and conforms to the four assumptions stated earlier, then it can be shown that our estimators are normally distributed, too, with population means equal to the parameters they are estimating:

$$a \text{ is } N[\alpha, \text{standard deviation of } a]$$

$$b \text{ is } N[\beta, \text{standard deviation of } b]$$

Figure 11.8 shows the probability distribution for our least-squares estimator b.

Both a and b are unbiased estimators, in that their population means are equal to α and β, respectively. The formulas for the standard deviations of a and b are too complicated to present here and unnecessary, too, since we let statistical software do the calculations for us. However, they do have four intuitively reasonable properties.

First, the standard deviations of these estimators are smaller if the standard deviation of the error term ϵ is small. If the values of ϵ are consistently close to zero, then the fitted line will be very close to the true relationship, regardless

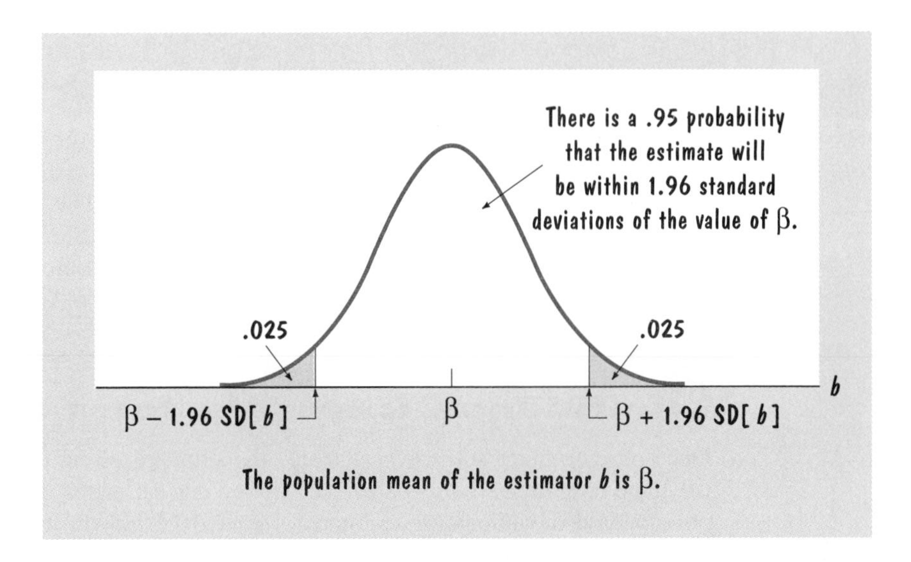

Figure 11.8 The distribution of the least-squares estimator b.

of whether the points happen to be slightly above or slightly below the line. If, however, the standard deviation of the error term is large, then the large positive and negative values of ϵ will cause the observations to be scattered widely about true relationship, creating considerable uncertainty about the values of the slope and intercept.

Second, the standard deviations of these estimators are smaller if there are a large number of observations n. With lots of data, the observations above and below the true relationship are almost certain to balance out, giving accurate estimates of both α and β. With only a handful of data, almost anything can—and probably will—happen.

Third, the standard deviations of these estimators are smaller if there is a large amount of variation in the values of x. If we observe substantial changes in x, we will have a good idea how x affects y. If x hardly varies, we will have little idea how changes in x affect y (and if we are very uncertain about the slope, we will also be very uncertain about the intercept).

Fourth, the standard deviation of our estimate of the intercept depends on how far the values of x are from zero. We will be very uncertain of the value of y when x is equal to zero if we never observe any values of x near zero.

To estimate the standard deviations of a and b, we need an estimate of the standard deviation of the error term ϵ, which is provided by the *standard error of estimate* (SEE),

$$\text{SEE} = \sqrt{\frac{\Sigma(y_i - \hat{y}_i)^2}{n - 2}} \tag{11.3}$$

The SEE is the square root of the average prediction error (but we divide by $n - 2$ rather than n, because the estimation of α and β uses up 2 degrees of freedom). As in earlier chapters, the term *standard error* is used to distinguish our sample estimate SEE from the population value σ. Similarly, the estimated standard deviations of our estimators a and b are called the *standard errors* of a and b.

Confidence Intervals for α and β

Once our computer software calculates the standard errors of a and b, we can calculate confidence intervals to accompany our estimates of α and β. Because our standard deviations are estimated, rather than known, we use the t distribution with $n - 2$ degrees of freedom.

After choosing a confidence level, such as 95 percent, we consult the t distribution in Table 4 to determine the value t^* that corresponds to this probability; then we set our confidence intervals for α and β equal to our estimates plus or minus the requisite number of standard errors:

$$a \pm t^*(\text{standard error of } a) \quad \text{and} \quad b \pm t^*(\text{standard error of } b) \tag{11.4}$$

For our plutonium example, the standard errors are

$$\text{Standard error of } a = 8.046 \quad \text{standard error of } b = 1.419$$

With 9 observations, we have $9 - 2 = 7$ degrees of freedom. Table 4 gives $t^* = 2.365$ for a 95 percent confidence interval, so that a 95 percent confidence interval for α is

$$\begin{aligned} 95\% \text{ confidence interval for } \alpha &= 114.71 \pm 2.365(8.046) \\ &= 114.71 \pm 19.03 \end{aligned}$$

and a 95 percent confidence interval for β is

$$\begin{aligned} 95\% \text{ confidence interval for } \beta &= 9.23 \pm 2.365(1.419) \\ &= 9.23 \pm 3.36 \end{aligned}$$

We are usually keenly interested in the estimate of the slope (since this measures the change in y associated with a 1-unit change in x) and not very interested in the intercept.

Hypothesis Tests

It is one short step from confidence intervals to hypothesis tests. The most common null hypothesis is that the value of β is 0:

$$H_0 : \beta = 0$$
$$H_1 : \beta \neq 0$$

If β does equal 0, then there is not a linear relationship between the explanatory variable and the dependent variable. Researchers generally specify their models precisely because they believe that there is such a relationship. Thus $\beta = 0$ is a straw hypothesis that the researcher hopes to be able to reject. This null hypothesis is two-sided unless the researcher can rule out positive or negative values before looking at the data.

As in earlier chapters, the null hypothesis is tested by seeing how many standard errors the estimate of β is from the value of β under the null hypothesis:

$$t = \frac{\text{estimated value of } \beta - \text{null hypothesis value of } \beta}{\text{standard error of estimate of } \beta} \tag{11.5}$$

Therefore:

For the null hypothesis $\beta = 0$, the t statistic simplifies to

$$t = \frac{b}{\text{standard error of } b} \tag{11.6}$$

The null hypothesis is rejected at a specified significance level if the t value is larger than the appropriate cutoff, using a t distribution with $n - 2$ degrees of freedom. For a test at the 5 percent level with a large number of observations, the cutoff is approximately 2. Thus the null hypothesis $\beta = 0$ is generally rejected at the 5 percent level (showing that x has a statistically significant effect on y) if the ratio of the absolute value of β to its standard error is larger than 2, thereby showing that the estimate of β is more than 2 standard errors from zero.

In our plutonium example, the estimate of β is $b = 9.23$, and the standard error of b is 1.42, implying a t value of

$$t = \frac{9.23}{1.42} = 6.51$$

The estimated slope is 6.51 standard deviations from zero, far larger than the 2.365 cutoff for a two-tailed test at the 5 percent level with $9 - 2 = 7$ degrees of freedom, and thus we reject the null hypothesis decisively. Statistical software shows the two-sided P value to be .0006.

Some researchers report their estimated coefficients along with the standard errors:

> Using 1965 data for nine Oregon counties, a least-squares regression of the cancer mortality rate (per 100,000 residents) on an index of exposure to Columbia River water (ranging from a low of 1.25 to a high of 11.64) yielded these estimates, with standard errors in parentheses:
>
> $$\underset{}{y} = \underset{(8.05)}{114.72} + \underset{(1.42)}{9.23x}$$

(Although the fitted equation gives the predicted value of y, rather than the observed value, research reports usually show the dependent variable as y, without a caret or other distinguishing mark.)

Most readers instinctively look at the ratio of the estimate of β to its standard error to see whether this ratio is larger than 2. This is such an ingrained habit that researchers often report t values along with their parameter estimates, to save readers the trouble of doing the arithmetic themselves. Because we have little interest in the intercept, some do not bother to report its t value:

> Using 1965 data for nine Oregon counties, a least-squares regression of the cancer mortality rate (per 100,000 residents) on an index of exposure to Columbia River water (ranging from a low of 1.25 to a high of 11.64) yielded these estimates, with the t value in brackets:
>
> $$y = 114.72 + \underset{[6.51]}{9.23x}$$

Another reporting format is to show the two-sided P value.

The reported t value (or two-sided P value) is for the null hypothesis $\beta = 0$, since our foremost concern is generally whether there is a statistically significant linear relationship between the explanatory variable and the dependent variable. However, this is not the only null hypothesis that may be of interest. An automaker may want to know, not whether a 1 percent increase in prices will affect the number of cars sold, but whether it will reduce sales by more or less than 1 percent. To test hypotheses other than $\beta = 0$, we need merely substitute the desired value into Eq. (11.5). Alternatively, we can see whether this value is inside a confidence interval for β, since we learned in Chap. 7 that values outside a 95 percent confidence interval will be rejected at the 5 percent level, while values inside will not.

In addition to asking whether the estimated effect of the explanatory variable on the dependent variable is statistically significant, we should ask if the effect is in the anticipated direction and whether the estimated size of the effect is substantial and plausible. Here, we anticipated that an increased exposure to radioactive waste would increase cancer mortality, and the estimated slope is indeed positive. If the estimate had turned out to be negative and statistically significant, the most likely explanation would be that an important variable had been omitted from our analysis and played havoc with our estimates. For example, it might have coincidentally turned out that those Oregon counties near the Columbia River happened to have relatively few elderly residents. A scatter diagram comparing exposure to Columbia River water and cancer mortality would then suggest that proximity to the river reduced cancer mortality, when it fact the harmful effects of the water were overwhelmed by this population's youthfulness. If we are able, on reflection, to identify such a confounding variable, then we can use the multiple-regression techniques discussed in more advanced courses to handle more than one explanatory variable—here, we could use both the exposure index and the age of the population as separate explanatory variables.

Another possible explanation for an estimate that has the wrong sign is that errors do happen. In Fig. 11.4, the value of y_1 might turn out to be substantially above the true relationship, while the values of y_2 and y_3 are substantially below, creating the illusion of a negative relationship between x and y. This is not likely, but unlikely things occasionally do occur. A final possibility is that our intuition was wrong. Perhaps exposure to radioactive waste actually reduces cancer mortality. This seems ludicrous, but empirical studies sometimes do reverse firmly held beliefs. Here, there is no need to search for explanations, because the estimated coefficient has the positive sign that we anticipated.

In addition to the right sign, we need to consider whether the estimated size of the coefficient is substantial and plausible. We cannot determine this merely by looking at 9.23 and judging this to be a large or small number. We have to provide some context by considering the magnitudes of variables x and y. The slope β tells us the effect on the dependent variable y of a 1-unit increase in x. (We say *1-unit* because an explanatory variable might be measured in hundredths of an inch or in trillions of dollars.) Here x is an exposure index, much like the consumer price index, that does not have natural units, like inches or dollars. To put into perspective the importance of a 1-unit change in an index, we need to take into account the scale of the index. The average value of x, the exposure index, is 4.62, and the average value of y, the cancer mortality rate, is 157.34. Remember, too, that (except for rounding error) the fitted line goes through the sample means:

$$\bar{y} = a + b\bar{x}$$
$$157.34 = 114.72 + 9.23(4.62)$$

At the sample means, a 1-unit increase in x is a 21.6 percent increase ($1/4.62 = .216$). Our 9.23 estimated coefficient implies that a 1-unit increase in x will increase y by 9.23, which is an additional 9.23 deaths from cancer per

100,000 residents. To put this in perspective, we can again convert to a percentage: At the mean, 9.23 additional deaths represents a 5.9 percent increase (9.23/157.34 = .059.) Thus we summarize our evaluation of the magnitude of the slope by observing that, at the mean, a 21.6 percent increase in exposure to radioactive waste is predicted to increase the annual number of deaths from cancer by 9.23 per 100,000, which represents a 5.9 percent increase. This estimate seems substantial and plausible.

Another way to put in perspective the magnitude of an estimated slope is to compare the predicted values of y for selected values of x. For instance, we can compare the predicted mortality rates at $x = 2$ (a relatively low exposure) and $x = 10$ (a relatively high exposure) by substituting these values into the least-squares equation:

$$x = 2: \qquad \hat{y} = 114.72 + 9.23(2) = 133.18$$

$$x = 10: \qquad \hat{y} = 114.72 + 9.23(10) = 207.03$$

An increase in exposure from $x = 2$ to $x = 10$ is predicted to increase cancer mortality by $207.03 - 133.18 = 73.85$ annual deaths per 100,000 residents. This figure is substantial and plausible.

Prediction Intervals for y

Regression models are usually intended for predictions, in either actual or hypothetical situations. We might want to predict what cancer mortality will be next year, or we might want to compare the estimated cost of stopping the leakage of radioactive wastes into the Columbia River with the predicted effect on mortality. Because the predicted value of the dependent variable y depends on our estimates of α and β, its accuracy is uncertain. Our parameter estimates depend on the error term values that happened to occur in the sample data, and so do predictions based on these estimates. If different values of ϵ had occurred, we would obtain somewhat different parameter estimates and therefore make somewhat different predictions.

All is not hopeless, however, because we can use an interval to quantify the precision of our prediction. For a selected value of the explanatory variable x^*, the predicted value of y is given by our least-squares equation:

$$\hat{y} = a + bx^* \tag{11.7}$$

After selecting a confidence level, such as 95 percent, we use a t distribution with $n - 2$ degrees of freedom to determine the appropriate value t^*, and we set our **prediction interval** for y equal to the predicted value plus or minus the requisite number of standard errors of the prediction error $y - \hat{y}$:

$$\hat{y} \pm t^*(\text{standard error of } y - \hat{y}) \tag{11.8}$$

The interpretation of a prediction interval is analogous to a confidence interval—for a 95 percent prediction interval, there is a .95 probability that the interval will encompass the actual value of y.

We can also consider the population mean value of y for a selected value of the explanatory variable x^*. This population mean of y is the average value of y that we would observe if the value of the explanatory variable were equal to x^*, over and over indefinitely, with random values for the error term ϵ. Since the population mean value of the error term is zero,

$$\text{Mean of } y = \alpha + \beta x^*$$

Therefore, we can estimate the population mean of the dependent variable with this equation:

$$\text{Estimate of mean of } y = a + bx^* \tag{11.9}$$

and we calculate a confidence interval for our estimate:

$$a + bx^* \pm t^*[\text{standard error of } (\alpha + \beta x^*)] \tag{11.10}$$

Our prediction of y in Eq. (1.7) and our estimate of the population mean of y in Eq. (11.9) are identical. However, the population mean of y is more certain than a single value, and thus a confidence interval for the population mean of y is narrower than is a prediction interval for y.

The formulas for the standard errors used in Eqs. (11.8) and (11.10) are very complicated, and, if possible, we should let statistical software do our calculations for us. Unfortunately, some software packages show prediction intervals for only the observed values of x. If we want prediction intervals for other values, we must do the calculations ourselves:

$$\text{Standard error of } y - \hat{y} = \text{SEE}\sqrt{1 + \frac{1}{n} + (x^* - \bar{x})^2\left(\frac{\text{standard error of } b}{\text{SEE}}\right)^2}$$

$$\text{Standard error of } (a + bx^*) = \text{SEE}\sqrt{\frac{1}{n} + (x^* - \bar{x})^2\left(\frac{\text{standard error of } b}{\text{SEE}}\right)^2}$$

These formidable equations do allow us to identify four sources of imprecise predictions, three of which are familiar from our earlier discussion of the sources of imprecise estimates of α and β. First, a large standard deviation of the error term ϵ will scatter the observed values of y, causing imprecise estimates. Second, a small number of observations makes our estimates less precise. Third, a small standard deviation of x undermines our efforts to measure the effects of changes in x on y. Fourth, our predictions are less reliable the farther from the average value of x is the value of x for which we are predicting y.

For our plutonium example, Table 4 gives $t^* = 2.365$ for a 95 percent interval with $9 - 2 = 7$ degrees of freedom. Statistical software gives the values of the standard errors:

$$\text{Standard error of } y - \hat{y} = 16.771$$
$$\text{Standard error of } (a + bx^*) = 9.219$$

Therefore,

95% prediction interval for y: $207.03 \pm 2.365(16.771) = 207.03 \pm 39.66$

95% confidence interval for $(\alpha + \beta x^*)$: $207.03 \pm 2.365(9.219) = 207.03 \pm 21.80$

As stated before, the value of y is less certain than is the population mean of y. Figure 11.9 shows how both of these intervals widen as we move away from the middle of the data, showing that we can be more confident of predictions near the center of the data and should be more cautious on the fringes of the data.

Predicting Old Faithful

In Example 2.4, a histogram of the time interval between eruptions of the Old Faithful geyser has two peaks and shows considerable variation in the time intervals, with some eruptions less than 50 minutes apart and some nearly 2

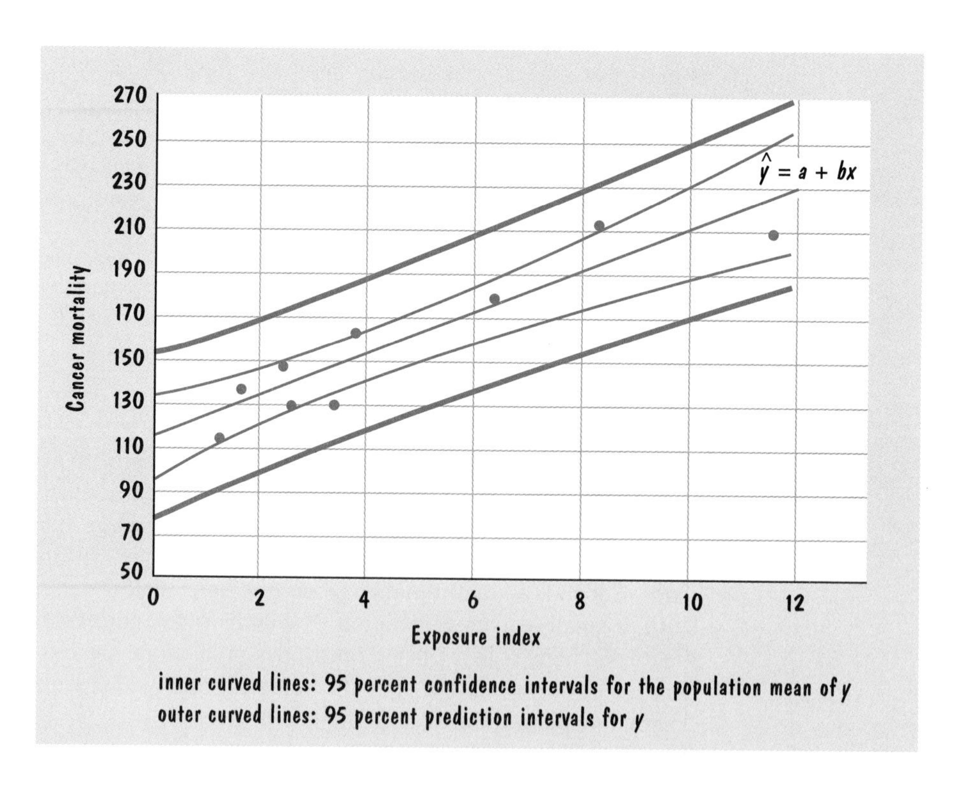

Figure 11.9 Gauging the precision of predictions.

hours apart. The average time between eruptions is 75.85 minutes, and if we used this figure to predict the next eruption, 76 percent of our forecasts would be off by more than 10 minutes and 23 percent would miss by more than 20 minutes.

The scatter diagram in Fig. 2.35 shows a positive correlation between the duration of an eruption and the time interval until the next eruption. We now have the tools to quantify this relationship in order to predict Old Faithful's eruptions. The figure below reproduces the scatter diagram in Fig. 2.35 and shows the least-squares line fit to these data. The equation for this line is

$$y = 34.549 + .2084x \qquad R^2 = .87$$

The standard error of the estimated coefficient of x is .0076, and the t value is 27.26, convincingly rejecting the null hypothesis that the coefficient is zero. The value of R^2 is .87, showing that 87 percent of the variation in the time intervals between Old Faithful's eruptions can be explained by the duration of the preceding eruption.

We can use the observed duration of an eruption and our estimated equation to predict the time interval until the next eruption. The confidence we have in this prediction is measured by a prediction interval. For example, the average duration of an eruption in these data is 198.21 seconds. For $x = 198.21$, the predicted value of y is 75.85 with a standard error of 6.0523, giving a 95

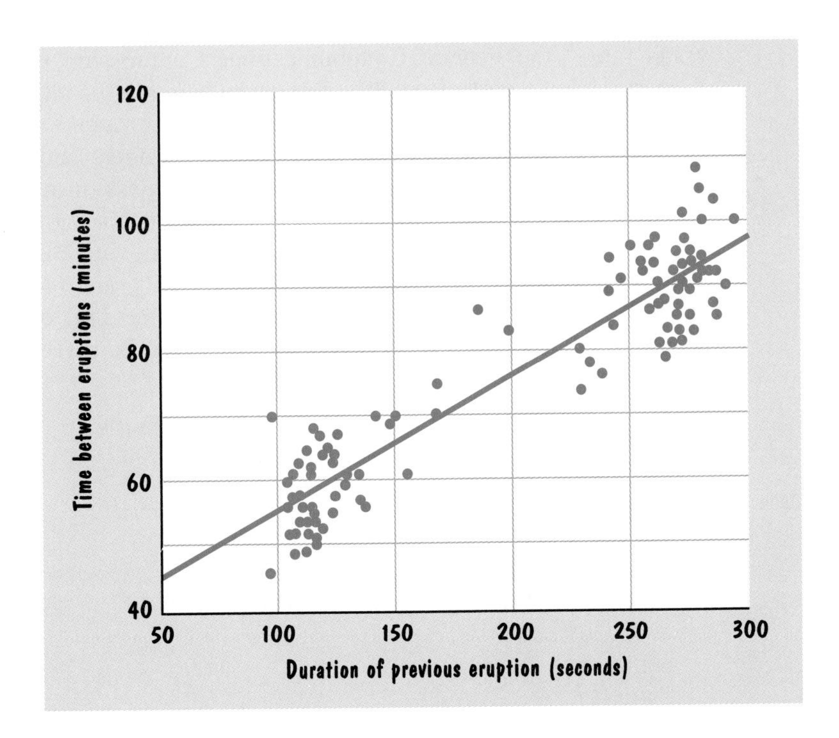

percent prediction interval of 75.81, plus or minus 11.84. There is a .95 probability that the actual eruption will be within 11.84 minutes of the predicted time. Yellowstone Park's rangers do in fact post their prediction of when each Old Faithful eruption will occur, based on the empirical relationship between the time interval and duration.

Estimating Western Union's Cost of Equity

There is a natural monopoly in the provision of water or electricity to households, in that it would be very inefficient to have adjacent homes connected to different suppliers of water or electricity. However, a single supplier of a vital product can exploit its monopolistic position by charging exorbitant prices. Recognizing this conflict between public efficiency and private exploitation, governments either run public utilities themselves or permit private monopolies to exist, but have public regulatory agencies set prices.

Regulators typically look at the firm's costs and set prices so that the rate of return on the firm's assets is comparable to that of similar business. As explained in Example 3.4, "Risk Has Its Rewards," relatively risky investments generally have relatively high average returns. Thus, regulatory agencies that are trying to identify comparable businesses often take into account a regulated utility's risk. For example, in a 1970 hearing to determine Western Union's telegraph rates,[8] the Federal Communications Commission wanted to estimate Western Union's cost of equity—the percentage rate of return that shareholders require on their stock. A stock's earnings-price ratio is often used for this purpose, but Western Union argued that its earnings had been temporarily depressed, making its earnings-price ratio meaningless. Instead, Western Union proposed using data for similar industries to estimate a regression model relating the cost of equity to risk, as measured by the variability of profits. (The earnings-price ratio was used for the cost of equity, and profit variability was calculated by dividing average profits into the standard deviation of profits about a trend line; both were multiplied by 100 to be percentages, rather than fractions.)

	PROFIT VARIABILITY	COST OF EQUITY
AT&T	3.6	9.9
ELECTRIC UTILITIES	7.1	10.4
INDEPENDENT TELEPHONES	10.6	10.4
GAS PIPELINES	11.9	11.4
GAS DISTRIBUTION	12.9	12.2
WATER UTILITIES	17.1	10.0
TRUCKERS	44.0	16.5
AIRLINES	76.0	18.9

To estimate the relationship between profit variability x and cost of equity y, an expert witness used the above data to estimate the regression equation

$$\hat{y} = 9.459 + 0.1312x$$
$$[18.6] \qquad [8.43]$$

with the t values shown in brackets. The estimated effect of profit variability on the cost of equity is, as expected, positive and highly statistically significant.

Because Western Union's profit variability was 27.0 percent, its cost of equity was estimated by calculating the predicted value of y for $x = 27.0$:

$$\hat{y} = 9.459 + .1312(27.0)$$
$$= 13.001$$

The standard error for this prediction is 1.083. With 8 observations, there are $n - 2 = 6$ degrees of freedom, and Table 4 shows that the appropriate t value for a 95 percent prediction interval is 2.447. Thus a 95 percent prediction interval for y at $x = 27.0$ is

$$13.001 \pm 2.447(1.083) = 13.001 \pm 2.650 \text{ or } 10.35 \text{ to } 15.65$$

Least-squares regression provided not only an estimate of Western Union's cost of equity, but also a means of quantifying the statistical uncertainty about this estimate.

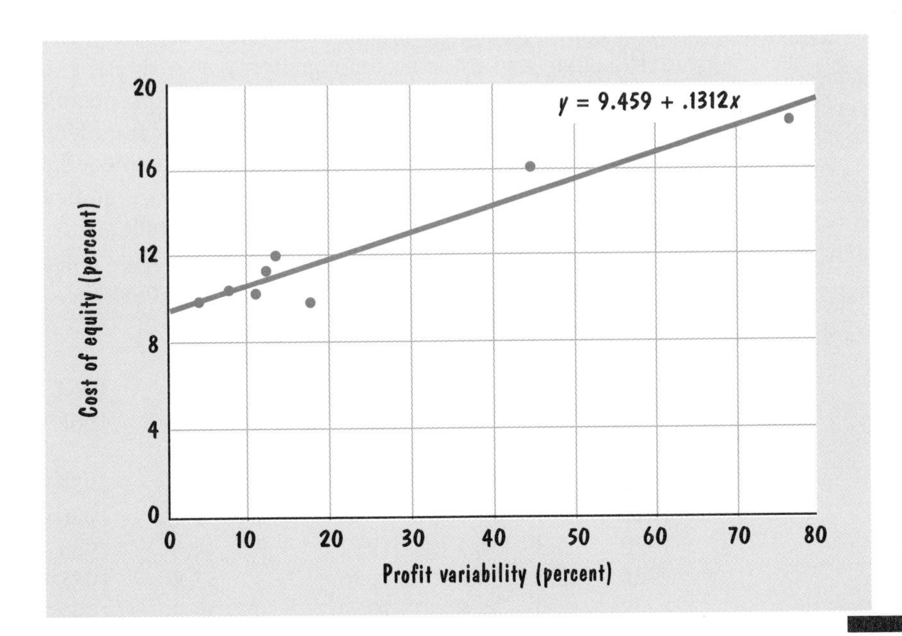

exercises

11.11 In the 1960s, data were collected for 21 countries on y = deaths from coronary heart disease per 100,000 persons aged 35 to 64 and x = annual per capita cigarette consumption.[9] Letting the horizontal axis go from 1000 to 4000 and the vertical axis go from 0 to 300, draw a scatter diagram and estimate the regression equation $y = \alpha + \beta x + \epsilon$ to see if you find a relationship that is positive, substantial, and statistically significant at the 5 percent level. Does the estimated value of β seem plausible?

	x	y		x	y
UNITED STATES	3900	259.9	GREECE	1800	41.2
CANADA	3350	211.6	AUSTRIA	1770	182.1
AUSTRALIA	3220	238.1	BELGIUM	1700	118.1
NEW ZEALAND	3220	211.8	MEXICO	1680	31.9
UNITED KINGDOM	2790	194.1	ITALY	1510	114.3
SWITZERLAND	2780	124.5	DENMARK	1500	144.9
IRELAND	2770	187.3	FRANCE	1410	144.9
ICELAND	2290	110.5	SWEDEN	1270	126.9
FINLAND	2160	233.1	SPAIN	1200	43.9
WEST GERMANY	1890	150.3	NORWAY	1090	136.3
NETHERLANDS	1810	124.7			

11.12 To investigate whether economic events influence presidential elections, draw a scatter diagram and estimate the equation $y = \alpha + \beta x + \epsilon$, using these data on x, the change in the unemployment rate during a presidential election year, and y, the percentage of the total vote for major-party presidential candidates received by the incumbent party. Does the estimated coefficient of x have a plausible sign and magnitude, and is it statistically significant at the 5 percent level? Does the intercept have a plausible magnitude? Test at the 5 percent level the null hypothesis that the intercept is 50. How close is the predicted value of y in 1996, when $x = -.2$, to the actual outcome, $y = 54.6$?

	x	y		x	y		x	y
1900	−1.5	53.2	1932	7.7	40.9	1964	−.5	61.3
1904	1.5	60.0	1936	−3.2	62.5	1968	−.2	49.6
1908	5.2	54.5	1940	−2.6	55.0	1972	−.3	61.8
1912	−2.1	54.7	1944	−.7	53.8	1976	−.8	48.9
1916	−3.4	51.7	1948	−.1	52.4	1980	1.3	44.7
1920	3.8	36.1	1952	−.3	44.6	1984	−2.1	59.2
1924	2.6	54.4	1956	−.3	57.8	1988	−.7	53.9
1928	.9	58.8	1960	.0	49.9	1992	.7	37.8

11.13 Shown below are data on the assets x (in billions of dollars) and profits y (in millions of dollars) of the 20 largest U.S. banks in 1973.[10] Draw a scatter diagram with assets on the horizontal axis and profits on the vertical axis. Does there seem to be a positive or negative relationship? The bank with $3.8 billion in assets and $13.8 million in profits was Franklin National Bank, which failed in 1974. Does its data point seem consistent or inconsistent with the general relationship between assets and profits shown in your scatter diagram? Now determine the least-squares estimates of the model $y = \alpha + \beta x + \epsilon$, and use these estimates to compare Franklin National's $13.8 million profit with that predicted by your model for a bank with $3.8 billion in assets. Is $13.8 million inside a 95 percent prediction interval?

ASSETS	PROFITS	ASSETS	PROFITS	ASSETS	PROFITS
49.0	218.8	13.4	60.9	7.2	32.2
42.3	265.6	13.2	144.2	6.7	42.7
36.3	170.9	11.8	53.6	6.0	28.9
16.4	85.9	11.6	42.9	4.6	40.7
14.9	88.1	9.5	32.4	3.8	13.8
14.2	63.6	9.4	68.3	3.4	22.2
13.5	96.9	7.5	48.6		

11.14 A study of 173,524 single births in Sweden between 1983 and 1987 obtained the following data on late fetal deaths (still birth at a gestation of 28 weeks or longer) and early neonatal deaths (during the first 6 days of life).[11]

MATERNAL AGE	BIRTHS	LATE FETAL DEATHS	EARLY NEONATAL DEATHS
20–24	70,557	251	212
25–29	68,846	253	177
30–34	26,241	125	78
35–39	6,811	34	33
40–52	1,069	7	5

Convert the late fetal and early neonatal data to rates per 1000 births by dividing the number of deaths by the number of births and multiplying the result by 1000. Draw a scatter diagram with the horizontal axis running from 20 to 55 and the vertical axis going from 2 to 7. Use least squares to estimate the regression model $y = \alpha + \beta x + \epsilon$, where y is the late fetal death rate and x is the maternal-age midpoint (22.5, 27.5, 32.5, 37.5, and 46.5). Now draw a scatter diagram with the same axes as before and estimate a similar model with y as the early neonatal death rate. Write a brief essay summarizing your findings.

11.15 A woman's chances of having a healthy baby are thought to be affected by the hormone androsenodione. A study[12] of 53 active, healthy women of normal weight used these data on androsenodione y and age x. Draw a scatter diagram and use least squares to estimate the model $y = \alpha + \beta x + \epsilon$, in order to see whether there is a positive or negative relationship between androsenodione and age. Is the relationship substantial and statistically significant at the 1 percent level?

x	y	x	y	x	y	x	y	x	y	x	y
33	273	43	176	84	84	34	137	32	130	42	140
75	45	73	137	53	56	29	140	38	140	38	172
40	196	34	109	35	154	40	165	73	74	58	81
39	175	31	172	42	108	44	91	61	35		
39	221	66	32	53	11	35	165	57	60		
56	102	22	227	61	84	63	84	62	84		
31	77	42	196	20	147	56	105	31	133		
29	165	56	74	46	116	55	116	66	63		
34	147	87	56	24	186	51	109	51	67		
40	116	28	158	39	123	47	98	20	189		

11.4 Correlation Coefficients

One measure of the success of a regression model is whether the estimate of β has a plausible magnitude and is statistically significant. Another is the size of the prediction errors—how close the data are to the least-squares line. Large prediction errors suggest that the model is of little use and may be fundamentally flawed. Small prediction errors indicate that the model may embody important insights.

Because we estimate the parameters by minimizing the sum of squared prediction errors, this sum is a logically consistent measure of the model's predictive prowess. But how do we judge whether a particular sum is large or small? A \$1 billion error in predicting the value of a \$1 trillion variable may be more impressive than a \$1 error in predicting a \$2 variable. We should also consider whether y fluctuates a lot or is essentially constant, because it is not at all difficult to predict the value of a variable that doesn't vary.

An appealing benchmark that takes these considerations into account is the **coefficient of determination R^2**, which compares the model's sum of the squared prediction errors with the sum of the squared deviations of y about its mean:

$$R^2 = 1 - \frac{\Sigma(y_i - \hat{y}_i)^2}{\Sigma(y_i - \bar{y})^2} \tag{11.11}$$

The statistical software used to estimate regression models invariably includes the calculated value of R^2 as part of the results. The value of R^2 cannot be less than 0 nor greater than 1. An R^2 close to 1 indicates that the model's prediction errors are minuscule in relation to the variation in y about its mean. An R^2 close to 0 indicates that the sum of the squared prediction errors is essentially equal to the variation of y about its mean—that the regression model is not an improvement over ignoring x and simply using the average value of y to predict y.

Mathematically, the sum of the squared deviations of y about its mean can be separated into the sum of squared deviations of the model's predictions about the mean and the sum of the squared prediction errors:

$$\begin{array}{ccccc} \Sigma(y_i - \bar{y})^2 & = & \Sigma(\hat{y}_i - \bar{y})^2 & + & \Sigma(y_i - \hat{y}_i)^2 \\ \text{Total sum of squares} & = & \text{Explained sum of squares} & + & \text{Unexplained sum of squares} \end{array}$$

The variation in the predictions about the mean is the "explained" sum of squares, because this is the variation in y that is attributed to the model. The prediction errors are the unexplained variation in y. Dividing through by the total sum of squares, we have

$$1 = \frac{\text{explained}}{\text{total}} + \frac{\text{unexplained}}{\text{total}}$$

Because R^2 is equal to 1 minus the ratio of the sum of squared prediction errors to the sum of squared deviations about the mean, we have

$$\begin{aligned} R^2 &= 1 - \frac{\text{unexplained}}{\text{total}} \\ &= \frac{\text{explained}}{\text{total}} \end{aligned}$$

Thus R^2 can be interpreted as the fraction of the variation in the dependent variable that is explained by the regression model. The Oregon plutonium example graphed in Figs. 11.1 and 11.6 has an R^2 of .86, showing that 86 percent of the variation in cancer mortality rates in these nine Oregon counties can be explained by differences in exposure to contaminated Columbia River water. The scatter diagram in Example 11.4 has an even higher R^2 of .92, with profit variability explaining 92 percent of the variation in these firms' cost of equity. The scatter diagram in Example 11.5, in sharp contrast, has an R^2 of .000004, showing that the rate of inflation explains virtually none of the annual variation in stock market returns.

The **correlation coefficient *R*** is equal to the square root of the coefficient of determination, R^2, and it can also be calculated directly from this formula:

$$R = \frac{1}{n-1} \frac{\Sigma(x_i - \bar{x})(y_i - \bar{y})}{(\text{standard deviation of } x)(\text{standard deviation of } y)} \tag{11.12}$$

The correlation coefficient is positive if the least-squares line has a positive slope and is negative if the least-squares line has a negative slope. The correlation coefficient equals 1 or −1 (and R^2 equals 1) if all the points lie on a

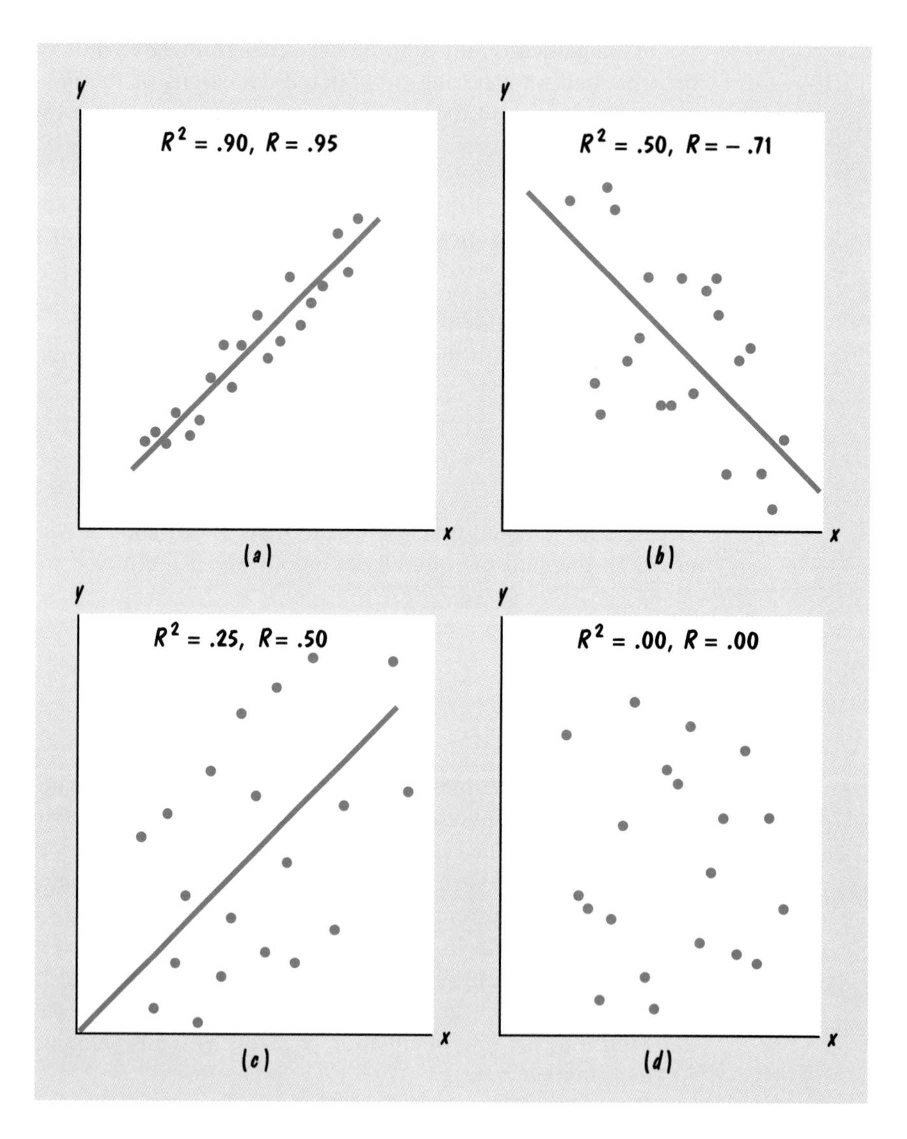

Figure 11.10 Four correlation coefficients.

straight line and equals 0 if a scatter diagram shows no linear relationship between x and y. (If there is no variation at all in x, so that the points lie on a vertical line, or no variation at all in y, so that the points lie on a horizontal line, then the correlation coefficient is defined as 0.)

Figure 11.10 gives four examples with correlation coefficients .95, −.71, .50, and .00. Figure 11.11 shows that a zero correlation coefficient does not rule out a perfect nonlinear relationship between two variables. In Fig. 11.11, y and x are exactly related by the equation $y^2 + x^2 = 1$. Yet, because there is no linear relationship between y and x, the correlation coefficient is 0. This is another compelling example of why we need to look at a scatter plot of our data.

Unlike the slope and intercept of a least-squares line, the value of the correlation coefficient does not depend on which variable is on the vertical axis and which is on the horizontal axis. Thus, it does not depend on which variable is considered the dependent variable, or indeed whether there is any causal relationship at all. For this reason, the correlation coefficient is often used to give a quantitative measure of the direction and degree of association between two variables that are related statistically, but not necessarily causally. For instance, Exercise 11.16 concerns the correlation between the IQ scores of 34 identical twins who were raised apart. We do not believe that either twin causes the other to have a high or low IQ score, but rather that both scores are influenced

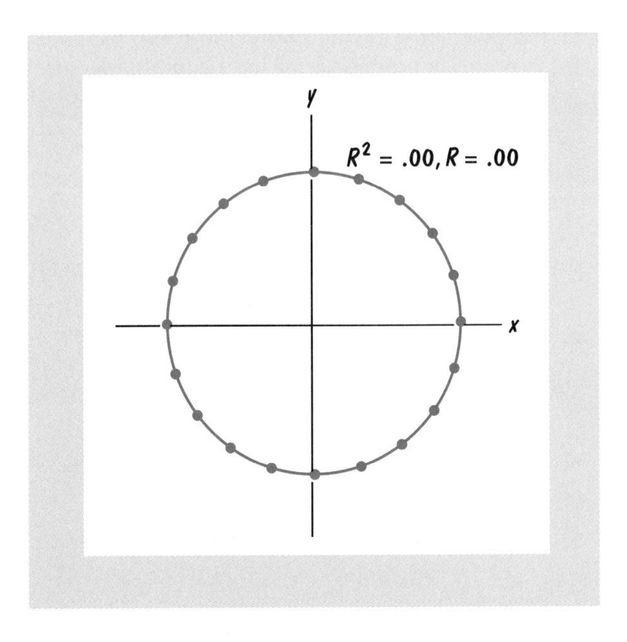

Figure 11.11 A zero correlation coefficient does not mean two variables are independent.

by common heredity factors; and we would like a statistical measure of the degree to which their scores are related statistically. The correlation coefficient does that.

The correlation coefficient also does not depend on the units in which the variables are measured—inches or centimeters, pounds or kilograms, degrees Celsius or Fahrenheit, thousands or millions. The correlation coefficient for the height and weight data in Example 11.2 would not be affected by converting the heights from inches to centimeters, or the weights from pounds to kilograms, or both.

A test of the null hypothesis that the population correlation coefficient is zero is computationally equivalent to a t test of the null hypothesis that $\beta = 0$ in the regression model. Because the correlation coefficient (and this t value) does not depend on which variable is labeled x and which is labeled y, this test is appropriate if either variable (or both) is normally distributed. Back when calculations were done by hand, Eq. (11.12) provided a simple, direct way to calculate the correlation coefficient. Nowadays, it is very easy to use the regression part of a statistics package to calculate the correlation coefficient, along with the t value and the P value for a statistical test of the equivalent null hypotheses that $\beta = 0$ and the population correlation coefficient is zero. In our Columbia River example, the .86 value of R^2 implies $R = \sqrt{.86} = .93$. The .0006 two-sided P value for testing the null hypothesis that $\beta = 0$ is also the two-sided P value for testing the null hypothesis that the population correlation coefficient between exposure and cancer mortality is zero.

Table 11.1 shows some of the output from the Minitab and SAS computer programs for a regression analysis of the Columbia River data. The first line of the Minitab output is the regression equation, using the variable names. Then

Table 11.1

Part of the Minitab and SAS regression results for the Columbia River data

MINITAB

```
THE REGRESSION EQUATION IS
MORT = 114.72 + 9.23 EXPOS

PREDICTOR        COEF     STDEV   T-RATIO      P
CONSTANT     114.7156    8.0457     14.26   .000
EXPOS          9.2315    1.4188      6.51   .000

S = 14.0099    R-SQ = 85.8%    R-SQ(ADJ) = 83.8%
```

SAS

```
                   Parameter  Standard        T for H0
Variable      DF    Estimate     Error  Parameter = 0  Prob > |T|
INTERCEP       1    114.7156    8.0457        14.2581      .00004
EXPOSURE       1      9.2315    1.4188         6.5066       .0006
```

we have the explanatory variables (predictors), with the intercept shown as constant. This table shows the estimated coefficients, the standard errors of the estimates, the t values, and the two-sided P values. The next line gives the standard error of estimate R^2 and R^2 adjusted for the degrees of freedom (which is beyond the scope of this book). The SAS output is quite similar, with slightly different labels. (The DF column is degrees of freedom.)

HOW TO DO IT

Least-Squares Estimation of a Simple Regression Model

We want to see if there is a relationship between exposure to radioactive waste and cancer mortality rates for nine Oregon counties near the Columbia River.

1. Specify the dependent variable y and the explanatory variable x for the linear regression model $y = \alpha + \beta x + \epsilon$. Here, we have cross-section data with y the cancer mortality rate and x the radioactive-waste exposure index.
2. A scatter diagram, with x on the horizontal axis and y on the vertical axis, provides a preliminary display of the data. Figure 11.1 shows our data.
3. Enter the observed values of y and x into a software program that will calculate the least-squares estimates of the model's parameters. Here are the estimates, with the standard errors in parentheses:

$$\begin{aligned} y = \underset{(8.05)}{114.72} + \underset{(1.42)}{9.23}x \end{aligned}$$

   Table 11.1 shows some Minitab and SAS computer output.
4. Check if the estimate of β has the anticipated sign and a plausible magnitude. Here, our estimate of β implies that, at the mean, a 21.6 percent increase in exposure to radioactive waste is predicted to increase the annual number of deaths from cancer by 9.23 per 100,000, which is a 5.9 percent increase.
5. To test the null hypothesis $\beta = 0$, we use the ratio of the estimate of β to its standard error: $t = b/(\text{standard error of } b)$ with $n - 2$ degrees of freedom. Here,

$$t = \frac{9.23}{1.42} = 6.51$$

With 9 − 2 = 7 degrees of freedom, statistical software shows the two-sided *P* value to be .0006, which decisively rejects the null hypothesis.

6. Confidence intervals for α and β can be obtained from their estimates and standard errors (and the *t* distribution with $n - 2$ degrees of freedom):

$$a \pm t^*(\text{standard error of } a) \quad \text{and} \quad b \pm t^*(\text{standard error of } b)$$

For our example, with 9 observations, we have 9 − 2 = 7 degrees of freedom, and Table 4 gives $t^* = 2.365$ for a 95 percent confidence interval. Using $b = 9.13$ and the 1.419 value for the standard error of *b*, we find that a 95 percent confidence interval for β is

$$9.13 \pm 2.365(1.419) = 9.23 \pm 3.36$$

7. The R^2 compares the model's sum of the squared prediction errors with the sum of the squared deviations of *y* about its mean; and R^2 can be interpreted as the fraction of the variation in *y* that is explained by the regression model. In our example, $R^2 = .86$, indicating that 86 percent of the variation in cancer mortality rates in these Oregon counties can be explained by differences in exposure to contaminated Columbia River water.
8. Computer software can also be used to predict the value of *y* for a specified value of *x*, using a prediction interval for *y* and a confidence interval for the population mean of *y* to quantify the precision of the prediction.

Are Stocks a Hedge against Inflation?

Inflation erodes the purchasing power of cash, bonds, and other assets with fixed dollar values. Consider, for example, $100 deposited in a bank for 1 year at a fixed 5 percent interest rate. At the end of the year, $100 will have grown to $105. However, if prices increase by 5 percent during the year, this $105 at the end of the year will buy no more than $100 would buy now. If prices increase by 10 percent, $105 at the end of the year buys 5 percent less than $100 does now.

Reluctant to settle for a 5 percent return during a 10 percent inflation, investors try to find investments that keep ahead of inflation, for example, shares of stock in firms that own land, buildings, and other physical assets that may appreciate in value during an inflation. To investigate whether stocks protect investors from inflation, the next figure shows a scatter diagram of the annual rate of inflation and annual stock returns (dividends plus capital gains) during the period from 1926 through 1994. (The peculiar 1926 starting date is used because this is as far back as these data go.)

Stock returns have averaged more than inflation (11.7 versus 3.0 percent), but bear little or no relation to the rate of inflation. Whether it is a year of deflation or double-digit inflation does not seem to affect the stock market's chances of collapse or boom. A least-squares regression of stock returns y on inflation x confirms this visual impression (the t values are in brackets):

$$\hat{y} = \underset{[3.36]}{11.72} - \underset{[.02]}{.009x} \qquad R^2 = .000004$$

The estimated coefficient of x is insubstantial, indicating that a 1 percentage point increase in the rate of inflation reduces stock returns by a minuscule .009 percentage point. The t value for this estimate is virtually zero and thus provides no reason why we should reject the null hypothesis that the coefficient of x is in fact zero. The R^2 is essentially zero, too, showing no correlation between inflation and stock returns. There is no evidence whatsoever in these data that the rate of inflation has a predictable effect on the stock market.

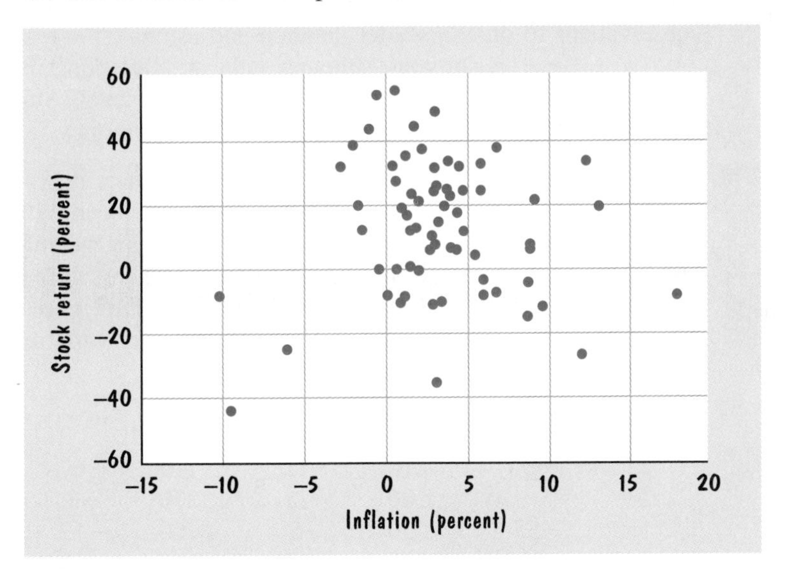

11.16 A classic study of the IQ scores of 34 identical twins who were raised apart used the intelligence scores that follow.[13] Draw a scatter diagram and calculate R^2 for a least-squares regression with the intelligence score of the second-born the dependent variable and the intelligence score of the first-born the explanatory variable. This R^2 was used to measure the influence of genes on intelligence, with $1 - R^2$ measuring the influence of environment. If 100 were added to all the first-born IQ scores and/or all the second-born IQ scores, how would R^2 be affected? Check your expectations by adding 100 to all the IQs and recalculating R^2.

FIRST	SECOND	FIRST	SECOND	FIRST	SECOND	FIRST	SECOND	FIRST	SECOND
22	12	30	26	30	34	6	10	41	41
36	34	29	35	27	24	23	21	19	9
13	10	26	20	32	18	38	27	40	38
30	25	28	22	27	28	33	26	12	9
32	28	21	27	22	23	16	28	13	22
26	17	13	4	15	9	27	25	29	30
20	24	32	33	24	33	4	2		

11.17 Table 3.2 gives the 1994 and 1995 returns from a sample of 17 mutual funds. Letting y be the fund's 1995 return and x be the 1994 return, use these 17 observations to draw a scatter diagram and estimate the regression equation $y = \alpha + \beta x + \epsilon$. Do your estimates indicate a relationship that is positive or negative? Statistically significant at the 5 percent level? Substantial? What conclusion regarding mutual fund performance do you draw from this exercise?

11.18 Tape recordings were made of 20 nonnative speakers of English reading a passage from the Test of Spoken English (TSE) developed by the Educational Testing Service.[14] These readings were scored independently by TSE professionals and by computer software developed at Rensselaer Polytechnic Institute to identify errors in spoken English. Letting the TSE score be the dependent variable and the computer score be the explanatory variable, draw a scatter diagram and determine the correlation coefficient.

TSE	COMPUTER	TSE	COMPUTER	TSE	COMPUTER	TSE	COMPUTER
220	−9	210	−12	110	−67	230	−5
280	−1	200	−18	230	−3	240	−8
150	−55	250	−5	200	−12	280	−5
240	−7	200	−14	190	−18	220	−11
180	−26	250	−6	160	−29	260	−5

11.19 Researchers examined 154,799 probability-of-precipitation forecasts made by 87 National Weather Service offices during the 2-year period from April 1977 to March 1979[15] These forecast probabilities were then divided into the 12 categories shown next, and for each category, the researchers tabulated the actual frequency of precipitation. (Both data sets are shown as percentages.) Draw a scatter diagram and use least squares to calculate the correlation coefficient between the forecast x and the actual y.

FORECAST	0	5	10	20	30	40	50	60	70	80	90	100
ACTUAL	1	3	8	18	30	40	50	60	70	79	88	91

11.20 A researcher compared the 1960 regular season and World Series batting averages of those players on the New York Yankees and Pittsburgh Pirates with at least 200 times at bat during the regular season.[16] For each team, see whether a player's regular season batting average x is a useful predictor of the player's World Series batting average y by drawing a scatter diagram and estimating the regression equation $y = \alpha + \beta x + \epsilon$. Are the teams' results similar or dissimilar?

	YANKEES			PIRATES	
	REGULAR SEASON	WORLD SERIES		REGULAR SEASON	WORLD SERIES
B. SKOWRON	.309	.375	D. GROAT	.325	.214
H. LOPEZ	.284	.429	R. CLEMENTE	.314	.310
R. MARIS	.283	.267	R. NELSON	.300	.333
Y. BERRA	.276	.318	H. SMITH	.295	.375
M. MANTLE	.275	.400	S. BURGESS	.294	.333
T. KUBEK	.273	.333	D. HOAK	.282	.217
G. MCDOUGALD	.258	.278	B. MAZEROSKI	.273	.320
B. RICHARDSON	.252	.367	B. SKINNER	.273	.200
B. CERV	.250	.357	G. CIMOLI	.267	.250
E. HOWARD	.245	.462	B. VIRDON	.264	.241
C. BOYER	.242	.250	D. STUART	.260	.150

11.5 Some Regression Pitfalls

Although regression is very elegant and powerful, it can also be misused and can give misleading conclusions. We consider several pitfalls to be avoided.

Careless Extrapolation

Regression models are sometimes used to predict outcomes far from the data that were used to estimate the model's parameters. For example, scientists who wanted to predict the earth's population and standard of living hundreds of years from now once estimated a model using data for the present and recent past and then extrapolated these calculations into the distant future. We should always proceed cautiously when making distant extrapolations. It is not enough to plug in a value for x and mechanically calculate the predicted value of y. At a minimum, we should calculate a 95 or 99 percent prediction interval. The width of this interval can provide a sobering indication of the fragility of our prediction. Prediction intervals widen as we move away from the center of the data, providing a fair warning that there is more uncertainty about predictions on the fringes of the data. If our prediction interval is 1.5 ± 232.6, we may realize that these data do not really tell us much about the dependent variable for this particular value of the explanatory variable.

Even the widening of prediction intervals may understate the unreliability of distant extrapolations, because linear models are seldom valid for any and all values of the explanatory variable. Linearity is a convenient assumption and a satisfactory approximation as long as we confine our attention to modest changes in the explanatory variable. Before using our model to predict the consequences of a huge change in the explanatory variable, we should think seriously about the range over which linearity is a reasonable assumption. A linear model assumes that the effect on y of a 1-unit change in x is a constant, β. For most phenomena, there comes a point at which the effect of x on y either becomes stronger or diminishes, and thereby violates the linearity assumption. The application of fertilizer increases crop yields, but eventually the effect gets smaller and smaller and then turns negative. Too much fertilizer damages crops.

Chapter 2 gave several examples of obviously ludicrous extrapolations. Here are two more. Example 11.2 estimated the relationship between male height and weight. An extrapolation without concern for eventual nonlinearities predicts a weight of zero for someone 33 inches tall and a negative weight for someone even shorter. Similarly, a researcher, tongue firmly in cheek, extrapolated data showing that automobile deaths declined after speed limits were reduced:

> To which Prof. Thirdclass of the U. of Pillsbury, stated that to reach a zero death rate on the highways, which was certainly a legitimate goal, we need only set a speed limit of zero mph. His data showed that death rates increased linearly with highway speed limits, and the line passing through the data points, if extended backwards, passed through zero at zero mph. In fact, if he extrapolated even further to negative auto speeds, he got negative auto deaths, and could only conclude, for his data, that if automobiles went backwards rather than forwards, lives would be created, not lost.[17]

Regression toward the Mean

Sir Francis Galton (1822–1911) was the first to apply the label *regression* to the least-squares estimation procedure. In a study of the relationship between the heights of parents and their children, Galton noticed a phenomenon that statisticians now call **regression toward the mean,** a tendency of extreme observations to be followed by more average values.[18] By purchasing family records, Galton obtained the heights of 205 pairs of parents and their 928 adult children. Because the average male height is about 8 percent larger than the average female height, he multiplied the female heights by 1.08 to make them comparable to the male heights. The heights of each set of parents were then averaged to give a "mid-parent height." The mid-parent heights were divided into nine categories, and the median heights of the children of parents in each category were computed. Figure 11.12 shows a least-squares line fit to these data.

Unusually tall parents tend to have children shorter than themselves, while very short parents usually have somewhat taller children. (The median height of the children of 72.5-inch parents is 72.2 inches; the median height of the children of 64.5-inch parents is 65.8 inches.) These data do not mean that soon everyone will be the same height; instead, they reflect the important regression-toward-the-mean phenomenon.

Our heights are influenced by the genes we inherit from our parents. For simplicity, we will call someone who, at conception, has a genetically predicted adult height of 6 feet 2 inches a person with "6-foot 2-inch genes." Because heights are affected by diet, exercise, and other environmental factors, our adult heights are not a perfect reflection of our genes and, hence, not a perfect predictor of the genetically expected height of our children. A person who is 6 feet 4 inches tall might have 6-foot 2-inch genes and experienced positive environmental influences

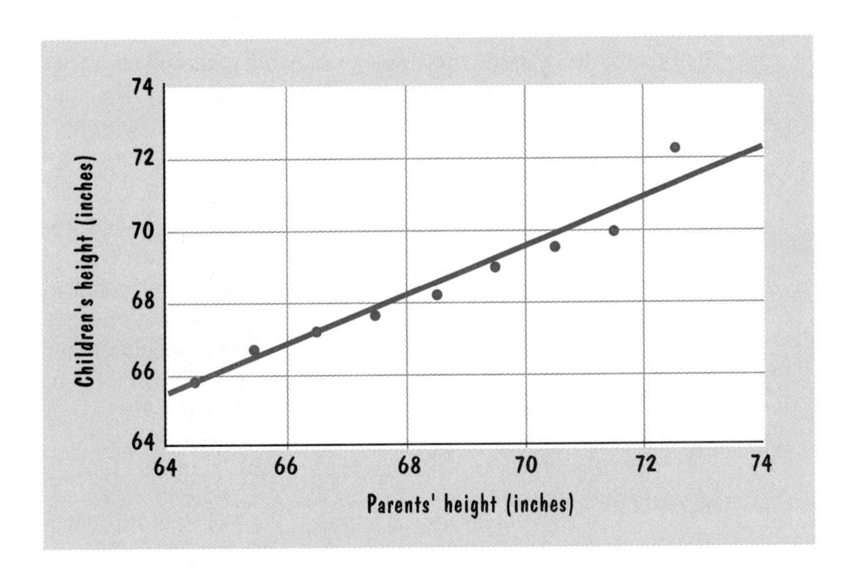

Figure 11.12 The heights of parents and their children.

or might have 6-foot 6-inch genes and had negative environmental factors. The former is more likely, simply because there are many more people with 6-foot 2-inch genes than with 6-foot 6-inch genes. Thus the observed heights of very tall parents are usually an overstatement of their genetic heights and thus of the expected heights of their children.

This reasoning does not imply that we will soon all be the same height. Indeed, we could just as well turn the argument upside down by noting that most very tall people had somewhat shorter parents, while most very short people had somewhat taller parents. Does this imply that heights are diverging? No, heights are neither converging nor diverging. There will always be unusually tall and unusually short people. What we must recognize is that heights are influenced by chance and that those who are unusually tall most likely had chance influences that pulled them above their genetically expected height. The regression-toward-the mean fallacy is to misinterpret the temporary nature of extreme observations as a trend.

Professors similarly observe that those who score highest on the midterm examination usually do not do quite as well on the final examination, while those who receive the lowest scores improve somewhat. Are students converging to a depressing mediocrity, with the weak students learning and the strong forgetting? Or, to turn the argument on its head, does the fact that the highest scorers on the final examination did not get the highest scores on the midterm show that scores are diverging from the mean? Two nos! The highest scores on any examination involve a bit of good luck, and lowest scores a bit of misfortune. Those students with the highest scores on any particular test are mostly above-average students who did unusually well because the questions asked happened to be ones that they were well prepared to answer. It is more probable that they are good students who did unusually well rather than great students who had an off day. The highest scorers on any given test most likely did not do as well on their last examination and will not do as well on their next examination.

Regression toward the mean is a subtle sort of data mining. If, before an examination, we select students at random, their average score on the examination will be an unbiased estimate of the population mean. But if, after the test, we identify those students who happened to have done exceptionally well, these are not a random sample. They were selected precisely because they had the highest scores. In any sample, the highest values are an overestimate of the population mean. For unbiased estimates, we need to use a random sample that is not based on the scores themselves.

This regression toward the mean has also been observed in IQ scores:

> 4 year olds with IQs of 120 typically have adult scores around 110. Similarly, 4 year olds with scores of 70 have an average adult score of 85. . . . This does not mean that there will be fewer adults than children

> with very high or very low IQs. While those who start out high or low will usually regress towards the mean, their places will be taken by others who started closer to the mean.[19]

Another example involves Air Force flight instructors who observed that very good landings were usually followed by landings that were not so good, while very poor landings were usually followed by somewhat better landings. Falling for the regression-toward-the-mean fallacy, the flight instructors reasoned that this pattern occurred because they had praised the good landings and harshly criticized the poor ones. Thus they concluded, contrary to well-accepted learning research, that praise is detrimental and severe criticism beneficial.[20]

An economic example is provided by a book with the provocative title *The Triumph of Mediocrity in Business*. The author discovered that businesses with exceptional profits in any given year tend to have smaller profits the following year, while firms with very low profits generally do somewhat better the next year. From this evidence, he concluded that strong companies were getting weaker, and the weak stronger, so that soon all will be mediocre. The author's fallacy is now obvious. A famous statistician explained it as follows:

> while [firms] at the margins . . . often go toward the center, those in the center of the group also go toward the margins. Some go up and some down; the average of the originally center group may, therefore, display little change, since positive and negative deviations cancel in averaging; while for an extreme group, the only possible motion is toward the center.[21]

A best-selling investments textbook makes this same error.[22] The author discusses a model of stock prices that assumes, "ultimately, economic forces will force the convergence of the profitability and growth rates of different firms." To support this assumption, he looked at the 20 percent of firms with the highest profit rates in 1966 and the 20 percent with the lowest profit rates. Fourteen years later, in 1980, the profit rates of both groups are more nearly average: "convergence toward an overall mean is apparent . . . the phenomenon is undoubtedly real." The phenomenon is regression toward the mean, and the explanation is statistical, not economic.

Correlation Is Not Causation

Sometimes we use regression models simply to quantify empirical relationships. We look at the relationship between midterm and final examination scores, not because we believe that midterm scores *cause* final test scores, but because we think that both reflect student ability and want to see how closely one predicts the other. We look at human heights, life expectancies, or atmospheric concentrations of carbon dioxide over time, not because we believe that the mere passage of time causes these things to change, but because we want to see if there have been important trends. In these cases, we should be careful to say that x and y are *associated,* or *correlated statistically,* without suggesting a causal relationship.

In other cases, a regression model reflects a belief that changes in the explanatory variable *cause* changes in the dependent variable. Our investigation of cancer mortality in nine Oregon counties was motivated by a belief that exposure to radioactive waste can cause cancer. If we plan to attach a causal interpretation to our results, it is sound practice to have this interpretation in mind when we specify the model and before we estimate it. However, this sound practice is not always followed, in that people often stumble upon unexpected correlations and then search for a plausible causal explanation. This is what Ed Leamer calls the *Sherlock Holmes inference,* and as pointed out in earlier chapters, it can lead to both useful and useless models.

There are three primary reasons why a causal interpretation of empirical correlations may be misleading: simple chance, reverse causation, and omitted factors. The first explanation, simple chance, refers to the fact that even unrelated data may, by coincidence, be statistically correlated. In the model $y = \alpha + \beta x + \epsilon$, we conclude that x has a statistically significant effect on y at the 5 percent level if there is less than a 5 percent chance that the estimate of β would be so far from zero, if β really were zero. What we must recognize is that even if β really is zero, there is a 5 percent chance that the data will happen by chance to indicate a statistically significant relationship. If, for example, we let x and y be artificial variables independently created from a table of random numbers, and if then we run a least-squares regression of y on x, there is a 5 percent chance of finding a relationship that is statistically significant at the 5 percent level. A hapless researcher who spends a lifetime looking at unrelated variables will find statistical significance on about 1 out of every 20 tries. We can hope that no one is so misguided as to study only unrelated variables, but we should also recognize that statistical correlation without a plausible explanation may be a coincidence—particularly if the researcher estimated dozens (or hundreds) of equations and reports only those few with the highest t values.

The remarkable correlation between the Super Bowl and the stock market is a clear example. Similarly, to illustrate the dangers of data mining, an economist once ran a large number of regressions and found an astounding correlation between the stock market and the number of strikeouts by the Washington Senators baseball team.[23] British researchers found a very high correlation between the annual rate of inflation and the number of dysentery cases in Scotland the previous year, and a spectacular correlation between Britain's price level and rainfall.[24] We should not be impressed with high, or even spectacular, values for the t statistic and R^2 unless there is also a logical explanation, and we should be particularly skeptical when there has been data mining.

The second reason why an observed correlation may be misleading is that perhaps it is y that determines x, rather than the other way round. Consider again the Columbia River data: for clarity, we let the variable E be the exposure index and C be cancer mortality. Our original regression, with cancer mortality as the dependent variable and exposure as the explanatory variable, gave these results:

$$C = \underset{[14.26]}{114.72} + \underset{[6.51]}{9.23}E \qquad R^2 = .858$$

with the t values given in brackets. Now, suppose that we had inadvertently reversed the dependent variable and the explanatory variable and estimated this equation:

$$E = \underset{[4.36]}{10.01} + \underset{[6.51]}{.093}C \qquad R^2 = .858$$

As stated earlier, R^2 and the t value for the slope are not affected by the reversal of the dependent variable and the explanatory variable. Either way, these statistics tell us that there is a highly significant correlation between exposure and cancer mortality. What they cannot tell us is which variable is causing changes in the other. For that, we need logical reasoning, not t values. Here, it is obvious that exposure to radioactive waste may increase cancer mortality, but that cancer mortality does not cause exposure to radioactive waste.

Other cases can be more subtle. Data from six large medical studies found that people with low cholesterol levels were more likely to die eventually of colon cancer; however, a later study indicated that the low cholesterol levels may have been caused by colon cancer that was in its early stages and therefore undetected.[25] For centuries, residents of New Hebrides believed that body lice made a person healthy. This folk wisdom was based on the observation that healthy people often had lice and unhealthy people usually did not. It was not the absence of lice that made people unhealthy, but the fact that unhealthy people often had fevers, which drove the lice away.

The third reason why impressive empirical correlations may be misleading is that some omitted factor may affect both x and y. In the regression model $y = \alpha + \beta x + \epsilon$, we assume that the omitted factors collected in the error term ϵ are uncorrelated with the explanatory variable x. If they are correlated, we may underestimate or overestimate the effect of x on y, or we may even identify an effect when there is none. Suppose that exposure to Columbia River water has no effect on cancer mortality, but that the state government had built elderly housing units all along the banks of the Columbia River. We would find a statistical correlation between proximity to the river and cancer mortality when the river was in fact benign. The observed correlation would be due to an omitted factor—the age of those living near the river.

This was not the case in Oregon, but we do need to be alert to the possible effects of omitted factors. A study once found that a certain state had an unusually high milk consumption and an abnormally high cancer rate, suggesting that drinking milk causes cancer. It turned out that the state had a relatively high cancer rate not because there were so many milk drinkers, but because many elderly people had retired to this state and elderly people often die of cancer. Similarly, Arizona has the highest death rate in the nation from bronchitis, emphysema, asthma, and other lung diseases, but this does not mean that Arizona's climate is especially hazardous to human lungs. On the contrary, doctors recommend that patients with lung problems move to Arizona for its beneficial climate. Many do move to Arizona and benefit from the dry, clean air. Although many eventually die of lung disease in Arizona, it is not due to the Arizona climate.

A less morbid example involves a reported positive correlation between stork nests and human births in northwestern Europe. Few would conclude that storks bring babies. A more logical explanation is that storks like to build their nests on buildings. Where there are more people, there are usually more human births and also more buildings on which storks can build nests.

Unfortunately, there are innumerable examples of statistical correlations that do not reflect causation. One safeguard is to use common sense in specifying models and to reexamine suspicious results with fresh data. The multiple-regression procedures developed in more advanced courses provide a way to deal with the confounding effects of omitted variables. For example, if we want to investigate the influence of a particular diet on death rates and know that death rates depend on the age of the population, then we can estimate a multiple-regression equation that includes several explanatory variables—diet, age, and anything else we believe to be important.

A particularly common source of coincidental correlations is that many variables are related to the size of the population and tend to increase over time as the population grows. If we pick two such variables at random, they may appear to be highly correlated when, in fact, they are both affected by a common omitted factor—the growth of the population. With only a small amount of data mining, I uncovered this simple example using annual data on the number of U.S. golfers and the nation's total number of missed workdays due to reported injury or illness (both in thousands):

	GOLFERS x	MISSED DAYS y
1960	4,400	370,000
1965	7,750	400,000
1970	9,700	417,000
1975	12,036	433,000
1980	13,000	485,000
1985	14,700	500,000

A least-squares regression gives

$$\hat{y} = \underset{[14.2]}{304{,}525} + \underset{[6.4]}{12.6x} \qquad R^2 = .91$$

with the t values in brackets. This relationship is highly statistically significant and very substantial—every additional golfer leads to another 12.6 missed days of work. It is semiplausible that people may call in sick in order to play golf (or that playing golf may cause injuries). But, in fact, most workers are not golfers, and most missed days are not spent playing golf or recovering from golf. The number of golfers and the number of missed days have both increased over time, not because one was causing the other, but because both were growing with the population.

A simple way to correct for the coincidental correlation caused by a growing population is to scale both sets of data by the size of the population, giving us per capita data. Here are the per capita data on golfers and missed days:

	GOLFERS PER CAPITA	MISSED DAYS PER CAPITA
1960	.02435	2.0479
1965	.03989	2.0586
1970	.04731	2.0336
1975	.05472	1.9684
1980	.05708	2.1295
1985	.06144	2.0896

There has been a large increase in the number of golfers per capita, while missed days per capita have no discernible trend. A least-squares regression gives

$$\hat{y} = \underset{[21.06]}{2.018} + \underset{[.39]}{.763x} \qquad R^2 = .04$$

Once we take into account the growing population by converting our data to per capita values, the coincidental correlation disappears, and we find that there is virtually no statistical relationship between the number of golfers and the number of missed workdays.

Do Champions Choke?

As of 1996, there had been 30 Super Bowls with only 6 teams able to win twice in a row. In baseball, of the 32 world champions from 1964 through 1996, only 4 have repeated. No professional basketball team repeated as champion between 1969 (Boston Celtics) and 1988 (Los Angeles Lakers). These data do not prove that champions choke. Instead, the data are an example of regression toward the mean. There are many teams capable of winning a championship, and which of these deserving teams ultimately wins is partly determined by luck—having few injuries and being the beneficiary of lucky bounces and questionable officiating. It is more likely that the winner is an above-average team that experienced good luck than an unbelievable team that survived bad luck. By definition, good luck cannot be counted on to repeat. The next year, another above-average team will probably get the breaks and win the championship.

There is plenty of evidence of regression toward the mean in sports.[26] Of those major league baseball teams that finish a season with winning records, two-thirds win fewer games the next season; of those teams with losing records, two-thirds do better the next season. Of those teams that win more than 100 out of 162 baseball games in a season, 90 percent do not do as well the next season. Of baseball players who bat over .300 in a season, 80 percent see their batting averages decline the following season. The regression-toward-the-mean fallacy is to conclude that the skills of good teams and players deteriorate. The correct conclusion is that those with the best performance in any particular season generally are not as skillful as their lofty records suggest. Most have had more good luck than bad, causing that season's record to be higher than that of the season before and higher than that of the season after—when their place at the top will be taken by others.

The Radiophobic Farmer

Darrell Huff tells the story of a farmer who claimed that his fruit trees were being damaged by radio waves from a nearby radio station. He put a wire fence around some of his trees to "shield" them from these radio waves, and, sure enough, they quickly recovered, while the unshielded trees continued to suffer. About the same time, many citrus trees throughout the country were threatened by the spread of little-leaf disease. Farmers in Texas found that a solution of iron sulfate cured this disease. However, it did not always work in Texas and almost never worked in Florida or California. Huff explains that these mysteries were solved when it was discovered that the real problem was a zinc deficiency in the soil:

> The radiophobic farmer's fence wire was galvanized; enough zinc washed off of the wire to give the trees the tiny trace of zinc they needed. The iron sulfate did nothing for the other trees, but the zinc-coated buckets in which it was carried saved them. In regions where buckets other than galvanized were used, the trees went right on ailing.[27]

11.21 Here are annual U.S. data on beer production x, in millions of barrels, and the number of married people y, in millions:

	1960	1965	1970	1975	1980	1985
x	95	108	135	158	193	200
y	84.4	89.2	95.0	99.7	104.6	107.5

Draw a scatter diagram and estimate the equation $y = \alpha + \beta x + \epsilon$ by least squares to see if there is a statistically significant relationship at the 5 percent level. If so, do you think that beer drinking leads to marriage, marriage leads to beer drinking, or what?

11.22 Shown below are some U.S. Department of Agriculture data on an index of the prices received by U.S. farmers and an index of the prices paid for producing farm products. Both indexes are scaled to equal 100 in 1977. Calculate the ratio y of prices received to prices paid; then draw a scatter diagram and estimate the equation $y = \alpha + \beta x + \epsilon$, where $x = 1950, 1955$, and so on. Interpret your estimate of β. At a 1 percent significance level, is there a statistically significant trend in this ratio? What is the predicted value of y when $x = 2050$?

	1950	1955	1960	1965	1970	1975	1980	1985	1990
PRICE RECEIVED	56	51	52	54	60	101	134	128	150
PRICE PAID	37	43	46	47	55	89	138	162	180

11.23 A meteorologist once observed that the fall price of corn is correlated with severity of hay fever (which is caused by ragweed). Does this mean that corn prices affect ragweed growth or that ragweed growth affects corn prices? Or is there another plausible explanation? Do you suppose that corn prices and hay fever are positively or negatively correlated?

11.24 In June 1989, *Sports Illustrated* reported that for Kevin Mitchell to threaten the single-season home run record, he "will have to avoid the traditional second-half power outage that plagues most fast starters."[28] According to *Sports Illustrated,* of the 39 batters who hit 20 or more home runs during the first half of the baseball season, only 2 hit 20 or more in the second half. How would a statistician explain this traditional power outage?

11.25 People who have a college education earn much more, on average, than people who do not have a college education. Here are 1992 census data on the mean income of U.S. citizens, 18 years old and older:

YEARS OF SCHOOLING	APPROXIMATE YEARS x	MEDIAN INCOME y
Less than 12	9	$10,272
12	12	16,284
13–15	14	18,540
16	16	29,868
More than 16	18	47,780

Use the approximate years-of-education numbers to draw a scatter diagram, and estimate the parameters of the regression equation $y = \alpha + \beta x + \epsilon$. Is the relationship between education and income substantial and statistically significant at the 5 percent level? What is the predicted income of a professional student who has 30 years of schooling? Do you believe it?

11.6 Some Regression Diagnostics*

Regression models assume that the model is correctly specified, in particular, that the error term ϵ has a constant standard deviation and that the values of ϵ are independent of one another. Violations of these assumptions do not bias the estimates of the coefficients of the explanatory variables, but their standard deviations are unnecessarily large and will be underestimated, causing us to overestimate the statistical significance of our results.

Nobel Laureate Paul Samuelson has forcefully advised forecasters to "always study your residuals." A study of a regression model's residuals (the deviations of the scatter points from the fitted line) can reveal outliers—perhaps an atypical observation or a typographical error—or violations of the assumptions that ϵ has a constant standard deviation and independent values. It is generally a good idea to use a scatter diagram to take a look at the data. Here is what might be discovered.

Looking for Outliers

In the early 1900s, researchers collected data for nine cigarette brands on each brand's percentage of total cigarette advertising expenditures and percentage of total cigarettes smoked by persons 12 to 18 years of age:[29]

	PERCENTAGE OF ADVERTISING	PERCENTAGE OF TEENAGE MARKET
MARLBORO	12.7	59.5
CAMEL	4.9	8.7

NEWPORT	4.7	11.1
SALEM	5.4	3.8
KOOL	4.8	4.3
BENSON & HEDGES	5.3	2.1
WINSTON	6.9	2.5
MERIT	6.3	.9
VIRGINIA SLIMS	4.8	2.2

A least-squares regression yielded a positive relationship with a t value of 5.83 and a two-sided P value of .001, leading to nationwide newspaper stories about the relationship between cigarette advertising and smoking by teens. However, the scatter diagram in Fig. 11.13 shows that these results are due entirely to the Marlboro outlier. For the eight brands other than Marlboro, there is a negative relationship between advertising and use, although it is not statistically significant at the 5 percent level. (The t value is 1.64, and the two-sided P value is .15.) The reader must decide whether the Marlboro outlier is decisive evidence or a misleading outlier. If we had not looked at a scatter diagram, we would not have known that this is the real issue here.

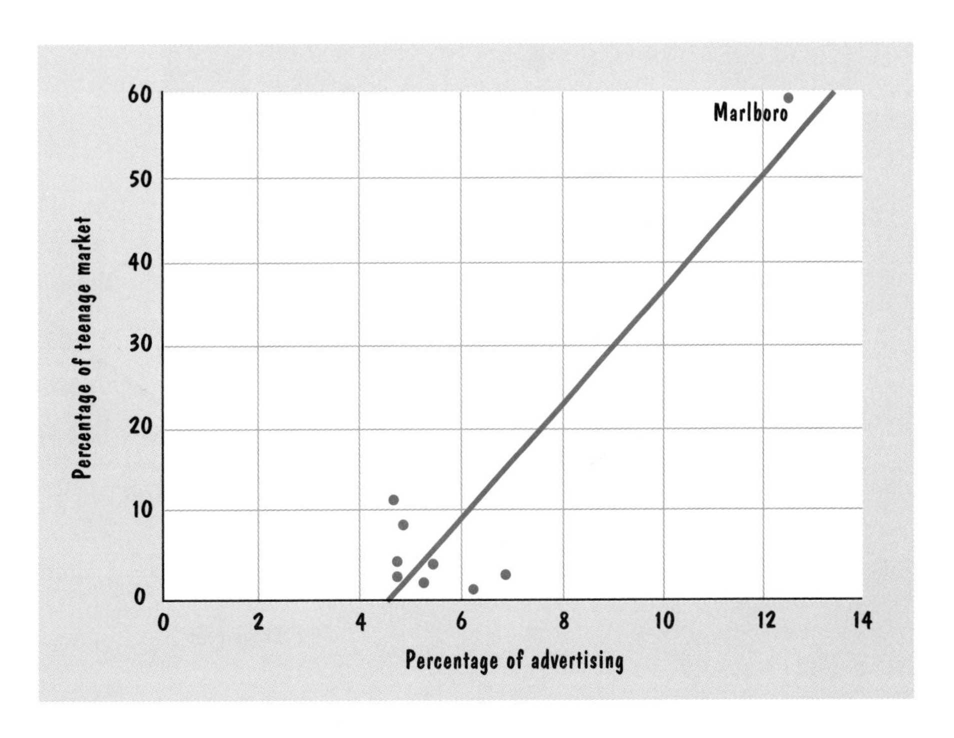

Figure 11.13 Cigarette advertising and the teenage market.

Sometimes we are interested not in how closely a line can be drawn to fit the data, but whether some of the observations deviate substantially from a line fit to the data. Figure 11.14 is an example. In reporting the votes cast for presidential candidates by each of the 50 states, the *Statistical Abstract* separates the United States into the nine geographic regions identified in this figure.[30] The horizontal axis shows the percentage of the total vote cast for the Democratic and Republican presidential candidates that was received by the Republican candidate, George Bush, when he ran against Michael Dukakis in 1988. The vertical axis shows the percentage George Bush received in 1992 when he ran against Bill Clinton. The nine data points are for the nine regions of the country.

Overall, Bush received 53.9 percent of the two-party vote in 1988 and only 46.5 percent in 1992. Although he received fewer votes in 1992 than in 1988 in all nine regions of the country, a line fit to these nine data points shows a statistical relationship between Bush's 1988 and 1992 percentages. Those four regions in which Bush was the strongest in 1988 were also the four regions where he was strongest in 1992 (South Atlantic, East South Central, West South Central, and Mountain). The three regions where Bush was the weakest in 1988 were also his weakest regions in 1992 (New England, Middle Atlantic,

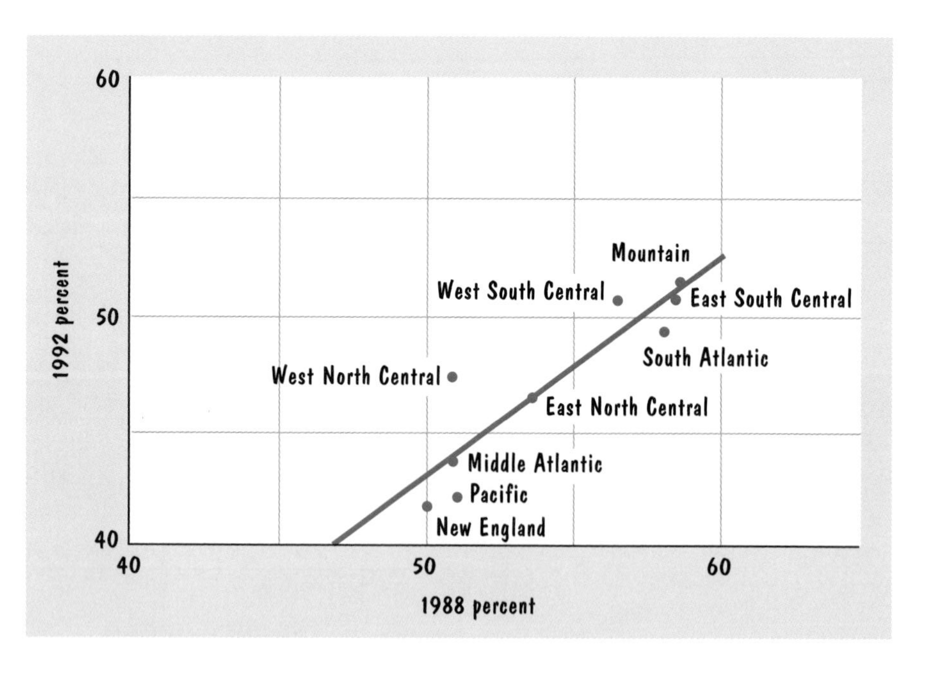

Figure 11.14 George Bush's percentage of the 1988 and 1992 presidential vote.

and Pacific). There is also one region that is noticeably off the line fit to these data—the West North Central region, which consists of Minnesota, Iowa, Missouri, North Dakota, South Dakota, and Nebraska, and Kansas. Given his performance elsewhere, Bush did much better in this region in 1992 than might have been expected. The point here is not that this outlier distorts the results, but that the very fact that it is an outlier is interesting.

Checking the Standard Deviation

Suppose that we are investigating the variation among cities in the annual sales of weekly newsmagazines. Believing that the sales y (in thousands) in a city depend primarily on the city's aggregate income x (in millions of dollars), we collect the following data for a sample of 10 cities:

SALES y	964	462	10,440	4,200	1,310	2,610	5,220	948	2,830	1,560
INCOME x	620	340	7,400	3,000	1,050	2,100	4,300	660	2,300	1,200

Least-squares estimation of the simple regression model yields

$$\underset{}{\hat{y}} = \underset{[.8]}{109.1} + \underset{[31.0]}{1.377x} \qquad R^2 = .992$$

where the t values are shown in brackets. The results seem very satisfactory. The relationship between income and sales is, as expected, positive and, as hoped, statistically significant. The value of R^2 is nearly 1.

Following Samuelson's advice, we plot the prediction errors in Fig. 11.15 with income on the horizontal axis. It appears from this scatter diagram that the residuals are larger for high-income cities than for low-income ones. On reflection, it is plausible that the errors in predicting total magazine sales in a large city will tend to be larger than the errors in predicting magazine sales in a small city. If so, the standard deviation is larger, too, violating our assumption that the standard deviation of the error term is constant.

A variety of formal statistical procedures for testing this assumption are described in more advanced courses on regression analysis. Here, we discuss one possible solution. The error term encompasses the effects of omitted variables, which may not have constant standard deviations—with cross-sectional data, because some observations (like large cities) have a different scale from others; with time-series data, because the scale of the omitted variables may increase over time. One solution is to respecify the model, rescaling the variables so that error term has a constant standard deviation. In our magazine sales example, the problem is apparently caused by the fact that some cities are much larger than others. To give the data a comparable scale, we can respecify

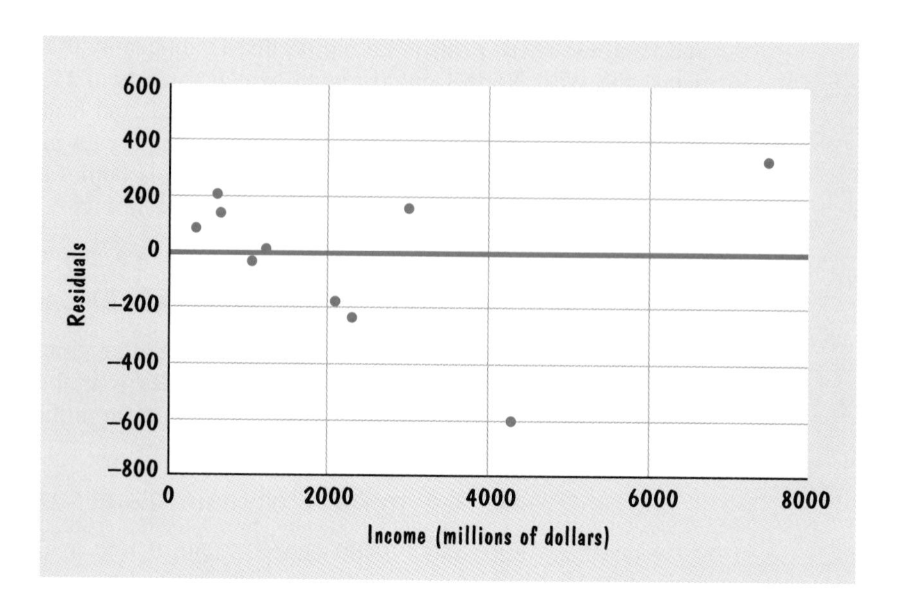

Figure 11.15 These sales residuals do not have a constant standard deviation.

the model in terms of per capita income and per capita magazine sales, using the following data on population p:

POPULATION p	40	20	400	150	50	100	200	30	100	50
PER CAPITA SALES y/p	24.1	23.1	26.1	28.0	26.2	26.1	26.1	31.6	28.3	31.2
PER CAPITA INCOME x/p	15.5	17.0	18.5	20.0	21.0	21.0	21.5	22.0	23.0	24.0

Least-squares estimation of a regression equation using per capita data yields

$$(y/p) = \underset{[2.3]}{10.207} + \underset{[3.8]}{.829}\,(x/p) \qquad R^2 = .641$$

with the t values in brackets. The plot of the residuals in Fig. 11.16 is a big improvement. Evidently, the problem was caused by the use of aggregate data for cities of very different sizes, and it was solved by a rescaling to per capita data.

Notice that this rescaling caused a substantial change in the estimated slope of the relationship. Our initial regression implies that a \$1 million increase in aggregate income will increase aggregate sales by 1.377 thousand—an additional 1.377 magazines per \$1000 of income. The rescaled regression gives an

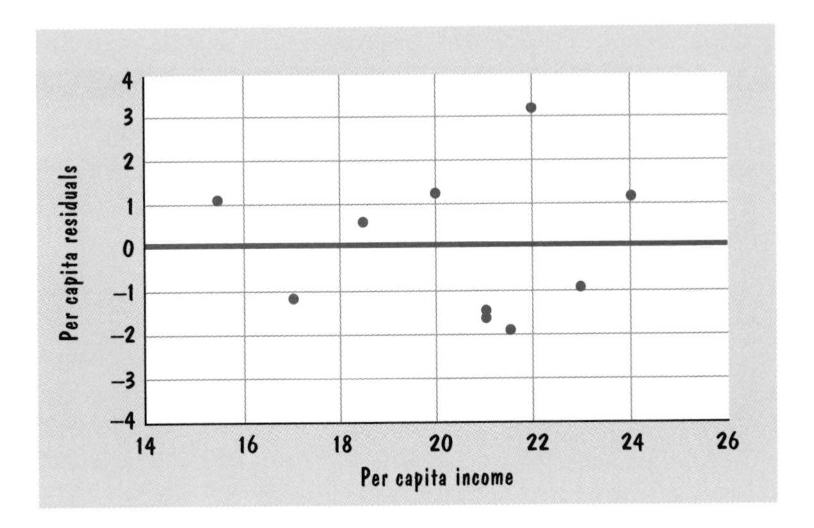

Figure 11.16 Per capita sales residuals look better.

estimate that is 40 percent lower: an additional .829 magazine per $1000 of income. If our rescaled estimates are correct, then our initial estimates were very misleading.

Some may be concerned by the fact that the rescaling caused a substantial decline in the t value and in R^2, suggesting that the rescaled model is not an improvement. However, we cannot compare these statistical measures directly because the explanatory variables are different. The first R^2 measures the accuracy in predicting aggregate sales; the second R^2 measures the accuracy in predicting per capita sales. If, as here, aggregate sales vary more than per capita sales, the per capita model could be more accurate and yet have the lower R^2!

To compare the predictive accuracy of these two models, we need to focus on a single variable—either aggregate sales or per capita sales. For the unscaled model, predicted aggregate sales are given by the regression equation; for the per capita model, aggregate sales can be predicted by multiplying predicted per capita sales by the city's population:

Aggregate model:	$\hat{y} = -109.1 + 1.377x$
Per capita model:	$\hat{y} = p\,(y/p)$
where	$(y/p) = 10.207 + .829\,(x/p)$

Table 11.2 shows these predictions. The per capita model generally has smaller prediction errors, often substantially so. The standard error of estimate (the

Table 11.2

Aggregate predictions of the unscaled and per capita models

	UNSCALED MODEL		PER CAPITA MODEL	
ACTUAL SALES	PREDICTED y	ERROR	PREDICTED $p(y/p)$	ERROR
964	744.8	219.2	922.3	41.7
462	359.2	102.8	486.0	−24.0
10,440	10,082.2	357.8	10,218.3	221.7
4,200	4,022.5	177.5	4,018.4	181.6
1,310	1,337.0	−27.0	1,380.9	−70.9
2,610	2,783.1	−173.1	2,761.9	−151.9
5,220	5,812.9	−592.9	5,606.6	−386.6
948	799.9	148.1	853.4	94.6
2,830	3,058.5	−228.5	2927.7	−97.7
1,560	1543.6	16.4	1,505.3	54.7

square root of the sum of the squared prediction errors, divided by $n - 2$) is 290.4 for the aggregate model and 188.3 for the per capita model. Despite its lower t value and R^2, the per capita model actually fits the data better than the aggregate model.

This example provides a good illustration not only of one possible technique for solving an unequal-standard-deviations problem, but also of the dangers of overrelying on t values and R^2 to judge a model's success. A successful model should have plausible estimates and residuals that are consistent with the model's assumptions.

Independent Error Terms

Sometimes, an examination of the residuals provides evidence that the values of the residuals are not independent of one another. Figure 11.17 shows one example of residuals with a definite pattern: consistently negative for low values of x, then consistently positive, and then negative again. The clumping of positive and negative values indicates that the residuals are not independent. In this case, as shown in Fig. 11.18, the explanation is that there is a pronounced nonlinear relationship between x and y, and the inappropriate use of a linear approximation gives the residuals a telltale pattern. The solution is to fit a nonlinear equation to these nonlinear data.

Figure 11.19 shows two examples of nonindependent residuals in regression models using time-series data. In these cases, the residuals have been plotted, not against the explanatory variable, but against time, to show the sequence in which the residuals appear. In the first case, the residuals are positively

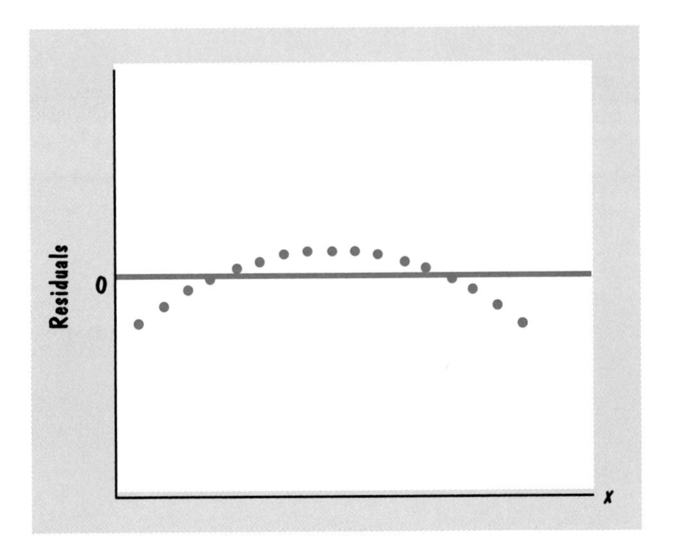

Figure 11.17 A suspicious residuals graph.

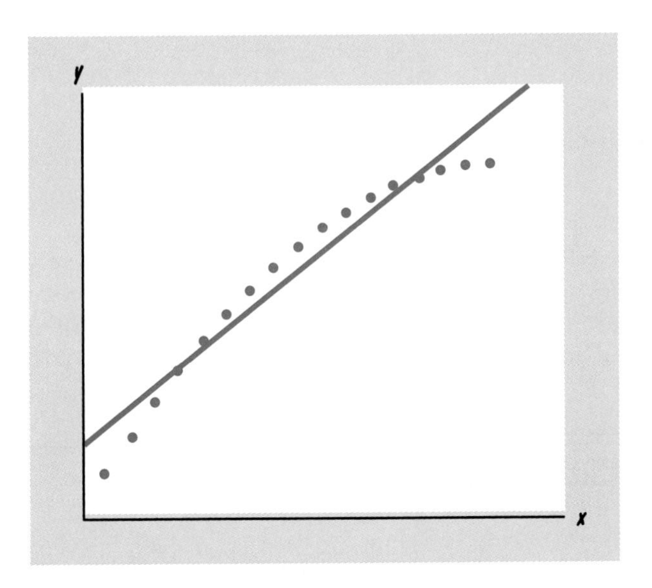

Figure 11.18 The solution to the suspicious residuals graph.

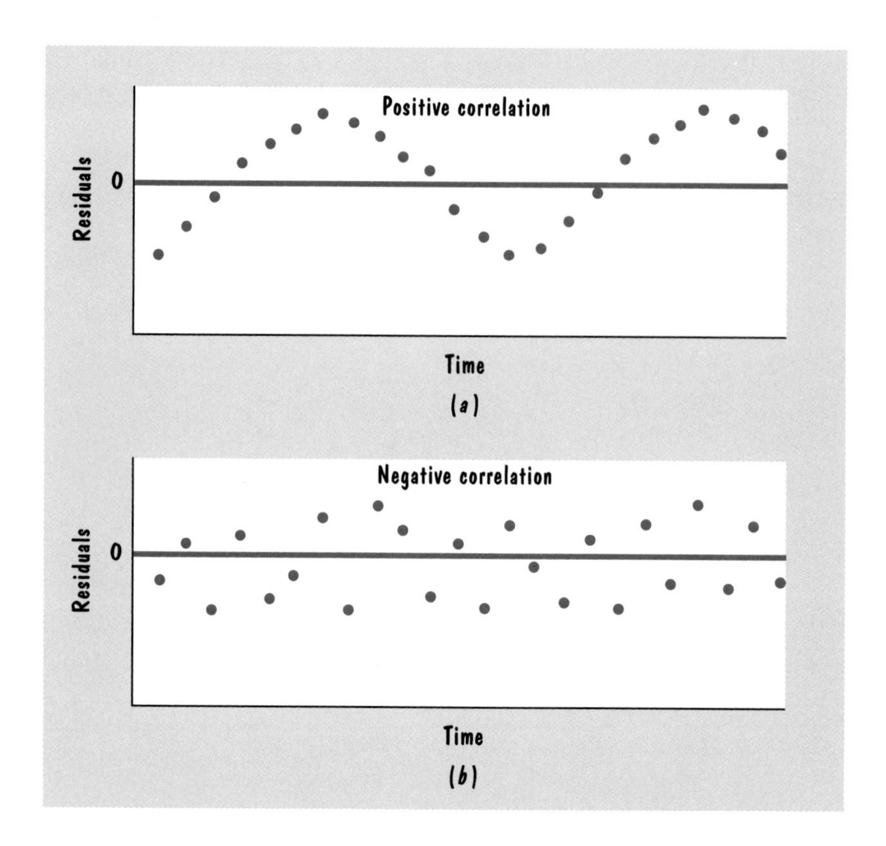

Figure 11.19 Residuals that are not independent.

correlated: Positive values tend to be followed by positive values and negative values by negative values. Negatively correlated residuals, in contrast, zigzag back and forth, alternating between positive and negative values. More advanced courses discuss a number of formal tests and possible cures.

exercises

11.26 Use these data on the average weight in pounds of U.S. males of various ages who are 68 inches tall to estimate the equation $y = \alpha + \beta x + \epsilon$. Plot the residuals and identify a pattern.

AGE x	20	30	40	50	60	70
WEIGHT y	157	168	174	176	174	169

11.27 Five plots in Georgia were studied over a 3-year period to see how tobacco yield y, in pounds per acre, is affected by the nitrogen percentage in the fertilizer x. Use these 15 observations to estimate the equation $y = \alpha + \beta x + \epsilon$. Is the relationship substantial and statistically significant at the 1 percent level? Now make a scatter diagram with the residuals on the vertical axis, x on the horizontal axis, and a horizontal line drawn at zero. What pattern do you see, and how do you explain it?

1924		1925		1926	
NITROGEN	YIELD	NITROGEN	YIELD	NITROGEN	YIELD
0	867	0	889	0	914
2	1094	2	1101	2	1092
3	1206	3	1180	3	1157
4	1281	4	1238	4	1224
5	1235	5	1237	5	1219

11.28 Below is a plot of the residuals from the least-squares estimation of a sample regression model. Which assumption appears to be violated?

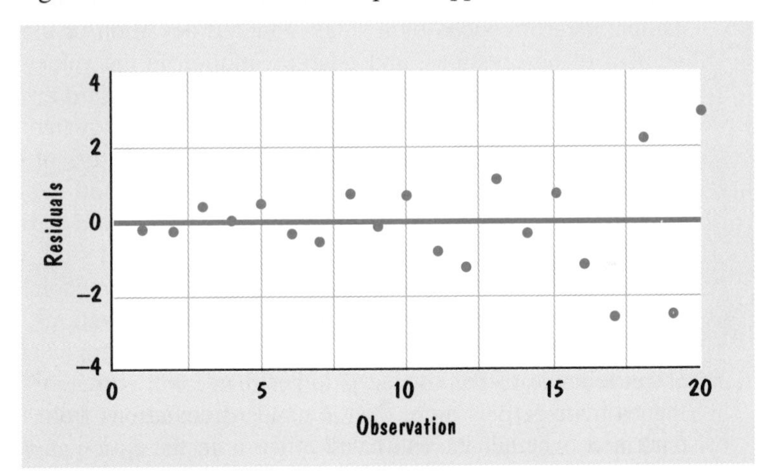

11.29 Below are data on x, the age (in months) at which 21 children said their first word, and y, their score on a later test of mental aptitude.[31] Estimate the equation $y = \alpha + \beta x + \epsilon$, and see whether there is a statistically significant relationship at the 1 percent level. Plot the residuals with x on the horizontal axis and a horizontal line drawn at zero. What pattern or other peculiarity do you see in the residuals?

x	y	x	y	x	y	x	y	x	y	x	y	x	y
15	95	9	91	18	93	20	94	10	83	10	100	17	121
26	71	15	102	11	100	7	113	11	84	12	105	11	86
10	83	20	87	8	104	9	96	11	102	42	57	10	100

summary

The simple linear regression model $y = \alpha + \beta x + \epsilon$ is a linear approximation to a possibly nonlinear relationship between the dependent variable y and the explanatory variable x. The parameter β tells us the change in y that accompanies a 1-unit change in x; the parameter α is the y intercept, and the error term ϵ encompasses the effects on y of influences other than x. We assume that ϵ is a normally distributed random variable with a zero mean and a constant standard deviation, and that the values of ϵ are independent of x and of one another.

The least-squares estimators a and b are the intercept and slope, respectively, of a line that minimizes the sum of the squared prediction errors for the observed values of y. These estimators are normally distributed with population means equal to the parameters they are estimating: a is $N[\alpha$, standard deviation of $a]$ and b is $N[\beta$, standard deviation of $b]$. The standard deviations of these estimators are reduced by a small standard deviation of the error term ϵ, a large number of observations, and a large variation in the values of x. The standard deviation of the error term is estimated by the standard error of estimate (SEE), which is essentially the square root of the average squared prediction error.

Using the SEE, we can compute the standard errors of our estimates of a and b and can calculate confidence intervals for α and β by using the t distribution with $n - 2$ degrees of freedom: $a \pm t^*$(standard error of a) and $b \pm t^*$(standard error of b). To test the null hypothesis $\beta = 0$, we use the t statistic $t = b/$(standard error of b) with $n - 2$ degrees of freedom. For a two-sided test at the 5 percent level with a large number of observations, we reject the null hypothesis and conclude that x has a statistically significant effect on y if the absolute value of the t value is larger than (approximately) 2, indicating that the estimate of β is more than 2 standard deviations from 0. We should also determine whether the estimated effect is in the anticipated direction and has a plausible magnitude.

For predicting the value of y for a particular value of x, we substitute the value of x into the least-squares estimated equation $y = a + bx$, quantifying the precision of this prediction with a prediction interval, which is analogous to a confidence interval for a parameter estimate. As with virtually all regression calculations, these computations are done for us by statistical software.

The coefficient of determination R^2 compares the model's sum of the squared prediction errors with the sum of the squared deviations of y about its mean:

$$R^2 = 1 - \frac{\Sigma(y_i - \hat{y}_i)^2}{\Sigma(y_i - \bar{y}_i)^2}$$

If the value of R^2 is close to 1, the model's prediction errors are minuscule in relation to the variation in y about its mean; if R^2 is close to 0, the model's prediction errors are no smaller than could be obtained by ignoring x and using the average value of y to predict y. The value of R^2 can be interpreted as the fraction of the variation in y that is explained by the regression model.

We should be wary of careless extrapolation—using the model to make predictions far from the available data. Not only do the prediction intervals widen, but also our linear approximation may become increasingly inaccurate. We should also be aware of the important regression-toward-the-mean phenomenon—the tendency of extreme observations to be followed by more average values. Finally, we need to realize that correlation does not prove causation. Only logical reasoning can do that. If x is correlated statistically with y, it may be a coincidence; or it may be because changes in x cause changes in y, changes in y cause changes in x, or some other factor causes changes in both x and y. We should be especially skeptical of results from data mining expeditions.

review exercises

11.30 Galton first noticed regression toward the mean in data on the diameters of parent and filial (offspring) sweet pea seeds.[32] He planted seeds of seven different diameters (in hundredths of an inch) and computed the average diameter of 100 seeds of their offspring:

PARENT	15	16	17	18	19	20	21
FILIAL	15.3	16.0	15.6	16.3	16.0	17.3	17.5

After you decide which is the dependent variable and which the explanatory variable, use least squares to estimate the regression equation $y = \alpha + \beta x + \epsilon$. Is there a statistically significant relationship at the 1 percent level? Plot these data in a scatter diagram with each axis going from 0 to 25, and draw in the least-squares line. What would the line look like if the diameters did not regress toward the mean?

11.31 The effective delivery and care of newborn babies can be aided by accurate estimates of birth weight. A study attempted to predict the birth weight of babies for 11 women whose pregnancies were judged to be abnormal by using echo-planar imaging (EPI), a form of magnetic resonance imaging, up to 1 week before delivery to estimate each baby's fetal volume.[33] Draw a scatter diagram, and use least-squares estimation of the regression model to see whether there is a statistically significant relationship at the 1 percent level between the child's estimated fetal volume x (in cubic decimeters) and the birth weight y (in kilograms). Summarize your findings (and be sure to interpret the estimated slope of your fitted line).

PATIENT	1	2	3	4	5	6	7	8	9	10	11
x	3.11	1.54	2.88	2.15	1.61	2.71	3.35	2.22	3.42	2.31	3.22
y	3.30	1.72	3.29	2.30	1.62	3.00	3.54	2.64	3.52	2.40	3.40

11.32 Shown below are the age x and bone density y of 25 females between the ages of 15 and 45 years who died between 1729 and 1852.[34] For a sample of modern-day British women of similar ages, the relationship between bone density and age is statistically significant at the 1 percent level, with an estimated slope of $-.658$. Draw a scatter diagram, and estimate the model $y = \alpha + \beta x + \epsilon$ by least squares. Now compare your results to those for modern-day British women.

x	y	x	y	x	y
15	104.66	28	88.92	38	111.49
17	90.68	29	108.49	39	88.82
17	105.28	30	84.06	41	93.89
19	98.55	34	85.30	43	128.99
23	100.52	35	101.66	44	111.39
26	126.60	35	91.41	45	114.60
27	82.09	35	91.30	45	100.10
27	114.29	36	75.78		
28	95.24	37	110.46		

11.33 Here are data from a 1993 statistics class showing each student's homework grade x and final examination grade y. Draw a scatter diagram, and estimate the model $y = \alpha + \beta x + \epsilon$ by least squares; give a one-paragraph summary of your findings.

x	y	x	y	x	y	x	y	x	y
76.3	80.5	14.6	68.0	92.5	88.0	92.9	90.5	85.8	82.5
93.3	90.0	74.2	84.0	40.0	75.5	87.9	83.5	87.5	87.5
92.5	95.5	3.8	33.0	92.9	83.0	27.9	67.5	75.4	85.0
68.8	82.5	80.4	79.0	73.3	80.5	86.7	86.0	35.0	73.5
96.7	92.0	9.2	53.0	96.3	92.0	45.0	82.0	87.9	88.0

11.34 Exercise 10.37 describes a Denver study of the relationship between climate and the age at which infants begin crawling. Shown next are their data on the average age (in weeks) when children born in different months began crawling and the average Denver temperature for the sixth month after birth.[35] In the linear regression model $y = \alpha + \beta x + \epsilon$, which of these variables would you use for y and x? After drawing a scatter diagram, estimate this equation by least squares and summarize your results.

BIRTH MONTH	CRAWLING AGE	TEMPERATURE
JANUARY	29.84	66
FEBRUARY	30.52	73
MARCH	29.70	72
APRIL	31.84	63
MAY	28.58	52
JUNE	31.44	39
JULY	33.64	33
AUGUST	32.82	30
SEPTEMBER	33.83	33
OCTOBER	33.35	37
NOVEMBER	33.38	48
DECEMBER	32.32	57

11.35 Some child-development researchers tabulated the amount of crying by 38 babies, aged 4 to 7 days, and compared these cry counts to each baby's measured IQ at 3 years of age.[36] Plot these points on a scatter diagram, and use least squares to estimate the regression line $y = \alpha + \beta x + \epsilon$, where y is IQ and x is the cry count. Is the relationship statistically significant at the 5 percent level? If there is a positive, statistically significant relationship, does this mean that you should make your baby cry?

CRY COUNT	IQ	CRY COUNT	IQ	CRY COUNT	IQ	CRY COUNT	IQ
9	103	13	162	17	94	22	135
9	119	14	106	17	141	22	157
10	87	15	112	18	109	23	103
10	109	15	114	18	109	23	113
12	94	15	133	18	112	27	108
12	97	16	100	19	103	30	155
12	103	16	106	19	120	31	135
12	119	16	118	20	90	33	159
12	120	16	124	20	132		
13	104	16	136	21	114		

11.36 The Australian Bureau of Meteorology calculates the southern oscillation index (SOI) from the monthly air pressure difference between Tahiti and Darwin, Australia. Negative values of the SOI indicate an El Niño episode; positive values indicate a La Niña episode. Since 1989, Australia's National Climate Center has used the SOI to predict rainfall 3 months in advance. Below are the Minitab computer results of a least-squares regression using annual data for 1948 through 1990 on the wheat yield relative to the long-run upward trend (positive values indicate unusually high yields, negative readings unusually low yields) and the average SOI reading for the 3 months of June, July, and August (Australia's winter).[37] In these data, the average value of wheat yield is .000, and the average SOI reading is .267. Write a brief summary of these results, being sure to mention whether El Niño or La Niña episodes are more often associated with high wheat yields.

```
THE REGRESSION EQUATION IS
WHEAT  =  -.0032  +  .0123  SOI
PREDICTOR         COEF      STDEV      T-RATIO          P
CONSTANT        -.0032      .0296        .1073      .9151
SOI              .0123      .0036       3.3664      .0016
S  =  .1943     R-SQ  =  21.6%    R-SQ(ADJ)  =  19.7%
```

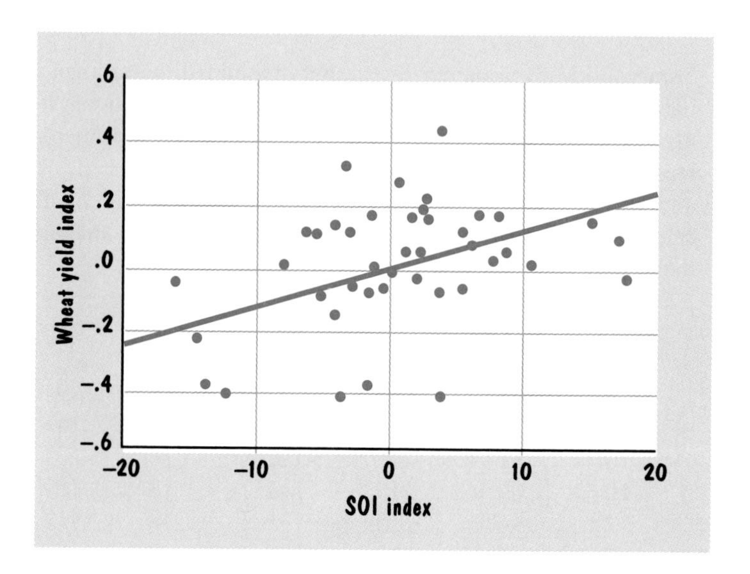

11.37 In the 1970s, researchers measured the pH of rain for 150 consecutive weeks in a wilderness area near Boulder, Colorado, and used these 150 observations to estimate the linear regression model where y = pH reading and x = week number (from 1 to 150).[38] Because pH uses a base-10 logarithmic scale, a difference of 1 represents a 10-fold difference in strength. A pH of 7 is considered neutral, 5 to 7 is weakly acidic, 2 to 5 is moderately acidic, and 0 to 2 is strongly acidic. Write a paragraph summarizing these results from the SAS computer program.

```
                    Parameter  Standard         T for H0
Variable    DF      Estimate      Error   Parameter  =  0   Prob > |T|
INTERCEP     1        5.4312      .1142           47.5587        .0000
WEEK         1        -.0053      .0013            4.0769        .0000
```

11.38 A farmer sold some California land that she had been farming for 20 years and bought some farmland in the midwest after investigating 20 different properties in five different states, spending one day at each property. Give a purely statistical explanation for why the property that appears to be the most attractive is probably not as good as it appears to be. Explain your reasoning carefully.

11.39 *Sports Illustrated*'s November 18, 1957, cover had a picture of the Oklahoma football team, which had not lost in 47 games, with the caption "Why Oklahoma is Unbeatable." The Saturday after this issue appeared, Oklahoma lost to Notre Dame 7 to 0, starting the legend of the *Sports Illustrated* cover jinx—that the performance of an individual or team usually declines after it is pictured on the cover of *Sports Illustrated.* Explain how regression toward the mean might be used to explain the *Sports Illustrated* cover jinx.

11.40 Below are decade averages of annual precipitation (in millimeters) in the contiguous United States.[39] Letting y be precipitation and x be the decade midpoints 1905, 1915, and so on, draw a scatter diagram and use least squares to estimate the equation $y = \alpha + \beta x + \epsilon$. Write a brief summary of your findings.

1900s	1910s	1920s	1930s	1940s	1950s	1960s	1970s	1980s
747.50	714.80	731.77	685.08	754.84	711.62	717.67	761.53	753.79

11.41 Here are data on inches of precipitation from 1961 to 1990 at the Los Angeles Civic Center during January (x) and during the rest of the year (y). To see how well the January precipitation predicts the rain during the rest of the year, draw a scatter diagram and use least squares to estimate the equation $y = \alpha + \beta x + \epsilon$. Summarize your findings.

x	y	x	y	x	y	x	y	x	y	x	y
1.28	4.55	.96	11.95	.43	8.83	.00	11.01	2.02	8.90	2.19	15.81
2.56	12.81	5.93	17.73	.00	6.54	2.84	12.13	2.17	12.24	1.39	7.72
.52	11.79	.90	6.68	4.39	13.06	7.70	22.87	6.49	27.55	1.65	8.33
1.43	6.55	14.94	11.38	8.35	8.34	6.59	10.41	.17	8.73	.73	3.83
.84	25.97	1.59	14.95	.12	10.58	7.50	18.83	.71	8.21	1.24	5.25

11.42 Explain why we could be skeptical of the implication that driver-training courses make people worse drivers: "A study comparing 34 drivers who had taken courses with 466 who had not, revealed that the driver-educated graduates actually had an accident rate 15% higher than their untrained counterparts."[40]

11.43 *U.S. News & World Report* once stated, "When a family gets a second or third car, you might expect that miles of driving per car would be reduced—but it doesn't work out that way. A study by economists of the Bureau of Public Roads indicates that families with only one car average 9,900 miles in a year but those with two or more average 10,000 miles for each car."[41] Does the purchase of a car make people feel they must drive more? Give an alternative explanation for the observed positive correlation between the number of cars a family has and the number of miles driven.

11.44 The Boston snow (B.S.) indicator uses the presence or absence of snow in Boston on Christmas Eve to predict the direction that stock prices will move the following year: "the average gain following snow on the ground was about 80% greater than for years in which there was no snow."[42] How do you explain the success of the B.S. indicator? Does snow on Christmas Eve affect the stock market, or does the stock market affect the weather?

11.45 Give a logical statistical explanation for the Cy Young jinx:

> Chicago White Sox pitcher LaMarr Hoyt thinks his 1983 American League Cy Young Award [for the best pitcher] may have been a jinx, and the four previous winners might agree with him. "I'll tell you, there have been a lot of times this season I've felt jinxed," said Hoyt. "A lot happens to me that is unexplainable." . . .
>
> Since Mike Flanagan of Baltimore won the award in 1979, all the AL winners have suffered disappointing seasons the following years. Flanagan has never won more than 16 games in a season since winning 23 in 1979. Steve Stone, the 1980 winner, retired less than two years later because of tendinitis. Rollie Fingers, the 1981 winner, and Pete Vuckovich, the 1982 winner, have been hampered with injuries since their big seasons.[43]

11.46 Rosenblatt and Cunningham found a strong positive correlation between family tension and the number of hours spent watching television.[44] Give a logical explanation other than that television shows tend to increase family tension.

11.47 A petition urging schools to offer and students to take Latin noted that "Seventy per cent of English words are of Latin origin, hence Latin students have larger English vocabularies. In fact, Latin students score 150 points higher on the verbal section of the SAT than their non-Latin peers." How might a skeptic challenge this argument?

11.48 Use the Vostok data in Exercise 2.13 to estimate the regression models $C = \alpha_1 + \beta_1 A + \epsilon_1$ and $A = \alpha_2 + \beta_2 D + \epsilon_2$, Write a brief essay summarizing your findings.

11.49 A psychologist once suggested this possible relationship between the IQs of children and of their parents:[45]

$$\text{Child IQ} - 100 = .50(\text{adult IQ} - 100)$$

What statistical phenomenon would explain this psychologist's use of the coefficient .5, rather than 1.0? A few years later, another researcher used the data that follows relating the mean IQ scores of children and of their fathers, grouped according to the father's occupational class, to estimate the regression equation $y = \alpha + \beta x + \epsilon$. What is suspicious about these data?

	FATHERS	CHILDREN
HIGHER PROFESSIONAL	139.7	120.8
LOWER PROFESSIONAL	130.6	114.7
CLERICAL	115.9	107.8
SKILLED	108.2	104.6
SEMISKILLED	97.8	98.9
UNSKILLED	84.9	92.6

11.50 After its staff ran more than 5000 miles in 30 different running shoes, *Consumer Reports* rated the shoes for overall performance on a scale of 0 to 100.[46] In the model $y = \alpha + \beta x + \epsilon$, where y = rating (0 to 100, with 100 best) and x = price (dollars), do you expect the value of β to be positive or negative? Explain your reasoning. Use the male and female data separately to draw scatter diagrams and to estimate this model by least squares. Briefly summarize your results.

FEMALE SHOES		MALE SHOES	
PRICE	RATING	PRICE	RATING
124	82	83	85
70	80	70	85
72	78	70	84
70	77	75	82
65	75	70	78
90	73	82	72
60	71	70	72
80	70	125	70
55	70	82	68
70	69	45	68
67	67	70	65
45	64	120	64
55	62	110	64
75	57	133	62
85	46	100	50

projects

For each of the following projects, type a report in ordinary English, using clear, concise, and persuasive prose. Any data that you collect for this project should be included as an appendix to your report. Data used in your report should be presented clearly and effectively.

11.1 Go to your campus bookstore, and select 30 new hardcover textbooks. (Be sure to explain how these books were selected.) For each book, record the number of pages and the price. Now use a scatter diagram to see if there appears to be a positive or negative relationship between the number of pages and the price. Calculate the correlation coefficient, and determine whether there is a statistically significant relationship between number of pages and price.

11.2 Go to a large bookstore that has a prominent display of best-selling fiction and nonfiction hardcover books. For each of these 2 categories, record the price and number of pages for at least 10 books and use these data to estimate the simple regression model for that category with the price the dependent variable and number of pages the explanatory variable. Are there any apparent outliers in your data?

11.3 Among students at your college, is there a statistical relationship between grade-point average and the number of siblings of a student?

11.4 For the 2 most recently completed major league baseball seasons, identify those players with at least 400 official times at bat in each season. Use the simple regression model to see how well each player's batting average in the most recent season is predicted by his batting average the preceding season. Do these results exhibit regression toward the mean?

11.5 Choose 10 recently completed full seasons of major league baseball. Do not include seasons that were shortened by strikes or other factors. For each season, identify those teams that won 90 or more games that season; also determine the average number of games these teams won during the season preceding their 90-win season and the average number of games they won during the season following their 90-win season. Now do the same for teams that won 100 or more games in a season.

11.6 After 3 years of major league play, a baseball player whose contract expires can declare himself a free agent at the end of the season and accumulate offers from any team. Determine the players who declared themselves free agents during a recent 5-year period, and compare their batting averages during the season in which they became free agents and during the next season.

11.7 Randomly select 25 mutual funds to investigate how well mutual fund performance is predicted by past performance.

11.8 Use the U.S. Department of Energy's monthly publication *Monthly Energy Review* to obtain annual U.S. data back to 1973 on total energy consumption (in quadrillion BU). The *Statistical Abstract of the United States* shows the U.S. population over this same period. Graph per capita energy consumption since 1973, and describe any patterns you see. To see whether there is a statistically significant trend, estimate the equation $y = \alpha + \beta x + \epsilon$, where y is per capita energy consumption and x is the year (1973, 1974, and so on).

11.9 Imagine that you are a U.S. government statistician in the 1860s. Use the census data for the years 1790 to 1860 to predict the U.S. population in 1930. To do so, estimate the equation $y = \alpha + \beta x + \epsilon$, where y is the population and x is the year (1790, 1800, and so on).

11.10 Has there been any long-term trend in voter turnout in U.S. presidential elections? Does the Democratic or Republican presidential candidate tend to do better when there is a heavy turnout?

11.11 Select an automobile model and year (at least 3 years old) that are of interest to you, for example, a 1993 Saab 900S convertible. Now find at least 30 of these cars that are for sale (from either dealers or private owners), and record the odometer mileage x and asking price y. As best you can, try to keep the cars as similar as possible. For example, ignore the car color, but do not mix together four-cylinder and six-cylinder cars or manual and automatic transmissions. Estimate the equation $y = \alpha + \beta x + \epsilon$, and summarize your results.

11.12 Find a local music store that has a section identifying the compact disks that are either the currently best-selling disks or the newest releases. For each disk, write down the total playing time and the price. Summarize your data for each of these variables, and estimate the correlation coefficient between them.

11.13 From either the police or district attorney's office in a nearby large metropolitan area, obtain data on the number of homicides each month for at least 5 years. From either a weather station or a newspaper, obtain data on the average high temperature for each of these months. Is there a statistical relationship between homicides and temperature?

11.14 In the most recent May issue of *Consumer Reports* (the car-buying guide), use the car facts-and-figures section to obtain data for at least 25 randomly selected passenger cars on the car's weight and overall gasoline mileage. Is there a statistical relationship between the two?

11.15 For each of the 50 states, calculate Bill Clinton's percentage of the 1992 votes cast for the Democratic and Republican presidential candidates; do not include the votes cast for other candidates. Do the same for the 1996 election. Is there a statistical relationship between these two sets of data? Are there any apparent outliers or anomalies?

11.16 An old Wall Street saying is, "As January goes, so goes the year." Use the simple regression model to see whether the February-through-December percentage change in the Dow Jones industrial average of stock prices is well predicted by the percentage change in January. Also use the simple regression model to see whether the annual percentage change in the Dow Jones industrial average of stock prices is well predicted by the percentage change the preceding year.

11.17 Ask 100 randomly selected students to tell you their heights, and the heights of both of their biological parents. Also note the gender of each student in your sample. Follow Galton's procedure for calculating mid-parent heights and then, for each gender, estimate the simple regression model with the student's height the dependent variable and the mid-parent height the explanatory variable. Is there evidence here of regression toward the mean?

11.18 Ask 100 randomly selected students to tell you their heights and weights. For each gender, estimate a simple regression model with weight the dependent variable and height the explanatory variable.

11.19 Ask a random sample of 100 college students to tell you their heights and grade-point averages. Use a procedure that will encourage honest answers to both questions. For each gender, draw a scatter diagram with heights on the horizontal axis and grades on the vertical axis. Does there seem to be a statistical relationship? Estimate the linear regression model for each gender and summarize your results.

11.20 Find the heights and weights of the quarterbacks, running backs, and wide receivers who have been inducted into professional football's hall of fame. For each of these 3 positions, estimate a linear regression model with height as the dependent variable and time as the explanatory variable and another linear regression model with weight as the dependent variable and time as the explanatory variable. In each case, let the time variable equal the year in which the player first played professional football.

11.21 Pick a date and approximate time of day (for example, 10:00 in the morning on April 1) for scheduling nonstop flights from an airport near you to at least a dozen large U.S. cities. By calling airlines, determine the cost of a coach seat on each of these flights and the distance covered by each flight. Use your data to estimate a simple regression model with ticket cost the dependent variable and distance the explanatory variable. Are there any outliers?

11.22 Find a local real-estate office that will let you browse through a recent multiple listing service (MLS) book on asking prices for houses that are for sale in your area. Use a random sample of at least 50 houses to estimate a simple regression model with asking price the dependent variable and square footage the explanatory variable.

11.23 Which is a better predictor of the day's high temperature: yesterday's high, the high 1 year ago, or the average high on this day for the past 5 years? To answer this question, randomly select at least 30 days of the year; for example, January 12, January 27, and so on. For each of these days, say January 12, determine the high temperature on the most recent January 12, the high temperature on January 11, the high temperature on January 12 a year earlier, and the average high temperature on January 12 for the 5 years preceding the most recent January 12. Use your data to estimate three simple regression models in order to answer the question posed in the first sentence of this project.

11.24 Find data for at least 25 countries on the rate of population growth and the level of per capita gross domestic product (GDP). Make sure that your GDP data are measured in a common currency, such as dollars. Use the simple regression model to see if there is a statistical relationship between population growth and GDP.

11.25 Collect data for at least 10 years on the cost of attending your college. Taking into account the increase in the overall price level during these years, use the simple regression model to see whether there has been a trend in the real cost of attending this college.

appendix tables

Table 1 Random digits

10 09 73 15 33	76 52 01 35 86	34 67 35 49 76	80 95 90 91 71	39 29 27 49 45
37 54 20 48 05	64 89 47 42 96	24 80 52 40 37	20 63 61 04 02	00 82 29 16 65
08 42 26 89 53	19 64 50 93 03	23 20 90 25 60	15 95 33 47 64	35 08 03 36 06
99 01 90 25 29	09 37 67 07 15	38 31 13 11 65	88 67 67 43 97	04 43 62 76 59
12 80 79 99 70	80 15 73 61 47	64 03 23 66 53	98 95 11 68 77	12 17 17 68 33
66 06 57 47 17	34 07 27 68 50	36 69 73 61 70	65 81 33 98 85	11 19 92 91 70
31 06 01 08 05	45 57 18 24 06	35 30 34 26 14	86 79 90 74 39	23 40 30 97 32
85 26 97 76 02	02 05 16 56 92	68 66 57 48 18	73 05 38 52 47	18 62 38 85 79
63 57 33 21 35	05 32 54 70 48	90 55 33 75 48	28 46 82 87 09	83 49 12 56 24
73 79 64 57 53	03 52 96 47 78	35 80 83 42 82	60 93 52 03 44	35 27 38 84 35
98 52 01 77 67	14 90 56 86 07	22 10 94 05 58	60 97 09 34 33	50 50 07 39 98
11 80 50 54 31	39 80 82 77 32	50 72 56 82 48	29 40 52 42 01	52 77 56 78 51
83 45 29 96 34	06 28 89 80 83	13 74 67 00 78	18 47 54 06 10	68 71 17 78 17
38 68 54 02 00	86 50 75 84 01	36 76 66 79 51	90 36 47 64 93	29 60 91 10 62
99 59 46 73 48	87 51 76 49 69	91 82 60 89 28	93 78 56 13 68	23 47 83 41 13
65 48 11 76 74	17 46 85 09 50	58 04 77 69 74	73 03 95 71 86	40 21 81 65 44
80 12 43 56 35	17 72 70 80 15	45 31 82 23 74	21 11 57 82 33	14 38 55 37 63
74 35 09 98 17	77 40 27 72 14	43 23 60 02 10	45 52 16 42 37	96 28 60 26 55
69 91 62 68 03	66 25 22 91 48	36 93 68 72 03	76 62 11 39 90	94 40 05 64 18
09 89 32 05 05	14 22 56 85 14	46 42 75 67 88	96 29 77 88 22	54 38 21 45 98
91 49 91 45 23	68 47 92 76 86	46 16 28 35 54	94 75 08 99 23	37 08 92 00 48
80 33 69 45 98	26 94 03 68 58	70 29 73 41 35	53 14 03 33 40	42 05 08 23 41
44 10 48 19 49	85 15 74 79 54	32 97 92 65 75	57 60 04 08 81	22 22 20 64 13
12 55 07 37 42	11 10 00 20 40	12 86 07 46 97	96 64 48 94 39	28 70 72 58 15
63 60 64 93 29	16 50 53 44 84	40 21 95 25 63	43 65 17 70 82	07 20 73 17 90
61 19 69 04 46	26 45 74 77 74	51 92 43 37 29	65 39 45 95 93	42 58 26 05 27
15 47 44 52 66	95 27 07 99 53	59 36 78 38 48	82 39 61 01 18	33 21 15 94 66
94 35 72 85 73	67 89 75 43 87	54 62 24 44 31	91 19 04 25 92	92 92 74 59 73
42 48 11 62 13	97 34 40 87 21	16 86 84 87 67	03 07 11 20 59	25 70 14 66 70
23 52 37 83 17	73 20 88 98 37	68 93 59 14 16	26 25 22 96 63	05 52 28 25 62
04 49 35 24 94	75 24 63 38 24	45 86 25 10 25	61 96 27 93 35	65 33 71 24 72
00 54 99 76 54	64 05 18 81 59	96 11 96 38 96	54 69 28 23 91	23 28 72 95 29
35 96 31 53 07	26 89 80 93 54	33 35 13 54 62	77 97 45 00 24	90 10 33 93 33
59 80 80 83 91	45 42 72 68 42	83 60 94 97 00	13 02 12 48 92	78 56 52 01 06
46 05 88 52 36	01 39 09 22 86	77 28 14 40 77	93 91 08 36 47	70 61 74 29 41

Table 1 Random digits (continued)

32 17 90 05 97	87 37 92 52 41	05 56 70 70 07	86 74 31 71 57	85 39 41 18 38
69 23 46 14 06	20 11 74 52 04	15 95 66 00 00	18 74 39 24 23	97 11 89 63 39
19 56 54 14 30	01 75 87 53 79	40 41 92 15 85	66 67 43 68 06	84 96 28 52 07
43 15 51 49 38	19 47 60 72 46	43 66 79 45 43	59 04 79 00 33	20 82 66 95 41
94 86 43 19 94	36 16 81 08 51	34 88 88 15 53	01 54 03 54 56	05 01 45 11 76
98 08 62 48 26	45 24 02 84 04	44 99 90 88 96	39 09 47 34 07	35 44 13 18 80
33 18 51 62 32	41 94 15 09 49	89 43 54 85 81	88 69 54 19 94	37 54 87 30 43
80 95 10 04 06	96 38 27 07 74	20 15 12 33 87	25 01 62 52 98	94 62 46 11 71
79 75 24 91 40	71 96 12 82 96	69 96 10 25 91	74 85 12 05 39	00 38 75 95 79
18 63 33 25 37	98 14 50 65 71	31 01 02 46 74	05 45 56 14 27	77 93 89 19 36
74 02 94 39 02	77 55 73 22 70	97 79 01 71 19	52 52 75 80 21	80 81 45 17 48
54 17 84 56 11	80 99 33 71 43	05 33 51 29 69	56 12 71 92 55	36 04 09 03 24
11 66 44 98 83	52 07 98 48 27	59 38 17 15 39	09 97 33 34 40	88 46 12 33 56
48 32 47 79 28	31 24 96 47 10	02 29 53 68 70	32 30 75 75 46	15 02 00 99 94
69 07 49 41 38	87 63 79 19 76	35 58 40 44 01	10 51 82 16 15	01 84 87 69 38

Reproduced by permission from the RAND Corporation, "A Million Random Digits," Glencoe, Illinois: 1955.

Table 2 Binomial probabilities

These are the probabilities of x successes in n Bernoulli trials, each with probability p of success: $P[x] = \binom{n}{x} p^x (1-p)^{n-x}$. For instance, with a $p = .20$ success probability, the probability of $x = 2$ successes in $n = 5$ trials is .2048. (*Note:* For $p > .5$, use the probability of $n - x$ successes with success probability $1 - p$. For instance, the probability of 4 successes in 6 trials when $p = .75$ is the same as the probability of 2 successes when $p = .25$; that is, .2966.)

		p										
n	*x*	.01	.05	.10	.15	.20	.25	.30	.35	.40	.45	.50
1	0	.9900	.9500	.9000	.8500	.8000	.7500	.7000	.6500	.6000	.5500	.5000
	1	.0100	.0500	.1000	.1500	.2000	.2500	.3000	.3500	.4000	.4500	.5000
2	0	.9801	.9025	.8100	.7225	.6400	.5625	.4900	.4225	.3600	.3025	.2500
	1	.0198	.0950	.1800	.2550	.3200	.3750	.4200	.4550	.4800	.4950	.5000
	2	.0001	.0025	.0100	.0225	.0400	.0625	.0900	.1225	.1600	.2025	.2500
3	0	.9703	.8574	.7290	.6141	.5120	.4219	.3430	.2746	.2160	.1664	.1250
	1	.0294	.1354	.2430	.3251	.3840	.4219	.4410	.4436	.4320	.4084	.3750
	2	.0003	.0071	.0270	.0574	.0960	.1406	.1890	.2389	.2880	.3341	.3750
	3	.0000	.0001	.0010	.0034	.0080	.0156	.0270	.0429	.0640	.0911	.1250
4	0	.9606	.8145	.6561	.5220	.4096	.3164	.2401	.1785	.1296	.0915	.0625
	1	.0388	.1715	.2916	.3685	.4096	.4219	.4116	.3845	.3456	.2995	.2500
	2	.0006	.0135	.0486	.0975	.1536	.2109	.2646	.3105	.3456	.3675	.3750
	3	.0000	.0005	.0036	.0115	.0256	.0469	.0756	.1115	.1536	.2005	.2500
	4	.0000	.0000	.0001	.0005	.0016	.0039	.0081	.0150	.0256	.0410	.0625
5	0	.9510	.7738	.5905	.4437	.3277	.2373	.1681	.1160	.0778	.0503	.0313
	1	.0480	.2036	.3280	.3915	.4096	.3955	.3602	.3124	.2592	.2059	.1563
	2	.0010	.0214	.0729	.1382	.2048	.2637	.3087	.3364	.3456	.3369	.3125
	3	.0000	.0011	.0081	.0244	.0512	.0879	.1323	.1811	.2304	.2757	.3125
	4	.0000	.0000	.0004	.0022	.0064	.0146	.0284	.0488	.0768	.1128	.1563
	5	.0000	.0000	.0000	.0001	.0003	.0010	.0024	.0053	.0102	.0185	.0313
6	0	.9415	.7351	.5314	.3771	.2621	.1780	.1176	.0754	.0467	.0277	.0156
	1	.0571	.2321	.3543	.3993	.3932	.3560	.3025	.2437	.1866	.1359	.0938
	2	.0014	.0305	.0984	.1762	.2458	.2966	.3241	.3280	.3110	.2780	.2344
	3	.0000	.0021	.0146	.0415	.0819	.1318	.1852	.2355	.2765	.3032	.3125
	4	.0000	.0001	.0012	.0055	.0154	.0330	.0595	.0951	.1382	.1861	.2344
	5	.0000	.0000	.0001	.0004	.0015	.0044	.0102	.0205	.0369	.0609	.0938
	6	.0000	.0000	.0000	.0000	.0001	.0002	.0007	.0018	.0041	.0083	.0156
7	0	.9321	.6983	.4783	.3206	.2097	.1335	.0824	.0490	.0280	.0152	.0078
	1	.0659	.2573	.3720	.3960	.3670	.3115	.2471	.1848	.1306	.0872	.0547
	2	.0020	.0406	.1240	.2097	.2753	.3115	.3177	.2985	.2613	.2140	.1641
	3	.0000	.0036	.0230	.0617	.1147	.1730	.2269	.2679	.2903	.2918	.2734
	4	.0000	.0002	.0026	.0109	.0287	.0577	.0972	.1442	.1935	.2388	.2734

Table 2 Binomial probabilities (continued)

		p										
n	*x*	.01	.05	.10	.15	.20	.25	.30	.35	.40	.45	.50
	5	.0000	.0000	.0002	.0012	.0043	.0115	.0250	.0466	.0774	.1172	.1641
	6	.0000	.0000	.0000	.0001	.0004	.0013	.0036	.0084	.0172	.0320	.0547
	7	.0000	.0000	.0000	.0000	.0000	.0001	.0002	.0006	.0016	.0037	.0078
8	0	.9227	.6634	.4305	.2725	.1678	.1002	.0576	.0319	.0168	.0084	.0039
	1	.0746	.2793	.3826	.3847	.3355	.2670	.1977	.1373	.0896	.0548	.0313
	2	.0026	.0515	.1488	.2376	.2936	.3115	.2965	.2587	.2090	.1569	.1094
	3	.0001	.0054	.0331	.0839	.1468	.2076	.2541	.2786	.2787	.2568	.2188
	4	.0000	.0004	.0046	.0185	.0459	.0865	.1361	.1875	.2322	.2627	.2734
	5	.0000	.0000	.0004	.0026	.0092	.0231	.0467	.0808	.1239	.1719	.2188
	6	.0000	.0000	.0000	.0002	.0011	.0038	.0100	.0217	.0413	.0403	.1094
	7	.0000	.0000	.0000	.0000	.0001	.0004	.0012	.0033	.0079	.0164	.0312
	8	.0000	.0000	.0000	.0000	.0000	.0000	.0001	.0002	.0007	.0017	.0039
9	0	.9135	.6302	.3874	.2316	.1342	.0751	.0404	.0207	.0101	.0046	.0020
	1	.0830	.2985	.3874	.3679	.3020	.2253	.1556	.1004	.0605	.0339	.0176
	2	.0034	.0629	.1722	.2597	.3020	.3003	.2668	.2162	.1612	.1110	.0703
	3	.0001	.0077	.0446	.1069	.1762	.2336	.2668	.2716	.2508	.2119	.1641
	4	.0000	.0006	.0074	.0283	.0661	.1168	.1715	.2194	.2508	.2600	.2461
	5	.0000	.0000	.0008	.0050	.0165	.0389	.0735	.1181	.1672	.2128	.2461
	6	.0000	.0000	.0001	.0006	.0028	.0087	.0210	.0424	.0743	.1160	.1641
	7	.0000	.0000	.0000	.0000	.0003	.0012	.0039	.0098	.0212	.0407	.0703
	8	.0000	.0000	.0000	.0000	.0000	.0001	.0004	.0013	.0035	.0083	.0176
	9	.0000	.0000	.0000	.0000	.0000	.0000	.0000	.0001	.0003	.0008	.0020
10	0	.9044	.5987	.3487	.1969	.1074	.0563	.0282	.0135	.0060	.0025	.0010
	1	.0914	.3151	.3874	.3474	.2684	.1877	.1211	.0725	.0403	.0207	.0098
	2	.0042	.0746	.1937	.2759	.3020	.2816	.2335	.1757	.1209	.0763	.0439
	3	.0001	.0105	.0574	.1298	.2013	.2503	.2668	.2522	.2150	.1665	.1172
	4	.0000	.0010	.0112	.0401	.0881	.1460	.2001	.2377	.2508	.2384	.2051
	5	.0000	.0001	.0015	.0085	.0264	.0584	.1029	.1536	.2007	.2340	.2461
	6	.0000	.0000	.0001	.0012	.0055	.0162	.0368	.0689	.1115	.1596	.2051
	7	.0000	.0000	.0000	.0001	.0008	.0031	.0090	.0212	.0425	.0746	.1172
	8	.0000	.0000	.0000	.0000	.0001	.0004	.0014	.0043	.0106	.0229	.0439
	9	.0000	.0000	.0000	.0000	.0000	.0000	.0001	.0005	.0016	.0042	.0098
	10	.0000	.0000	.0000	.0000	.0000	.0000	.0000	.0000	.0001	.0003	.0010
11	0	.8953	.5688	.3138	.1673	.0859	.0422	.0198	.0088	.0036	.0014	.0005
	1	.0995	.3293	.3835	.3248	.2362	.1549	.0932	.0518	.0266	.0125	.0054
	2	.0050	.0867	.2131	.2866	.2953	.2581	.1998	.1395	.0887	.0513	.0269
	3	.0002	.0137	.0710	.1517	.2215	.2581	.2568	.2254	.1774	.1259	.0806
	4	.0000	.0014	.0158	.0536	.1107	.1721	.2201	.2428	.2365	.2060	.1611

Table 2 Binomial probabilities (continued)

		p										
n	*x*	.01	.05	.10	.15	.20	.25	.30	.35	.40	.45	.50
	5	.0000	.0001	.0025	.0132	.0388	.0803	.1321	.1830	.2207	.2360	.2256
	6	.0000	.0000	.0003	.0023	.0097	.0268	.0566	.0985	.1471	.1931	.2256
	7	.0000	.0000	.0000	.0003	.0017	.0064	.0173	.0379	.0701	.1128	.1611
	8	.0000	.0000	.0000	.0000	.0002	.0011	.0037	.0102	.0234	.0462	.0806
	9	.0000	.0000	.0000	.0000	.0000	.0001	.0005	.0018	.0052	.0126	.0269
	10	.0000	.0000	.0000	.0000	.0000	.0000	.0000	.0002	.0007	.0021	.0054
	11	.0000	.0000	.0000	.0000	.0000	.0000	.0000	.0000	.0000	.0002	.0005
12	0	.8864	.5404	.2824	.1422	.0687	.0317	.0138	.0057	.0022	.0008	.0002
	1	.1074	.3413	.3766	.3012	.2062	.1267	.0712	.0368	.0174	.0075	.0029
	2	.0060	.0988	.2301	.2924	.2835	.2323	.1678	.1088	.0639	.0339	.0161
	3	.0002	.0173	.0852	.1720	.2362	.2581	.2397	.1954	.1419	.0923	.0537
	4	.0000	.0021	.0213	.0683	.1329	.1936	.2311	.2367	.2128	.1700	.1208
	5	.0000	.0002	.0038	.0193	.0532	.1032	.1585	.2039	.2270	.2225	.1934
	6	.0000	.0000	.0005	.0040	.0155	.0401	.0792	.1281	.1766	.2124	.2256
	7	.0000	.0000	.0000	.0006	.0033	.0115	.0291	.0591	.1009	.1489	.1934
	8	.0000	.0000	.0000	.0001	.0005	.0024	.0078	.0199	.0420	.0762	.1208
	9	.0000	.0000	.0000	.0000	.0001	.0004	.0015	.0048	.0125	.0277	.0537
	10	.0000	.0000	.0000	.0000	.0000	.0000	.0002	.0008	.0025	.0068	.0161
	11	.0000	.0000	.0000	.0000	.0000	.0000	.0000	.0001	.0003	.0010	.0029
	12	.0000	.0000	.0000	.0000	.0000	.0000	.0000	.0000	.0000	.0001	.0002
13	0	.8775	.5133	.2542	.1209	.0550	.0238	.0097	.0037	.0013	.0004	.0001
	1	.1152	.3512	.3672	.2774	.1787	.1029	.0540	.0259	.0113	.0045	.0016
	2	.0070	.1109	.2448	.2937	.2680	.2059	.1388	.0836	.0453	.0220	.0095
	3	.0003	.0214	.0997	.1900	.2457	.2517	.2181	.1651	.1107	.0660	.0349
	4	.0000	.0028	.0277	.0838	.1535	.2097	.2337	.2222	.1845	.1350	.0873
	5	.0000	.0003	.0055	.0266	.0691	.1258	.1803	.2154	.2214	.1989	.1571
	6	.0000	.0000	.0008	.0063	.0230	.0559	.1030	.1546	.1968	.2169	.2095
	7	.0000	.0000	.0001	.0011	.0058	.0186	.0442	.0833	.1312	.1775	.2095
	8	.0000	.0000	.0001	.0001	.0011	.0047	.0142	.0336	.0656	.1089	.1571
	9	.0000	.0000	.0000	.0000	.0001	.0009	.0034	.0101	.0243	.0495	.0873
	10	.0000	.0000	.0000	.0000	.0000	.0001	.0006	.0022	.0065	.0162	.0349
	11	.0000	.0000	.0000	.0000	.0000	.0000	.0001	.0003	.0012	.0036	.0095
	12	.0000	.0000	.0000	.0000	.0000	.0000	.0000	.0000	.0001	.0005	.0016
	13	.0000	.0000	.0000	.0000	.0000	.0000	.0000	.0000	.0000	.0000	.0001
14	0	.8687	.4877	.2288	.1028	.0440	.0178	.0068	.0024	.0008	.0002	.0001
	1	.1229	.3593	.3559	.2539	.1539	.0832	.0407	.0181	.0073	.0027	.0009
	2	.0081	.1229	.2570	.2912	.2501	.1802	.1134	.0634	.0317	.0141	.0056
	3	.0003	.0259	.1142	.2056	.2501	.2402	.1943	.1366	.0845	.0462	.0222
	4	.0000	.0037	.0349	.0998	.1720	.2202	.2290	.2022	.1549	.1040	.0611

Table 2 Binomial probabilities (continued)

		p										
n	*x*	.01	.05	.10	.15	.20	.25	.30	.35	.40	.45	.50
	5	.0000	.0004	.0078	.0352	.0860	.1468	.1963	.2178	.2066	.1701	.1222
	6	.0000	.0000	.0013	.0093	.0322	.0734	.1262	.1759	.2066	.2088	.1833
	7	.0000	.0000	.0002	.0019	.0092	.0280	.0618	.1082	.1574	.1952	.2095
	8	.0000	.0000	.0000	.0003	.0020	.0082	.0232	.0510	.0918	.1398	.1833
	9	.0000	.0000	.0000	.0000	.0003	.0018	.0066	.0183	.0408	.0762	.1222
	10	.0000	.0000	.0000	.0000	.0000	.0003	.0014	.0049	.0136	.0312	.0611
	11	.0000	.0000	.0000	.0000	.0000	.0000	.0002	.0010	.0033	.0093	.0222
	12	.0000	.0000	.0000	.0000	.0000	.0000	.0000	.0001	.0005	.0019	.0056
	13	.0000	.0000	.0000	.0000	.0000	.0000	.0000	.0000	.0001	.0002	.0009
	14	.0000	.0000	.0000	.0000	.0000	.0000	.0000	.0000	.0000	.0000	.0001
15	0	.8601	.4633	.2059	.0874	.0352	.0134	.0047	.0016	.0005	.0001	.0000
	1	.1303	.3658	.3432	.2312	.1319	.0668	.0305	.0126	.0047	.0016	.0005
	2	.0092	.1348	.2669	.2856	.2309	.1559	.0916	.0476	.0219	.0090	.0032
	3	.0004	.0307	.1285	.2184	.2501	.2252	.1700	.1110	.0634	.0318	.0139
	4	.0000	.0049	.0428	.1156	.1876	.2252	.2186	.1792	.1268	.0780	.0417
	5	.0000	.0006	.0105	.0449	.1032	.1651	.2061	.2123	.1859	.1404	.0916
	6	.0000	.0000	.0019	.0132	.0430	.0917	.1472	.1906	.2066	.1914	.1527
	7	.0000	.0000	.0003	.0030	.0138	.0393	.0811	.1319	.1771	.2013	.1964
	8	.0000	.0000	.0000	.0005	.0035	.0131	.0348	.0710	.1181	.1647	.1964
	9	.0000	.0000	.0000	.0001	.0007	.0034	.0116	.0298	.0612	.1048	.1527
	10	.0000	.0000	.0000	.0000	.0001	.0007	.0030	.0096	.0245	.0515	.0916
	11	.0000	.0000	.0000	.0000	.0000	.0001	.0006	.0024	.0074	.0191	.0417
	12	.0000	.0000	.0000	.0000	.0000	.0000	.0001	.0004	.0016	.0052	.0139
	13	.0000	.0000	.0000	.0000	.0000	.0000	.0000	.0001	.0003	.0010	.0032
	14	.0000	.0000	.0000	.0000	.0000	.0000	.0000	.0000	.0000	.0001	.0005
	15	.0000	.0000	.0000	.0000	.0000	.0000	.0000	.0000	.0000	.0000	.0000
16	0	.8515	.4401	.1853	.0743	.0281	.0100	.0033	.0010	.0003	.0001	.0000
	1	.1376	.3706	.3294	.2097	.1126	.0535	.0228	.0087	.0030	.0009	.0002
	2	.0104	.1463	.2745	.2775	.2111	.1336	.0732	.0353	.0150	.0056	.0018
	3	.0005	.0359	.1423	.2285	.2463	.2079	.1465	.0888	.0468	.0215	.0085
	4	.0000	.0061	.0514	.1311	.2001	.2252	.2040	.1553	.1014	.0572	.0278
	5	.0000	.0008	.0137	.0555	.1201	.1802	.2099	.2008	.1623	.1123	.0667
	6	.0000	.0001	.0028	.0180	.0550	.1101	.1649	.1982	.1983	.1684	.1222
	7	.0000	.0000	.0004	.0045	.0197	.0524	.1010	.1524	.1889	.1969	.1746
	8	.0000	.0000	.0001	.0009	.0055	.0197	.0487	.0923	.1417	.1812	.1964
	9	.0000	.0000	.0000	.0001	.0012	.0058	.0185	.0442	.0840	.1318	.1746
	10	.0000	.0000	.0000	.0000	.0002	.0014	.0056	.0167	.0392	.0755	.1222
	11	.0000	.0000	.0000	.0000	.0000	.0002	.0013	.0049	.0142	.0337	.0667
	12	.0000	.0000	.0000	.0000	.0000	.0000	.0002	.0011	.0040	.0115	.0278
	13	.0000	.0000	.0000	.0000	.0000	.0000	.0000	.0002	.0008	.0029	.0085
	14	.0000	.0000	.0000	.0000	.0000	.0000	.0000	.0000	.0001	.0005	.0018

Table 2 Binomial probabilities (continued)

		p										
n	*x*	.01	.05	.10	.15	.20	.25	.30	.35	.40	.45	.50
	15	.0000	.0000	.0000	.0000	.0000	.0000	.0000	.0000	.0000	.0001	.0002
	16	.0000	.0000	.0000	.0000	.0000	.0000	.0000	.0000	.0000	.0000	.0000
17	0	.8429	.4181	.1668	.0631	.0225	.0075	.0023	.0007	.0002	.0000	.0000
	1	.1447	.3741	.3150	.1893	.0957	.0426	.0169	.0060	.0019	.0005	.0001
	2	.0117	.1575	.2800	.2673	.1914	.1136	.0581	.0260	.0102	.0035	.0010
	3	.0006	.0415	.1556	.2359	.2393	.1893	.1245	.0701	.0341	.0144	.0052
	4	.0000	.0076	.0605	.1457	.2093	.2209	.1868	.1320	.0796	.0411	.0182
	5	.0000	.0010	.0175	.0668	.1361	.1914	.2081	.1849	.1379	.0875	.0472
	6	.0000	.0001	.0039	.0236	.0680	.1276	.1784	.1991	.1839	.1432	.0944
	7	.0000	.0000	.0007	.0065	.0267	.0668	.1201	.1685	.1927	.1841	.1484
	8	.0000	.0000	.0001	.0014	.0084	.0279	.0644	.1134	.1606	.1883	.1855
	9	.0000	.0000	.0000	.0003	.0021	.0093	.0276	.0611	.1070	.1540	.1855
	10	.0000	.0000	.0000	.0000	.0004	.0025	.0095	.0263	.0571	.1008	.1484
	11	.0000	.0000	.0000	.0000	.0001	.0005	.0026	.0090	.0242	.0525	.0944
	12	.0000	.0000	.0000	.0000	.0000	.0001	.0006	.0024	.0081	.0215	.0472
	13	.0000	.0000	.0000	.0000	.0000	.0000	.0001	.0005	.0021	.0068	.0182
	14	.0000	.0000	.0000	.0000	.0000	.0000	.0000	.0001	.0004	.0016	.0052
	15	.0000	.0000	.0000	.0000	.0000	.0000	.0000	.0000	.0001	.0003	.0010
	16	.0000	.0000	.0000	.0000	.0000	.0000	.0000	.0000	.0000	.0000	.0001
	17	.0000	.0000	.0000	.0000	.0000	.0000	.0000	.0000	.0000	.0000	.0000
18	0	.8345	.3972	.1501	.0536	.0180	.0056	.0016	.0004	.0001	.0000	.0000
	1	.1517	.3763	.3002	.1704	.0811	.0338	.0126	.0042	.0012	.0003	.0001
	2	.0130	.1683	.2835	.2556	.1723	.0958	.0458	.0190	.0069	.0022	.0006
	3	.0007	.0473	.1680	.2406	.2297	.1704	.1046	.0547	.0246	.0095	.0031
	4	.0000	.0093	.0700	.1592	.2153	.2130	.1681	.1104	.0614	.0291	.0117
	5	.0000	.0014	.0218	.0787	.1507	.1988	.2017	.1664	.1146	.0666	.0327
	6	.0000	.0002	.0052	.0301	.0816	.1436	.1873	.1941	.1655	.1181	.0708
	7	.0000	.0000	.0010	.0091	.0350	.0820	.1376	.1792	.1892	.1657	.1214
	8	.0000	.0000	.0002	.0022	.0120	.0376	.0811	.1327	.1734	.1864	.1669
	9	.0000	.0000	.0000	.0004	.0033	.0139	.0386	.0794	.1284	.1694	.1855
	10	.0000	.0000	.0000	.0001	.0008	.0042	.0149	.0385	.0771	.1248	.1669
	11	.0000	.0000	.0000	.0000	.0001	.0010	.0046	.0151	.0374	.0742	.1214
	12	.0000	.0000	.0000	.0000	.0000	.0002	.0012	.0047	.0145	.0354	.0708
	13	.0000	.0000	.0000	.0000	.0000	.0000	.0002	.0012	.0045	.0134	.0327
	14	.0000	.0000	.0000	.0000	.0000	.0000	.0000	.0002	.0011	.0039	.0117
	15	.0000	.0000	.0000	.0000	.0000	.0000	.0000	.0000	.0002	.0009	.0031
	16	.0000	.0000	.0000	.0000	.0000	.0000	.0000	.0000	.0000	.0001	.0006
	17	.0000	.0000	.0000	.0000	.0000	.0000	.0000	.0000	.0000	.0000	.0001
	18	.0000	.0000	.0000	.0000	.0000	.0000	.0000	.0000	.0000	.0000	.0000

Table 2 Binomial probabilities (continued)

		p										
n	*x*	.01	.05	.10	.15	.20	.25	.30	.35	.40	.45	.50
19	0	.8262	.3774	.1351	.0456	.0144	.0042	.0011	.0003	.0001	.0000	.0000
	1	.1586	.3774	.2852	.1529	.0685	.0268	.0093	.0029	.0008	.0002	.0000
	2	.0144	.1787	.2852	.2428	.1540	.0803	.0358	.0138	.0046	.0013	.0003
	3	.0008	.0533	.1796	.2428	.2182	.1517	.0869	.0422	.0175	.0062	.0018
	4	.0000	.0112	.0798	.1714	.2182	.2023	.1491	.0909	.0467	.0203	.0074
	5	.0000	.0018	.0266	.0907	.1636	.2023	.1916	.1468	.0933	.0497	.0222
	6	.0000	.0002	.0069	.0374	.0955	.1574	.1916	.1844	.1451	.0949	.0518
	7	.0000	.0000	.0014	.0122	.0443	.0974	.1525	.1844	.1797	.1443	.0961
	8	.0000	.0000	.0002	.0032	.0166	.0487	.0981	.1489	.1797	.1771	.1442
	9	.0000	.0000	.0000	.0007	.0051	.0198	.0514	.0980	.1464	.1771	.1762
	10	.0000	.0000	.0000	.0001	.0013	.0066	.0220	.0528	.0976	.1449	.1762
	11	.0000	.0000	.0000	.0000	.0003	.0018	.0077	.0233	.0532	.0970	.1442
	12	.0000	.0000	.0000	.0000	.0000	.0004	.0022	.0083	.0237	.0529	.0961
	13	.0000	.0000	.0000	.0000	.0000	.0001	.0005	.0024	.0085	.0233	.0518
	14	.0000	.0000	.0000	.0000	.0000	.0000	.0001	.0006	.0024	.0082	.0222
	15	.0000	.0000	.0000	.0000	.0000	.0000	.0000	.0001	.0005	.0022	.0074
	16	.0000	.0000	.0000	.0000	.0000	.0000	.0000	.0000	.0001	.0005	.0018
	17	.0000	.0000	.0000	.0000	.0000	.0000	.0000	.0000	.0000	.0001	.0003
	18	.0000	.0000	.0000	.0000	.0000	.0000	.0000	.0000	.0000	.0000	.0000
	19	.0000	.0000	.0000	.0000	.0000	.0000	.0000	.0000	.0000	.0000	.0000
20	0	.8179	.3585	.1216	.0388	.0115	.0032	.0008	.0002	.0000	.0000	.0000
	1	.1652	.3774	.2702	.1368	.0576	.0211	.0068	.0020	.0005	.0001	.0000
	2	.0159	.1887	.2852	.2293	.1369	.0669	.0278	.0100	.0031	.0008	.0002
	3	.0010	.0596	.1901	.2428	.2054	.1339	.0716	.0323	.0123	.0040	.0011
	4	.0000	.0133	.0898	.1821	.2182	.1897	.1304	.0738	.0350	.0139	.0046
	5	.0000	.0022	.0319	.1028	.1746	.2023	.1789	.1272	.0746	.0365	.0148
	6	.0000	.0003	.0089	.0454	.1091	.1686	.1916	.1712	.1244	.0746	.0370
	7	.0000	.0000	.0020	.0160	.0545	.1124	.1643	.1844	.1659	.1221	.0739
	8	.0000	.0000	.0004	.0046	.0222	.0609	.1144	.1614	.1797	.1623	.1201
	9	.0000	.0000	.0001	.0011	.0074	.0271	.0654	.1158	.1597	.1771	.1602
	10	.0000	.0000	.0000	.0002	.0020	.0099	.0308	.0686	.1171	.1593	.1762
	11	.0000	.0000	.0000	.0000	.0005	.0030	.0120	.0336	.0710	.1185	.1602
	12	.0000	.0000	.0000	.0000	.0001	.0008	.0039	.0136	.0355	.0727	.1201
	13	.0000	.0000	.0000	.0000	.0000	.0002	.0010	.0045	.0146	.0366	.0739
	14	.0000	.0000	.0000	.0000	.0000	.0000	.0002	.0012	.0049	.0150	.0370
	15	.0000	.0000	.0000	.0000	.0000	.0000	.0000	.0003	.0013	.0049	.0148
	16	.0000	.0000	.0000	.0000	.0000	.0000	.0000	.0000	.0003	.0013	.0046
	17	.0000	.0000	.0000	.0000	.0000	.0000	.0000	.0000	.0000	.0002	.0011
	18	.0000	.0000	.0000	.0000	.0000	.0000	.0000	.0000	.0000	.0000	.0002
	19	.0000	.0000	.0000	.0000	.0000	.0000	.0000	.0000	.0000	.0000	.0000
	20	.0000	.0000	.0000	.0000	.0000	.0000	.0000	.0000	.0000	.0000	.0000

Table 2 Binomial probabilities (continued)

		p										
n	*x*	.01	.05	.10	.15	.20	.25	.30	.35	.40	.45	.50
25	0	.7778	.2774	.0718	.0172	.0038	.0008	.0001	.0000	.0000	.0000	.0000
	1	.1964	.3650	.1994	.0759	.0236	.0063	.0014	.0003	.0000	.0000	.0000
	2	.0238	.2305	.2659	.1607	.0708	.0251	.0074	.0018	.0004	.0001	.0000
	3	.0018	.0930	.2265	.2174	.1358	.0641	.0243	.0076	.0019	.0004	.0001
	4	.0001	.0269	.1384	.2110	.1867	.1175	.0572	.0224	.0071	.0018	.0004
	5	.0000	.0060	.0646	.1564	.1960	.1645	.1030	.0506	.0199	.0063	.0016
	6	.0000	.0010	.0239	.0920	.1633	.1828	.1472	.0908	.0442	.0172	.0053
	7	.0000	.0001	.0072	.0441	.1108	.1654	.1712	.1327	.0800	.0381	.0143
	8	.0000	.0000	.0018	.0175	.0623	.1241	.1651	.1607	.1200	.0701	.0322
	9	.0000	.0000	.0004	.0058	.0294	.0781	.1336	.1635	.1511	.1084	.0609
	10	.0000	.0000	.0000	.0016	.0118	.0417	.0916	.1409	.1612	.1419	.0974
	11	.0000	.0000	.0000	.0004	.0040	.0189	.0536	.1034	.1465	.1583	.1328
	12	.0000	.0000	.0000	.0000	.0012	.0074	.0268	.0650	.1140	.1511	.1550
	13	.0000	.0000	.0000	.0000	.0003	.0025	.0115	.0350	.0760	.1236	.1550
	14	.0000	.0000	.0000	.0000	.0000	.0007	.0042	.0161	.0434	.0867	.1328
	15	.0000	.0000	.0000	.0000	.0000	.0002	.0013	.0064	.0212	.0520	.0974
	16	.0000	.0000	.0000	.0000	.0000	.0000	.0004	.0021	.0088	.0266	.0609
	17	.0000	.0000	.0000	.0000	.0000	.0000	.0001	.0006	.0031	.0115	.0322
	18	.0000	.0000	.0000	.0000	.0000	.0000	.0000	.0001	.0009	.0042	.0143
	19	.0000	.0000	.0000	.0000	.0000	.0000	.0000	.0000	.0002	.0013	.0053
	20	.0000	.0000	.0000	.0000	.0000	.0000	.0000	.0000	.0000	.0001	.0016
	21	.0000	.0000	.0000	.0000	.0000	.0000	.0000	.0000	.0000	.0000	.0004
	22	.0000	.0000	.0000	.0000	.0000	.0000	.0000	.0000	.0000	.0000	.0001
30	0	.7397	.2146	.0424	.0076	.0012	.0002	.0000	.0000	.0000	.0000	.0000
	1	.2242	.3389	.1413	.0404	.0093	.0018	.0003	.0000	.0000	.0000	.0000
	2	.0328	.2586	.2277	.1034	.0337	.0086	.0018	.0003	.0000	.0000	.0000
	3	.0031	.1270	.2361	.1703	.0785	.0269	.0072	.0015	.0003	.0000	.0000
	4	.0002	.0451	.1771	.2028	.1325	.0604	.0208	.0056	.0012	.0002	.0000
	5	.0000	.0124	.1023	.1861	.1723	.1047	.0464	.0157	.0041	.0008	.0001
	6	.0000	.0027	.0474	.1368	.1795	.1455	.0829	.0353	.0115	.0029	.0006
	7	.0000	.0005	.0180	.0828	.1538	.1662	.1219	.0652	.0263	.0081	.0019
	8	.0000	.0001	.0058	.0420	.1106	.1593	.1501	.1009	.0505	.0191	.0055
	9	.0000	.0000	.0016	.0181	.0676	.1298	.1573	.1328	.0823	.0382	.0133
	10	.0000	.0000	.0004	.0067	.0355	.0909	.1416	.1502	.1152	.0656	.0280
	11	.0000	.0000	.0001	.0022	.0161	.0551	.1103	.1471	.1396	.0976	.0509
	12	.0000	.0000	.0000	.0006	.0064	.0291	.0749	.1254	.1474	.1265	.0806
	13	.0000	.0000	.0000	.0001	.0022	.0134	.0444	.0935	.1360	.1433	.1115
	14	.0000	.0000	.0000	.0000	.0007	.0054	.0231	.0611	.1101	.1424	.1354

Table 2 Binomial probabilities (continued)

		p										
n	*x*	.01	.05	.10	.15	.20	.25	.30	.35	.40	.45	.50
	15	.0000	.0000	.0000	.0000	.0002	.0019	.0106	.0351	.0783	.1242	.1445
	16	.0000	.0000	.0000	.0000	.0000	.0006	.0042	.0177	.0489	.0953	.1354
	17	.0000	.0000	.0000	.0000	.0000	.0002	.0015	.0079	.0269	.0642	.1115
	18	.0000	.0000	.0000	.0000	.0000	.0000	.0005	.0031	.0129	.0379	.0806
	19	.0000	.0000	.0000	.0000	.0000	.0000	.0001	.0010	.0054	.0196	.0509
	20	.0000	.0000	.0000	.0000	.0000	.0000	.0000	.0003	.0020	.0088	.0280
	21	.0000	.0000	.0000	.0000	.0000	.0000	.0000	.0001	.0006	.0034	.0133
	22	.0000	.0000	.0000	.0000	.0000	.0000	.0000	.0000	.0002	.0012	.0055
	23	.0000	.0000	.0000	.0000	.0000	.0000	.0000	.0000	.0000	.0003	.0019
	24	.0000	.0000	.0000	.0000	.0000	.0000	.0000	.0000	.0000	.0001	.0006
	25	.0000	.0000	.0000	.0000	.0000	.0000	.0000	.0000	.0000	.0000	.0001
40	0	.6690	.1285	.0148	.0015	.0001	.0000	.0000	.0000	.0000	.0000	.0000
	1	.2703	.2706	.0657	.0106	.0013	.0001	.0000	.0000	.0000	.0000	.0000
	2	.0532	.2777	.1423	.0365	.0065	.0009	.0001	.0000	.0000	.0000	.0000
	3	.0068	.1851	.2003	.0816	.0205	.0037	.0005	.0001	.0000	.0000	.0000
	4	.0006	.0901	.2059	.1332	.0475	.0113	.0020	.0003	.0000	.0000	.0000
	5	.0000	.0342	.1647	.1692	.0854	.0272	.0061	.0010	.0001	.0000	.0000
	6	.0000	.0105	.1068	.1742	.1246	.0530	.0151	.0031	.0005	.0000	.0000
	7	.0000	.0027	.0576	.1493	.1513	.0857	.0315	.0080	.0015	.0002	.0000
	8	.0000	.0006	.0264	.1087	.1560	.1179	.0557	.0179	.0040	.0006	.0001
	9	.0000	.0001	.0104	.0682	.1386	.1397	.0849	.0342	.0095	.0018	.0002
	10	.0000	.0000	.0036	.0373	.1075	.1444	.1128	.0571	.0196	.0047	.0008
	11	.0000	.0000	.0011	.0180	.0733	.1312	.1319	.0838	.0357	.0105	.0021
	12	.0000	.0000	.0003	.0077	.0443	.1057	.1366	.1090	.0576	.0207	.0051
	13	.0000	.0000	.0001	.0029	.0238	.0759	.1261	.1265	.0827	.0365	.0109
	14	.0000	.0000	.0000	.0010	.0115	.0488	.1042	.1313	.1063	.0575	.0211
	15	.0000	.0000	.0000	.0003	.0050	.0282	.0774	.1226	.1228	.0816	.0366
	16	.0000	.0000	.0000	.0001	.0019	.0147	.0518	.1031	.1279	.1043	.0572
	17	.0000	.0000	.0000	.0000	.0007	.0069	.0314	.0784	.1204	.1205	.0807
	18	.0000	.0000	.0000	.0000	.0002	.0029	.0172	.0539	.1026	.1260	.1031
	19	.0000	.0000	.0000	.0000	.0001	.0011	.0085	.0336	.0792	.1194	.1194
	20	.0000	.0000	.0000	.0000	.0000	.0004	.0038	.0190	.0554	.1025	.1254
	21	.0000	.0000	.0000	.0000	.0000	.0001	.0016	.0097	.0352	.0799	.1194
	22	.0000	.0000	.0000	.0000	.0000	.0000	.0006	.0045	.0203	.0565	.1031
	23	.0000	.0000	.0000	.0000	.0000	.0000	.0002	.0019	.0106	.0362	.0807
	24	.0000	.0000	.0000	.0000	.0000	.0000	.0001	.0007	.0050	.0210	.0572
	25	.0000	.0000	.0000	.0000	.0000	.0000	.0000	.0003	.0021	.0110	.0366
	26	.0000	.0000	.0000	.0000	.0000	.0000	.0000	.0001	.0008	.0052	.0211
	27	.0000	.0000	.0000	.0000	.0000	.0000	.0000	.0000	.0003	.0022	.0109
	28	.0000	.0000	.0000	.0000	.0000	.0000	.0000	.0000	.0001	.0008	.0051
	29	.0000	.0000	.0000	.0000	.0000	.0000	.0000	.0000	.0000	.0003	.0021

Table 2 Binomial probabilities (continued)

		p										
n	*x*	**.01**	**.05**	**.10**	**.15**	**.20**	**.25**	**.30**	**.35**	**.40**	**.45**	**.50**
	30	.0000	.0000	.0000	.0000	.0000	.0000	.0000	.0000	.0000	.0001	.0008
	31	.0000	.0000	.0000	.0000	.0000	.0000	.0000	.0000	.0000	.0000	.0002
	32	.0000	.0000	.0000	.0000	.0000	.0000	.0000	.0000	.0000	.0000	.0001
50	0	.6050	.0769	.0052	.0003	.0000	.0000	.0000	.0000	.0000	.0000	.0000
	1	.3056	.2025	.0286	.0026	.0002	.0000	.0000	.0000	.0000	.0000	.0000
	2	.0756	.2611	.0779	.0113	.0011	.0001	.0000	.0000	.0000	.0000	.0000
	3	.0122	.2199	.1386	.0319	.0044	.0004	.0000	.0000	.0000	.0000	.0000
	4	.0015	.1360	.1809	.0661	.0128	.0016	.0001	.0000	.0000	.0000	.0000
	5	.0001	.0658	.1849	.1072	.0295	.0049	.0006	.0000	.0000	.0000	.0000
	6	.0000	.0260	.1541	.1419	.0554	.0123	.0018	.0002	.0000	.0000	.0000
	7	.0000	.0086	.1076	.1575	.0870	.0259	.0048	.0006	.0000	.0000	.0000
	8	.0000	.0024	.0643	.1493	.1169	.0463	.0110	.0017	.0002	.0000	.0000
	9	.0000	.0006	.0333	.1230	.1364	.0721	.0220	.0042	.0005	.0000	.0000
	10	.0000	.0001	.0152	.0890	.1398	.0985	.0386	.0093	.0014	.0001	.0000
	11	.0000	.0000	.0061	.0571	.1271	.1194	.0602	.0182	.0035	.0004	.0000
	12	.0000	.0000	.0022	.0328	.1033	.1294	.0838	.0319	.0076	.0011	.0001
	13	.0000	.0000	.0007	.0169	.0755	.1261	.1050	.0502	.0147	.0027	.0003
	14	.0000	.0000	.0002	.0079	.0499	.1110	.1189	.0714	.0260	.0059	.0008
	15	.0000	.0000	.0001	.0033	.0299	.0888	.1223	.0923	.0415	.0116	.0020
	16	.0000	.0000	.0000	.0013	.0164	.0648	.1147	.1088	.0606	.0207	.0044
	17	.0000	.0000	.0000	.0005	.0082	.0432	.0983	.1171	.0808	.0339	.0087
	18	.0000	.0000	.0000	.0001	.0037	.0264	.0772	.1156	.0987	.0508	.0160
	19	.0000	.0000	.0000	.0000	.0016	.0148	.0558	.1048	.1109	.0700	.0270
	20	.0000	.0000	.0000	.0000	.0006	.0077	.0370	.0875	.1146	.0888	.0419
	21	.0000	.0000	.0000	.0000	.0002	.0036	.0227	.0673	.1091	.1038	.0598
	22	.0000	.0000	.0000	.0000	.0001	.0016	.0128	.0478	.0959	.1119	.0788
	23	.0000	.0000	.0000	.0000	.0000	.0006	.0067	.0313	.0778	.1115	.0960
	24	.0000	.0000	.0000	.0000	.0000	.0002	.0032	.0190	.0584	.1026	.1080
	25	.0000	.0000	.0000	.0000	.0000	.0001	.0014	.0106	.0405	.0873	.1123
	26	.0000	.0000	.0000	.0000	.0000	.0000	.0006	.0055	.0259	.0687	.1080
	27	.0000	.0000	.0000	.0000	.0000	.0000	.0002	.0026	.0154	.0500	.0960
	28	.0000	.0000	.0000	.0000	.0000	.0000	.0001	.0012	.0084	.0336	.0788
	29	.0000	.0000	.0000	.0000	.0000	.0000	.0000	.0005	.0043	.0208	.0598
	30	.0000	.0000	.0000	.0000	.0000	.0000	.0000	.0002	.0020	.0119	.0419
	31	.0000	.0000	.0000	.0000	.0000	.0000	.0000	.0001	.0009	.0063	.0270
	32	.0000	.0000	.0000	.0000	.0000	.0000	.0000	.0000	.0003	.0031	.0160
	33	.0000	.0000	.0000	.0000	.0000	.0000	.0000	.0000	.0001	.0014	.0087
	34	.0000	.0000	.0000	.0000	.0000	.0000	.0000	.0000	.0000	.0006	.0044
	35	.0000	.0000	.0000	.0000	.0000	.0000	.0000	.0000	.0000	.0002	.0020
	36	.0000	.0000	.0000	.0000	.0000	.0000	.0000	.0000	.0000	.0001	.0008
	37	.0000	.0000	.0000	.0000	.0000	.0000	.0000	.0000	.0000	.0000	.0003
	38	.0000	.0000	.0000	.0000	.0000	.0000	.0000	.0000	.0000	.0000	.0001

Table 3 Standardized normal distribution

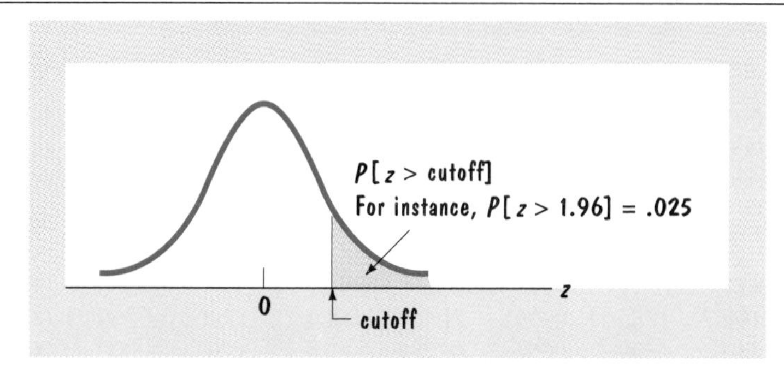

CUTOFF	.00	.01	.02	.03	.04	.05	.06	.07	.08	.09
.0	.5000	.4960	.4920	.4880	.4840	.4801	.4761	.4721	.4681	.4641
.1	.4602	.4562	.4522	.4483	.4443	.4404	.4364	.4325	.4286	.4247
.2	.4207	.4168	.4129	.4090	.4052	.4013	.3974	.3936	.3897	.3859
.3	.3821	.3783	.3745	.3707	.3669	.3632	.3594	.3557	.3520	.3483
.4	.3446	.3409	.3372	.3336	.3300	.3264	.3228	.3192	.3156	.3121
.5	.3085	.3050	.3015	.2981	.2946	.2912	.2877	.2843	.2810	.2776
.6	.2743	.2709	.2676	.2643	.2611	.2578	.2546	.2514	.2483	.2451
.7	.2420	.2389	.2358	.2327	.2296	.2266	.2236	.2206	.2177	.2148
.8	.2119	.2090	.2061	.2033	.2005	.1977	.1949	.1922	.1894	.1867
.9	.1841	.1814	.1788	.1762	.1736	.1711	.1685	.1660	.1635	.1611
1.0	.1587	.1562	.1539	.1515	.1492	.1469	.1446	.1423	.1401	.1379
1.1	.1357	.1335	.1314	.1292	.1271	.1251	.1230	.1210	.1190	.1170
1.2	.1151	.1131	.1112	.1093	.1075	.1056	.1038	.1020	.1003	.0985
1.3	.0968	.0951	.0934	.0918	.0901	.0885	.0869	.0853	.0838	.0823
1.4	.0808	.0793	.0778	.0764	.0749	.0735	.0722	.0708	.0694	.0681
1.5	.0668	.0655	.0643	.0630	.0618	.0606	.0594	.0582	.0571	.0559
1.6	.0548	.0537	.0526	.0516	.0505	.0495	.0485	.0475	.0465	.0455
1.7	.0446	.0436	.0427	.0418	.0409	.0401	.0392	.0384	.0375	.0367
1.8	.0359	.0352	.0344	.0336	.0329	.0322	.0314	.0307	.0301	.0294
1.9	.0287	.0281	.0274	.0268	.0262	.0256	.0250	.0244	.0239	.0233
2.0	.0228	.0222	.0217	.0212	.0207	.0202	.0197	.0192	.0188	.0183
2.1	.0179	.0174	.0170	.0166	.0162	.0158	.0154	.0150	.0146	.0143
2.2	.0139	.0136	.0132	.0129	.0125	.0122	.0119	.0116	.0113	.0110
2.3	.0107	.0104	.0102	.0099	.0096	.0094	.0091	.0089	.0087	.0084
2.4	.0082	.0080	.0078	.0075	.0073	.0071	.0069	.0068	.0066	.0064
2.5	.0062	.0060	.0059	.0057	.0055	.0054	.0052	.0051	.0049	.0048
2.6	.0047	.0045	.0044	.0043	.0041	.0040	.0039	.0038	.0037	.0036
2.7	.0035	.0034	.0033	.0032	.0031	.0030	.0029	.0028	.0027	.0026
2.8	.0026	.0025	.0024	.0023	.0023	.0022	.0021	.0021	.0020	.0019
2.9	.0019	.0018	.0017	.0017	.0016	.0016	.0015	.0015	.0014	.0014
3.0	.0013	.0013	.0013	.0012	.0012	.0011	.0011	.0011	.0010	.0010

Table 3 Standardized normal distribution (continued)

CUTOFF	.00	.01	.02	.03	.04	.05	.06	.07	.08	.09
4.0	.000 031 7									
5.0	.000 000 287									
6.0	.000 000 000 987									
7.0	.000 000 000 001 28									
8.0	.000 000 000 000 001									

Table 4 Student's *t* distribution

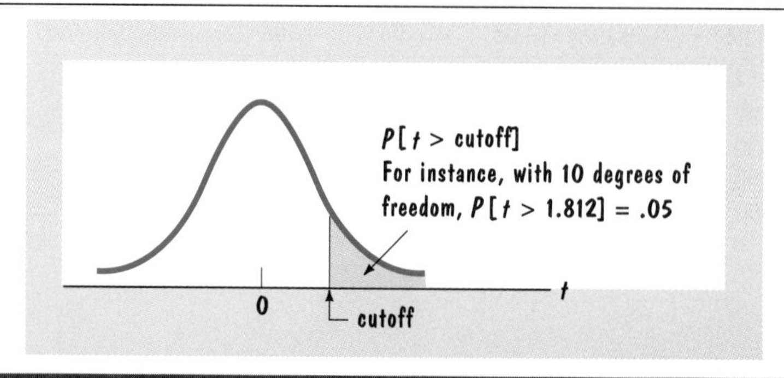

	PROBABILITY OF A *t* VALUE LARGER THAN INDICATED CUTOFF				
DEGREES OF FREEDOM	.10	.05	.025	.01	.005
1	3.078	6.314	12.706	31.821	63.657
2	1.886	2.920	4.303	6.965	9.925
3	1.638	2.353	3.182	4.541	5.841
4	1.533	2.132	2.776	3.747	4.604
5	1.476	2.015	2.571	3.365	4.032
6	1.440	1.943	2.447	3.143	3.707
7	1.415	1.895	2.365	2.998	3.499
8	1.397	1.860	2.306	2.896	3.355
9	1.383	1.833	2.262	2.821	3.250
10	1.372	1.812	2.228	2.764	3.169
11	1.363	1.796	2.201	2.718	3.106
12	1.356	1.782	2.179	2.681	3.055
13	1.350	1.771	2.160	2.650	3.012
14	1.345	1.761	2.145	2.624	2.977
15	1.341	1.753	2.131	2.602	2.947
16	1.337	1.746	2.120	2.583	2.921
17	1.333	1.740	2.110	2.567	2.898
18	1.330	1.734	2.101	2.552	2.878
19	1.328	1.729	2.093	2.539	2.861
20	1.325	1.725	2.086	2.528	2.845
21	1.323	1.721	2.080	2.518	2.831
22	1.321	1.717	2.074	2.508	2.819
23	1.319	1.714	2.069	2.500	2.807
24	1.318	1.711	2.064	2.492	2.797
25	1.316	1.708	2.060	2.485	2.787

Table 4 Student's t distribution (continued)

DEGREES OF FREEDOM	PROBABILITY OF A t VALUE LARGER THAN INDICATED CUTOFF				
	.10	**.05**	**.025**	**.01**	**.005**
26	1.315	1.706	2.056	2.479	2.779
27	1.314	1.703	2.052	2.473	2.771
28	1.313	1.701	2.048	2.467	2.763
29	1.311	1.699	2.045	2.462	2.756
30	1.310	1.697	2.042	2.457	2.750
40	1.303	1.684	2.021	2.423	2.704
60	1.296	1.671	2.000	2.390	2.660
120	1.289	1.661	1.984	2.358	2.626
∞	1.282	1.645	1.960	2.326	2.576

Table 5 The *F* distribution (5 percent probabilities)

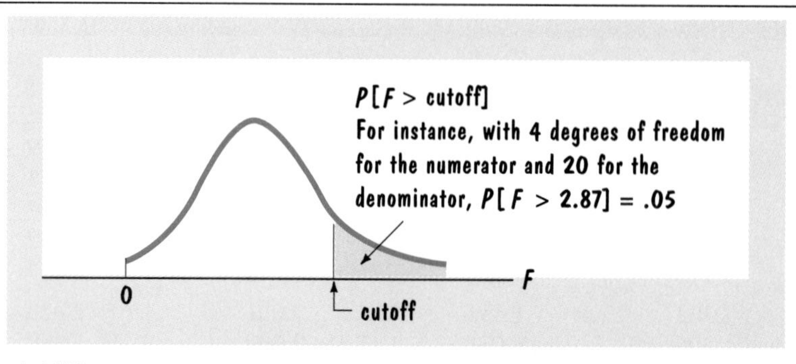

Cutoffs for 5 percent probabilities

DEGREES OF FREEDOM FOR DENOMINATOR	DEGREES OF FREEDOM FOR NUMERATOR 1	2	3	4	5	6	7	8	9	10
1	161	200	216	225	230	234	237	239	241	242
2	18.5	19.0	19.2	19.2	19.3	19.3	19.4	19.4	19.4	19.4
3	10.1	9.55	9.28	9.12	9.01	8.94	8.89	8.85	8.81	8.79
4	7.71	6.94	6.59	6.39	6.26	6.16	6.09	6.04	6.00	5.96
5	6.61	5.79	5.41	5.19	5.05	4.95	4.88	4.82	4.77	4.74
6	5.99	5.14	4.76	4.53	4.39	4.28	4.21	4.15	4.10	4.06
7	5.59	4.74	4.35	4.12	3.97	3.87	3.79	3.73	3.68	3.64
8	5.32	4.46	4.07	3.84	3.69	3.58	3.50	3.44	3.39	3.35
9	5.12	4.26	3.86	3.63	3.48	3.37	3.29	3.23	3.18	3.14
10	4.96	4.10	3.71	3.48	3.33	3.22	3.14	3.07	3.02	2.98
11	4.84	3.98	3.59	3.36	3.20	3.09	3.01	2.95	2.90	2.85
12	4.75	3.89	3.49	3.26	3.11	3.00	2.91	2.85	2.80	2.75
13	4.67	3.81	3.41	3.18	3.03	2.92	2.83	2.77	2.71	2.67
14	4.60	3.74	3.34	3.11	2.96	2.85	2.76	2.70	2.65	2.60
15	4.54	3.68	3.29	3.06	2.90	2.79	2.71	2.64	2.59	2.54
16	4.49	3.63	3.24	3.01	2.85	2.74	2.66	2.59	2.54	2.49
17	4.45	3.59	3.20	2.96	2.81	2.70	2.61	2.55	2.49	2.45
18	4.41	3.55	3.16	2.93	2.77	2.66	2.58	2.51	2.46	2.41
19	4.38	3.52	3.13	2.90	2.74	2.63	2.54	2.48	2.42	2.38
20	4.35	3.49	3.10	2.87	2.71	2.60	2.51	2.45	2.39	2.35
25	4.24	3.39	2.99	2.76	2.60	2.49	2.40	2.34	2.28	2.24
30	4.17	3.32	2.92	2.69	2.53	2.42	2.33	2.27	2.21	2.16
40	4.08	3.23	2.84	2.61	2.45	2.34	2.25	2.18	2.12	2.08
60	4.00	3.15	2.76	2.53	2.37	2.25	2.17	2.10	2.04	1.99
120	3.92	3.07	2.68	2.45	2.29	2.18	2.09	2.02	1.96	1.91
∞	3.84	3.00	2.60	2.37	2.21	2.10	2.01	1.94	1.88	1.83

The F distribution (1 percent probabilities)

Cutoffs for 1 percent probabilities

	DEGREES OF FREEDOM FOR NUMERATOR									
DEGREES OF FREEDOM FOR DENOMINATOR	1	2	3	4	5	6	7	8	9	10
1	4052	5000	5403	5625	5764	5859	5928	5982	6022	6056
2	98.5	99.0	99.2	99.2	99.3	99.3	99.4	99.4	99.4	99.4
3	34.1	30.8	29.5	28.7	28.2	27.9	27.7	27.5	27.3	27.2
4	21.2	18.0	16.7	16.0	15.5	15.2	15.0	14.8	14.7	14.5
5	16.3	13.3	12.1	11.4	11.0	10.7	10.5	10.3	10.2	10.1
6	13.7	10.9	9.78	9.15	8.75	8.47	8.26	8.10	7.98	7.87
7	12.2	9.55	8.45	7.85	7.46	7.19	6.99	6.84	6.72	6.62
8	11.3	8.65	7.59	7.01	6.63	6.37	6.18	6.03	5.91	5.81
9	10.6	8.02	6.99	6.42	6.06	5.80	5.61	5.47	5.35	5.26
10	10.0	7.56	6.55	5.99	5.64	5.39	5.20	5.06	4.94	4.85
11	9.65	7.21	6.22	5.67	5.32	5.07	4.89	4.74	4.63	4.54
12	9.33	6.93	5.95	5.41	5.06	4.82	4.64	4.50	4.39	4.30
13	9.07	6.70	5.74	5.21	4.86	4.62	4.44	4.30	4.19	4.10
14	8.86	6.51	5.56	5.04	4.70	4.46	4.28	4.14	4.03	3.94
15	8.68	6.36	5.42	4.89	4.56	4.32	4.14	4.00	3.89	3.80
16	8.53	6.23	5.29	4.77	4.44	4.20	4.03	3.89	3.78	3.69
17	8.40	6.11	5.19	4.67	4.34	4.10	3.93	3.79	3.68	3.59
18	8.29	6.01	5.09	4.58	4.25	4.01	3.84	3.71	3.60	3.51
19	8.19	5.93	5.01	4.50	4.17	3.94	3.77	3.63	3.52	3.43
20	8.10	5.85	4.94	4.43	4.10	3.87	3.70	3.56	3.46	3.37
25	7.77	5.57	4.68	4.18	3.86	3.63	3.46	3.32	3.22	3.13
30	7.56	5.39	4.51	4.02	3.70	3.47	3.30	3.17	3.07	2.98
40	7.31	5.18	4.31	3.83	3.51	3.29	3.12	2.99	2.89	2.80
60	7.08	4.98	4.13	3.65	3.34	3.12	2.95	2.82	2.72	2.63
120	6.85	4.79	3.95	3.48	3.17	2.96	2.79	2.66	2.56	2.47
∞	6.63	4.61	3.78	3.32	3.02	2.80	2.64	2.51	2.41	2.32

Table 6 Chi-square distribution

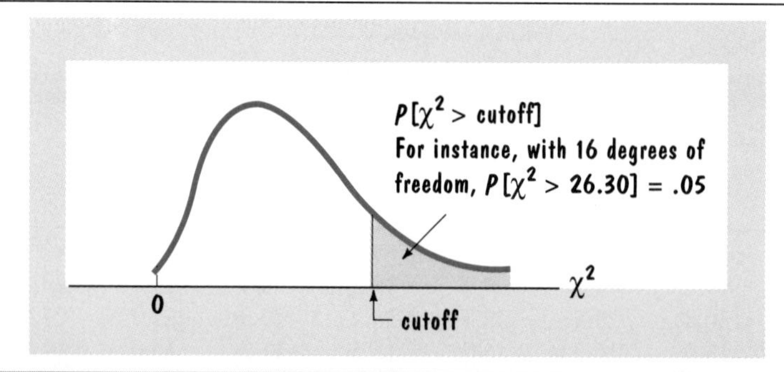

DEGREES OF FREEDOM	PROBABILITY OF A CHI-SQUARE VALUE LARGER THAN INDICATED CUTOFF				
	.10	.05	.025	.01	.005
1	2.71	3.84	5.02	6.63	7.88
2	4.61	5.99	7.38	9.21	10.60
3	6.25	7.81	9.35	11.34	12.84
4	7.78	9.49	11.14	13.28	14.86
5	9.24	11.07	12.83	15.09	16.75
6	10.64	12.59	14.45	16.81	18.55
7	12.02	14.07	16.01	18.48	20.28
8	13.36	15.51	17.53	20.09	21.95
9	14.68	16.92	19.02	21.67	23.59
10	15.99	18.31	20.48	23.21	25.19
11	17.28	19.68	21.92	24.73	26.76
12	18.55	21.03	23.34	26.22	28.30
13	19.81	22.36	24.74	27.69	29.82
14	21.06	23.68	26.12	29.14	31.32
15	22.31	25.00	27.49	30.58	32.80
16	23.54	26.30	28.85	32.00	34.27
17	24.77	27.59	30.19	33.41	35.72
18	25.99	28.87	31.53	34.81	37.16
19	27.20	30.14	32.85	36.19	38.58
20	28.41	31.41	34.17	37.57	40.00
21	29.62	32.67	35.48	38.93	41.40
22	30.81	33.92	36.78	40.29	42.80
23	32.01	35.17	38.08	41.64	44.18
24	33.20	36.42	39.36	42.98	45.56
25	34.38	37.65	40.65	44.31	46.93

Table 6 Chi-square distribution (continued)

DEGREES OF FREEDOM	PROBABILITY OF A CHI-SQUARE VALUE LARGER THAN INDICATED CUTOFF				
	.10	**.05**	**.025**	**.01**	**.005**
26	35.56	38.89	41.92	45.64	48.29
27	36.74	40.11	43.19	46.96	49.64
28	37.92	41.34	44.46	48.28	50.99
29	39.09	42.56	45.72	49.59	52.34
30	40.26	43.77	46.98	50.89	53.67
40	51.81	55.76	59.34	63.69	66.77
50	63.17	67.50	71.42	76.15	79.49
60	74.40	79.08	83.30	88.38	91.95
70	85.53	90.53	95.02	100.43	104.22
80	96.58	101.88	106.63	112.33	116.32
90	107.57	113.15	118.14	124.12	128.30
100	118.50	124.34	129.56	135.81	140.17

references

chapter 1 introduction

1. For example, Edward O. Thorpe, *Beat the Dealer,* rev. ed., New York: Random House, 1966; Maxine Brady, *The Monopoly Book,* New York: David McKay, 1976; Oswald Jacoby, *Oswald Jacoby on Poker,* New York: Doubleday, 1981; George Lindsey, "An Investigation of Strategies in Baseball," *Operations Research,* 1963, vol. 11, pp. 477–501; and Shaul P. Ladany and Robert E. Machol, *Optimal Strategies in Sports,* Amsterdam: North-Holland, 1977.

chapter 2 displaying data

1. U.S. Department of Commerce, *Statistical Abstract of the United States,* Washington: Government Printing Office (GPO), 1996, Table 415.
2. *Statistical Abstract of the United States,* 1996, Table 467.
3. Harold R. Jacobs, *Mathematics: A Human Endeavor,* San Francisco: W. H. Freeman, 1982, p. 562.
4. Tracy Tseng, Marina Apotovsky, and Karen Crowley, "Family Plans of Pomona Seniors," Pomona College, Claremont, CA, Fall 1996.
5. *Statistical Abstract of the United States,* 1996, Tables 27, 1184.
6. James Gleick, "Hole in Ozone over South Pole Worries Scientists," *The New York Times,* July 29, 1986, p. C1.
7. A. Quetelet, Lêttres S.A.R. le Duc Régnant de Saxe-Cobourg et Gotha, *Sur la Théorie des Probabilités, Appliqué aux Sciences Morales et Politiques,* Brussels: M. Hayes, 1846, p. 400.
8. Edward R. Tufte, *The Visual Display of Quantitative Information,* Cheshire, CT: Graphics Press, 1983.
9. Arthur H. Miller, Edie N. Goldenberg, and Lutz Erbring, "Type-Set Politics: Impact of Newspapers on Public Confidence," *American Political Science Review,* 1979, vol. 73, pp. 67–84.
10. Jacobs, *Mathematics: A Human Endeavor,* p. 539, reproduces this graph with the label "The Simple Truth."
11. John Tukey, "Some Graphic and Semigraphic Displays," in T. A. Bancroft (ed.), *Statistical Papers in Honor of George W. Snedcor,* Ames, IA: Iowa State University Press, 1972, p. 296.
12. For a thorough explanation of stem-and-leaf diagrams and other tools of exploratory data analysis, see John Tukey, *Exploratory Data Analysis,* Reading, MA: Addison-Wesley, 1977.
13. Frederick Mosteller and David L. Wallace, *Inference and Disputed Authorship: The Federalist,* Reading, MA: Addison-Wesley, 1964.
14. Philip J. Boland and Michael Proschan, "The Use of Statistical Evidence in Allegations of Exam Cheating," *Chance,* Summer 1990, pp. 10–14.
15. Rich Hutchinson, Yellowstone National Park's research geologist, kindly provided these data.
16. *Collier's* April 24, 1953 newspaper advertisement. From Darrell Huff, *How to Lie with Statistics,* New York: W. W. Norton, 1954, p. 65.
17. William Playfair, *The Commercial and Political Atlas,* London: John Stockdale, 1786, pp. 3–4.
18. These data are from various issues of *The Wall Street Journal.*
19. *Trenton Times,* March 24, 1989; cited in Chamont Wang, *Sense and Nonsense of Statistical Inference,* New York: Marcel Dekker, 1993, p. 170.
20. Robert Stansfield, *The New York Times,* February 21, 1971.
21. Tom Petruno, "Getting a Clue on When to Buy and When to Bail Out," *Los Angeles Times,* May 16, 1990.
22. Thomas L. Friedman, "That Numberless Presidential Chart," *The New York Times,* August 2, 1981; this article also compares the President's

graphic with a figure with an origin and labeled axes.

23. National Science Foundation, *Science Indicators, 1974,* Washington: General Accounting Office, 1976, p. 15.
24. *Washington Post,* 1976.
25. Associated Press, 1974, from David S. Moore, *Statistics: Concepts and Controversies,* New York: W. H. Freeman, 1979, p. 151.
26. Mark Twain, *Life on the Mississippi,* New York: Harper, 1874.
27. *Time,* February 11, 1980, p. 3.
28. W. Vance Grant and C. George Lind, *Digest of Education Statistics, 1979,* U.S. Department of Health, Education, and Welfare, National Center for Education Statistics, 1979, p. 97.
29. Edward R. Tufte, *The Visual Display of Quantitative Information,* Cheshire, CT: Graphics Press, 1983, p. 118.
30. Ivan Valiela, Paulette Peckol, Charlene D'Avanzo, Kate Lajtha, James Kremer, W. Rockwell Geyer, Ken Foreman, Douglas Hersh, Brad Seely, Tatsu Isajl, and Rick Crawford "Hurricane Bob on Cape Cod," *American Scientist,* March–April 1996, pp. 154–165; and personal correspondence with Prof. Valiela.
31. U.S. Department of Commerce, *Statistical Abstract of the United States,* Washington: GPO, various issues.
32. *Statistical Abstract of the United States,* various issues.
33. J. M. Barnola, D. Raynaud, C. Lorius, and Y. S. Korotkevich, "Historical CO_2 Record from the Vostok Ice Core," in T. A. Boden, D. P. Kaiser, R. J. Sepanski, and F. W. Stoss (eds.), *Trends '93: A Compendium of Global Change,* Oak Ridge, TN: Carbon Dioxide Information Analysis Center, Oak Ridge National Laboratory, 1993, pp. 7–10.
34. Brian J. Whipp and Susan A. Ward, "Will Women Soon Outrun Men?" *Nature,* January 2, 1992, p. 25.
35. These data are based on R. Doll, "Etiology of Lung Cancer," *Advances in Cancer Research,* 1955, pp. 1–50, and were used in the Report of the Advisory Committee to the Surgeon General, *Smoking and Health,* Washington: U.S. Department of Health, Education, and Welfare, Public Health Service, 1964, p. 176.
36. *Statistical Abstract of the United States,* 1996, Table 263, p. 172.
37. These data are from Wiesenberger Investment Companies Service, *Investment Companies,* New York: Arthur Wiesenberger & Co., 1987, pp. 154–157.
38. "Running Shoes," *Consumer Reports,* May 1995, pp. 313–317.
39. *The Way Things Work Book of the Computer: An Illustrated Encyclopedia of Information Science, Cybernetics, and Data Processing,* New York: Simon & Schuster, 1974.
40. Neva Kerbeshian, "The Sophomore Slump," Pomona College, Claremont, CA, spring 1992.
41. R. J. Hoyle, "Decline of Language as a Medium of Communication," in George H. Scherr (ed.), *The Best of the Journal of Irreproducible Results,* New York: Workman, 1983, pp. 134–135.
42. "Air Travel Boom Makes Agents Fume," *The New York Times,* August 8, 1978.
43. National Institute on Drug Abuse, National Household Survey, *Statistical Abstract of the United States,* Washington: GPO, 1995, p. 143. I closed the oldest interval, 35–64, which is open-ended in the *Statistical Abstract.*
44. Tushar Atre, Kristine Auns, Kurt Badenhausen, Kevin McAuliffe, Christopher Nikolov, and Michael K. Ozanian, "The High-Stakes Game of Team Ownership," *Financial World,* May 20, 1996, 52–68.
45. David Frum, *The New York Times,* August 14, 1995; reprinted in *Reader's Digest,* December 1995, p. 123.
46. "What We Earn," *The Honolulu Advertiser,* March 12, 1995.
47. These data are from "Getting Calcium without the Fat," *Consumer Reports,* May 1996, pp. 48–49.
48. Adapted from a graph in the Business section of the *Los Angeles Times,* January 28, 1997. The original graph had these labels on the horizontal axis: 9:30–10, 11, noon, 1, 2, 3, 3–5:15.

49. "Inflation to Smile about," *Los Angeles Times,* January 11, 1988.

50. Jackie Spinner, "Appealing to the Time Deprived," *Washington Post,* December 18, 1995, p. B1.

51. Christopher B. Utgaard, "A Statistical Analysis of Pomona College Relationships," Pomona College, Claremont, CA, Spring 1994.

52. Richard Benedetto and Susan Page, "Poll: 51% Hold Poor Opinion of First Lady," *USA Today,* January 17, 1996.

53. Belinda Lees, Theya Molleson, Timothy R. Arnett, and John C. Stevenson, "Differences in Proximal Femur Bone Density over Two Centuries," *The Lancet,* March 13, 1993, pp. 673–675. I am grateful to them for sharing their raw data with me.

54. "The Merrill Lynch Performance Difference," advertisement in *The Wall Street Journal,* January 9, 1996.

55. Lynn Rapaport, "The Cultural and Material Reconstruction of the Jewish Communities in the Federal Republic of Germany," *Jewish Social Studies,* Spring 1987, pp. 137–154.

56. Jose P. Peixoto and M. Ali Kettani, "The Control of the Water Cycle," *Scientific American,* April 1973, pp. 46–61.

57. Ann Landers, *Cape Cod Times,* June 9, 1996.

chapter 3 **summarizing data**

1. William A. Spurr and Charles P. Bonini, *Statistical Analysis for Business Decisions,* rev. ed., Homewood, IL: Irwin, 1973, p. 219.

2. Quoted in Jerry Kirshenbaum, "Scorecard," *Sports Illustrated,* February 1984, p. 16.

3. F. Y. Edgeworth, "The Choice of Means," *Philosophical Magazine,* 1887, p. 269.

4. Arnold Barnett, "How Numbers Can Trick You," *Technology Review,* October 1994, p. 42.

5. P. J. Bickel, E. A. Hammel, and J. W. O'Connell, "Sex Bias in Graduate Admissions: Data from Berkeley," *Science,* February 7, 1975.

6. Dean E. Murphy, "Wondering Why the Rainfall Isn't Normal? Maybe Because It Never Is," *Los Angeles Times,* February 24, 1993.

7. Ann Landers, *Boston Globe,* August 14, 1976.

8. S. M. Stigler, "Do Robust Estimators Work with Real Data?" *Annals of Statistics,* 1977, vol. 4, pp. 1055–1078. As explained by Stigler, Cavendish made an additional six measurements, one of which was misrecorded, before he replaced the suspension wire and made the 13 measurements shown in this exercise.

9. Marilyn vos Savant, "Ask Marilyn," *Parade,* July 25, 1993.

10. Francis Galton, *Natural Inheritance,* London: Macmillan, 1889.

11. These data are from "Can Fast Food be Good Food?" *Consumer Reports,* August 1994, pp. 493–498.

12. These data are from U.S. Geological Survey, *The Yellowstone Catalog,* Washington: U.S. Government Printing Office, 1996.

13. William D. Nordhaus, "Expert Opinion on Climatic Change," *American Scientist,* January/February 1994, pp. 45–51. The original graph was vertical, rather than horizontal.

14. These data are from Tushar Atre, Kristine Auns, Kurt Badenhausen, Kevin McAuliffe, Christopher Nikolov, and Michael K. Ozanian, "The High-Stakes Game of Team Ownership," *Financial World,* May 20, 1996, pp. 52–68.

15. *Newsweek,* January 16, 1967, p. 6.

16. Milt Freudenheim, "Survey Finds Health Costs Rose in '95," *The New York Times,* January 30, 1996; Ellen Neuborne, "Health Care Costs Slowing for Employers," *USA Today,* January 30, 1996, p. B1; Ron Winslow, "Employee Health-Care Costs Are Steady," *The Wall Street Journal,* January 30, 1996, p. A2.

17. Mary Ann Mason, *The Equality Trap,* New York: Simon & Schuster, 1988, p. 19.

18. Victor R. Fuchs, "Sex Differences in Economic Well-Being," *Science,* April 1986, p. 460.

19. W. Allen Wallis and Harry V. Roberts, *Statistics: A New Approach,* Glencoe, IL: Free Press, 1956, p. 83.

20. *The New York Times,* February 22, 1933.

21. Susan Milton, "Wellfleet the Victim in Statistical Murder Mystery," *Cape Cod Times,* December 12, 1994.

22. H. Schmid, "Kaposi's Sarcoma in Tanzania: A Statistical Study of 220 Cases," *Tropical Geographical Medicine,* 1973, vol. 25, pp. 266–276.
23. David Dempsey, "In and out of Books," *The New York Times Book Review,* April 29, 1951, p. 8.
24. *Markey vs. Tenneco Oil Company,* 707 *F.2d* 172 (5th Cir. 1983).
25. *Hamilton vs. United States,* 309 *F. Supp.* 468 (S.D.N.Y. 1969).
26. Claremont Courier, *College New Mag,* November 1989.
27. John M. Chambers, *Graphical Methods for Data Analysis,* Belmont, CA: Wadsworth International Group, 1983, p. 350.
28. Joel E. Cohen, "An Uncertainty Principle in Demography and the Unisex Issue," *The American Statistician,* February 1986, pp. 32–39.
29. John A. Johnson, "Sharing Some Ideas," *Cape Cod Times,* July 12, 1984.
30. David Barnes, *Statistics As Proof,* Boston: Little, Brown, 1983, p. 18.
31. Stigler, "Do Robust Estimators Work with Real Data?"
32. Edward Yardeni, *Money & Business Alert,* Prudential-Bache Securities, November 11, 1988.
33. Robert M. Wood, "Giant Discoveries of Future Science," *Virginia Journal of Science,* 1970, vol. 21, pp. 169–177.
34. Gertrude E. Pearse, "On Corrections for the Moment-Coefficients of Frequency Distributions When There Are Infinite Ordinates at One or Both of the Terminals of the Range," *Biometrika,* vol. 20A, 1928, pp. 314–355.
35. Letter to Marilyn vos Savant, "Ask Marilyn," *Parade Magazine,* June 11, 1995.
36. Samantha Woirol, "A Change in Perspective: Average Class Size," Pomona College, spring 1993.

chapter 4 **producing data**

1. R. Clay Sprowls, "The Admissibility of Sample Data into a Court of Law: A Case History," *UCLA Law Review,* vol. 4, 1957, pp. 222–232.
2. *Physicians Union vs. Miller,* 675 *F.2d* 151 (7th Circuit 1982).
3. This example is discussed by John Neter "How Accounts Save Money by Sampling," in Judith Tanur, Frederic Mosteller, William H. Kruskal, Richard F. Link, Richard S. Pieters, and Gerald R. Rising, *Statistics: A Guide to the Unknown,* San Francisco: Holden-Day, 1972, pp. 203–211.
4. Eugene Carlson, "A. C. Nielsen Took the Measure of the Crowd," *The Wall Street Journal,* May 16, 1989.
5. Many of these details are from Michael Wheeler, *Lies, Damn Lies, and Statistics,* New York: Norton, 1976, chap. 10.
6. R. Ruggles and H. Brodie, "An Empirical Approach to Economic Intelligence in World War II," *Journal of the American Statistical Association,* vol. 42, 1947, pp. 72–91.
7. This example is from Tom Moore, who uses it as a laboratory assignment in his introductory physics course at Pomona College.
8. Catherine M. Viscoli, Mark S. Lachs, and Ralph I. Horowitz, "Bladder Cancer and Coffee Drinking: A Summary of Case-Control Research" *The Lancet,* June 5, 1993, pp. 1432–1437.
9. Associated Press, "College Broadens Not Just Mind, Study Says," *Cape Cod Times,* August 10, 1984.
10. H. S. Diehl, A. B. Baker, and D. W. Cowan, "Cold Vaccines," *Journal of the American Medical Association,* September 24, 1938, vol. 111, pp. 1168–1173.
11. L. L. Miao, "Gastric Freezing: An Example of the Evaluation of Medical Therapy by Randomized Clinical Trials," in J. P. Bunker, B. A. Barnes, and F. Mosteller (eds.), *Costs, Risks and Benefits of Surgery,* New York: Oxford University Press, 1977, pp. 198–211.
12. Advertisement for Seldane, *Sports Illustrated,* March 23, 1996.
13. F. J. Roethlisberger and W. J. Dickson, *Management and the Worker,* Cambridge, MA: Harvard University Press, 1946.
14. Robert Rosenthal and Kermit L. Fode, "The Effect of Experimenter Bias on the Performance of the Albino Rat," *Behavioral Science,* 1963, vol. 8, pp. 183–189.

15. Arnold Barnett, "How Numbers Can Trick You," *Technology Review,* October 1994, p. 40.

16. For more details, see Paul Meier, "The Biggest Public Health Experiment Ever: The 1954 Field Trials of the Salk Poliomyelitis Vaccine," in Judith Tanur, Frederic Mosteller, William H. Kruskal, Richard F. Link, Richard S. Pieters, and Gerald R. Rising, *Statistics: A Guide to the Unknown,* San Francisco: Holden-Day, 1972, pp. 2–13.

17. John Snow, *On the Mode of Communication of Cholera,* 2d ed., London: John Churchill, 1855.

18. Associated Press, "Anger Doubles Risk of Attack of Heart Disease Victims," *The New York Times,* March 19, 1994.

19. W. Allen Wallis and Harry V. Roberts, *Statistics: A New Approach,* New York: Free Press, 1956, pp. 479–480.

20. The RAND Corporation, *A Million Random Digits with 100,000 Normal Deviates,* New York: Free Press, 1955.

21. The details of this case are from Hans Zeisel, "Dr. Spock and the Case of the Vanishing Women Jurors," *University of Chicago Law Review,* Fall 1969, pp. 1–18.

22. Many of the details in this example are from Michael Wheeler, *Lies, Damn Lies, and Statistics,* New York: Norton, 1976, pp. 257–259.

23. Alan Abelson, "Up & Down Wall Street," *Barrons',* September 17, 1990.

24. Gina Kolata, "In Shuffling Cards, 7 Is Winning Number," *The New York Times,* January 9, 1990, p. C1.

25. "Down on the Farm," *Newsweek,* May 2, 1949, pp. 47–48.

26. L. L. Bairds, *The Graduates,* Princeton, NJ: Educational Testing Service, 1973.

27. *Civil Liberties,* December 1952.

28. Naomi Wolf, *Beauty Myth,* New York: Anchor Books, 1992, p. 67.

29. M. C. Bryson, "The Literary Digest: Making of a Statistical Myth," *The American Statistician,* 1976, vol. 30, pp. 184–185, argues that nonresponse bias was more important than the *Literary Digest*'s selection bias.

30. Don A. Dilman, "Token Financial Incentives and the Reduction of Nonresponse Error in Mail Surveys," unpublished, 1996.

31. John Brehm, *The Phantom Respondents: Opinion Surveys and Political Representation,* Ann Arbor, MI: University of Michigan Press, 1993.

32. Brehm, p. 38. Brehm found a difference in telephone surveys and personal interviews: Blacks and those with relatively little education are underrepresented in telephone surveys and overrepresented in face-to-face interviews.

33. Earl W. Kintner, *A Primer on the Law of Deceptive Practices,* New York: Macmillan, 1971, p. 153.

34. "Scorecard," *Sports Illustrated,* August 27, 1984, p. 14.

35. F. F. Stephan and P. J. McCarthy, *Sampling Opinions,* New York: John Wiley & Sons, 1958, p. 286.

36. Wheeler, *Lies, Damn Lies, and Statistics,* p. 88.

37. Wheeler, p. 90.

38. *U.S. News & World Report,* October 21, 1968, p. 28.

39. Kevin Goldman, "As Ad Clutter Becomes Tradition at Super Bowl, Viewer Recall Slips," *The Wall Street Journal,* February 3, 1993.

40. Darrell Huff, *How to Lie with Statistics,* New York: Norton, 1954, p. 24.

41. The numbers and quotations in this example are from Richard Boudreaux and Marjorie Miller, "Sandinistas Conclude They Lost Touch with Populace," *Los Angeles Times,* March 4, 1990.

42. E. F. Loftus and J. C. Palmer, "Reconstruction of Automobile Destruction: An Example of the Interaction between Language and Memory," *Journal of Verbal Learning and Verbal Behavior,* 1974, vol. 13, pp. 585–589.

43. "World Briefs," *The Gazette* (Montreal), May 20, 1994, p. A7.

44. Cited in Lucy Horwitz and Lou Ferleger, *Statistics for Social Change,* Boston: South End Press, 1980, pp. 181–182.

45. Wheeler, *Lies, Damn Lies, and Statistics,* p. 15.

46. Frederick Raymond of New York, cited in "(Crab)Grass Roots: Questionnaires Sent by Congress Tap Voter Vitriol," *The Wall Street Journal,* August 21, 1975.

47. Darrell Huff, *How to Take a Chance,* New York: Norton, 1959, pp. 115–117.

48. Betsy Morris, "In This Taste Test, the Loser Is the Taste Test," *The Wall Street Journal,* June 3, 1987.

49. Michael J. McCarthy, "New Coke Gets a New Look, New Chance," *The Wall Street Journal,* March 7, 1990.

50. Michael Kelly, "Clinton's Escape Clause," *The New Yorker,* October 24, 1994, pp. 42–53.

51. Michael Kelly, "You Say You Want a Revolution," *The New Yorker,* November 21, 1994, pp. 56–65.

52. Wheeler, *Lies, Damn Lies, and Statistics,* p. 289.

53. Ellen Hume and Peter Gumbel, "Gorbachev Impresses Group of Missourians; He Is High in Poll, too," *The Wall Street Journal,* December 4, 1987.

54. U.S. Department of Commerce, *Statistical Abstract of the United States,* Washington: GPO, 1981, table 202, p. 123.

55. "More Children Equal Less Divorce," *Look,* February 13, 1951, p. 80.

56. David Fay Smith and John Kochevar, "How Computers Are Changing Lives," *Dial,* June 1984, pp. 26–33, 55.

57. "SAT Coaching Disparaged," *The New York Times,* February 10, 1988, section II, p. 8.

58. "Coke-Pepsi Slugfest," *Time,* July 26, 1976, pp. 64–65.

59. *Cape Cod Times,* August 28, 1984.

60. Vernita Ediger, "Making Math a Little Less Problematic," *San Francisco Chronicle,* July 8, 1992.

61. *Boykin vs. Georgia Power,* 706 *F. 2d* 1384, 32 *FEP Cases* 25 (5th Cir. 1983).

62. Cynthia Crossen, "Studies Galore Support Products and Positions, but Are They Reliable?" *The Wall Street Journal,* November 14, 1991.

63. Marcia Peoples Halio, "Student Writing: Can the Machine Maim the Message?" *Academic Computing,* January 1990, pp. 16–19, 45.

64. *Chicago Daily News,* April 8, 1955.

65. Crossen, "Studies Galore Support Products and Positions, but Are They Reliable?"

chapter 5 **explaining data**

1. Geoffrey D. Bryant and Geoffrey R. Norman, "Expressions of Probability: Words and Numbers," letter to the *New England Journal of Medicine,* February 14, 1980, p. 411; also see George A. Diamond and James S. Forrester, "Metadiagnosis," *The American Journal of Medicine,* July 1983, pp. 129–137.

2. Leo Gould, *You Bet Your Life,* Hollywood, CA: Marcel Rodd, 1946.

3. An early report is by Hans Ulrich Balthasar, Roberto A. A. Boschi, and Michael M. Menke, "Calling the Shots in R&D," *Harvard Business Review,* May–June 1978, pp. 151–160.

4. Matthew Rose, "When Tilting at Windmills Gets Boring, Impossible Races Will Do," *The Wall Street Journal,* February 15, 1996.

5. Edward Yardeni, *Money & Business Alert,* Prudential-Bache, February 4, 1987, p. 1.

6. Elaine M. Sloand, Elisabeth Pitt, Robert J. Chiarello, and George J. Nemo, "HIV Testing: State of the Art," *Journal of the American Medical Association,* 1991, vol. 266, pp. 2861–2866.

7. Clement McQuaid (ed.), *Gambler's Digest,* Northfield, IL: Digest Books, 1971, pp. 24–25.

8. David Eddy, "Probabilistic Reasoning in Clinical Medicine: Problems and Opportunities," in Daniel Kahneman, Paul Slovak, and Amos Tversky, *Judgment under Uncertainty: Heuristics and Biases,* Cambridge, England: Cambridge University Press, 1982, pp. 249–267.

9. *The Rocky Mountain News,* November 27, 1968.

10. Quoted in Darrell Huff, *How to Take a Chance,* New York: Norton, 1959, p. 110.

11. Norman Dash, *Great Betting Systems,* Los Angeles: Price/Stern/Soan and Ravenna Books, 1968, pp. 155–156.

12. Tom Clancy, "In a Frigate's Combat Center, Time and Information Run Out Quickly," *Los Angeles Times,* May 20, 1987.

13. Quoted in McQuaid, *Gambler's Digest,* p. 287.

14. Computed from G. Bateman, "On the Power Function of the Longest Runs as a Test for Randomness in a Sequence of Alternatives," *Biometrika,* 1948, vol. 35, pp. 97–112. For some discussion of a basketball study, see Robert L. Wardrop, "Simpson's Paradox and the Hot Hand in Basketball," *The American Statistician,* February 1995, pp. 24–28.

15. Tim Molloy, "Gatemen, Athletics Cape League Picks," *Cape Cod Times,* July 19, 1991.

16. *Charlotte, West Virginia, Gazette,* July 29, 1987.

17. Marilyn vos Savant, "Ask Marilyn," *Parade,* July 12, 1992, p. 6.

18. Edgar Allan Poe, "The Mystery of Marie Roget," *Lady's Companion,* 1842.

19. McQuaid, *Gambler's Digest,* p. 287.

20. H. P. Bowditch, "The Growth of Children," Report of the Board of Health of Massachusetts, MA: Board of Health, 1877, VIII.

21. These data were provided by Shaun Johnson of Australia's National Climate Centre, Bureau of Meteorology. The index is SOI = $10(x - \bar{x})/s$, where x is the air pressure difference in the current month, $\bar{x}$ is the particular month's historical average value, and s is the standard deviation of this month's historical values.

22. Karl Pearson and Alice Lee, "On the Laws of Inheritance in Man," *Biometrika,* 1903, vol. 2, p. 364.

23. D. D. Kosambi, "Scientific Numismatics," *Scientific American,* 214, February 1966, pp. 102–111.

24. Harold Jacobs, *Mathematics: A Human Endeavor,* San Francisco: W. H. Freeman, 1982, p. 570.

25. Dean E. Murphy, "Wondering Why the Rainfall Isn't Normal? Maybe Because It Never Is," *Los Angeles Times,* February 24, 1993.

26. J. E. Vader, "Mr. Bases Loaded," *Sports Illustrated,* April 24, 1989, p. 73.

27. Mark Fischetti, "You Said a Mouthful, Dick, Baby," *Sports Illustrated,* March 19, 1990.

28. Charles McCarry, *The Secret Lovers,* New York: E. P. Dutton, 1977, p. 100.

29. Gould, *You Bet Your Life,* p. 80.

30. David Lykken, "Polygraph Interrogation," *Nature,* February 23, 1984, pp. 681–684. Another careful study of six polygraph experts found that 18 of 50 innocent people were classified as liars, while 12 of 50 confessed thieves were judged truthful: Benjamin Kleinmuntz and Julian J. Szucko, "A Field Study of the Fallibility of Polygraph Lie Detection," *Nature,* March 29, 1984, pp. 449–450.

31. A. Barnett, I. Greenberg, and R. Marchol, "Hinckley and the Chemical Bath," *Interfaces,* 1984, vol. 14, pp. 48–52.

32. *A. B. G. Instrument & Engineering, Inc. vs. United States* (1979), cited in David Barnes, *Statistics as Proof,* Boston: Little, Brown, 1983, p. 261.

33. Marilyn vos Savant, "Ask Marilyn," *Parade Magazine,* July 1, 1990.

34. Melvin Durslag, "After 51 Years, It's Time to Say Goodbye," *Los Angeles Times,* May 22, 1991.

35. E. B. Grosfils and J. W. Head, "The Timing of Giant Radiating Dike Swarm Emplacement on Venus: Implications for Resurfacing of the Planet and Its Subsequent Evolution," *Journal of Geophysical Research,* February 1996.

36. This was a question asked of Marilyn vos Savant, "Ask Marilyn," *Parade Magazine,* August 20, 1995.

chapter 6 estimation

1. Arthur Conan Doyle, *The Sign of Four,* London: Spencer Blackett, 1890.

2. This example is from W. Allen Wallis and Harry V. Roberts, *Statistics: A New Approach,* New York: Free Press, 1956, p. 471.

3. *Reserve Mining Co. vs. EPA* (1975), cited in David W. Barnes, *Statistics as Proof,* Boston: Little, Brown, 1983, p. 244.

4. Robert J. Samuelson, "The Strange Case of the Missing Jobs," *Los Angeles Times,* October 27, 1983.

5. James T. McClave and P. George Benson, *Statistics for Business and Economics,* 2d ed., San Francisco: Dellen, 1982, p. 279.

6. E. S. Pearson and N. W. Please, "Relation between the Shape of Population Distribution and the Robustness of Four Simple Test Statistics," *Biometrika,* vol. 62, 1975, pp. 223–241; Harry O. Poston, "The Robustness of the One-Sample *t*-Test over the Pearson System," *Journal of Statistical Computation and Simulation,* 1979, pp. 133–149.

7. Samantha Barryessa Good, "The Pomona-Pitzer Women's Volleyball 1992 Season," Pomona College, Claremont, CA, December 1992.

8. S. M. Stigler, "Do Robust Estimators Work with Real Data?" *Annals of Statistics,* 1977, pp. 1055–1078.

9. Stigler.

10. Stephen M. Stigler, "Eight Centuries of Sampling Inspection," *Journal of the American Statistical Association,* September 1977, pp. 493–500.

11. Jerry E. Bishop, "Statisticians Occupy Front Lines in Battle over Passive Smoking," *The Wall Street Journal,* July 28, 1993.

12. E. J. Dionne, Jr., "Poll Shows where Blacks Stand," *San Francisco Chronicle,* July 9, 1992, p. A4.

13. Ellen Hume and Peter Gumbel, "Gorbachev Impresses Group of Missourians; He Is High in Poll, too," *The Wall Street Journal,* December 4, 1987.

14. "Who's Hassled, Who Complains," *Working Woman,* January 1992, p. 11.

15. "How They Scored, How They're Eating," *Consumer Reports,* May 1990, p. 325.

16. Anthony Lewis, "Chief Justice Bird: Calm at the Center," *The New York Times,* October 23, 1986.

17. *U.S. vs. 43 1/2 Gross of Rubber Prophylactics, etc.* 65 *F. Supp.* 534 (D. Minn. 1946).

18. "The Science of Polling," *Newsweek,* September 28, 1992, pp. 38–39.

19. Margaret Harris, Caroline Yeeles, Joan Chasin, and Yvonne Oakley, "Symmetries and Asymmetries in Early Lexical Comprehension and Production," *Journal of Child Language,* February 1995, pp. 1–18.

20. Cathy Lynn Grossman and Carrie Dowling, "Vacation in New York? No Way, Guidebook Told," *USA Today,* May 4, 1995, p. 1.

21. Adapted from Fred Grieman, first hour examination, Chemistry la, Pomona College, fall 1993.

22. G. Invernizzi, "II Tempo Ritrovato," *L' Espresso,* January 10, 1993, p. 139.

23. Jaime Diaz, "Perils of Putting," *Sports Illustrated,* February 1989, pp. 76–78.

24. J. K. Chambers, "Dialect Acquisition," *Language,* 1992, vol. 68, pp. 673–705.

25. Todd Austin, "Roshambo," Pomona College, Claremont, CA, spring 1990.

26. Dina Colman, "How Can I Write My Paper?" Pomona College, Claremont, CA, 1987.

27. H. G. Reza, "New Study Indicates Wide Sex Harassment in Navy," *Los Angeles Times,* February 10, 1992.

28. Charles Kenney, *Boston Globe Magazine,* August 30, 1987.

29. Jay Sala, "A Statistical Test of the Life Cycle Theory of Consumption," Pomona College, Claremont, CA, fall 1990.

30. Daniel Matsuda, "Effect of the Backboard on Shooting Accuracy," Pomona College, Claremont, CA, spring 1990.

31. Paul Farmer, "Make a Wish! (Your Candle Will Go Out in 715 Seconds.)" Pomona College, Claremont, CA, spring 1990.

32. Ellen Goodman, "The Last Power Struggle," *San Francisco Chronicle,* May 19, 1992.

33. Peter Joe, "Typo Probabilities," Pomona College, Claremont, CA, spring 1990.

34. David Kahn, "Modern Cryptology," *Scientific American,* July 1966, pp. 38–46. Also see Henry Beker and Fred Piper, *Cipher Systems,* New York: John Wiley & Sons, 1982, esp. p. 25 and app. 1.

35. Clark Jensen, "Hometown Refs," Pomona College, Claremont, CA, spring 1990.

36. Barnes, *Statistics as Proof,* p. 35.

37. *U.S. vs. United Shoe Machinery Company,* 110 *F. Supp.* 295 (D. Mass. 1953).

38. Jennifer Kurz, Cabell Westbrook, and Jim Wright, "Sleeping Habits of Pomona Students," Pomona College, Claremont, CA, fall 1996.

39. Charles Seiter, "Forecasting the Future," *MacWorld,* September 1993, p. 187.

40. Tracy Hobbs, Pomona College, Claremont, CA, 1987.

chapter 7 **hypothesis tests**

1. Edward Leamer, "Let's Take the Con out of Econometrics," *American Economic Review,* March 1983, p. 36.
2. P. A. Mackowiak, S. S. Wasserman, and M. M. Levine, "A Critical Appraisal of 98.6°F, the Upper Limit of the Normal Body Temperature, and Other Legacies of Carl Reinhold August Wunderlich," *Journal of the American Medical Association,* September 23/30, 1992, pp. 1578–1590. I am grateful to Steven Wasserman for sharing the data with me.
3. R. A. Fisher, "The Arrangement of Field Experiments," *Journal of the Ministry of Agriculture of Great Britain,* 1926, vol. 8, p. 504.
4. Arthur Melton, "Editorial," *Journal of Experimental Psychology,* 1962, p. 553–557.
5. For a discussion of such systems, see Timothy O. Bakke, "Body Language Security Systems," *Popular Science,* June 1986, pp. 76–78, 112–113.
6. Harley Tinkham, "Morning Briefing," *Los Angeles Times,* November 26, 1983.
7. Otto Klineberg, *Negro Intelligence and Selection Migration,* New York: Columbia University Press, 1935.
8. P. K. Whelpton and Clyde V. Kiser, *Social and Psychological Factors Affecting Fertility,* vol. 1, New York: Milbank Memorial Fund, 1950, p. 109.
9. Sheldon Blackman and Don Catalina, "The Moon and the Emergency Room," *Perceptual and Motor Skills,* 1973, pp. 624–626.
10. Pamela Weintraub, "Preschool?" *Omni,* August 1989, pp. 34–38.
11. E. Yehle, "Accuracy in Clerical Work," in Victor Lazzaro (ed.), *Systems and Procedures: A Handbook for Business and Industry,* 2d ed., Englewood Cliffs, NJ: Prentice-Hall, 1968.
12. Kenneth B. Clark and Mamie P. Clark, "Racial Identification and Preference in Negro Children," in Eleanor E. Macoby, Theodore M. Newcomb, and Eugene L. Hartley (eds.), *Readings in Social Psychology,* 3d ed., New York: Holt, Rinehart, and Winston, 1958, pp. 602–611.
13. Lee Salk, "The Role of the Heartbeat in the Relations between Mother and Infant," *Scientific American,* May 1973, pp. 26–29.
14. R. A. Fisher, *Experiments in Plant Hybridization,* Edinburgh: Oliver and Boyd, p. 53. This book includes Mendel's original paper and comments by Fisher based on a paper he wrote in 1936.
15. Jack McCallum, "Out of Joint," *Sports Illustrated,* February 13, 1995, pp. 44–53.
16. Andy Jacobson, "That's the Way the Ball Bounces," Pomona College, Claremont, CA, April 1992.
17. Thomas H. Maugh, II, "Breast Cancer Survival Rate May Double, Study Finds," *Los Angeles Times,* May 14, 1990.
18. Jeremy Lopez, "The Penalty Kick," Pomona College, Claremont, CA, April 1992.
19. F. Arcelus and A. H. Meltzer, "The Effect of Aggregate Economic Variables on Congressional Elections," *American Political Science Review,* 1975, vol. 69, pp. 1232–1239.
20. William Feller, "Are Life Scientists Overawed by Statistics?" *Scientific Research,* February 3, 1969, pp. 24–29.
21. T. D. Sterling, "Publication Decisions and Their Possible Effects on Inferences Drawn from Tests of Significance—or Vice Versa," *Journal of the American Statistical Association,* 1959, vol. 54, pp. 30–34.
22. Lloyd Shearer, *Parade,* October 8, 1972. For additional evidence see Dennis D. Miller, "Is It Height or Sex Discrimination?" *Challenge,* September/October 1986, pp. 59–61. On the other hand, an article in the October 1986 issue of *Pediatrics* suggests that height and IQ are related.
23. Sue Avery, "Market Investors Will Be High on Redskins Today," and "Morning Briefing: Wall Street 'Skinnish' on Big Game," *Los Angeles Times,* January 22, 1983.
24. James Bates, "Reality Wears Loser's Jersey in Super Bowl Stock Theory," *Los Angeles Times,* January 22, 1989.
25. Cited by Martin Gardner, *Fads and Fallacies in the Name of Science,* New York: Dover, 1957, p. 305.

26. Warner-Lambert Co. 86 *F.T.C.* 1398 (1975), modified 562 *F.2d* 749 (D.C. Cir. 1977), cert. denied, 435 *U.S.* 950 (1978).

27. *Sporting Edge,* 1988.

28. Gail Howard, *State Lotteries: How to Get in It . . . and How to Win It!,* 5th ed., Ben Buxton, 1986.

29. Frederick Mosteller, Robert E. K. Rourke, and George B. Thomas, Jr., *Probability with Statistics Applications,* Reading, MA: Addison-Wesley, 1961, p. 17.

30. "Scorecard," *Sports Illustrated,* January 5, 1987, p. 10.

31. "News Briefs," *Cape Cod Times,* July 5, 1984.

32. Francis Iven Nye, *Family Relationships and Delinquent Behavior,* New York: John Wiley & Sons, 1958, p. 29.

33. Floyd Norris, "Predicting Victor in Super Bowl," *The New York Times,* January 17, 1989.

34. Beatrice E. Garcia, "Will It Be Friday the 13th, Part 6? Study Ties 3 in Year to Recession," *The Wall Street Journal,* November 13, 1987.

35. "Science at the EPA," *The Wall Street Journal,* October 2, 1985.

36. L. A. Chung, "San Francisco Fire Commission Pressured on Hiring," *San Francisco Chronicle,* July 11, 1992.

37. Seymour Siwoff, Steve Hirdt, and Peter Hirdt, *The New Elias Baseball Analyst,* New York: Macmillan, 1985.

38. Tom Purcell, "Free Throws and Rebounds from Misses," Pomona College, Claremont, CA, December 1989.

39. Ted Lai, "Hockey," Pomona College, Claremont, CA, spring 1991.

40. Georgette Jasen, "Investment Dartboard: Pros Top Darts in Selecting Stock Winners," *The Wall Street Journal,* September 13, 1995, p. C1.

41. Andrew Jacobson, "Is the Mail Finally Here?" Pomona College, Claremont, CA, spring 1990.

42. Jeffrey Laderman, "Insider Trading," *Business Week,* April 29, 1985, pp. 78–92.

43. David Van Oppen, Pomona College, Claremont, CA, 1987.

44. Anonymous, "Pickles and Humbug," *The Journal of Irreproducible Results: Selected Papers,* Chicago Heights, Illinois: *Journal of Irreproducible Results,* 1976, p. 107.

45. William G. Baxt, *Annals of Internal Medicine,* December 1991; cited in Michael Waldholz, "Computer 'Brain' Outperforms Doctors in Diagnosing Heart Attack Patients," *The Wall Street Journal,* December 2, 1991.

46. JS&A advertisement, *Discoveries,* United Airlines, June 1986.

47. Elizabeth Cudd, Sarah Cunningham, Erica Green, and Ty Schultz, "Your Weight, Sir?" Pomona College, Claremont, CA, spring 1996.

48. *Sports Illustrated,* January 30, 1984.

49. *The New York Times,* January 18, 1976.

50. *Allen vs. Prince George's County,* MD 538 *F. Supp.* 833 (1982), affirmed 737 *F.2d* 1299 (4th Cir. 1984).

51. Steven J. Milloy, "The EPA's Houdini Act," *The Wall Street Journal,* August 8, 1996.

52. D. Allison, S. Heshka, D. Sepulveda, and S. Heymsfield, "Counting Calories—Caveat Emptor," *Journal of the American Medical Association,* 1993, vol. 270, pp. 1454–1456.

chapter 8 comparing two samples

1. David E. Harrington, "Economic News on Television," *Public Opinion Quarterly,* 1989, pp. 17–40, and personal correspondence.

2. See Harry O. Poston, "The Robustness of the Two-Sample *t* Test over the Pearson System," *Journal of Statistical Computation and Simulation,* 1978, pp. 295–311; Paul Leaverton and John J. Birch, "Small Sample Power Curves for the Two Sample Location Problem," *Technometrics,* 1969, vol. 11, pp. 299–307; and D. J. Best and J. C. W. Rayner, "Welch's Approximate Solution for the Behrens-Fisher Problem," *Technometrics,* 1987, vol. 29, pp. 205–210.

3. Charles W. Dunnett, "Drug Screening: The Never-Ending Search for New and Better Drugs," in Judith M. Tanur, Frederic Mosteller, William H. Kruskal, Richard F. Link, Richard S. Pieters, and Gerald R. Rising, *Statistics: A Guide to the Unknown,* San Francisco: Holden-Day, 1972, pp. 23–33.

4. Lipid Research Clinic Program, "The Lipid Research Clinics Coronary Primary Prevention

Trial Results," *Journal of the American Medical Association,* January 20, 1984, pp. 351–374.

5. Ronald Frank and William Massy, "Shelf Position and Space Effects on Sales," *Journal of Marketing Research,* February 1970, pp. 59–66.

6. For example, William Beyer (ed.), *Handbook of Tables for Probability and Statistics,* 2d ed., Cleveland, OH: Chemical Rubber Company, 1972.

7. For example, Beyer.

8. *Motor Vehicles Manufacturers Association vs. EPA,* 768, *F.*2d 385, D.C. Cir. 1985.

9. Marion Gillim, "Physical Measurements of Mount Holyoke College Freshmen in 1918 and 1943," *Journal of the American Statistical Association,* March 1944, vol. 39, pp. 53–56.

10. John Harris, "El Niño and Precipitation in Southern California," Pomona College, Claremont, CA, April 1992.

11. H. S. Diehl, A. B. Baker, and D. W. Cowan, "Cold Vaccines," *Journal of the American Medical Association,* September 24, 1938, pp. 1168–1173.

12. Sarah Wales, "Will You Hold, Please?" Pomona College, Claremont, CA, spring 1993.

13. Byron Wm. Brown, Jr., and Myles Hollander, *Statistics: A Biomedical Introduction,* New York: John Wiley & Sons, 1977, p. 87.

14. Joanne Simpson, Anthony Olsen, and Jane C. Eden, "A Bayesian Analysis of a Multiplicative Treatment Effect in Weather Modification," *Technometrics,* vol. 17, 1975, pp. 161–166.

15. An easily read account is in *Consumer Reports,* February 1976, pp. 68–70. There were actually 1000 volunteers at the beginning of this experiment, but 200 dropped out.

16. J. R. Berrueta-Clement, L. Schweinhart, W. W. Barnett, W. Epstien, and D. Weikart, *Changed Lives: The Effects of the Perry Preschool Program on Youths through Age 19,* Monographs of the High/Scope Educational Research Foundation, no. 8, Ypsilanti, MI: High/Scope Educational Research Foundation, 1984.

17. John L. Coulehan, Keith S Reisinger, Kenneth D. Rogers, and Daniel W. Bradley, "Vitamin C Prophylaxis in a Boarding School," *New England Journal of Medicine,* vol. 290, January 1974, pp. 6–10.

18. The Steering Committee of the Physicians' Health Study Research Group, "Preliminary Report: Findings from the Aspirin Component of the Ongoing Physicians' Health Study," *New England Journal of Medicine,* January 28, 1988, pp. 262–264.

19. *Elison vs. City of Knoxville 33 FEP Cases* 1141 (E.D. Tenn. 1983).

20. Jeremy Lopez, "The Penalty Kick," Pomona College, Claremont, CA, April 1992.

21. Arlie Russell Hochschild, *Second Shift,* New York: Viking, 1989, p. 242, citing Louis Harris and Associates, "Families at Work," *General Mills American Family Report,* 1980–81.

22. John Leaning, "Barnstable Auto Fatalities Drop 600%," *Cape Cod Times,* August 7, 1986.

23. Julie Scheer, Pomona College, Claremont, CA, 1987.

24. Niels Ehlers, "On Corneal Thickness and Intraocular Pressure, II," *Acta Ophthalmologica,* 1970, vol. 48, pp. 1107–1112.

25. Diana May Lin, "The Model Minority Myth," Pomona College, Claremont, CA, April 1992.

26. D. Weiss, B. Whitten, and D. Leddy, "Lead Content of Human Hair (1871–1971)," *Science,* 1972, vol. 178, pp. 69–70.

27. Kristin Fix, "Track Surfaces: Are They Significant?" Pomona College, Claremont, CA, fall 1990.

28. Russell Green, "GPAs of Single and Double Majors at Pomona College," Pomona College, Claremont, CA, April 1992.

29. Sheryl F. Kelsey, *Circulation,* March 1993: cited in Elyse Tanouye, "Women Face Much Higher Risk of Dying after Angioplasty than Men, Study Says," *The Wall Street Journal,* March 9, 1993.

30. Derek Churchill, "Heavy Packs and Sore Knees," Pomona College, Claremont, CA, spring 1993.

31. Louis Freedberg, "Survey Finds Fewer Blacks, Men in Teaching," *San Francisco Chronicle,* July 7, 1992, p. A2.
32. Frank J. Massey, Jr., "Vasectomy and Health," *Journal of the American Medical Association,* August 24, 1984, pp. 1023–1029.
33. S. N. Frankel and J. E. Alman, *Journal of Oral Therapeutic Pharmacology,* 1969, pp. 443–449.
34. G. R. Lindsey, "A Scientific Approach to Strategy in Baseball," in Shaul P. Ladany and Robert E. Machol, *Optimal Strategies in Sports,* Amsterdam: North-Holland, 1977, pp. 1–30.
35. Stanley Coren and Diane F. Halpern, "Left-handedness: A Marker for Decreased Survival Fitness," *Psychological Bulletin,* January 1991, pp. 90–106.
36. Tracy Tseng, Marina Apotovsky, and Karen Crowley, "Family Plans of Pomona Seniors," Pomona College, Claremont, CA, fall 1996.
37. Nichole M. Christian, "Call for Bright Yellow Firetrucks Has Many Firefighters Seeing Red," *The Wall Street Journal,* June 26, 1995.
38. S. W. Prayitno, M. Addy, and W. G. Wade, "Does Gingivitis Lead to Periodontis in Young Adults?" *The Lancet,* August 21, 1993, pp. 471–472.
39. Richard Hudson, "Discussion Note: About 37% of Word-Tokens Are Nouns," *Language,* 1994, vol. 70, pp. 331–339.
40. Elizabeth Cudd, Sarah Cunningham, Erica Green, and Ty Schultz, "Your Weight, Sir?" Pomona College, Claremont, CA, spring 1996.
41. Susan Inman, "Do Baseball Players Perform Differently in Day and Night Games?" Pomona College, Claremont, CA, fall 1995.
42. Elizabeth Cudd, Sarah Cunningham, Erica Green, and Ty Schultz, "Who's the Real Low-Price Leader?" Pomona College, Claremont, CA, spring 1996.
43. Stephanie Shideler, "A Comparison of Elite Male and Female Runners' Velocities," Pomona College, Claremont, CA, fall 1995.
44. Craig J. Newschaffer, Trudy L. Bush, and William E. Hale, "Aging and Total Cholesterol Levels: Cohort, Period, and Survivorship Effects," *American Journal of Epidemology,* vol. 136, no. 1, 1992, pp. 23–31.
45. I am indebted for these data to Suzanne Thompson, professor of psychology at Pomona College.
46. Daniel Kahneman and Amos Tversky, "Choices, Values, and Frames," *American Psychologist,* vol. 39, April 1984, pp. 341–350.
47. Don A. Dilman, "Token Financial Incentives and the Reduction of Nonresponse Error in Mail Surveys," unpublished, 1996.
48. George De Leon, "The Baldness Experiment," *Psychology Today,* vol. 11, October 1977, pp. 62–66.
49. Joy Reid, "Using the Writer's Workbench in Composition Teaching and Testing," in *Technology and Language Testing,* Charles W. Stansfield, editor, Washington, D.C.: TESOL, 1986, pp. 167–188.
50. Marilyn Chase, "Old Heart Drug Works as Well as Costly TPA in Study," *The Wall Street Journal,* March 8, 1980.
51. Karl P. Koenig and John Masters, "Experimental Treatment of Habitual Smoking," *Behaviour Research and Therapy,* vol. 3, December 1965, pp. 235–243.
52. Kelly Redfield, "The Sophomore Shift," Pomona College, Claremont, CA, fall 1990.
53. E. Cameron and L. Pauling, "Supplemental ascorbate in the supportive treatment of cancer: re-evaluation of prolongation of survival times in terminal human cancer," *Proceedings of the National Academy of Science USA,* 1978, pp. 4538–4542.

chapter 9 using ANOVA to compare several means

1. T. R. Karl, D. R. Easterling, R. W. Knight, and P. Y. Hughes, "U.S. National and Regional Temperature Anomalies," in T. A. Boden, D. P. Kaiser, R. J. Sepanski, and F. W. Stoss, (eds.), *Trends '93: A Compendium of Global Change,* Oak Ridge, TN: Carbon Dioxide Information Analysis Center, Oak Ridge National Laboratory, 1993, pp. 686–736.
2. Elisabeth Street and Mavis B. Carroll, "Preliminary Evaluation of a New Food Product,"

in Judith M. Tanur, Frederic Mosteller, William H. Kruskal, Richard F. Link, Richard S. Pieters, and Gerald R. Rising, *Statistics: A Guide to the Unknown,* San Francisco: Holden-Day, 1972, pp. 220–229.

3. Mark Gibbons and Patrick Hess, "Day of the Week Effects and Asset Returns," *Journal of Business,* October 1981, pp. 579–595. The calculations in the text are not exactly right, because holidays left an unequal number of observations for each day of the week. If the underlying data are entered in a statistical software package, it will make the necessary adjustments.
4. The article that the subjects read was Darhl M. Pedersen, Sheila Keithly, and Karie Brady, "Effects of an Observer on Conformity to Handwashing Norm," *Perceptual and Motor Skills,* 1986, pp. 169–170; the researcher was Jarrod Kula, "The Effects of Music on Studying and Reading Comprehension," Pomona College, Claremont, CA, fall 1993.
5. These data from the Federal Trade Commission were kindly provided by Lauren McIntyre, North Carolina State University.
6. Luke Sobota, "Do Office Hours Vary by Division?" Pomona College, Claremont, CA, fall 1995.
7. J. R. Stroop, "Studies of Interference in Serial Verbal Reactions," *Journal of Experimental Psychology,* December 1935, vol. 18, pp. 643–662.
8. These data were kindly provided by Deborah Burke, Professor of Psychology at Pomona College, Claremont, CA.
9. Betsy Smith, "Faculty Response to ESL Writing," Ph.D. dissertation, Claremont Graduate School, Claremont, CA, 1996.
10. M. H. Criqui and Brenda L. Ringel, "Does Diet or Alcohol Explain the French Paradox?" *The Lancet,* December 24/31, 1994, pp. 1719–1723; and personal correspondence with Michael Criqui.
11. O. L. Davies, *The Design and Analysis of Industrial Experiments,* London: Oliver and Boyd, 1956, p. 164.
12. I am grateful to Ian Simmonds and Kevin Keay, School of Earth Sciences, University of Melbourne, for sharing their results with me.
13. These data were provided by the wonderful Statlib Project run by the Statistics Department at Carnegie Mellon University.
14. A. Tubb, A. J. Parker, and G. Nickless, "The Analysis of Romano-British Pottery by Atomic Absorption Spectrophotometry," *Archaeometry,* 1980, vol. 22, pp. 153–171.
15. A. Thomson and R. Randall-Maciver, *Ancient Races of the Thebaid,* Oxford: Oxford University Press, 1905.
16. June Webb, "College Mathematics Achievement and Logical Reasoning," *Perceptual and Motor Skills,* August 1985, vol. 61, pp. 15–21.
17. Robert Moody, Dennis Dusevich, and Chris Cain, "Upscale Cookies—A Good Value or a Good Gimmick?" statistics paper, Pomona College, Claremont, CA, Spring 1997.
18. Danya Dumbrill, "She Never Studies and Still Gets A's," Pomona College, Claremont, CA, Spring 1991.
19. Cletus C. Coughlin, K. Alec Chrystal, and Geoffrey E. Wood, "Protectionist Trade Policies: A Survey of Theory, Evidence, and Rationale," *Federal Reserve Bank of St. Louis Review,* January/February 1988, pp. 12–26.
20. Chester I. Bliss, *Statistics in Biology,* New York: McGraw-Hill, 1967, p. 351.

chapter 10 chi-square tests for categorical data

1. Tyson Thomas, "The Magic of Three," Pomona College, Claremont, CA, December 1989.
2. Christopher Robson and Brendan M. Walsh, "Alphabetical Voting: A Study of the 1973 General Election in the Republic of Ireland," Dublin: The Economic and Social Research Institute, June 1973.
3. Wei Hopeman, "47?" Pomona College, Claremont, CA, spring 1990.
4. P. L. Yu, C. Wrather, and G. Kozmetsky, "Auto Weight and Public Safety: A Statistical Study of Transportation Hazards," Research Report

233, Austin, TX: Center for Cybernetic Studies, University of Texas, 1975.

5. Wen Yen, "M&Ms' Home Runs," Pomona College, Claremont, CA, April 1992.

6. David P. Phillips, "Deathday and Birthday: An Unexpected Connection," in Judith Tanur, Frederic Mosteller, William H. Kruskal, Richard F. Link, Richard S. Pieters, and Gerald R. Rising, *Statistics: A Guide to the Unknown,* San Francisco: Holden-Day, 1972, pp. 52–65.

7. Lucy Hovis, "Statistics and Stock Prices," Pomona College, Claremont, CA, spring 1990.

8. Francis Iven Nye, *Family Relationships and Delinquent Behavior,* New York: John Wiley & Sons, 1958, p. 37.

9. Joan R. Rosenblatt and James J. Filliben, "Randomization and the Draft Lottery," *Science,* vol. 171, 1971, pp. 306–308. Interestingly, there was also a draft lottery in 1940, in which the capsules were not very well mixed; see the statement by Samuel Stouffer and Walter Bartky in *Chicago Tribune,* November 2, 1940, p. 4.

10. Christopher Silagy, David Mant, Godfrey Fowler and Mark Lodge, "Meta-analysis of Efficacy of Nicotine Replacement Therapies in Smoking Cessation," *The Lancet,* January 15, 1994, pp. 139–142.

11. Gerald A. Hudgens, Victor H. Denenberg, and M. X. Zarrow, "Mice Reared with Rats: Effects of Preweaning and Postweaning Social Interactions upon Adult Behavior," *Behaviour,* 1968, vol. 30, pp. 259–274.

12. Peter Redfield, *The Titanic and the California,* New York: John Day, 1965.

13. Fleur Templeton, "The Taming of the Flu among Newborns," *Business Week,* June 1, 1992, p. 85.

14. Kathleen Doheny, "Big Feet Make for Bad Fit," *San Francisco Chronicle,* June 29, 1992.

15. J. David Erickson, "The Secondary Sex Ratio in the United States, 1969–1971: Race, Parental Ages, Birth Order, Paternal Education, and Legitimacy," *Annals of Human Genetics,* 1976, vol. 40, pp. 205–212. Because race matters, too, these data are for nonblack babies.

16. Jane Luh, "The Effect of Birth Order on Academic Performance," Pomona College, Claremont, CA, spring 1990.

17. Karl Pearson, "On the Laws of Inheritance in Man, II," *Biometrika,* 1904, vol. 3, pp. 131–190.

18. *Contra Costa Times,* May 27, 1992, p. 2B.

19. *Bazemore vs. Friday* 751 *F.2d* 662, 36 *FEP Cases* 834 (4th Cir. 1984).

20. Phil Taylor, "Spur of the Moment," *Sports Illustrated,* March 7, 1994, pp. 58–60.

21. Gardy Laguerre, "Seles' First Serves," Pomona College, Claremont, CA, spring 1993.

22. Daniel Matsuda, "Effect of the Backboard on Shooting Accuracy," Pomona College, Claremont, CA, spring 1990.

23. G. R. Lindsey, "A Scientific Approach to Strategy in Baseball," in Shaul P. Ladany and Robert E. Machol, *Optimal Strategies in Sports,* Amsterdam: North-Holland, 1977, pp. 1–30.

24. Warren E. Miller, Arthur H. Miller, and Edward J. Schneider, *American National Election Studies Data Sourcebook, 1952–1978,* Cambridge, MA: Harvard University Press, 1980, p. 268.

25. H. C. White, "Cause and Effect in Social Mobility Tables," *Behavioral Science,* 1963, vol. 8, pp. 14–27.

26. Centers for Disease Control and Prevention of the U.S. Department of Health and Human Services, *HIV/AIDS Surveillance Report,* 1995.

27. Donald Erlandson, "Billiard Ball Falls on the Break," Pomona College, Claremont, CA, fall 1995.

28. L. Smith, S. Folkard, and C. J. M. Poole, "Increased Injuries on the Night Shift," *The Lancet,* October 22, 1994, pp. 1137–1139.

29. Dennis Lendrum, "Should John McEnroe Grunt?" *New Scientist,* July 21, 1983, pp. 188–189. Professor Lendrum was kind enough to send me his original data.

30. Heidi Cobar, "How Strong are Cupid's Arrows?" Pomona College, Claremont, CA, December 1992.

31. The female data are from Audra Timmins, "Which Side to Block Out On?" Pomona College, Fall 1990; the male data are from Andy Jacobson, "That's the Way the Ball Bounces," Pomona College, Claremont, CA, April 1992.
32. John R. Emshwiller, "Car Makers Face Lawsuits Alleging Rear Seat Belts Aren't Safe Enough," *The Wall Street Journal,* January 6, 1988.
33. Matt Markatos, "The Effects of Pronoun Choice on Subjective Gender Interpretation," Pomona College, Claremont, CA, spring 1993.
34. John Van Hook, "Free Throw Percentages and Heights of Basketball Players," Pomona College, Claremont, CA, spring 1994.
35. Erica Green, Liz Cudd, Ty Schultz, and Sarah Cunningham, "Which Cookies Are the Best?," Pomona College, Claremont, CA, spring 1996.
36. Janette B. Benson, "Season of Birth and Onset of Locomotion: Theoretical and Methodological Implications," *Infant Behavior and Development,* 1993, pp. 69–81.
37. Eric Castellanos, "The Neatness Factor," Pomona College, Claremont, CA, fall 1990.
38. Aubrey Milunsky, et al, "Maternal Heat Exposure and Neural Tube Defects," *Journal of the American Medical Association,* August 19, 1992, pp. 882–885.

chapter 11 simple regression

1. Robert Cunningham Fadeley, "Oregon Malignancy Pattern Physiographically Related to Hanford, Washington, Radioisotope Storage," *Journal of Environmental Health,* 1965, pp. 883–897.
2. T. R. Karl, D. R. Easterling, and P. Ya. Groisman, "United States Historical Climatology Network—National and Regional Estimates of Monthly and Annual Precipitation," in T. A. Boden, D. P. Kaiser, R. J. Sepanski, and F. W. Stoss (eds.), *Trends '93: A Compendium of Global Change,* Carbon Dioxide Information Analysis Center, Oak Ridge National Laboratory, Oak Ridge, TN, 1993, pp. 830–905.
3. Johann Heinrich Lambert, *Beyträge zum Gebrauche de Mathematik und deren Anwendung,* Berlin, 1765, as quoted in Laura Tilling, "Early Experimental Graphs," *British Journal for the History of Science,* 1975, vol. 8, pp. 193–213.
4. Frederick E. Croxton and Dudley J. Cowdon, *Applied General Statistics,* 2d ed., Englewood Cliffs, NJ: Prentice-Hall, 1955, pp. 451–454.
5. Peter Passell, "Probability Experts May Decide Vote in Philadelphia," *The New York Times,* April 11, 1994, p. A10.
6. Ivan Valiela, Paulette Peckol, Charlene D'Avanzo, Kate Lajtha, James Kremer, W. Rockwell Geyer, Ken Foreman, Douglas Hersh, Brad Seely, Tatsu Isaji, and Rick Crawford, "Hurricane Bob on Cape Cod," *American Scientist,* March–April 1996, pp. 154–165; and personal correspondence with Prof. Valiela. Some of these data are the median values of ranges.
7. J. W. Kuzma and R. J. Sokel, "Maternal Drinking Behavior and Decreased Intrauterine Growth," *Alcoholism: Clinical and Experimental Research,* 1982, vol. 6, pp. 396–401.
8. Re Western Union Tel. Co., No. 18935, F.C.C., August 12, 1970.
9. Restéard Mulcahy, J. W. McGilvray, and Noel Hickey, "Cigarette Smoking Related to Geographic Variations in Coronary Heart Disease Mortality and to Expectations of Life in the Two Sexes," *American Journal of Public Health,* 1970, vol. 60, pp. 1515–1521.
10. These data are from D. E. Booth, *Regression Methods and Problem Banks,* Arlington, MA: COMAP, 1986.
11. Sven Cnattingius, Michele R. Forman, Heinz W. Berendes, and Leena Isotalo, "Delayed Childbearing and Risk of Adverse Perinatal Outcome," *Journal of the American Medical Association,* August 19, 1992, vol. 268, pp. 886–890.
12. Frances E. Purifoy, Lambert H. Koopmans, and Ronald W. Tatum, "Steroid Hormones and Aging: Free Testosterone Concentration, Testosterone and Androsenodione in Normal

Females Age 20–87 Years," *Human Biology,* May 1980, pp. 181–191.

13. James Shields, *Monozygotic Twins,* London: Oxford University Press, 1962; three similar, separate studies by Cyril Burt all reported the same value of R^2 (.594)! A logical explanation is that the data were flawed; see Nicholas Wade, "IQ and Heredity: Suspicion of Fraud Beclouds Classic Experiment," *Science,* 1976, vol. 194, pp. 916–919.

14. Garry Molholt and Art M. Presler, "Correlation between Human and Machine Ratings of Test of Spoken English Reading Passages," in Charles W. Stansfield (ed.), *Technology and Language Testing,* Washington: TESOL, 1986, pp. 111–128.

15. Allan H. Murphy and Robert L. Winkler, "Probability Forecasting in Meteorology," *Journal of the American Statistical Association,* September 1984, pp. 489–500.

16. Scott Vanourek, "The 1960 Yankee Sluggers—Grace under Pressure or Luck?" Pomona College, Claremont, CA, spring 1990.

17. Charles Osterberg, "Unsafe at Zero Speed," *Journal of Irreproducible Results,* September/October 1983, p. 19.

18. Francis Galton, "Regression towards Mediocrity in Hereditary Stature," *Journal of the Anthropological Institute,* 1886, pp. 246–263.

19. Christopher Jencks, Marshall Smith, Henry Acland, Mary Jo Bane, Daivd Cohen, Herbert Gintis, Barbara Heyns, and Stephen Michelson, *Inequality: A Reassessment of the Effect of Family and Schooling in America,* New York: Basic Books, 1972, p. 59.

20. Amos Tversky and Daniel Kahneman, "On the Psychology of Prediction," *Psychological Review,* 1973, vol. 80, pp. 237–251.

21. Harold Hotelling, review of Horace Secrist, "The Triumph of Mediocrity in Business," *Journal of the American Statistical Association,* 1933, vol. 28, pp. 463–465; Secrist and Hotelling debated this further in the 1934 volume of this journal, pp. 196–199.

22. William F. Sharpe, *Investments,* 3d ed., Englewood Cliffs, NJ: Prentice-Hall, 1985, p. 430.

23. Lawrence S. Ritter and William F. Silber, *Principles of Money, Banking, and Financial Markets,* New York: Basic Books, 1986, p. 533.

24. John Llewellyn and Roger Witcomb, letters to *The Times,* London, April 4–6, 1977; and David Hendry, quoted in *The New Statesman,* November 23, 1979, pp. 793–795.

25. Judith R. Glynn, "A Question of Attribution," *The Lancet,* August 28, 1993, pp. 530–532.

26. These baseball statistics are from Bill James, Esquire's 1981 Baseball Forecast, *Esquire,* April 1981, vol. 95, pp. 106–113.

27. Darrell Huff, *How to Take a Chance,* New York: Norton, 1959, p. 141.

28. Peter Gammons, "Inside Baseball," *Sports Illustrated,* June 19, 1989, p. 68.

29. Richard W. Pollay, S. Siddarth, Michael Siegel, Anne Haddix, Robert K. Merritt, Gary A. Giovino, and Michael P. Eriksen, "The Last Straw? Cigarette Advertising and Realized Market Shares among Youths and Adults, 1979–1993," *Journal of Marketing,* April 1996, vol. 60, pp. 1–16.

30. U.S. Department of Commerce, *Statistical Abstract of the United States,* Washington: GPO, 1995, p. 273.

31. Max R. Mickey, Olive Jean Dunn, and Virginia Clark, "Note on the Use of Stepwise Regression in Detecting Outliers," *Computers and Biomedical Research,* July 1967, pp. 105–111.

32. Galton, "Regression towards Mediocrity in Hereditary Stature."

33. Philip N. Baker, Ian R. Johnson, Penny A. Gowland, Jonathan Hykin, Paul R. Harvey, Alan Freeman, Valerie Adams, Brian S. Worthington, and Peter Mansfield, "Fetal Weight Estimation by Echo-Planar Magnetic Resonance Imaging," *The Lancet,* March 12, 1994, vol. 343, pp. 644–645.

34. Belinda Lees, Theya Molleson, Timothy R. Arnett, and John C. Stevenson, "Differences in Proximal Femur Bone Density over Two Centuries," *The Lancet,* March 13, 1993, pp. 673–675; I am grateful to them for sharing their raw data with me.

35. Janette B. Benson, "Season of Birth and Onset of Locomotion: Theoretical and Methodological Implications," *Infant Behavior and Development,* 1993, pp. 69–81.

36. S. Karelitz, V. R. Fisichelli, J. Costa, R. Kavelitz, and L. Rosenfeld, "Relation of Crying in Early Infancy to Speech and Intellectual Development at Age Three Years," *Child Development,* 1964, vol. 35, pp. 769–777.

37. The data were provided by Shaun Johnson of Australia's National Climate Centre, Bureau of Meteorology.

38. William M. Lewis, Jr., and Michael C. Grant, "Acid Precipitation in the Western United States," *Science,* January 11, 1980, vol. 207, pp. 176–177. This SAS output is estimated based on the summary provided in the paper.

39. Karl, Easterling, and Groisman, "United States Historical Climatology Network."

40. *Time,* November 3, 1967, p. 49.

41. *U.S. News & World Report,* March 4, 1968, p. 86.

42. David Upshaw of Drexel Burnham Lambert, quoted in John Andrew, "Some of Wall Street's Favorite Stock Theories Failed to Foresee Market's Slight Rise in 1984," *The Wall Street Journal,* January 2, 1985.

43. "Hoyt says Cy Young Awards a Jinx," *Cape Cod Times,* August 4, 1984.

44. P. C. Rosenblatt and M. R. Cunningham, "Television Watching and Family Tensions," *Journal of Marriage and the Family,* 1976, vol. 38, pp. 105–111.

45. This case is described in D. D. Dorfman, "The Cyril Burt Question: New Findings," *Science,* 1978, vol. 201, pp. 1177–1186.

46. "Running Shoes," *Consumer Reports,* May 1995, pp. 313–317.

brief solutions to odd-numbered exercises

chapter 2

2.1 The percentages are 43.68 for walking, 34.40 for gym, 16.58 for jogging, and 5.34 for aerobic. A bar graph of these percentages is identical to the bar graph in Fig. 2.1, except for the labels on the vertical axis.

2.3

a. The mobile-home percentages are 4.7 northeast, 18.1 midwest, 63.8 south, and 13.4 west.

b. The population percentages are 19.6 northeast, 23.5 midwest, 35.0 south, and 21.9 west.

c. The south's mobile home sales are disproportionately large in comparison to its population, and the other three regions are disproportionately small, particularly the northeast.

2.5 By (roughly) doubling not only the height, but also the width of the pictures as we move from Boston to Philadelphia to New York, the area quadruples—making Philadelphia seem 4 times as big as Boston and New York 4 times as big as Philadelphia.

2.7 The histogram heights (the relative frequencies divided by the interval widths) for the four intervals are .0367, .0433, .0133, and .0067. The center of the data is in the interval from 10 to 20 inches, with a range of 0 to 40 inches. The data are skewed to the right (positively skewed).

2.9 By omitting most of the vertical axis, the bar for the first quarter of 1953 is drawn nearly 4 times the height of the bar for the first 9 months of 1952, even though the estimated average sales in the first quarter of 1953 were only 3.2 percent higher than in the first 9 months of 1952.

2.11 This graph shows quite clearly the long-term decline in the birthrate during the 20th century, from around 130 at the beginning of the century to around 70 at the end, as well as the baby dearth during the Great Depression in the 1930s and the baby boom following World War II.

2.13 A time-series graph shows a peak 120,000 years ago, followed by a long-term decline over the next 80,000 years—a decline that ended 40,000 years ago, with atmospheric carbon dioxide 36 percent below the earlier peak (178.5/280.0 = .64). There has been a marked increase during the past 20,000 years, but not quite back to the peak 120,000 years ago.

2.15 The division of each time (in minutes) into 42,195 meters gives the velocity in meters per minute, for example, 299.043 for females and 332.682 for males in the 1980s. Fitting two straight lines to a time-series graph suggests that women will pass men in the 1990s. By extrapolating the female line backward, the implied velocity in 1910 was negative, indicating that they would have been running backward and never finished a marathon! Clearly, there are dangers to incautious extrapolation.

2.17 In each of these cases, the relationship is not perfect, but reflects a general tendency. The dependent variable is given first in these answers.

a. Because genes are inherited, the height of the son depends (positively) on the father's height.

b. Because most women tend to have their first child when they are relatively young, the age of the oldest child depends (positively) on the age of the mother.

c. Because male hair tends to thin with age, the number of scalp hairs depends (negatively) on age.

d. Humans are fully grown well before the age of 30; heights do decline a bit when people are elderly, but this is well after the age of 40. Thus, between the ages of 30 and 40, there should not be any statistical relationship between age and height.

e. Because the chances of death rise with age, the cost of life insurance depends (positively) on age.

2.19 There seems to be little, if any, relationship between shoe price and rating. (In fact, a line fit to these points—by using the procedures described in Chap. 11—has a slight negative slope!)

2.21 The figure depends on your judgment about first-year salaries. The horizontal axis should show salaries, and the vertical axis should show the density (relative frequency divided by interval width).

2.23 The bar charts indicate that English and German are the most similar, while Arabic words tend to have more syllables than either, and Japanese words have more syllables on average and also a greater diversity.

2.25 In nearly one-half the cases, the batting average declined by .00 to .03 point; the remaining observations are very roughly symmetric about this interval, with no decline larger than .12 and no increase larger than .06.

2.27 Histogram heights can be obtained by dividing each frequency by the width of the interval; for example, in the 0-to-19 age group, .032/20 = .16 for black females and .044/20 = .22 for black males. Black females tend to live longer than black males.

2.29 Per capita cigarette consumption is the explanatory variable and is consequently on the horizontal axis. Death from coronary heart disease is the dependent variable and is consequently on the vertical axis. There seems to be a positive relationship.

2.31 I would draw a histogram for the number of smokers and either a bar graph or line graph for the percentage of persons in each group who smoked (a histogram would be inappropriate for the rate data).

2.33 The first and second years shown (1967 and 1973) are 6 years apart, and the second and third (1973 and 1980) are 7 years apart; however, the third and fourth (1980 and 1993) are 13 years apart. By omitting an intervening year, such as 1986, the graphic gives a misleading impression of a great surge in the number of households earning $100,000 or more. The inclusion of a 1986 figure, perhaps 4.1 million, would make the increase seem steadier. In addition, these data should be adjusted for the increase in prices and population during this period.

2.35 All the bars are the same width, even though the income intervals are not equally wide. This makes it appear that the middle class is a very narrow band. Also, by taking the middle three intervals, this definition of the middle class has 34.9 percent of the families in the upper class and only 16.8 percent of the families in the lower class. It would be more logical to put the middle class in the middle of the data, so that the numbers of people in the upper class and lower class are roughly equal.

2.37 By reducing both the height and the width by 50 percent, the area of the 1978 dollar is only one-fourth that of the 1968 dollar. Our attention is drawn to the area, not the height, and the decline in purchasing power is consequently overly dramatic.

2.39 Darryl Strawberry was paid more in real terms, since his salary was 50 times Babe Ruth's salary and prices increased by only a factor of 9. If Darryl Strawberry had been paid an amount with the same purchasing power as Ruth's 1931 salary, his 1991 salary would have been only $80,000(136.2/15.2) = $716,842.

2.41 The caption purports to tell us the "portion of merchandise purchased through catalogues," but the actual figure gives the number of adults who purchased items by catalog. The *portion of merchandise* should be measured by the value of the items purchased (or perhaps the number of items purchased), rather than by the number of people who purchased items. More people may have bought clothing than bought electronic items, but the value or number of electronic items purchased might be far larger.

Also, the lengths of the bars are not consistently proportional to the values; in particular, the gap between 31.6 and 32.4 is much too large. It appears that lengthening the 31.6 bar would make the bars roughly proportional.

2.43 There are equal spaces between the months, but the months are not evenly spaced over the calendar.

2.45 All the graphs have been scaled to begin and end at the same visual points on the graph. Thus all four funds appear at first glance to have done equally well, when in fact Technology A and the Constellation Fund did much better than Capital Fund A and Foreign Fund I.

2.47

a. Looking vertically, the lowest time in seconds is the fastest: 1984.

b. This winning time was 212.53 seconds (3 minutes 32.53 seconds).

c. The winning times have been steadily declining over time, although in fits and starts.

d. The winning times declined more rapidly before World War II (between 1836 and 1936) than afterward.

2.49 We need to adjust these data for the number of drivers or the number of miles driven.

chapter 3

3.1 The mean is 4.6, the median is 2.4, and the mode is −6.3. This modal value is misleading as a summary description because these happened to be United Airlines' 2 worst years.

3.3 The mean is $76,000, and the median is $25,000. This exercise illustrates the general principle that the median is less sensitive than the mean to measurement errors.

3.5 This professor must have been below average at Yale, but above average at Harvard.

3.7 The mean is 5.483; the median is 5.46; there are two modes, at 5.29 and 5.34; and the 10 percent trimmed mean is 5.474. The mean is closest to the value 5.517 that is now accepted as the density of earth.

3.9 Chris's low (outlier) midterm score gives the midterm the lower mean and the higher standard deviation. If Jessica's score had been 9 points lower—9 points closer to the mean—this would have reduced the mean (by almost a full point) and reduced the standard deviation, too.

3.11 The actual mean is 2.91, and the actual standard deviation is .77. You can provide reasonable estimates by looking at the center of the data and using Chebyshev's inequality or the one-or-two-standard-deviations rules of thumb for bell-shaped histograms.

3.13 If the responses ranged from 0 to 30 with a mean and median below 5, the mean should have been pulled above the median by the 30 (and by any other high answers). The graph, however, is drawn with the mean below the median. The legends are reversed, in that the diamond is the median and the circle is the mean.

3.15 The mean is 11.65, and the standard deviation is 3.67. A value of 23 is 3.09 standard deviations above the mean. The median is 10.9, the first quartile is 8.5, the third quartile is 14.75, and the interquartile range is 6.25. A value of 23 was more than 1.5 times the interquartile range above the median (although not quite more than 1.5 times the interquartile range above the third quartile).

3.17 The price of coffee increased by 209.9 percent; the price of ice cream, by 125.9 percent.

3.19 In a tribe of 8000, an incidence of 1.3 cases per 10,000 is exactly 1 case!

3.21 The increase in total costs is a weighted average of the increases in itemized costs. Thus, a weighted average of numbers that range from 6 to 12 percent will itself be in the range of 6 to 12 percent. Book costs could not have increased by 33 percent unless some component of book costs increased by at least 33 percent.

3.23 The respective means are 7.1, 10.2, 10.3, and 13.5; the standard deviations are 2.8, 12.5, 12.2, and 15.6. Treasury bills have been the safest; corporate stock had the highest average return and the highest standard deviation, too.

3.25 Each is a mode, the most common type of house. The mean is unlikely to be a whole number, and neither the mean nor the median could be used for air conditioning, which does not have a numerical value.

3.27 To estimate the amount wagered daily, we divide the three-digit wagers over the 3-day period by 3 and the two-digit wagers over the 2-day period by 2. This procedure gives an estimate of $2443.03.

3.29 Some of the younger members must have died.

3.31 Because the median is the middle value, 49 percent of the class could have low SAT scores without affecting the median. To evaluate Poch's conclusion, we need to know the distribution of SAT scores—in particular, the distribution among whites and nonwhites.

3.33 The respective means are 64.25, 64.89, 69.15, and 70.72. The medians are 65.00, 65.00, 68.50, and 71.00. The standard deviations are 1.87, 2.79, 3.22, and 2.36. Box plots confirm visually that the basses and tenors are generally taller than the sopranos and altos, and that there is more dispersion in the alto and tenor heights than in that of the sopranos and basses.

3.35 He is confusing the rate of inflation with the price level. The rate of inflation slowed, but prices did not fall (and will not unless there is deflation).

3.37 If the price index was 100 in 1967, the base year, and 116 in 1970, the index increased by 16 percent, not 116 percent, from 1967 to 1970.

3.39 The median is somewhere in the second interval, from $25 million to $49.9 million. None of the observations are less than $0, but some are far above $100 million. The mean is pulled far above the median by the relatively few giant banks.

3.41 A drop of 48 percent reduces the Dow average to 198.2084; a subsequent 48 percent increase raises the Dow to 293.35, some 23 percent below the initial level.

3.43 This histogram is U-shaped. The mean degree of cloudiness is 6.11, which is a potentially misleading description of a typical Greenwich day, because the U-shaped frequency distribution says that the days are usually at one of the extremes—very cloudy or virtually cloudless.

3.45 The distribution of gifts is no doubt asymmetric because gifts cannot be less than zero and can be more than twice the mean (or median). Those gifts that are much larger than others pull the mean above the median. Thus the college's use of the median decreased the threshold for a free book.

3.47 The number of deaths by category does vary by race. One way to demonstrate this is to calculate the white and black fractions by category; for example, white males had 86.5 percent of fatal motor vehicle accidents, but only 47.2 percent of homicide deaths.

3.49 The U.S. resident who sells a car after 3¼ years may do so in order to buy a new, improved model, not because the car has worn out. The car may well last 11 years or longer, although with a succession of owners. Similarly, the fact that a Volvo lasts an average of 11 years does not mean that someone in Sweden only buys a car once every 11 years. There is no doubt a market for used cars in Sweden, just as there is in the United States.

chapter 4

4.1 The cost is the testing of samples, and the benefit is the timely detection of deterioration in the production process, before large numbers of defective products are produced and shipped to irate customers.

4.3 Their statisticians advised them to take relatively larger samples from the population of expensive waybills because sampling errors here would be more expensive. A 1 percent error on a $50 waybill is 100 times more expensive than a 1 percent error on a $.50 waybill.

4.5 The weakness of a people meter is that it relies on the viewer to enter information about who is watching television; whoopee cushions do this automatically. However, people may avoid the cushions if they are uncomfortable; people may switch chairs; and people may sit in a chair with no cushion or on the floor.

4.7 We do not know that Edison and Ford actually used *Ray's Arithmetics,* and even if they did, there is no evidence that it was *Ray's Arithmetics* that made them geniuses or that their era had an unusually large number of geniuses. Every era has geniuses, no matter what the textbooks.

4.9 These numbers should be compared to the total number of people in the military and in the civilian population. In addition, those who fought in the war were presumably younger and healthier than most civilians.

4.11 We do not know if the women at these two colleges are of comparable ability, if the women took similar courses, and if these colleges' grading practices are similar.

4.13 The patients may have been drawn from all over the world, but there is no assurance that they are in proportion to their shares of the entire population.

4.15 Members of the ACLU are not a random sample of Chicago citizens, but rather are those most believing that free speech should seldom be stifled by censorship.

4.17 Perhaps January, coming right after food-filled holidays and New Year's resolutions, is not a typical month, but instead the primary month in which diet articles appear. We also need some measure of the size of the popular press in 1979 and 1989.

4.19 Some people will step forward (maybe more than once) for famine relief but try to remain uncounted for military service and tax collection.

4.21 Only a bit more than 10 percent responded, and they may have been disproportionately fervent on this issue. The real silent majority is the 90 percent who did not bother to respond.

4.23 It is similar to the question "Have you stopped beating your wife yet?" How can a husband who does not beat his wife answer this loaded question? Here, if you think Reagan is doing a marvelous job, how should you answer?

4.25 These are not otherwise identical groups. Former smokers are undoubtedly not a random sample of the population, but are more likely to be those with health problems who have been advised to quit smoking because of their health problems.

4.27 The article implies that having more children reduces the likelihood of divorce; but the causation may well run the other way round—the longer you stay married, the more likely you are to have children.

4.29 Those who chose to take an SAT preparation course were self-selected, not a random selection from potential applicants to Harvard. Students who were confident that they were going to do very well (based on their scores on the PSAT and maybe the SAT itself) were less likely to spend money on an SAT preparation course.

4.31 Those who wanted a cheap date in Burbank, California, are not a random sample of all U.S. movie and television viewers—too young, too poor, too Californian.

4.33 Many people may associate Q with unpopular (or at least unusual) words. There is no real reason why the glasses should have any visible markings; there could be a small label on the underside of each cup to remind the statistician which is which.

4.35 A college does not have the current addresses of all its 1980 graduates, and those with missing addresses are, on average, more likely to be people who do not have stable careers and are not earning large incomes. Also, graduates may be more likely to respond to such a survey if they are doing well financially, and respondents may exaggerate their income in the hope of impressing their classmates. On the other hand, some alumni may understate their income in the hope of discouraging fund-raising solicitations.

4.37 This is not a poll of a random sample of readers who are asked to watch a movie and rate it, but a tabulation of readers who paid to see a movie, presuming that they would like it, and then wanted to share their opinions with the world. A certain movie may appeal to very few people but may receive a high rating because the only people who vote on it are the fans who paid to see it. An alternative problem is that some of the votes may have been cast by people who have not seen the movie they rated, but were encouraged to vote so that it would receive a high, or low, rating.

4.39 People who do not own foreign cars are more likely than the general population to not like foreign cars.

4.41 There is likely to be self-selection or nonresponse bias, in that students with relatively high SAT scores are more likely to report their scores to Bowdein than are students with relatively low scores. Thus (assuming that some of these below-average nonreporting students are admitted) the average of the reported Bowdein SAT scores will be higher than the average SAT scores of all Bowdein students.

4.43 It is implausible that the bed causes death; rather, people who are ill stay in bed. To reach the conclusion offered, we need a control group that is otherwise identical to the treatment group but does not go to bed.

4.45 Those who buy coffee in a New York City 7-Eleven are surely not a random sample of U.S. voters. Also, the fat-congressman caricature no doubt influenced their cup choice.

4.47 People who have visited France for pleasure more than once during the past 2 years are likely to have had good experiences, while those who are making their first visit or have visited once and never returned were deliberately excluded from the sample.

4.49 The commercial does not say that 70 percent of the cars made in 1974 are still on the road. Perhaps its cars last only 3 years, but this company has been growing rapidly and 70 percent of its sales have been within the past 3 years. Also, the quality of the company's cars may have changed over time.

chapter 5

5.1 9/47 = .1915.

5.3 4/96 = .15.

5.5 This probability must be subjective.

5.7 Of these DUI arrests, the probability that a randomly selected person under 21 is male is .948. The gender and age of DUI arrests are largely, but not completely, independent.

5.9 Dark clothing is intuitively unwise, but these data to not prove it. The data are P[dark clothes if victim] = .8 and P[light clothes if victim] = .2 while the relevant comparison is P[victim if dark clothes] versus P[victim if light clothes]. We need to know how common dark clothing is among pedestrians.

5.11 In the binomial model, each trial has two possible outcomes, with the probabilities of success and failure constant, irrespective of the outcomes of other trials.

- **a.** Yes, the probability of a head stays at .5 regardless of previous flips.
- **b.** No, the probabilities are not the same for each batter or for one batter against different pitchers.
- **c.** No, the probabilities vary from hour to hour and are not independent, since similar birds tend to flock together.
- **d.** No, the probabilities vary from election to election.

5.13 Analytically, the six peach pits are equivalent to six coins fairly flipped. The probability of five points is .03125; the probability of one point is .1875.

5.15 Using the binomial distribution with $n = 8$ and $p = .75$ or $p = .90$, the respective probabilities of $P[x > 5]$ are .6787 for $p = .75$ and .9619 for $p = .90$.

5.17 The phrase *used my statistical miles* evidently refers to the law-of-averages fallacy: Because he has flown so many miles without incident, he is due for a crash. There is no persuasive reason why his past flights have any effect on the safety of future flights.

5.19 Edgar Allen Poe, despite acknowledging that the past should have no bearing on future outcomes of dice rolls, nonetheless believed in the fallacious law of averages.

5.21 Probably the simplest test is to ask whether the variable could, in theory, be measured to an unlimited number of decimal places. Thus the number of females is discrete, while time is continuous.

- **a.** Discrete
- **b.** Continuous
- **c.** Continuous
- **d.** Continuous
- **e.** Discrete
- **f.** Discrete

5.23 To convert from a raw score x to a standardized score z, we subtract the mean (75) from the raw score and then divide this difference by the standard deviation (10). The standardized scores are 1.9, 1.4, .8, .6, .3, .1, −.1, −.4, and −.8. A raw score of 90 corresponds to a z value of 1.5.

5.25 The height of a histogram bar is equal to the relative frequency divided by the interval width. Since the interval widths are all 1 inch in this exercise, the height of each histogram bar is equal to the relative frequency. The histogram does look like a normal distribution.

5.27 $a = -2.00$, $b = .00$, and $c = .025$.

5.29 The probabilities are .4013, .0168, and .003.

5.31 The z value is 5.12, making the probability very close to 0.

5.33 The z value is 16.0, making the probability very close to 0.

5.35 The probability is .124.

5.37 Not all occupations encompass an equal number of people, and, in addition, the nephew may have a good idea (not a completely random guess) of what he wants to do.

5.39 Nineteen percent of those persons with brain atrophy are schizophrenic. This 19 percent figure is much larger than the 1.5 percent of the general population who are schizophrenic; however, it is far from certain that a person with brain atrophy is schizophrenic.

5.41 Hank Stram believes in the law of averages; I do not.

5.43

- **a.** The binomial probability is .4040.
- **b.** This is not an unlikely defiance of the percentages.
- **c.** These binomial probabilities assume that the six foul shots are independent, each with .375 probability of success. In reality, his chances may drop when he becomes fatigued at the end of a game and may vary with the pressure and emotions of the moment.

5.45 If the dice rolls are independent, there is no "continuity of the dice." Nick should go to the washroom.

5.47 $P[x > 865.6] = .0091$ and $P[x < 602.2] = .0122$. (In each case, there is about a 1-in-100 chance, which is consistent with each having occurred once in these 90 years.)

5.49 The z value is 1.60, and this fraction is .0548.

chapter 6

6.1 These probabilities are .0228 and .00003.

6.3 A 99 percent confidence interval is 186,270 $\pm$ .448.

6.5 The upper and lower limits on confidence intervals are generally equidistant from the mean.

6.7 A random sample of 60,000 people has very little sampling error, even if the population is infinite.

6.9 A 90 percent confidence interval is 68.3 $\pm$ 4.14; a 99 percent confidence interval would be wider than a 90 percent confidence interval.

6.11 A 99 percent confidence interval is 299,756.2 $\pm$ 63.0

6.13 The 95 percent confidence intervals are -1.000 ± 2.361 for 1994 and 27.766 ± 3.397 for 1995.

6.15 A 95 percent confidence interval is 19.05 $\pm$ 2.46. It does not seem reasonable to interpret these data as a random sample from a population.

6.17 A 95 percent confidence interval has a .014 margin for sampling error.

6.19 The 19-in-20 reference is to a 95 percent confidence interval, which has a .025 margin for sampling error, slightly less than the 3 percent figure cited.

6.21 A 95 percent confidence interval is .15 $\pm$ .023; and a 99 percent confidence interval is .15 $\pm$ .030.

6.23 A 95 percent confidence interval is .0735 $\pm$ .0333.

6.25 A 95 percent confidence interval is 79.7 $\pm$ 57.7. This confidence interval is very wide because the sample is small and the standard deviation is large. We do not know if these children were randomly selected from the population of all British children, or if British children are representative of all children.

6.27

- **a.** The 95 percent confidence intervals are 15.71635 $\pm$ .03217 and 15.68134 $\pm$ .02806. These estimates appear to be consistent in that the confidence intervals overlap.
- **b.** Janet's estimate is more precise because (with seven measurements instead of five) the standard deviation of the sample mean is smaller.
- **c.** Sally's technique is more precise, because the standard deviation of a single estimate is smaller.

6.29 The 95 percent confidence intervals are .589 $\pm$ .051 and .168 $\pm$ .057. The binomial model might be inappropriate because the success probability varies from golfer to golfer. Also, the results are not independent if golfers are affected by their successes or failures.

6.31 If four bolts are samples, the control engineer can set the upper and lower control limits at 1.000 $\pm$ .0644; if nine bolts are sampled instead of four, the upper and lower control limits are 1.000 $\pm$ .0429. The control limits narrow because the larger the sample, the more certain it is that the sample mean will be close to its mean value.

6.33 A 95 percent confidence interval is .29 $\pm$ .10. The smallest value gives 266,000 missing books, an incredible 44.3 books per person.

6.35 By using a conservative value $p = .5$ for estimating the standard deviation, a 95 percent confidence interval for a sample of 400 gives a margin for sampling error of ±.049. It is puzzling that a sample of 600 (with a 4 percent margin for sampling error) is automatically acceptable, while a sample of 400 (with a 5 percent margin) is automatically unacceptable. The usefulness of the poll should depend on its purpose and results, not just the margin for error.

6.37 The 95 percent confidence intervals are .700 ± .082, .517 ± .089, and .442 ± .089.

6.39 An estimate of 1000 fish in the lake is consistent with our estimate that 10 percent of the fish in the lake are tagged.

6.41 A 95 percent confidence interval is .1094 ± .0795. If a letter of the alphabet has a disproportionately large number of typographical errors, this suggests an awkward layout of the keyboard or perhaps a mechanical malfunction.

6.43 A 95 percent confidence interval is 6.19 ± .24. There is a .95 probability that a confidence interval constructed in this fashion will include the true value of the mean prediction of the population.

- **a.** No. This says nothing about how accurate or inaccurate the economic forecasters are.
- **b.** No. If anything, we might estimate that approximately 95 percent of the forecasts are in an interval equal to our estimate of the mean plus or minus 2 standard deviations of the individual observations: 6.19 ± 1.96(.686).
- **c.** No. There would be a .95 probability of including the population mean, which is not necessarily 6.19.

6.45 With $n = 234$, a 95 percent confidence interval is .406 ± .063; with $n = 45$, a 95 percent confidence interval is .406 ± .143. It is not certain that $n = 234$ is appropriate, because a particular company probably tends to buy several identical machines. The selection of a company that has five shoe-cementing machines is not really five independent observations.

As a statistician employed by United Shoe, you could argue that there had been a change in the company's market share between 1947 and 1949, or you could note that, on average, 1 out of every 20 times, a 95 percent confidence interval will exclude the true value of the parameter being estimated. It is not that surprising that one of these 30 confidence intervals excludes the firm's estimate.

6.47 A 95 percent confidence interval is 5.875 ± .347.

6.49 There is a .68 probability that a normally distributed random variable will be within 1 standard deviation of its expected value.

chapter 7

7.1 A formal statistical test would require an estimate of the standard deviation and information about the number of observations; perhaps the difference between 13.6 and 18 yards can be explained by sampling error. I would also like to see who the opponents were; perhaps Texas A&M had tougher away games than home games.

7.3 The sample mean would be drawn from a normal distribution with a mean of 87 and a standard deviation of 3.875. The z value is −5.93, and the P value is virtually 0. Perhaps the blacks who migrated first were more intelligent than those who migrated later, or perhaps the New York environment increased IQ scores.

7.5 The t value is −2.457, and the one-sided P value is .0133.

7.7 For 1994, the t value is −2.065, and the two sided P value is .056; for 1995, the t value is −6.004, and the two-sided P value is .000018.

7.9 The binomial distribution with $p = .5$ and $n = 20$ gives a probability of .0059.

7.11 The z value is 2.613, which is just barely statistically significant at the 1 percent level, since the two-sided P value is .00898.

7.13 Her results are substantial and highly statistically significant, with a z value of 9.80 and a P value very close to 0.

7.15

- **a.** The null hypothesis is that the probability of selecting a Spanish-surnamed person for jury duty is $p = .79$.
- **b.** For the null hypothesis $p = .79$, the observed proportion $x/n = .39$ gives a z value of 29.0.
- **c.** The z value is based on a normal approximation to the binomial distribution, which is more accurate as the sample gets larger.
- **d.** Two standard deviations is appropriate for a two-tailed test at the 5 percent significance

level; 3 standard deviations is statistically significant at an even lower level (less than 1 percent).

7.17 For precision-made dice, the z value is 1.27 and the two-sided P value is .2040, which is not statistically significant at the 1 percent level. For the inexpensive dice, the z value is 15.62, and the two-sided P value is virtually 0, decisively rejecting the null hypothesis. However, the sample proportions are so close to .5, particularly with the precision-made dice, that even if statistically significant, they would be of little or no practical importance.

7.19 With no logical reason for the quarterback's name to influence the outcome of a football game, I would explain the J theory as data mining—an explanation supported by the observation that the New York Giants with Phil Simms as quarterback won the Super Bowl in 1987.

7.21 This is obviously data mining. These particular six numbers were not selected randomly, but were carefully picked from among the millions of possible comparisons in this table.

7.23 This is undoubtedly a coincidence uncovered by data mining. The behavior of the Dow and the names of the cities have no effect on the outcome of a football game. They used the end of November because this is the starting time (of many they tried) that worked the best.

7.25 The t value is $-.911$, and the two-sided P value is .366, far from the .01 needed to reject the null hypothesis at the 1 percent level.

7.27 The z value is 9.32, and the P value is less than .0000000001. To determine how aggressively this fire department is moving to hire minorities and females, we need to know how many of the 1448 firefighters have been hired since 1988 and the proportions of these new hires that are minorities and females.

7.29 The binomial distribution gives a two-sided P value of .0784, which is not quite statistically significant at the 5 percent level.

7.31 The binomial distribution gives P values of .0650 and .3073.

7.33 The results are statistically significant (for a one-tailed test) at the 5 percent level, with the binomial distribution giving a P value of .0376.

7.35 The binomial distribution implies a .0433 probability of 5 or more correct out of 20. The probability that 2 or more people out of 23 will do so well is also given by the binomial distribution, this time with $p = .0433$ and $n = 23$, and works out to be .2626. There is a substantial chance of at least two people getting five or more correct by luck alone; that fact that they did not repeat their success suggests that it was indeed luck the first time.

7.37 If almost everyone has eaten pickles, then almost everyone who has cancer (or anything else) will have eaten pickles. The probability that a cancer victim has eaten pickles is not at all the same as the probability that a person who eats pickles will get cancer (which should be compared to the probability that a person who does not eat pickles will get cancer).

7.39 In each case, the z value is so large that the P value is virtually zero.

a. $z = 15.0$
b. $z = 7.66$
c. $z = 28.87$
d. $z = 26.7$

7.41 The binomial distribution gives the exact probability $P[x > 34] = .00054$, so that the two-sided P value is .00108. Unless this experiment was poorly designed, this very low P value is a persuasive reason to reject the null hypothesis.

7.43 Blackmun's logic relies on the presumption (evidently supported by statistical studies as well as common sense) that an increase in jury size reduces the chances of a conviction. Therefore, the risk of convicting an innocent person increases as the jury size decreases, and the risk of not convicting a guilty person increases as the jury size decreases.

7.45

a. The null hypothesis is that pay raises are independent of race.
b. This probability is .0000000000391.
c. This probability is .337.
d. Data that are divided into eight job categories might not show statistical significance in any of the categories because each of the eight samples is too small, even though there is a statistically significant relationship when the data are aggregated.

e. Data that are divided into eight job categories might show statistical significance in some categories in favor of blacks and in other regions in favor of whites, differences that cancel when the data are aggregated.

7.47 All three statements are wrong.

a. This statement confuses $P[A \text{ if } B]$ with $P[B \text{ if } A]$. The test shows that if the cold vaccine is ineffective, there is a .08 probability of observing a difference this large (or larger) between the fractions of each group that caught colds.

b. Statistical significance is not the same as practical importance. We cannot tell the estimated size of the effect from the P value.

c. Not being able to reject the null hypothesis is not the same as proving the null hypothesis to be true. The correct interpretation is given in (*a*).

7.49 The two-sided P value is .6072 for the national products, .0063 for the regional products, and .0078 for the local products.

chapter 8

8.1 The t values are 5.956, 6.684, and 3.306, with two-sided P values of .000000006, .0000000000, and .0010. The 99 percent confidence intervals are 3.10 ± 1.34, 9.60 ± 3.70, and 1.40 ± 1.09. The differences seem substantial, representing 1 additional inch of height (about 2 percent), nearly a 10-pound increase in weight (about 8 percent), and about a 5 percent increase in strength.

8.3 The t value is 6.46, and the two-sided P value is .0000000001. Despite the overwhelming statistical significance of this result, the doctors apparently felt that having, on average, .5 fewer colds per year was not worth the trouble and expense of taking the vaccine.

8.5 The t value is 1.046, and the two-sided P value is .30, which is not low enough to reject at the 5 percent level the null hypothesis that the population mean difference is zero.

8.7 The sum of the signed ranks is 66, and the z value is 1.232, giving a two-sided P value of .2180.

8.9 For the fatal heart attack data, the z value is 2.712 and the two-sided P value is .0067; for the nonfatal heart attack data, the z value is 4.409 and the two-sided P value is .0000105. In both cases, the observed differences are substantial and statistically significant. The test was stopped because the researchers were persuaded by the evidence and did not want the control group to suffer more heart attacks.

8.11 The z value is 1.72, and the two-sided P value is .0854. The observed difference is not quite statistically significant at the 5 percent level.

8.13 There was an 83.33 percent decline in automobile fatalities between 1980 and 1985. For statistical significance, the respective t values are 2.897 and 4.001, giving two-sided P values of .0442 and .0162. In both cases, the observed difference between 1980–1982 and 1983–1985 is statistically significant at the 5 percent level, since the two-sided P values are less than .05.

8.15 They are matched pairs in that even better than comparing the eyes in two seemingly similar persons, we compare the eyes in the same person. The t value is -1.053, and the two-sided P value is .327, which is not nearly statistically significant at the 1 percent (or 5 percent) level.

8.17 For each of these four tests, the t value is well above 2 and thus rejects the null hypothesis at the 5 percent level. The t values are 2.69, 6.31, 7.15, and 5.31, with respective two-sided P values of .0093, .0000039, .0000032, and .000123.

8.19 The t value is 1.17, which is not statistically significant at the 5 percent level. Statistical software shows the degrees of freedom to be 33.6 and the two-sided P value to be .250.

8.21 The two sample success proportions are not substantially different: .4902 versus .5217. The z value is .355, which is far less than 1.96 needed for statistical significance with a two-tailed test at the 5 percent level. The two-sided P value is .722.

8.23 The t value is 2.08, barely large enough to show statistical significance at the 5 percent level. With 8 degrees of freedom, the two-sided P value is .041. To evaluate whether the observed difference is substantial, we can consider the fact that the .383 ratio is 3.5 percent larger than the .370 ratio.

8.25 The ratio of the white/black passing rates for whites is .681, which is less than .80 and thus indicates a substantial difference in selection rates.

The difference in passing rates is statistically significant at the 1 percent level, with $z = 3.757$ and the two-sided P value equal to .000173.

8.27 The t value is 2.158, and the two-sided P value is .0316. This is (barely) statistically significant at the 5 percent level, and it does seem substantial (although not overwhelming), representing an 11 percent drop in cavities.

8.29 The center of the female data appears more compact, but overall the distributions seem roughly similar and not very asymmetric. The t value is 1.785, and with 42.5 degrees of freedom, the two-sided P value is .0816. These data do not reject at the 5 percent level the null hypothesis that males and females have the same average body temperature.

8.31 The t value is .294, and with an estimated 57.3 degrees of freedom, the two-sided P value is .7698. The observed difference of .185 child is not substantial or even remotely statistically significant.

8.33 The plaque and gingivitis data are numerical data, while the periodontis data are categorical. For the plaque and gingivitis data, the t values are -19.49 and -13.39, and the P values are virtually zero. For the periodontis data, the z value is .315, giving a two-sided P value of .753, which is far from statistical significance. The difference in the sample proportions does not seem substantial either: .039 versus .042. These data suggest that while people of high socioeconomic status with substantial access to dental care may have less plaque and gingivitis, they do not necessarily have less periodontis.

8.35 The z value is 2.428, and the two-sided P value is .0152. Thus the observed difference is statistically significant at the 5 percent level, but not at the 1 percent level. The difference (.412 versus .105) is substantial.

8.37 The t value is 2.674, giving a two-sided P value of .0092.

8.39 The t value is 10.521, and the two-sided P value is less than .0000000001, decisively rejecting the null hypothesis.

8.41 The sample success proportions are .4600 and .8798, a difference that seems substantial. For a statistical test, the z value is 8.666 and the P value is less than .0000000001. The observed difference is highly statistically significant.

8.43 The t value is .578, and with 33.8 degrees of freedom, the two-sided P value is .5670.

8.45 For the number of words in the essay, the t value is -4.576; with 126.4 degrees of freedom, the two-sided P value is .000012. For the average sentence length, the t value is 2.308; with 124.2 degrees of freedom, the two-sided P value is .0226. Both differences are statistically significant at the 5 percent level, although the number of words more so than the sentence length. The differences also seem substantial, again more so with the number of words. For the descriptive essay, students tended to write shorter essays but use longer sentences.

8.47 The t value is a minuscule .073, which provides no evidence whatsoever against the null hypothesis; the two-sided P value is .94.

8.49 The sums of the ranks are 203 for the stomach cancer patients and 97 for the breast cancer patients. The z value is 2.346, giving a two-sided P value of .019.

chapter 9

9.1 Because of inevitable sampling error (the luck of the draw in choosing random samples), the means of samples drawn from a single population will vary from sample to sample. The same is true of samples drawn from different populations that have the same mean and standard deviation.

9.3 The t value is 1.685, and the two-sided P value is .1084.

9.5 Yes, the visual appearance of the four box plots is generally consistent with the seasonal standard deviations in Table 9.2. The variation in the data is largest for winter and smallest for summer. Fall and spring are in the middle with roughly equal standard deviations; although fall has the larger overall range, spring has the larger interquartile range.

9.7 There is relatively little variation within each group, but there are considerable differences among the group means. The F value is much larger than 1, providing strong evidence against the null hypothesis that these data came from populations with the same mean.

9.9 The t value is 1.93, and the F value is 3.73. Except for rounding error, the F value is equal to the t value squared.

9.11 The means are 2.569, 2.850, and 3.186. The F value is 5.947, and the P value is .0070. The null hypothesis is that the population mean GPAs are the same for these three categories. In fact, the higher the predicted GPA, the higher is the average first-year GPA—and the differences are statistically significant at the 1 percent level.

9.13 The F value is 16.508, and the P value is .00005. The observed differences among these groups—with the high-nicotine category having the highest carbon monoxide content and the low-nicotine category having the lowest carbon monoxide content decline—are highly statistically significant.

9.15 There were highly statistically significant differences among the final examination scores for the four sections (P value = .003). We cannot tell if these differences were also substantial without examining the average scores for the sections.

9.17 The F value is 3.59, and the P value is .038. The 95 percent confidence intervals are 2.67 ± .42, 3.07 ± .64, and 2.05 ± .51. In this study, those in the Indo-European group had the lowest average score and thus were judged the most proficient in their grammatical abilities.

9.19 Coronary heart disease tends to be lower in countries where the citizens eat more vegetables. However, the F value is .865 and the P value is .43, not nearly low enough to reject the null hypothesis at the 5 percent level.

9.21 The F value is 6.55, which is larger than the 5.95 cutoff for a 1 percent test; the P value is .0070. These data reject the null hypothesis that the fabrics are equally durable.

9.23 The average sales for troops 287 and 319 are nearly identical, while troop 403 sold substantially more boxes. The observed differences are statistically significant at the 5 percent level, with an F value of 4.020 and a P value of .0287.

9.25 The null hypothesis is that there is no difference in the population means. Corporate stock had the highest average return, but the observed differences among the returns are not statistically significant at the 5 percent level; the F value is .446, and the P value is .640.

9.27 The F value is 55.80, and the P value is less than .0000000001. The observed differences among these heights reject decisively the null hypothesis that the population mean heights are the same for these four groups of singers. The 95 percent confidence intervals are 64.25 ± .83, 64.89 ± .84, 69.15 ± 1.11, and 70.72 ± .80. There is substantial overlap in the soprano and alto confidence intervals and in the tenor and bass intervals, but the soprano and alto intervals are far from the tenor and bass intervals. This make sense, since the sopranos and altos are female, while the tenors and basses are male.

9.29 The observed differences are statistically significant at the 5 percent level, with an F value of 34.53 and a P value of .0004. The 95 percent confidence intervals are 12.56 ± .84, 18.18 ± 1.41, and 17.32 ± 1.41. The Island Thorns and Ashley Rails intervals overlap substantially, but neither overlaps the Llanederyn confidence interval.

9.31 The mean hours of sleep are lower for these first-year and third-year students than for the second-year and fourth-year students. However, the observed differences are not statistically significant at the 5 percent level, since the F value is 2.061 and the P value is .1216.

9.33 SnackWell had the highest average rating (6.1) and Chips Ahoy the lowest (4.4), a difference that seems substantial. The observed differences among the ratings of the three brands are statistically significant at the 1 percent level, with a P value of .0004. An examination of the 95 percent confidence intervals shows considerable overlap between Pepperidge Farms and SnackWell, a slight overlap between Pepperidge Farms and Chips Ahoy, and no overlap between SnackWell and Chips Ahoy. These college students do not seem to like Chips Ahoy.

9.35 The females in this sample studied more, on average, than did the males, but with a P value of .312, the observed difference is not nearly statistically significant at the 5 percent level.

9.37 The free-trade countries had the highest average growth rate, the moderate-restrictions countries the next highest, and the strong-restrictions countries the lowest average growth rate. The differences seem substantial and are statistically significant at the 1 percent level, with an *F* value of 8.655 and a *P* value of only .0008.

9.39 The *F* value is 294.17, and the *P* value is less than .0001. The 95 percent confidence intervals are 652.40 ± 7.59, 648.50 ± 7.59, 558.10 ± 7.59, and 552.70 ± 7.59. The female sophomore and senior confidence intervals overlap; the male sophomore and senior confidence intervals do not overlap each other or the female confidence intervals.

chapter 10

10.1 The chi-square value is 135.6, which is larger than the 21.67 cutoff for a 1 percent test and thus decisively rejects the null hypothesis at the 1 percent level. Statistical software shows the *P* value to be .0000007.

10.3 With 5 degrees of freedom, the cutoff for each test at the 1 percent level is 15.09.

a. The chi-square value is 128.74; these data decisively reject the null hypothesis.

b. The chi-square value is 16.84; these data barely reject the null hypothesis at the 1 percent level.

10.5 The binomial distribution gives the probabilities under the null hypothesis. The chi-square value is 7.77, and the *P* value is .17, not low enough to reject the null hypothesis at the 5 percent level.

10.7 There were far more women, somewhat more children, and far fewer men saved than would be true with independence. (Overall, 74 percent of adult women, 18 percent of adult men, and 52 percent of children were saved.) The chi-square value is 360.7, which decisively rejects the null hypothesis. Statistical software shows the *P* value to be .000001.

10.9 Overall, 73 percent experienced foot pain. If there were no relationship between shoe size and foot pain, we would expect 73 percent of those in each shoe-size category to experience foot pain. These data indicate that women with large feet are much more likely to experience foot pain. The chi-square statistic is 7.55, which exceeds the 5.99 cutoff for a test at the 5 percent level. (Statistical software shows the *P* value to be .023.) These data reject at the 5 percent level the null hypothesis that foot pain is unrelated to shoe size.

10.11 The value of the chi-square statistic is 3.37, which is far smaller than the 9.21 value required to reject the null hypothesis at the 1 percent level. If the null hypothesis is true, there is a .17 probability of this large a chi-square value.

10.13 The chi-square value is 105.32, which overwhelmingly rejects the null hypothesis at the 1 percent level, since the cutoff is 9.21 for a test at the 1 percent level.

10.15 The chi-square value is 38.6; with 3 degrees of freedom, there is a .05 chance of a chi-square value above 7.81. Thus these data reject at the 5 percent level the null hypotheses that sex and grades are unrelated. (Statistical software shows the *P* value to be .0000003.) The superior performance of these female students in these courses is too large to be explained by chance alone.

10.17 The chi-square value is 8.42; with 2 degrees of freedom, the probability of this large a chi-square value, if the null hypothesis is true, is .015. The observed relationship is statistically significant at the 5 percent level. There are several reasons why we should treat our results cautiously. First, it is not clear why a player's hair color should affect the outcome of a basketball game, although perhaps there is some emotional effect on the player, his teammates, or the opposition. The observed relationship may simply reflect the results of some energetic data mining. Second, we do not know that the caliber of the opposition was the same (or that the hair color was chosen randomly); perhaps Rodman chose red or blue hair when the Spurs were playing the best teams. Third, the expected values are so small that a chi-square approximation may be inaccurate.

10.19 The chi-square value is 17.15; these data reject the null hypothesis at the 5 percent level, since the cutoff for a 5 percent test is 5.99. Statistical software shows that if the null hypothesis is true, the probability of such a large chi-square value is .0002.

10.21 The chi-square value is 935.46, which is far above the 9.21 cutoff required for a 1 percent test. These data reject the null hypothesis. Statistical software shows that if the null hypothesis is true, the probability of such a large chi-square value is .000002.

10.23 The chi-square value is 2.14, which is less than the 6.63 cutoff for a 1 percent test. These data do not reject the null hypothesis of independence. (Statistical software shows the *P* value to be .144.) Tripling her data triples her chi-square value. This is not a logical procedure. It is like flipping a coin once, obtaining a head, and then assuming that 1000 flips would yield 1000 heads.

10.25 For the three categories, 3/15 of the balls are in the corners, 6/15 are adjacent, and 6/15 are other. Far more corner balls were sunk and far fewer in the other category than would be expected if each ball were equally likely to be sunk. The chi-square value is 28.55, which decisively rejects the null hypothesis, since the cutoffs are 5.99 for a test at the 5 percent level and 9.21 for a test at the 1 percent level. (Statistical software shows that the *P* value is .0000007.)

10.27 The chi-square value is 14.39, and the *P* value is .025—which is statistically significant at the 5 percent level, but not at the 1 percent level.

10.29 The value of the chi-square statistic is 32.94, which decisively rejects the null hypothesis. With 2 degrees of freedom, the cutoffs are 5.99 at the 5 percent level and 9.21 at the 1 percent level. Statistical software shows that the *P* value is .0000064.

10.31 The chi-square value is 20.17, and the *P* value is .043. The cutoff for a test at the 5 percent level is 19.68. Thus these data barely reject at the 5 percent level the null hypothesis that all birth months are equally likely. Although the 21 observations in November are nearly twice the expected value under the null hypothesis that November births have a 1/12 probability, the other months are much closer to what we expect if all birth months are equally likely. Perhaps this researcher looked at the data before deciding to separate November for special attention.

10.33 The chi-square value is 12.972, and the *P* value is .011. These data indicate that a person wearing only a lap belt is more likely to be killed in a frontal crash than is a person wearing a lap-and-shoulder belt or a person wearing no restraint at all. Our calculated *P* value is only a rough approximation because some of the expected values are quite small.

10.35 The chi-square value is 25.32, and the *P* value is .000045; these data decisively reject the null hypothesis that free-throw percentage and height are unrelated.

10.37 The chi-square value is 10.337, and the *P* value is .01591. Of those babies born in the winter and spring, a relatively large number began crawling before 30.89 weeks; of those born in the summer and fall (especially the fall), a relatively large number began crawling after 30.89 weeks. The observed differences are statistically significant at the 5 percent level, but not at the 1 percent level.

10.39 The z value is 1.732, and the two-sided *P* value is 2(.0416) = .083, which is not quite low enough for statistical significance at the 5 percent level. Not surprisingly, the chi-square test and the difference-in-means test agree with each other. In fact, the chi-square value is the square of the z value, and the *P* values are identical.

chapter 11

11.1

- **a.** Batting average depends (positively) on practice.
- **b.** Attendance depends (negatively) on temperature.
- **c.** Average speed depends (negatively) on congestion.
- **d.** Miles per gallon depends (negatively) on weight.
- **e.** Price depends (negatively) on mileage.

11.3

- **a.** Bill
- **b.** Elaine
- **c.** Final
- **d.** Homework
- **e.** Positively related

11.5 If the more experienced get higher ratings, a scatter diagram will be positively sloped; if the less experienced get higher ratings, it will be negatively sloped; and if there is no relationship, the scatter diagram will show no pattern. Such relationships are imperfect because of variations from professor to professor and class to class—other factors matter, too.

11.7 The dependent variable is cricket chirps, since temperatures may influence cricket behavior but cricket behavior does not affect the temperature. A scatter diagram shows an approximately linear relationship, and the estimated regression equation shows a positive relationship between temperature and chirping frequency: $y = -.3092 + .2119x$. The .2119 estimate of β implies that a 1° increase in the temperature increases the number of cricket chirps per second by .2119.

11.9 The least-squares line is $y = -6.33 + .170x$. The .170 estimated value of β implies than an increase in wind velocity of 1 mile per hour increases the percentage of trees killed or severely damaged by .170.

11.11 The least-squares estimated equation is $y = 29.456 + .0557x$. The estimated relationship is positive and statistically significant at the 5 percent level (the t value of 4.3 is far larger than the 2.093 cutoff for a two-tailed test at the 5 percent level). To evaluate whether the coefficient is substantial, consider the fact that an increase in per capita cigarette consumption of 1000 (roughly the difference between Sweden and Iceland or between Iceland and Australia) is predicted to increase annual deaths from coronary heart disease per 100,000 persons aged 35 to 64 by 55.7, that is, an additional 557 people per 1 million or 5570 per 10 million. This seems to be a substantial (and plausible) number.

11.13 Franklin National Bank has the smallest profits and the second-smallest assets, but otherwise does not appear to be inconsistent with the positive relationship between profits and assets shown by the other points. However, a closer inspection reveals that Franklin's income was $13 million below the line, which is a large amount for such a small bank. A least-squares regression of profits on assets gives $y = 7.57 + 4.99x$, and a 95 percent prediction interval for $x = 3.8$ (Franklin National's assets) is $y = 26.5 \pm 55.3$. The prediction is nearly twice the size of Franklin's actual profits of $13.8 million, but the prediction interval is quite wide.

11.15 The least-squares estimated equation is $y = 229.17 - 2.32x$. There is a negative relationship, and the t value of 6.70 has a two-sided P value of .000000016. Thus the relationship is highly statistically significant. To gauge whether the estimated effect is substantial, note that a 10-year increase in age from 30 to 40 will reduce the predicted level of androsenodione by 23.2, which is about 14 percent. We do not have enough information here to determine whether such a decrease has a substantial effect on the probability of having a healthy child.

11.17 The estimated equation is $y = 28.18 + .406x$. The slope of this line is positive, but not statistically significant at the 5 percent level since the two-sided P value is .273. The estimated coefficient does seem substantial, with each percentage point increase in the 1994 return predicted to increase the 1995 return by .406 percentage point. The low value of R^2 reinforces our conclusion that, overall, these data do not demonstrate a reliable relationship between a fund's return one year and its return the next year.

11.19 The least-squares estimated equation is $y = .180 + .966x$. The R^2 value is .995, and the correlation coefficient is the square root of R^2, or .9975.

11.21 The estimated equation is $y = 66.47 + .204x$. The relationship is highly statistically significant, but correlation is not necessarily causation. Most likely, both x and y have grown over time as the population has increased; thus instead of x causing y or y causing x, it is population growth that causes both.

11.23 Corn prices and ragweed growth are negatively related, because favorable weather conditions increase the growth of both ragweed and corn, increasing the incidence of hay fever and lowering corn prices.

11.25 The least-squares line is $y = -29{,}598.33 + 3923.70x$. Because the t value of 4.25 is larger than 3.182 (the cutoff for a 5 percent test with 3 degrees of freedom), there is a statistically significant relationship between education and income. (The two-sided P value is .024.) On average, an extra year of education is associated with an additional $3923.70 in annual income. These data show that the better-educated earn more than the less-educated. But the data do not prove that a particular person would earn more income if he or she were better educated; that is, they do not prove that it is the education that *causes* the higher income. Undoubtedly, those who continue their education are, on average, brighter and more motivated than those who stop—qualities that will help make them successes regardless of their education.

The predicted income of someone who went to school for 30 years is $88,113, which is not very believable. The professional student is most likely averse to work and unlikely to be successful at it, if he or she ever leaves the security of school.

11.27 The estimated equation is $y = 926.20 + 72.405x$, indicating that there is a positive relationship between nitrogen and yield. The coefficient value of 72.405 predicts that a 1 percentage point increase in nitrogen (from 2 to 3 percent, for example) will increase crop yield by 72.405 pounds, about 7 percent. This seems substantial. The t value of 10.51 shows the effect to be highly statistically significant. A scatter diagram of the residuals reveals that the prediction errors are negative for both large and small values of x and positive for the middle values, suggesting that the relationship between nitrogen and crop yield is nonlinear.

11.29 The estimated equation is $y = 109.87 - 1.13x$. With 19 degrees of freedom, the cutoff for a two-tailed test at the 1 percent level is $t = 2.861$; thus the observed relationship is statistically significant at the 1 percent level. (The two-sided P value is .002.) An examination of a plot of the residuals shows that the observation $x = 42$, $y = 57$ is an outlier, with the actual value of y far larger than the predicted value.

11.31 The least-squares estimated relationship is $y = .1192 + 1.0315x$. The two-sided P value is .00000002, which is decisively statistically significant at the 1 percent level. The 1.0315 estimated slope of the regression line implies that a baby who has a 1 cubic decimeter larger estimated fetal volume is predicted to have a 1.0315-kilogram larger birth weight.

11.33 The estimated equation is $y = 52.79 + .398x$. There is a substantial and highly statistically significant relationship, with the variation in homework scores explaining 78 percent of the variation in final examination scores. The t value of 8.95 is far beyond the 2.807 cutoff for a two-tailed test of $\beta = 0$ at the 1 percent level. Statistical software shows that the two-sided P value is .000000006.

11.35 The estimated equation is $y = 91.27 + 1.49x$. Because the t value of 3.1 is larger than 2.0 (the cutoff for a two-tailed t test at the 5 percent level), the relationship is statistically significant at the 5 percent level. The two-sided P value is .004.

Presumably, it is not crying that makes babies smart, and making your baby cry will not improve the child's IQ score. Instead, there are more fundamental causes that apparently tend to make some children smarter than others and also cry more. Crying is an *indicator* of intelligence, but not its cause.

11.37 The relationship is strongly statistically significant. With 148 degrees of freedom, the cutoff for statistical significance at the 1 percent level is a t value of approximately 2.6. Here the t value is 4.08, and the two-sided P value is less than .0001. The predicted pH level at the beginning of this 3-year period was 5.42, and at the end it was 4.64, representing nearly a 10-fold increase in acidity, which changed the rain pH from weakly acidic to moderately acidic.

11.39 Those particular individuals or teams that appear on the cover of *Sports Illustrated* are not a random sample. They have typically done something exceptional recently—perhaps won the World Series, a major tennis tournament, or 47 football games in a row. Such accomplishments are more likely to have involved good luck than bad and are not likely to continue indefinitely.

11.41 The least-squares line is $y = 9.95 + .626x$. With 28 degrees of freedom, the cutoff for statistical significance at the 5 percent level is a t value of 2.048, making the result barely statistically significant at the 5 percent level. (The two-sided P value is .0494). The estimates imply that a 1-inch increase in January precipitation (a 34 percent increase over the 2.92-inch average) is associated with a .626 percent increase in precipitation during the remainder of the year (a 5.3 percent increase over the 11.78-inch average).

11.43 No doubt, families buy an additional car because they drive more, not the other way round.

11.45 I suspect that there may be some careful data selection here, a suspicion fueled by the fact that we are only told of the past four American League pitchers who won the Cy Young Award, not the ones before that or any National League winners. Even if this were a comprehensive tabulation of Cy Young Award winners, baseball pitchers do have good and bad years, with lucky bounces and injuries, and any year in which a pitcher wins the Cy Young Award is almost certainly an atypically good year. It would be extraordinary for a pitcher to be the best in the league while having a below-average year. After a pitcher wins the Cy Young Award, there is nowhere to go but down. The so-called Cy Young jinx is an extreme example of regression toward the mean.

11.47 The argument quoted here assumes that it is the taking of Latin that causes the high SAT scores; it may well be that the students who take Latin are above-average students with above-average language skills who would have scored 150 points above mean even if they had not taken Latin.

11.49 The .5 coefficient, rather than 1.0, reflects the regression-toward-the-mean phenomenon: The IQs of the children of parents with relatively high (or low) IQs tend to be more nearly average. The later researcher's data are suspect because the fit is almost perfect (that is, it is too good to be true). Not only is the value of R^2 virtually 1, but also a rearrangement of the estimated equation shows the estimated coefficients to be almost exactly those suggested by the earlier psychologist.

index

Table 3 Standardized normal distribution

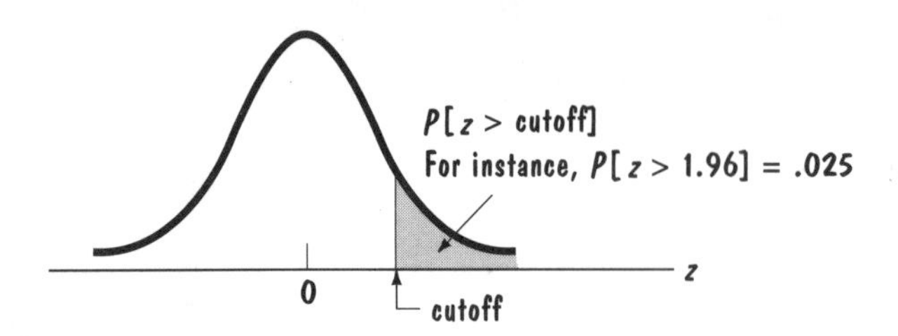

CUTOFF	.00	.01	.02	.03	.04	.05	.06	.07	.08	.09
.0	.5000	.4960	.4920	.4880	.4840	.4801	.4761	.4721	.4681	.4641
.1	.4602	.4562	.4522	.4483	.4443	.4404	.4364	.4325	.4286	.4247
.2	.4207	.4168	.4129	.4090	.4052	.4013	.3974	.3936	.3897	.3859
.3	.3821	.3783	.3745	.3707	.3669	.3632	.3594	.3557	.3520	.3483
.4	.3446	.3409	.3372	.3336	.3300	.3264	.3228	.3192	.3156	.3121
.5	.3085	.3050	.3015	.2981	.2946	.2912	.2877	.2843	.2810	.2776
.6	.2743	.2709	.2676	.2643	.2611	.2578	.2546	.2514	.2483	.2451
.7	.2420	.2389	.2358	.2327	.2296	.2266	.2236	.2206	.2177	.2148
.8	.2119	.2090	.2061	.2033	.2005	.1977	.1949	.1922	.1894	.1867
.9	.1841	.1814	.1788	.1762	.1736	.1711	.1685	.1660	.1635	.1611
1.0	.1587	.1562	.1539	.1515	.1492	.1469	.1446	.1423	.1401	.1379
1.1	.1357	.1335	.1314	.1292	.1271	.1251	.1230	.1210	.1190	.1170
1.2	.1151	.1131	.1112	.1093	.1075	.1056	.1038	.1020	.1003	.0985
1.3	.0968	.0951	.0934	.0918	.0901	.0885	.0869	.0853	.0838	.0823
1.4	.0808	.0793	.0778	.0764	.0749	.0735	.0722	.0708	.0694	.0681
1.5	.0668	.0655	.0643	.0630	.0618	.0606	.0594	.0582	.0571	.0559
1.6	.0548	.0537	.0526	.0516	.0505	.0495	.0485	.0475	.0465	.0455
1.7	.0446	.0436	.0427	.0418	.0409	.0401	.0392	.0384	.0375	.0367
1.8	.0359	.0352	.0344	.0336	.0329	.0322	.0314	.0307	.0301	.0294
1.9	.0287	.0281	.0274	.0268	.0262	.0256	.0250	.0244	.0239	.0233
2.0	.0228	.0222	.0217	.0212	.0207	.0202	.0197	.0192	.0188	.0183
2.1	.0179	.0174	.0170	.0166	.0162	.0158	.0154	.0150	.0146	.0143
2.2	.0139	.0136	.0132	.0129	.0125	.0122	.0119	.0116	.0113	.0110
2.3	.0107	.0104	.0102	.0099	.0096	.0094	.0091	.0089	.0087	.0084
2.4	.0082	.0080	.0078	.0075	.0073	.0071	.0069	.0068	.0066	.0064
2.5	.0062	.0060	.0059	.0057	.0055	.0054	.0052	.0051	.0049	.0048
2.6	.0047	.0045	.0044	.0043	.0041	.0040	.0039	.0038	.0037	.0036
2.7	.0035	.0034	.0033	.0032	.0031	.0030	.0029	.0028	.0027	.0026
2.8	.0026	.0025	.0024	.0023	.0023	.0022	.0021	.0021	.0020	.0019
2.9	.0019	.0018	.0017	.0017	.0016	.0016	.0015	.0015	.0014	.0014
3.0	.0013	.0013	.0013	.0012	.0012	.0011	.0011	.0011	.0010	.0010
4.0	.000 031 7									
5.0	.000 000 287									
6.0	.000 000 000 987									
7.0	.000 000 000 001 28									
8.0	.000 000 000 000 001									